The Neuronal Environment

Contemporary Neuroscience

The Neuronal Environment: Brain Homeostasis in Health and Disease, edited by **Wolfgang Walz,** 2002

Neurotransmitter Transporters: Structure, Function, and Regulation, 2/e, edited by **Maarten E. A. Reith,** 2002

Pathogenesis of Neurodegenerative Disorders, edited by **Mark P. Mattson**, 2001

Stem Cells and CNS Development, edited by **Mahendra S. Rao,** 2001

Neurobiology of Spinal Cord Injury, edited by **Robert G. Kalb** and **Stephen M. Strittmatter,** 2000

Cerebral Signal Transduction: From First to Fourth Messengers, edited by **Maarten E. A. Reith,** 2000

Central Nervous System Diseases: Innovative Animal Models from Lab to Clinic, edited by **Dwaine F. Emerich, Reginald L. Dean, III,** and **Paul R. Sanberg,** 2000

Mitochondrial Inhibitors and Neurodegenerative Disorders, edited by **Paul R. Sanberg, Hitoo Nishino,** and **Cesario V. Borlongan,** 2000

Cerebral Ischemia: Molecular and Cellular Pathophysiology, edited by **Wolfgang Walz**, 1999

Cell Transplantation for Neurological Disorders, edited by **Thomas B. Freeman** and **Håkan Widner,**1998

Gene Therapy for Neurological Disorders and Brain Tumors, edited by **E. Antonio Chiocca** and **Xandra O. Breakefield,** 1998

Highly Selective Neurotoxins: Basic and Clinical Applications, edited by **Richard M. Kostrzewa,** 1998

Neuroinflammation: Mechanisms and Management, edited by **Paul L. Wood,** 1998

Neuroprotective Signal Transduction, edited by **Mark P. Mattson,** 1998

Clinical Pharmacology of Cerebral Ischemia, edited by **Gert J. Ter Horst** and **Jakob Korf,** 1997

Molecular Mechanisms of Dementia, edited by **Wilma Wasco** and **Rudolph E. Tanzi,** 1997

Neurotransmitter Transporters: Structure, Function, and Regulation, edited by **Maarten E. A. Reith,** 1997

Motor Activity and Movement Disorders: Research Issues and Applications, edited by **Paul R. Sanberg, Klaus-Peter Ossenkopp,** and **Martin Kavaliers,** 1996

Neurotherapeutics: Emerging Strategies, edited by **Linda M. Pullan** and **Jitendra Patel,** 1996

Neuron–Glia Interrelations During Phylogeny: II. Plasticity and Regeneration, edited by **Antonia Vernadakis** and **Betty I. Roots,** 1995

Neuron–Glia Interrelations During Phylogeny: I. Phylogeny and Ontogeny of Glial Cells, edited by **Antonia Vernadakis** and **Betty I. Roots,** 1995

The Biology of Neuropeptide Y and Related Peptides, edited by **William F. Colmers** and **Claes Wahlestedt,** 1993

The Neuronal Environment

Brain Homeostasis in Health and Disease

Edited by

Wolfgang Walz

Department of Physiology, University of Saskatchewan, Saskatoon, Saskatchewan, Canada

Humana Press **Totowa, New Jersey**

999 Riverview Drive, Suite 208
Totowa, New Jersey 07512

www.humanapress.com

This publication is printed on acid-free paper. ∞
ANSI Z39.48-1984 (American Standards Institute) Permanence of Paper for Printed Library Materials.

Production Editor: Diana Mezzina

Cover Illustration: Figure 9 from Chapter 4, "Transmitter-Receptor Mismatches in Central Dopamine, Serotonin, and Neuropeptide Systems," *Further Evidence for Volume Transmission*, by A. Jensson, L. Descarries, V. Cornea-Hébert, M. Riad, D. Vergé, M. Bancila, L. F. Agnati, and K. Fluxe.

Cover design by Patricia Cleary.

For additional copies, pricing for bulk purchases, and/or information about other Humana titles, contact Humana at the above address or at any of the following numbers: Tel.: 973-256-1699; Fax: 973-256-8341; E-mail: humana@humanapr.com; or visit our Website: www.humanapress.com.

Printed in the United States of America. 10 9 8 7 6 5 4 3 2 1

Library of Congress Cataloging in Publication Data

The neuronal environment: brain homeostasis in health and diease/edited by
Wolfgang Walz
p. cm.--(Contemporary neuroscience)
Includes bibliographical references and index.
ISBN : 0-89603-882-3 (alk. paper)
1. Neurons--Physiology. 2. Homeostasis. 3. Neuroglia. 4. Brain--Metabolism. 5. Blood-brain barrier.
I. Walz, Wolfgang. II. Series.

QP363.N47758 2002
612.8'2--dc21

2001039827

Preface

To function properly, neurons cannot tolerate fluctuations of their local environmental variables. This mainly results from their high degree of specialization in synaptic integration and action potential conduction. Even small changes of certain extracellular ion concentrations, as well as in the dimensions of the extracellular space, alter ion channel kinetics in such a way as to distort the information represented by the nerve impulses. Another potential problem is the huge consumption of glucose and oxygen by neurons caused by the heavy compensatory ion pumping used for counteracting passive ion flux. This problem is compounded by the low glucose storage capacity of the neurons. A complicated structure surrounds the neurons to sustain the required level of metabolites and to remove waste products.

The Neuronal Environment: Brain Homeostasis in Health and Disease examines the function of all the components involved, including their perturbation during major disease states, and relates them to neuronal demands. The two introductory chapters focus on neuronal requirements. The dependence of their excitability on external factors that accumulate in the extracellular space, as well as their varying demands for energy metabolites, are described. Following that, the close interaction of neurons with elements of their microenvironment is illustrated. The extracellular space is no longer seen as a passive constituent of the CNS, but as a separate compartment in its own right, as a communication channel, and an entity that reacts with plastic changes in its size that will affect the concentrations of all its contents. Astrocytes participate in many neuronal processes, particularly in the removal of excess waste and signal substances, the supply of energy metabolites, and the modulation of synaptic transmission. In addition to their homeostatic role, astrocytes are now seen as an active partner involved in synaptic transmission between neurons. The classical example of a close relationship of neurons with a component of their environment is, of course, their relationship with the surrounding myelin sheath. This speeds up action potential conduction, but is itself a potential source of problems in various disease states. In the last few years new imaging techniques have demonstrated a close coupling between local blood flow and neuronal activity, and several theories have been put forward to explain these interactions. The special status of the brain in having its own insulated circulation system—the cerebrospinal fluid contained in the ventricles and ducts—is also underlined. The brain is the only organ that is protected from fluctuations of blood-borne chemicals by the existence of the blood–brain barrier. However, windows exist in this barrier in the form of the circumventricular organs that allow direct two-way communication between neurons and blood constituents. Finally, despite their protection and insulation, the neurons are accessible to the immune system. Resident macrophages and invasion by blood-borne immune cells that cross the endothelial cell barrier enable an immune reaction to take place. This complex interaction of neurons with their immediate environment is integral to the tasks that the neurons must perform to ensure that the organism can cope with its environmental challenges. Most diseases originating in the brain start in these accessory systems of the neuronal microenvironment and affect neurons only second hand. Therefore, understanding the elements of the neuronal environment and the interactions with neurons, and with each other, is crucial in understanding the development and impact of most brain diseases.

All the authors contributing to *The Neuronal Environment: Brain Homeostasis in Health and Disease* have made an attempt not only to explain the normal functioning of these accessory elements, but also their involvement in major diseases. Therefore, this book not only addresses researchers, graduate students, and educators who want to understand the complex environment of neurons, but also health professionals who need to know more about the normal homeostatic role of the neuronal environment to follow disease patterns.

Wolfgang Walz

Contents

Contributors

LUIGI FRANCESCO AGNATI, *Department of Human Physiology, University of Modena, Modena, Italy*

JAMES W. ANDERSON, *Department of Physiology, Queen's University, Kingston, Ontario, Canada*

MIRCEA BANCILA, *Laboratoire de Neurobiologie de Signaux Intercellulaires, Institut des Neurosciences, Université Pierre et Marie Curie, Paris, France*

GIORGIO CARMIGNOTO, *Department of Experimental Biomedical Sciences, University of Padova, Padova, Italy*

SOPHIE CHABOT, *Department of Oncology and Clinical Neurosciences, University of Calgary, Calgary, Canada*

VIRGINIA CORNEA-HÉBERT, *Département de Pathologie et Biologie Cellulaire, Université de Montréal, Montréal, Canada*

MARC DEL BIGIO, *Department of Pathology, Health Sciences Centre and University of Manitoba, Winnipeg, Canada*

LAURENT DESCARRIES, *Département de Pathologie et Biologie Cellulaire, Université de Montréal, Montréal, Canada*

STEPHEN DOMBROWSKI, *Department of Neurosurgery, Cleveland Clinic Foundation, Cleveland, OH*

ANDREAS FAISSNER, *Laboratoire de Neurobiologie du Developpment et de la Regeneration, Strasbourg, France*

ALASTAIR V. FERGUSON, *Department of Physiology, Queen's University, Kingston, Ontario, Canada*

KJELL FUXE, *Department of Neuroscience, Karolinska Institute, Stockholm, Sweden*

JEREMY GARWOOD, *Laboratoire de Neurobiologie du Developpment et de la Regeneration, Strasbourg, France*

ALBERT GJEDDE, *The Pathophysiology and Experimental Tomography Center, Aarhus General Hospital, Aarhus C, Denmark*

NICOLAS HECK, *Centre National De la Recherche Scientifique, Strasbourg, France*

DAMIR JANIGRO, *Division of Cerebrovascular Research, Department of Neurosurgery, Cleveland Clinic Foundation, Cleveland, OH*

ANDERS JANSSON, *Department of Neuroscience, Division of Cellular and Molecular Neurochemistry, Karolinska Institute, Stockholm, Sweden*

Bernhard H.J. Juurlink, *Department of Anatomy and Cell Biology, University Saskatchewan, Saskatoon, Canada*

Hiroyuki Kato, *Department of Neurology and Neuroendovascular Therapy, Tohoku University School of Medicine, Sendai, Japan*

Richard F. Keep, *Departments of Surgery and Physiology, University of Michigan, Ann Arbor, MI*

Jeffery D. Kocsis, *Neuroscience Research Center, Department of Veterans Affairs Medical Center, Yale University School of Medicine, West Haven, CT*

Imad Najm, *Department of Neurosurgery, Cleveland Clinic Foundation, Cleveland, OH*

Takanori Oikawa, *Department of Neurology, Tohoku University School of Medicine, Sendai, Japan*

Mustapha Riad, *Departement de Pathologie et Biologie Cellulaire, Universite de Montreal, Montreal, Canada*

Franck Rigato, *Centre Natioanl De la Recherche Scientifique, Strasbourg, France*

Avital Schurr, *Department of Anesthesiology, University of Louisville, School of Medicine, Louisville, KY*

Eva Syková, *Department of Neuroscience, Institute of Experimental Medicine, Academy of Sciences, Prague, Czech Republic*

Daniel Vergé, *Laboratoire de Neurobiologie de Signaux Intercellulaires, Institut des Neurosciences, Université Pierre et Marie Curie, Paris, France*

Wolfgang Walz, *Department of Physiology, University of Saskatchewan, Saskatoon, Canada*

Roy O. Weller, *Department of Microbiology and Pathology, Southhampton General Hospital, Southampton, UK*

V. Wee Yong, *Departments of Oncology and Clinical Neurosciences, University of Calgary, Calgary, Canada*

Micaela Zonta, *Department of Experimental Biomedical Sciences, University of Padova, Padova, Italy*

I

Neuronal Activity and Its Dependence on the Microenvironment

1
Central Nervous System Microenvironment and Neuronal Excitability

Stephen Dombrowski, Imad Najm, and Damir Janigro

1. INTRODUCTION

The biological cell membrane, the interface between the cell and its environment, is a complex biochemical entity, one of whose major jobs is to allow or impede transport of specific substances in one direction or another. A related major job of the cell membrane is the maintenance of chemical gradients, particularly electrochemical gradients, across the plasma membrane. These gradients can be of high specificity (e.g., sodium vs. potassium ions), and of great functional significance (e.g., in the production of action potentials in nerve and muscle cells) *(1)*.

The separation of intra- and extracellular compartments by lipidic bilayers is one of the crucial steps in evolution. One of the consequences of this partition is the significant difference in the cytosol and extracellular contents of cells. Furthermore, cells with different functions tend to have different intracellular composition, and cellular elements from different tissues are exposed to extracellular media of different chemical nature. In addition to a variety of nutrients and growth factors, the extracellular milieu also contains molecules that either promote cell differentiation (e.g., adhesion molecules) or survival (growth factors), as well as ions constituting the basis of electrical activity (or silence) of mammalian cells. Granting that appropriate control of the composition of the extracellular space significantly impacts the cytosolic content, and vice versa, change in the intracellular components of central nervous system (CNS) cells impacts the composition of extracellular fluids. The dynamic process involved in the maintenance of the composition of intra- and extracellular ingredients is called "homeostasis."

The general design used for the separation of intracellular and extracellular space has also been used during the evolution to maintain the nervous system of vertebrates, isolated, at least in part, from systemic influences. Therefore, a double bilayer, similar to the lipophilic barrier isolating the cytoplasm from the external milieu and formed by brain microvascular endothelial cells [the blood–brain barrier (BBB)], separates the CNS from the blood, in vertebrates.

From a neuroscientist's point of view, the fact that the neuronal extracellular milieu composition is controlled by such a complex cascade of serially occurring events best illustrates the relevance of controlled neuronal activity to ensure the organism's

From: *The Neuronal Environment: Brain Homeostasis in Health and Disease*
Edited by: W. Walz © Humana Press Inc., Totowa, NJ

Table 1
Examples of Homeostatic Mechanisms in CNS and Their Possible Involvement in Pathogenesis

	Mechanisms involved	Cell types involved	Pathology	Refs.
Barrier function	BBB	Endothelium	Brain tumors	
	Choroid plexus	Neuroepithelium	Stroke	
	Brain–CSF barrier	Pia–glia	Hypertension	*(90)*
			Alzheimer's	*(91)*
	P-glycoprotein	Endothelium	Epilepsy	
Transport of nutrients and neurotransmitters	Glucose transport GLUT1–GLUT3	Endothelium Astrocytes	GLUT1 deficiency Epilepsy Alzheimer's	
	Amino acid transport	Neurons		*(92–96)*
		Glia		*(43,45,46,97,98)*
	GLAST	Endothelium		
Ion homeostasis	Na^+/K^+-ATPase	Neurons	Epilepsy	
		Glia	Vascular dementia	
		Endothelium		
	Inward rectifier	Astrocytes		
Metabolic control of CNS function	Autoregulation Systemic influences	Vascular smooth muscle; glia, neurons	Head injury	*(99,100)*

survival (Table 1). The following paragraphs summarize some of the most relevant mechanisms involved in the regulation of neuronal excitability by factors present in the extracellular milieu.

2. CELLULAR CORRELATES OF BRAIN HOMEOSTASIS

2.1. *Neuroglia*

The necessity for tight control of the composition of brain extracellular fluids is in part a consequence of the evolutionary push for miniaturization of the cellular components of the CNS (neurons and glia), paralleled by the need to produce ultrafast signaling at the neuronal synapse, and to allow comparably fast neural transmissions along axons. Functional compromise between a high velocity of neuronal computation and reduced size of the neuron-to-neuron axonal wiring has been reached, in vertebrates, by ensheathing the axons by a myelin isolator produced by oligodendroglia, allowing for so-called "saltatory conductance" *(2)*. One of the clear advantages of this design is that the myelin sheath occupies much less volume than an equally conductive axon with a much larger diameter would occupy (for mathematical modeling and other biophysical considerations, *see* ref. *3*).

Miniaturization of the vertebrate CNS occurred as a consequence of the necessity to protect the brain and spinal cord with a bony structure, limiting the overall volume available for cellular expansion. A consequence of this limiting factor is that the extracellular space in the brain is very small, amplifying the concentration changes occurring across the plasmalemma surrounding the cells *(4)*. The size of the extracellular space is not homogeneous, and regional differences have been found, even within the contiguous CA1 and CA3 hippocampal regions *(5)*. The possibility that these regional variations also relate to different glial subpopulations within the hippocampus has been proposed *(6)*.

Finally, in an attempt to further minimize the cellular number and volume of the CNS, the lymphatic drainage apparatus has been sacrificed, leaving the composition of extracellular fluids in the brain at the mercy of the brain cells themselves. The subsequent necessity to shield the central nervous system from uncontrolled systemic influences, and in order to minimize the extravasation of potentially noxious or osmotically active molecules from the blood, is perhaps the best-understood reason for the creation of the blood–brain-barrier *(7–9)*. Similarly, the requirement for an extralymphatic mechanism of clearance and homeostasis constitutes the teleonomic reason for the numeric preponderance of glial cells in the mammalian central nervous system. These glia are directly responsible for the control of the composition of the extracellular space.

Glial cells themselves do not constitute a homogeneous population, and at least three classes of glial cells have been described. Oligodendroglia are primarily responsible for the production of myelin, which isolates axons, leaving unsheathed segments with high densities of sodium and potassium channels *(10,11)*. Astrocytes are present in both gray and white matter of the CNS, and are perhaps the most numerous subpopulation of glial cells. Astrocytes are involved in a number of different processes, including the control of ionic homeostasis, control of neuronal metabolism, as well as maintenance of blood–brain barrier integrity *(12–19)*; recent evidence also suggests that they may actively participate in synaptic transmission *(20–23)*. Microglia are the cellular substrates of the neuroimmune response. Their possible role in the homeostasis of CNS extracellular fluids is not known, but these cells express ion channels involved in the control of potassium homeostasis performed by astrocytes *(24,25)*.

2.2. *Vascular Endothelium and Smooth Muscle*

In addition to parenchymally located glial cells, at least two additional cell types participate in the process of the control of the composition of the extracellular space in the brain: the cellular elements constituting intraparenchymal vessels, the endothelial cells lining the intraluminal portion of blood vessels, and only cellular element constituting the BBB at the capillary level; and vascular smooth muscle, the final effectors responsible for the control of cerebral perfusion.

There are numerous ways by which these vascular elements may cooperate with parenchymal glia toward the maintenance of a stable extracellular milieu. BBB endothelial cells are believed to control ionic homeostasis, by preventing equalization of plasma levels of ions with those present in the cerebral spinal fluid *(26–28)*. Part of this process is energy-dependent, and directly impacts the ionic homeostasis for potassium ions (*see* Subheading 3.).

Vascular smooth muscle are also indirectly involved in the control of brain homeostasis, since their powerful effect on the control of cerebral perfusion will be the final determinant of the amount of oxygen and glucose delivered to the brain, as well as to the level of "cleansing" by cerebral blood flow of potential noxious metabolites produced by neural activity. The control of cerebral circulation is mostly independent of extrinsic neuronal influences *(29)*. Both capillary function and the amount of blood perfusing the brain parenchyma are directly proportional to the metabolic activity of neuronal cells, a phenomenon called "autoregulation," which appears to depend on a number of different mechanisms, including nitric oxide, adenosine, potassium, and pH *(30–35)*.

Finally, vascular (endothelial cells and vascular smooth muscle) and parenchymal (neurons and glia) cells cooperate closely, and directly influence each other's development. The best-understood mechanism of this tight cell-to-cell interaction is perhaps the ontogenesis of the blood–brain barrier, a phenomenon directly dependent on the presence of abluminal glial endfeet, which transmit as-yet unknown signals to neighboring endothelial cells *(17,36,37)*. This example clearly illustrates one of the unique mechanisms by which the central nervous system parenchyma influences the cerebral vasculature, without involvement of signals generated distally, a feature that is common in the systemic circulation, where barrier function is not crucial, because of the presence of lymphatic drainage. Note that this general difference does not apply to highly specialized peripheral systems, such as the testicle, where active barrier function is bestowed upon capillary endothelial cells *(38)*.

3. BASIC ELECTROPHYSIOLOGY AS RELEVANT TO EXTRACELLULAR SPACE (ECS) HOMEOSTASIS

Electrical phenomena occur whenever charges of opposite sign are separated or moved in a given direction. Static electricity is the accumulation of electric charge: An electric current results when these charges flow across a permissive material, called a "conductor." An ion current is a particular type of current carried by charges present on atoms or small molecules flowing in aqueous solution. Separation of charges in an aqueous solution can be achieved by inserting an impermeable membrane in the solution itself. In mammalian cells, these membranes coincide with the plasma membrane, and its lipophylic composition ensures a remarkable level of electrical isolation for cells and tissues. Excitable cells, as well as most nonexcitable cells, are characterized by an asymmetric distribution of electrical charges across the plasma membrane.

The control of the distribution of electrical charges across the plasma membrane is an energy-consuming process. A significant portion of this homeostatic control involves the tight regulation of sodium and potassium gradients. The molecular mechanism responsible is the so-called "Na^+/K^+-adenosine triphosphatase (ATPase)," an ubiquitous enzyme whose activity is highly dependent on intra-cellular levels of ATP. It is clear that even minimal changes in the availability of energy substrates (ATP) will cause significant changes in the resting potential of the cells. It is well understood that intracellular Na^+ concentrations are controlled by the exchange of three Na^+ against two K^+, an electrogenic mechanism that contributes substantially to the regulation of resting membrane potential (RMP) in both neurons and glia. The activity of this enzyme depends, in addition to availability of ATP, on internal Na^+ and external K^+ concentrations, and, since $[K^+]$ (and, to a lesser extent, $[Na^+]$) are the main ionic mechanism of the generation of a stable resting membrane potential *(39)*, it becomes obvious that energy supply, ionic homeostasis, and the control of RMP are closely interconnected mechanisms. Because the probability of neuronal firing depends to a large extent on the transmembrane voltage, the link between ionic homeostasis and neuronal excitability becomes evident.

Neuronal cells use a single type of long distance signaling strategy, based on the propagation of all-or-nothing action potentials. Sodium action potentials, such as those recorded in axons or cell bodies, are relatively invariant in normal tissue, and thus the shape and duration of these electrical signals does not vary significantly within the nervous system. Calcium action potentials are similarly predictable, but the underlying ionic mechanism can be complex, depending on the cell type, and on the topographic location within the cell. The terms "sodium action potential" and "calcium action potential" refer to the initial (depolarizing) phase of these rapid membrane polarity changes, and, although genetic or molecular alteration of I_{Na} and I_{Ca} can significantly affect neuronal firing and, ultimately, CNS/peripheral nervous system neurophysiology, gross changes in neuronal excitability may also result by altering the repolarization phase of individual action potentials because of the dramatic changes in extracellular potassium concentrations that accompany neuronal firing, and the consecutive feedback effect of $[K^+]_{out}$ on neuronal resting membrane potential (Fig. 1).

From a functional standpoint, the genesis of fast, sodium action potentials is a hallmark of neuronal function, to the degree that during neurophysiological recordings, presence or absence of Na^+ spikes is frequently used to determine the neuronal or glial cell type *(40–42)*. Recently, this notion has been challenged, and glial action potentials have been reported with increasing frequency *(43–46)*. These responses, however, usually appear to be associated with pathologic conditions (brain tumors, epilepsy), and the old perception that neuronal cells are the exclusive tenants of sufficient I_{Na} density, to promote active responses, is still generally accepted.

Although it is obvious that any significant ionic flux across neuronal membranes will invariably lead to changes in the extracellular/intercellular milieu composition,

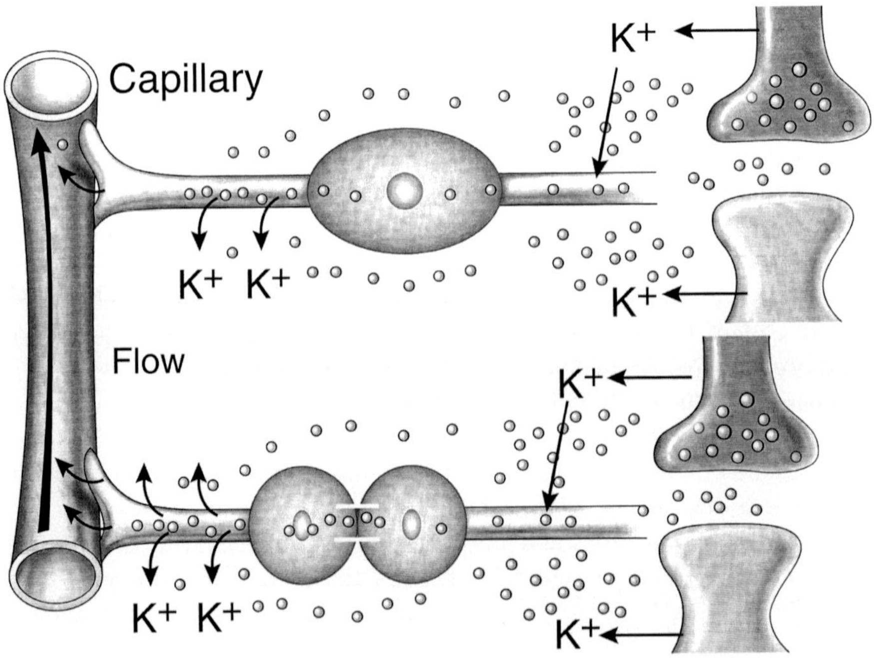

Fig. 1. Diagrammatic representations of potassium fluxes into the CNS. This scheme is based on original spatial buffering concepts described by Orkand *(46a)*, as well as from results inferred from experiments on isolated cells and BVs isolated from the brain *(30)*. The depolarization of pre- and postsynaptic terminals depicted in the right side of the picture causes opening of voltage-dependent potassium channels in neurons. Activation of outward potassium currents causes large potassium fluxes from the cytoplasm to the ECS. Although a fraction of excess potassium ions may directly return into the neuronal cell by active transport via Na^+/K^+-ATPase (not shown in figure), additional uptake of potassium occurs, under most conditions, by voltage-dependent uptake into astrocytic endfeet. Fluxes of potassium through the glial syncytium may then lead either to return of K^+ into the ECS surrounding the neurons, or, perhaps, under more extreme conditions, to release of excess potassium into the blood stream by glial endfeet. The top part of the figure represents the passage of potassium across one single astrocyte, characterized by a cell body and endfeet surrounding a BV, as well as an ensheathment of synaptic terminals. The bottom part of the figure refers to a more common situation, in which multiple glial cells are coupled by gap junctions *(6)*. Gap junction expression is altered in epileptic tissue *(45)*.

the following subheading describes in some detail only mechanisms involved in the control of K^+ homeostasis, because K^+ ions have historically been linked to strict control of neuronal excitation by their profound effect on neuronal resting potential and synaptic transmission. Recent evidence from the author's laboratory also suggests that failure of K^+ homeostasis by glial cells may lead to abnormal extracellular fluid composition and a propensity to seizures.

4. POTASSIUM HOMEOSTASIS

Potassium channels are present in virtually every animal cell type, and serve a variety of functions. Historically, these ubiquitous ionic mechanisms were associated with control of cell resting potential and, after the discovery of sodium action potentials, the repolarization phase leading to the recovery of pre-action potential RMPs. Potassium channels belong to a large and complex group that can be divided into functional, structural, or molecular families. Voltage-dependent K channels (K_V) are constituted of six transmembrane regions (S1–S6) and a P or H5 segment between S5 and S6; the selectivity filter contains a specific sequence (glycine-tyrosine-glycine); the voltage sensor consists of positively charged amino acids in the S4 region. Inward rectifier potassium channels (K_{IR}) are distantly related to the voltage-dependent family, and are made of four subunits, each consisting of two transmembrane segments (M1 and M2) and a P or H5 segment located between. These channels do not allow passage of current at positive potentials. The voltage-dependency of the K_V channels depends on the presence of a voltage sensor, but inward rectification is achieved by voltage-dependent blockade of the intracellular portion of the channel pore by cytosolic cations *(47,48)*. Opening of the channels may be achieved by G protein-coupled mechanisms (as in the GIRK subfamily), or by metabolic changes (intracellular ATP, K_{IR} 6.1, or ATP-sensitive potassium channels) *(49–51)*.

4.1. Extracellular Space Composition and Regulation of Neuronal Excitability

Central nervous system astrocytes are strategically located in proximity to excitable neurons, and are sensitive to changes in extracellular ion composition that follow neuronal activity (*see* diagram in Fig. 1). Several lines of evidence suggest that brain glial cells support the homeostatic regulation of the neuronal microenvironment. In cortical regions, glial cells participate in the genesis of the extracellular field potential changes associated with neuronal depolarization and efflux of potassium in the extracellular space *(52–54)*.

Several mechanisms have been proposed to explain how astrocytes sense and react to changes in extracellular potassium concentrations, following both normal and abnormal neuronal activity. As summarized in the previous paragraphs, neuronal excitability is regulated by a complex interaction of voltage-dependent ion currents and synaptically mediated excitatory and inhibitory potentials. In principal neocortical or hippocampal neurons, depolarizing ion conductances involved in action potential generation, are regulated primarily by the voltage-dependent activation/inactivation properties of Na^+ and Ca^{2+} channels; inward Na^+ and Ca^{2+} fluxes also underlie the generation of excitatory postsynaptic potential (EPSPs). Termination of these depolarizing potentials occurs by the voltage- and calcium-dependent activity of intrinsic potassium conductances, and by activation of interneurons, which release inhibitory neurotransmitters to produce inhibitory postsynaptic potentials (IPSPs): The latter are mediated by postsynaptic activation of chloride and potassium currents.

Although I_{Na}, I_{Ca}, and I_{EPSP} are, under physiological conditions, independent of modest changes in the driving force for the permeant ions (since E_{Na} and E_{Ca} are remote with respect to cell resting potential), both repolarizing potassium and IPSP conductances are critically affected by even modest changes in cell RMP, $[K^+]_{out}$ and $[Cl]_{in}/[Cl]_{out}$. Thus, ionic changes directly associated with excitatory (depolarizing) activity seem to impact minimally ionic homeostasis, but repolarization and inhibition are powerful modulators of $[K^+]_{out}$, $[Cl^-]_{in/out}$, and so on. As a consequence, failure to control potassium and chloride homeostasis is likely to promote neuronal excitability by decreasing the efficacy of repolarizing K currents and IPSPs.

The concentration of potassium in the ECS (K_{ECS}), in the mammalian CNS, increases measurably (from 3 to ~4 m*M*) during physiological stimulation; to a larger extent (up to 12 m*M*), during seizures or direct, synchronous stimulation of afferent pathways; and to exceedingly high values (>30 m*M*), during anoxia or spreading depression *(6,41,55,56)*. Despite these rapid and large changes in K_{ECS}, K^+ values return to normal levels in a short period of time. Several mechanisms have been proposed to explain the rapid clearance of K^+ from the ECS, including uptake by glia, passive diffusion, and neuronal reuptake. Experiments have suggested that glial uptake plays a pivotal role under conditions in which there is massive K_{out} accumulation *(41,53,55,57)*. $[K]_{out}$ can also be redistributed through active removal by blood flow, or by passive diffusion through the ECS *(58)*; however, these mechanisms alone are not fast enough to account for the rapid K^+ removal from the ECS seen under experimental conditions.

4.2. Astrocytes and Buffering of ECS Potassium

Glial involvement in CNS potassium homeostasis has been long suspected, but never unequivocally demonstrated in the mammalian CNS. Two different hypotheses have been formulated: K^+ may accumulate directly into astrocytes, and increased local concentrations of K_{ECS} may be buffered through glial cells by current-carried transport mechanisms. The combination of potassium uptake into glial cells, immediately followed by redistribution through electronically coupled glial gap junctions ("spatial buffering" *[59–65]*) provides a valid working hypothesis to explain some of the features of K^+ movements in the extracellular space. The spatial buffering mechanism rests on the following hypotheses: Glial RMP closely follows E_K (i.e., glial cells are selectively and exclusively permeant to K^+); and glial cells form a topographically complex syncytium, by virtue of their tight electrotonic coupling via gap junctions. Both of these hypotheses have been experimentally challenged. A clear correlation between astrocyte RMP and $[K^+]_{out}$ has been described in virtually every glial cell type studied, but RMP more positive than those predicted by a Nernstian behavior have been frequently reported (for discussion, *see* ref. *65*). The deviation of glial RMP from those predicted by Nernstian behavior has been attributed to one or more of the following: the electrogenicity of the Na^+/K^+-ATPase pump; a persistent sodium conductance activated at cell resting potential *(66)*; chloride currents.

A modification of spatial buffering has been described for retinal Müller cells (potassium siphoning) *(67)*. This process is characterized by a topographic segregation of high conductance zones in the plasma membrane. Thus, a large density of potassium channels is localized in the cell region where extracellular space accumulation occurs, and distally, at the glial endfeet, where potassium excretion into the ECS occurs. No significant K fluxes are possible in the central region of the cell, where no K^+ removal or excretion occurs. A similar mechanism could explain several features of extracellular potassium dynamics in cortical structures, and preliminary evidence *(65)*, supporting a preferential distribution of potassium channels in cortical astrocyte membrane has been recently provided; double recordings from neighboring astrocytes demonstrated that a heterogeneous expression of inward rectifier and outward rectifier channels is present in these cells. The proposed model of potassium movements in these syncytia of neocortical astrocytes shared characteristics of both spatial buffering (influx of potassium driven by E_m and E_K) and siphoning (segregated expression of inward rectifier and outward rectifier channels).

4.3. Astrocytic Ion Currents and Potassium Homeostasis

For the purpose of this minireview, we will describe in some detail the endowment of potassium channels expressed by glial cells in the CNS. Their role in the control of extracellular K homeostasis is also briefly summarized.

Glial cells express a variety of potassium currents, but the expression of these currents seems to depend on the glial cell type considered (microglia vs astrocyte vs oligodendroglia), as well as on the experimental conditions used to determine potassium channel expression. Culturing of glia in vitro dramatically changes the expression of a variety of glial specific markers, including ion channels. The relevance of these expression changes is presently unknown. However, since exposure to serum (or, conversely, serum deprivation) plays a major role in cell differentiation, it is possible that, under physiological conditions, the serum-free CNS environment may act as a maturation factor for astrocytic differentiation. When and where brain homeostasis fails, as the result of opening of the blood–brain barrier, serum proteins will extravasate into the CNS: This may have profound effects on glial ion channel expression, and play a major pathogenetic role.

It has been known for a long time that the predominant ion channel mechanism expressed in astrocytes is the so-called "inward rectifier potassium channel." The properties of these channels are consistent with the mechanisms involved in the simultaneous control of RMP and voltage-dependent uptake of potassium from the extracellular space. The coupling of these channels to this dual control mechanism justifies, in part, the old spatial-buffering theory proposed many years ago by Orkand (1986). The mechanism of potassium entry into the cell is consistent with the biophysical properties of inwardly rectifying potassium channels *(68,69)*, but it is still unclear how the spatial redistribution of potassium ions occurs, and what mechanisms are used by astrocytes to return potassium ion to the extracellular space. The presence of both inward and outward currents on astrocytes, if topographically segregated as shown in retinal astrocytes *(67,70)*, would account for both uptake and redistribution of potassium across cortical structures.

4.3.1. Potassium Uptake via Voltage-Dependent Currents

Glial cells lack regenerative, AP-like responses. However, glial cells express a number of voltage-, second messenger-, and agonist-operated channels *(71–73)*. Potassium channels are the most common electrophysiological feature of both cultured and *in situ* astrocytes, and can be categorized as follows: channels that allow inward, but not outward, current flow (inward rectifiers [K_{IR}]); channels that allow outward, but not inward, current flow (delayed rectifier [I_{DR}]; transient outward current [I_A]); channels that are opened by intracellular calcium ($I_{K(Ca)}$). Glial potassium channels differ in their sensitivity to blockers: Inward rectifiers are blocked by submillimolar concentrations of external Cs^+ and Ba^{2+}; outward I_{DR} and I_A are both sensitive to tetraethylammonium and 4-aminopyridine, but I_A blockade by tetraethylammonium requires high concentrations. Recently, a mixed-cation channel (I_{ha}), permeant to K^+ and Na^+, has

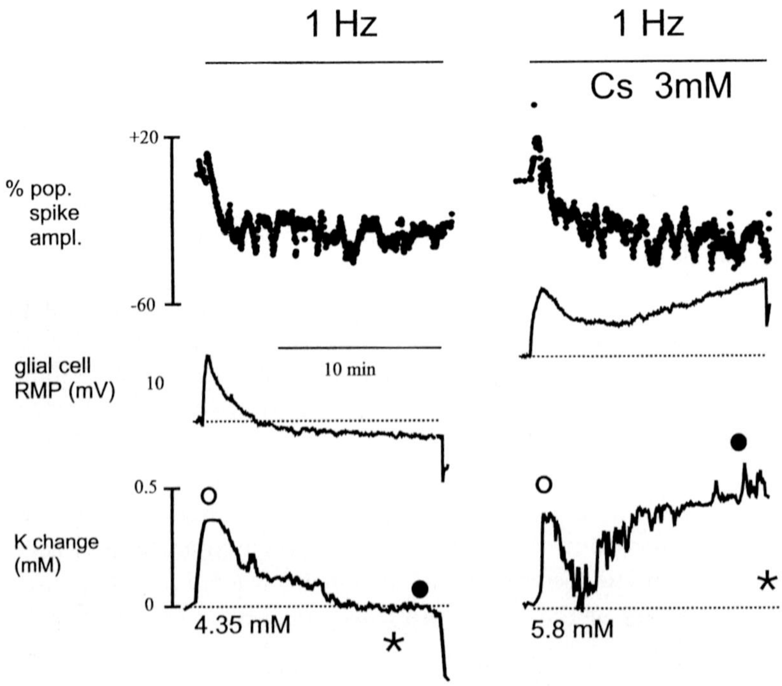

Fig. 2. Changes in neuronal activity, EC potassium, and glial resting potential, after chemical ablation of spatial buffering by cesium. The left panel shows the size of EC field potentials recorded in the CA1 region of the hippocampus during a 1 Hz stimulation trial of 15 min in duration. The traces below refer to the IC recording, by patch clamp, of glial cell resting potential during the same period of time, as well as changes in EC potassium. Note that decreased field potential amplitude does not necessarily correlate with either glial or EC potassium changes. After exposure of the cells to a saturating concentration of cesium, the EC field potential response was not significantly altered, suggesting that direct neuronal effects were absent. However, profound changes in basal concentration of EC potassium (indicated by the millimole values under the bottom traces), as well as glial resting potential, occurred. (Reproduced with permission from ref. *41*.)

been described in cultured astrocytes *(74)*. A cardiac-type outward potassium current has been described in *in situ* glia *(25)*; this current (I_{HERG}) seems to be involved in potassium homeostasis in cooperation with K_{IR}. A direct demonstration that K_{IR}, I_{HERG}, or I_{ha} play a role in spatial K^+ buffering is still lacking, but evidence from both in vivo and in vitro studies has demonstrated proepileptogenic neuronal changes after application of mostly glia-specific potassium channel blockers (*[6,25,41]*; *see also* Subheading 4.3.2., and Figs. 2–6).

Voltage-dependent, tetrodotoxin-sensitive and -insensitive sodium channels are also expressed in both cultured and *in situ* glial cells *(75)*. Although astrocytes are incapable of producing action potential-like responses (but *see* ref. *73*), possibly because of the low Na^+ current densities in these cells, a role for Na^+ channels in spatial buffering has been proposed. According to this hypothesis Na^+ influx sustains the Na^+/K^+-ATPase pump, resulting in net K^+ uptake. Finally, calcium channels are represented sparingly in glial cells, and require either neuronal or otherwise-differentiating factors for

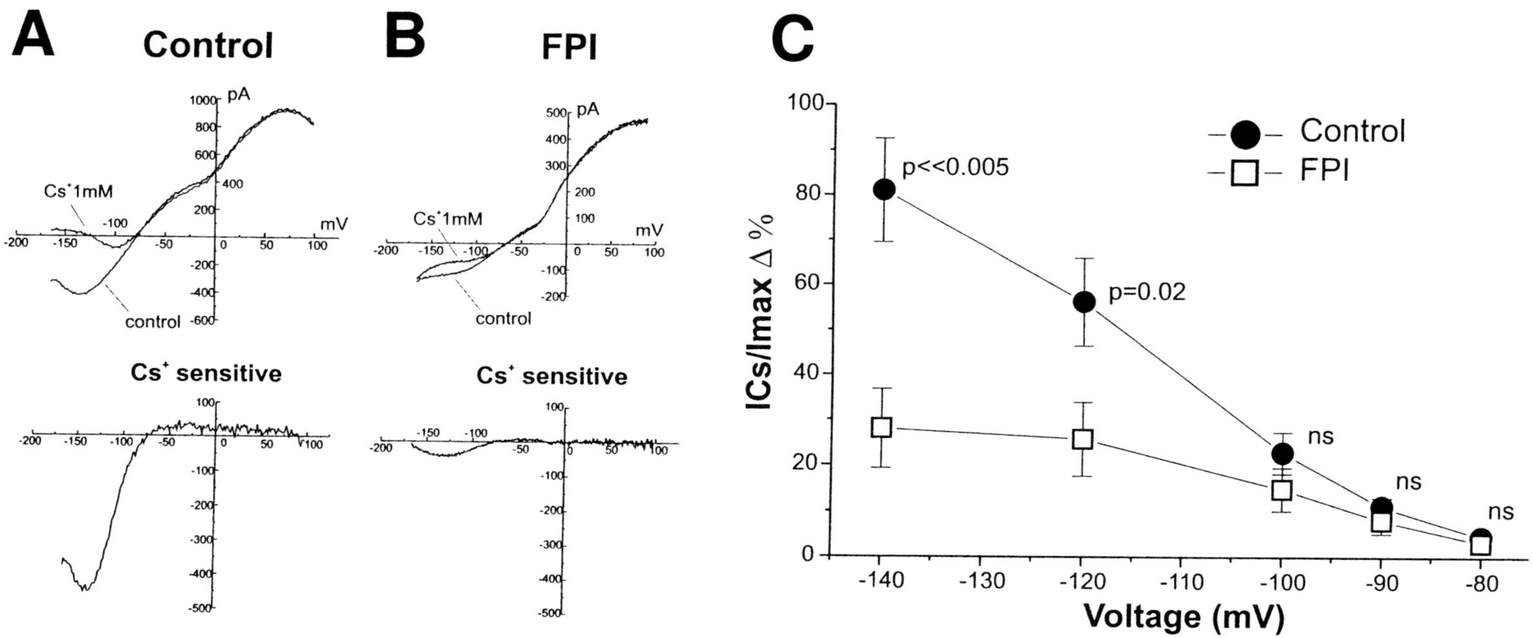

Fig. 3. Astrocytes lose inward K^+ currents after in vivo traumatic brain injury. Whole-cell voltage-clamped cells in control hippocampal slices exhibited large Cs^+-sensitive currents (**A**, **top** and **bottom panel**), and were characterized by a large Cs^+-sensitive component. In contrast, cells in post-FPI hippocampal slices displayed little Cs^+-sensitivity (**B**, **top** and **bottom panel**), and showed a decreased Cs^+-sensitive component of the whole-cell inward currents. **(C)** The percentage of Cs^+-sensitive currents (I_{Cs}) for glia in normal and post-FPI hippocampus is shown for membrane potentials from –140 to –80 mV. Voltage commands consisted of ramps from –170 to +100 mV, over 750 ms, from holding potential of –70 mV. (Reprinted with permission from ref. *79*.)

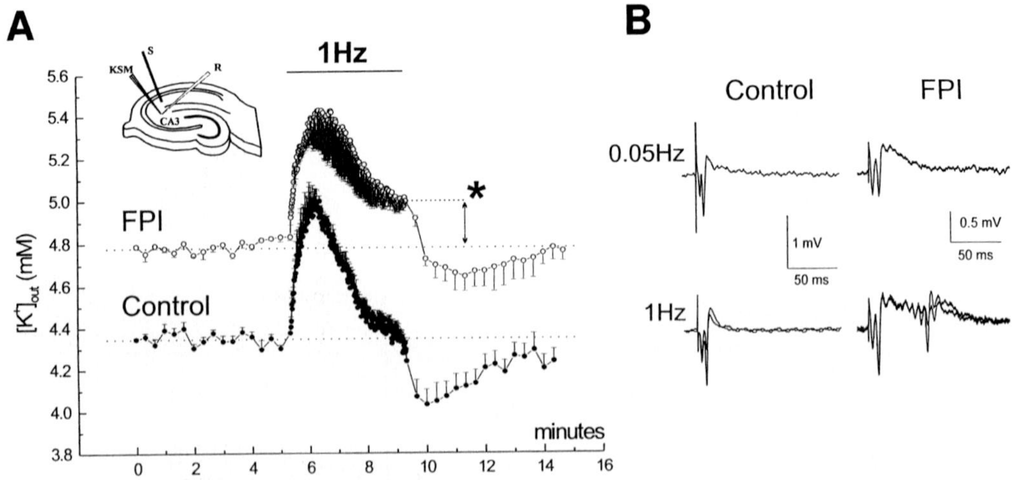

Fig. 4. Neuronal stimulation induces abnormal accumulation of EC K^+ and burst discharge in slices from post-traumatic rats. Field electrode and KSM were placed in CA3 stratum radiatum. A stimulating electrode was placed in CA2 stratum radiatum. K^+ activity recordings were performed during 0.05- and 1-Hz antidromic stimulation. **(A)** Control slices (filled circles) had a basal $[K^+]_{out}$ similar to that of bathing a CSF. Antidromic stimulation, at 1 Hz for 4 min, induced a transient elevation of $[K^+]_{out}$ to about 5 m*M*, and its recovery toward baseline values within the fourth minute. During the following 0.05 Hz, $[K^+]_{out}$ transiently decreased to about 4 m*M*, then recovered. Post-FPI slices (empty circles) had elevated basal $[K^+]_{out}$ during stimulation at 0.05 Hz. When the high-frequency stimulation was performed, $[K^+]_{out}$ transiently increased to 5.4 m*M*, then decreased to 5 ± 0.05 m*M*, without reaching the baseline value (asterisk, $p < 0.001$). During the following 0.05 Hz, $[K^+]_{out}$ transiently decreased to ~4.7 m*M*. **(B)** Post-FPI CA3 develops frequency-dependent afterdischarges for antidromic stimulation. In control, only a small fraction of slices developed afterdischarges during antidromic 1-Hz stimulation (28%, 2/7 slices). Post-FPI slices showed a higher excitability. They did not display afterdischarges during 0.05-Hz stimulation, but afterdischarges appeared during 1-Hz stimulation (80%, 8/10 slices). (Reprinted with permission from ref. *79*.)

expression *(71,72)*; whether I_{Ca} can be recorded from *in situ* hippocampal astrocytes is still unknown, but release of calcium from intracellular stores, in response to neurotransmitters acting on astrocytes, has been clearly demonstrated. Relevant to spatial buffering, micromolar $[Ca^{2+}]_i$ can cause opening of $I_{K(Ca)}$, and may thus participate in the generation of outward potassium fluxes.

4.3.2. K⁺ Buffering by Furosemide-Sensitive Na,K,2Cl Cotransporter

Under conditions involving high levels of neuronal activity (e.g., seizures), $[K]_{out}$ accumulation is accompanied by cell swelling. The swelling that accompanies epileptiform neuronal discharge results from excess activity of ionic mechanisms normally involved in the control of ECS homeostasis. One of several proposed mechanisms associated with cell swelling, the Na,K,2Cl co-transporter, is also believed to participate in uptake of K^+ into glia. This transporter is blocked by the general diuretic, furosemide. Treatment of epileptic hippocampal slices (treated with bicuculline, 0 Ca^{2+}, or 4-aminopyridine) with furosemide has been shown to inhibit spontaneous burst discharge. It has been hypothesized *(76)* that this mechanism was related to furosemide

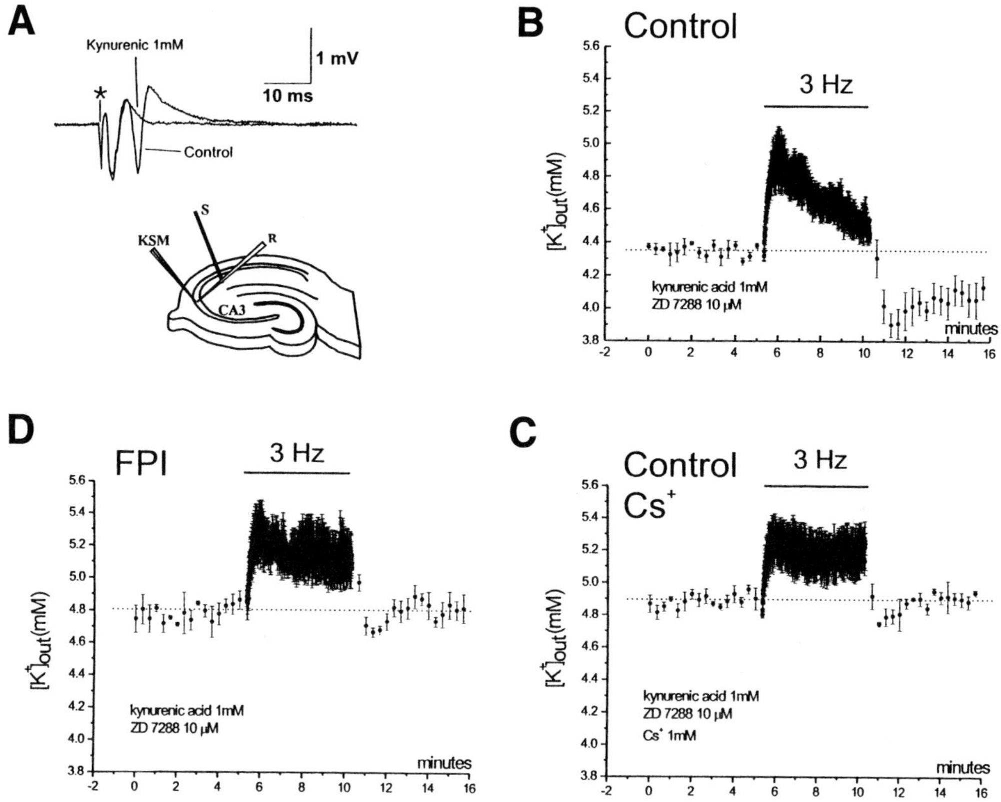

Fig. 5. Abnormal accumulation of EC potassium is caused by impaired glial homeostasis. In naïve slices, 3-Hz antidromic stimulation induced a modest K rise to 4.9 ± 0.1 m*M*. At the end of the 5-min period, $[K^+]_{out}$ was 4.5 ± 0.05 m*M*. In the following 5 min of stimulation at 0.05 Hz, $[K^+]_{out}$ reached the value of 3.9 ± 0.1 m*M* (**B**). (**C**) Cs^+ (1 m*M*), added to the control bath solution, increased baseline $[K^+]_{out}$ to 4.9 m*M*. As expected, blockade of potassium uptake into glia caused exaggerated potassium transients. These were identical to those recorded from Cs^+-free post-traumatic slices.

blockade of the swelling induced by large ionic (and water) shifts that accompany Na,K,2Cl co-transporter activity.

4.3.3. K^+ Buffering by Na^+/K^+-ATPase

Neuronal K^+ reuptake, and part of the hyperpolarizing undershoot that follows the action potential is mediated in part by an energy-dependent process that requires Na^+/K^+-ATPase activity. Similarly, extracellular potassium accumulation into glia may depend on energy-dependent processes.

Na^+/K^+-ATPase activity is regulated by both $[Na^+]_i$ and $[K]_{out}$. Thus, extracellular potassium increases, or Na^+ influx will cause activation of this electrogenic uptake mechanism. As a result, glial cells will accumulate potassium and extrude Na^+, the net result being a hyperpolarization. Whether Na^+/K^+-ATPase-dependent potassium uptake plays any role in K buffering is still controversial, partially because selective pharmacological blockade of the glial pump has been unavailable.

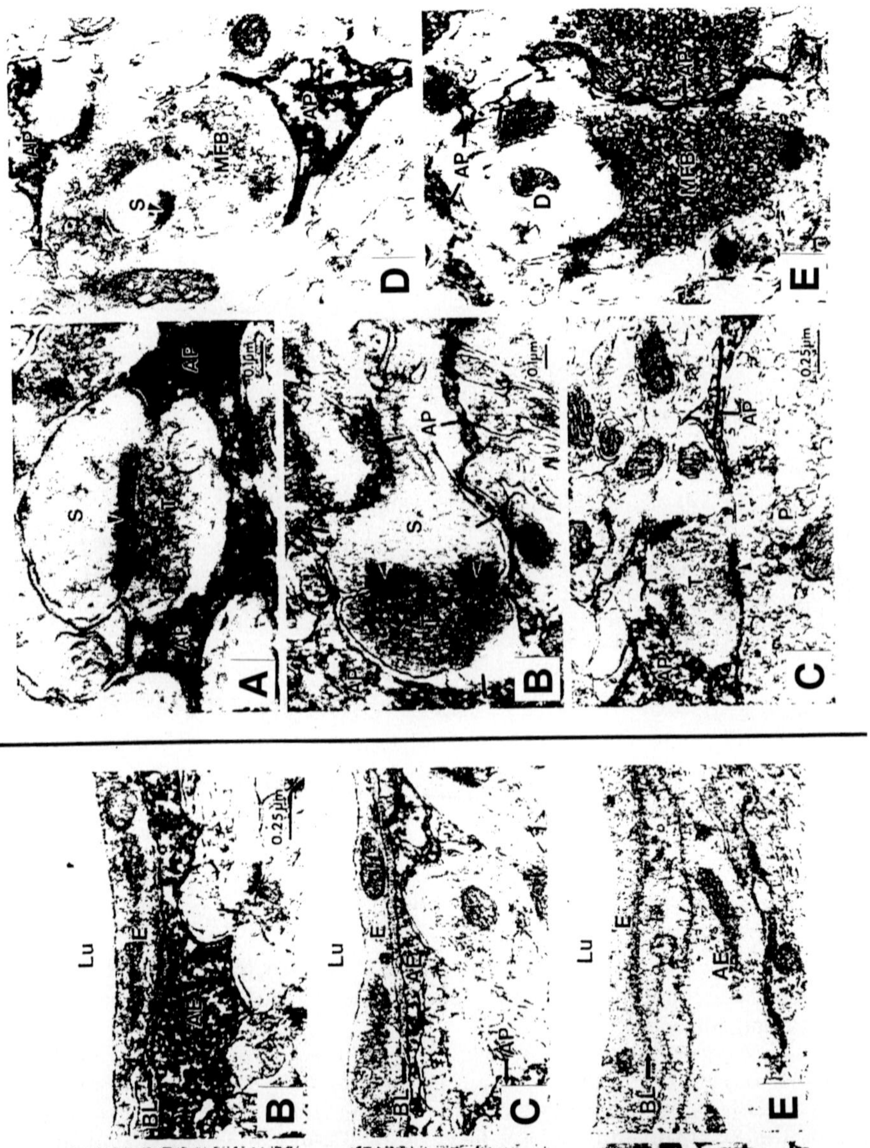

Fig. 6

To summarize this brief synopsis on K^+ buffering by glia: At least three independent mechanisms of potassium uptake into astrocytes exist; those mediated by voltage-dependent ion channel mechanisms are relatively ATP- and energy-independent, but require a negative RMP (and, thus, indirectly, ATP hydrolysis) and high conductance to potassium ions; transporters (such as the furosemide-sensitive transporter) are also energy-independent, but their function results in cellular swelling and subsequent shrinkage of the extracellular space, a condition known to cause hyperexcitability/synchronization; redistribution of excess $[Na]_i$ will eventually require some energy. Finally, energy-dependent pumps cause electrogenic potassium influx, which, under normal ionic conditions, leads to hyperpolarization. Under extreme conditions, such as synchronous neuronal activity during epileptic seizure, and if glial K^+ uptake mechanisms are operational one at a time (one K-uptake mechanism per each individual astrocyte), these mechanisms are likely to fail, as the result of glial depolarization (voltage-dependent uptake mechanisms), cell swelling/ECS shrinkage (Na,K,2Cl co-transporter), or energy failure (Na^+/K^+-ATPase). However, it is known that, even at exceedingly high $[K]_{out}$ (such as during spreading depression), potassium homeostasis still occurs; indeed, extracellular potassium levels are quickly returned to physiological levels following seemingly overwhelming $[K]_{out}$ increases (>40 m*M*). It seems reasonable to assume that all these different pathways for K^+ uptake exist and cooperate in maintenance of ionic homeostasis. However, it is still unclear whether these mechanisms are expressed in all astrocytes, or whether segregation of different K-uptake mechanisms takes place. This question is important, because different brain regions display different levels of sensitivity to insults, such as head injury, ischemia/anoxia, and seizures, which cause, or derive from, perturbation of ionic homeostasis.

5. ION HOMEOSTASIS AND NEUROLOGICAL DISEASE

Substantial progress has been made in the understanding of the pathophysiology and mechanisms involved in the attenuation of brain homeostasis. In many diseases that

---

Fig. 6. *(previous page)* Immunocytochemical (ICC) localization of ERG channel protein in hippocampal astrocytes. **Left panel:** Electron microscopy of biocytin-filled astrocytes (**A** and **C**) is shown for comparison with ICC of ERG (C1) channel protein in astrocytes (**B** and **D**). Note the similar morphological features of hippocampal astrocytes of biocytin-filled and ERG-immunopositive cell bodies (**A** and **B**) and their processes (*arrows*, **C** and **D**). ERG immunoreactivity was confined within the cell body cytoplasm of astrocytes, and within large primary astrocytic processes, and on small, branched processes within the neuropil. Note that an oligodendrocyte (*O*) was immunonegative (**A**). Pyramidal cells did not show ERG immunoreactivity. **Right panel:** ICC localization of ERG channel protein in astrocytes surrounding blood vessels in hippocampal CA1 (**A** and **D**); comparison with biocytin-filled cells (**B** and **C**). **(A)** Low-power magnification of an ERG imunoreactive astrocyte process forming an astrocytic endfoot (AE) around a capillary (BV). The endothelial cell (E) of the capillary wall does not show ERG immunoreactivity. As comparison, the capillary wall from a biocytin-filled astrocytic endfoot is shown in **B**. **(D)** ICC localization of ERG channel protein in astrocytes surrounding blood vessels in hippocampal CA3 subregion; note the immunonegative basal lamina (BL). **(C)** Comparison with the capillary wall of biocytin-filled astrocytes. **(E)** Specificity of the ERG antibody is demonstrated in a control section following preabsorption of the primary antibody. (Reprinted with permission from ref. *25*.)

affect the brain, the astrocytes and cerebral endothelium play an active part in the disease process, with ionic homeostasis becoming disrupted, or modified, in such a way that there is a dramatic increase in tissue osmolarity, followed by increased water content. Thus, a notable consequence of homeostatic failure is cerebral edema. Brightman et al. *(77)*, in the 1960s, classified brain edema into two major types: vasogenic, or "wet" edema; and cellular cytotoxic, or "dry" edema. The extravasation of plasma proteins, resulting from a weakened BBB, causes vasogenic edema. Increased BBB permeability may also be the outcome of enhanced pinocytotic activity. Changes in cyclic nucleotides, arachidonic acid, histamine, and other mediators may contribute to the local changes in BBB permeability. Once established, vasogenic edema can spread under the influence of hydrostatic and oncotic pressure. In cytotoxic edema, the primary target is the intracellular metabolic machinery (i.e., ATP-dependent sodium pump) and/or metabolic substrates. The mechanisms involved in the generation of cytotoxic edema may lead to glial swelling, changes in BBB function, and, finally, production of vasogenic edema.

Cerebral edema is commonly associated with acute episodes of neurological disease (e.g., head injury), but it is also recognized that less-dramatic increases in CNS water content may chronically and regionally accompany persistent conditions, such as multiple sclerosis, Alzheimer's dementia, and perhaps some forms of epilepsy. Here are reviewed only a few examples of experimental evidence linking changes in potassium homeostasis to acute or chronic neurological disease. Table 1 summarizes other homeostatic mechanisms that may impact transmitters, or other molecules and ions.

5.1. Astrocytes and Epilepsy

Given the considerations earlier in this chapter, it is evident that failure of potassium homeostasis may become epileptogenic. Whether this concept applies only to experimental reality, or extends to the true etiology of the disease, is unknown. Indirect evidence linking potassium uptake failure to seizures is, however, accumulating. Potassium channel mechanisms that are believed to be involved in astrocytic function are altered in conditions such as human epilepsy. Several investigators *(43,46,78,79)* have clearly demonstrated deficits of inward rectifier occurrence in reactive astrocytes, in both human and animal models of epilepsy. These reactive glia were found in animal models of neuronal injury, as well as in resections from cortical structures from human epileptic patients affected by epilepsy refractory to drug treatment.

The fact that conditions such as epilepsy are associated with loss of an important mechanism of potassium homeostasis by astrocytes has led to the hypothesis that perhaps ablation of voltage-dependent potassium uptake by glia could lead to neuronal hyperexcitability/synchronization. This has been shown indirectly by two different models. In the first set of experiments, it has been shown that blockade of extracellular potassium uptake into hippocampal astrocytes by millimolar concentration of cesium, causes profound changes in the dynamics of extracellular potassium, as well as important changes in neuronal excitability and synchronization. Furthermore, these manipulations had a profound effect on synaptic plasticity, of as-yet unknown significance for pathological changes *(41)*. Figure 2 shows some of the results of these experiments.

Taken together, these results suggested that chemical ablation of potassium uptake into glia may be epileptogenic. Numerous issues, however, complicate the interpretation of these results, such as possible effects of Cs^+ on neurons, or on channels other than K_{IR} (e.g., I_{ha} or I_h *[74,80–82]*). Further evidence to support the hypothesis that the effects of acutely applied Cs^+ were indeed mediated by glia came from a separate set of experiments, in which it was shown that traumatic brain injury (TBI), induced in an animal model, decreases inward rectifier channel expression in glia. These functional expression changes were comparable to those found in resections from epilepsy patients, and, quantitatively, here virtually undistinguishable from the effects of cesium applied in vitro (Figs. 3–5). This is of relevance, since TBI causes propensity toward unusual hyperexcitability, and may be proepileptogenic *(83,84)*. Thus, these results, combined with knowledge of the glial changes occurring in human epileptic tissue, strongly suggested an etiological, and perhaps temporal, link between an initial astrocytic deficit, leading to loss of homeostatic control, exaggerated $[K^+]_{out}$ transients, and proepileptogenic changes in neuronal function.

Consistent with the idea that loss of inward rectification/potassium uptake mechanisms, leading to neurological disease, may greatly affect neuronal function is the fact that potassium transients, as well as changes in neuronal excitability observed in post-traumatic brains, lacking inward rectification expression in glia, were comparable to proepileptogenic changes observed in naïve slices treated with cesium. The results of these experiments are shown in Figs. 4–6.

Finally, it has been recently shown that "channelopaties," which may be involved in seizure disorders, affect the expression of ion channels usually associated with cardiac function *(85,86)*, and responsible for the so-called "long QT syndrome" *(85)*. Long QT syndrome genes appear to be frequently associated with various forms of human epilepsy *(87–89)*. In the CNS, the gene product responsible for one form of long QT is found primarily in glia (Fig. 6); furthermore, blockade of this current with antiarrhythmic drugs specific for I_{HERG} led to neuronal excitability changes identical to those obtained after exposure of hippocampal slices to Cs^+, or after TBI in vivo *(25)*.

6. CONCLUSIONS

Neuronal excitability is controlled by a variety of factors, including intrinsic and extrinsic mechanisms. The fact that the CNS milieu is normally shielded from systemic influences makes neuronal activity mostly independent of systemic influences. However, failure of these barrier mechanisms, and/or failure of internal mechanisms of ionic homeostasis, may lead to chronic neurological diseases, such as epilepsy. Increasing evidence has linked failure of astrocytic function, particularly uptake of potassium, to epileptic disorders, and experimentally induced defects that are proepileptogenic have also been found in human epileptic tissue.

ACKNOWLEDGMENTS

The authors' work reported here was supported by NIH-1R29 HL51614, NIH-2RO1 HL51614, and NIH-RO1 NS38195. We wish to acknowledge the experimental and conceptual help of many colleagues during the past several years, including Drs. P. A. Schwartzkroin, J. Wenzel, R. D'Ambrosio, A. Emmi, G. Maccaferri, E. Guatteo, S. Grady, and D. Maris.

REFERENCES

1. Hille, B. (1992) in *Ion Channels in Excitable Membranes*, Sinauer Associates, Sunderland, MA.
2. Huxley, A. F. and Stampfli, R. (1949) Evidence for saltatory conductance in peripheral myelinated nerve. *J. Physiol.* (Lond.) **108,** 315–339.
3. Jack, J. J. B., Noble, D., and Tsien, R. W. (1975) *Electric Current Flow in Excitable Cells*, Oxford University Press, New York.
4. Nicholson, C. and Sykova, E. (1998) Extracellular space structure revealed by diffusion analysis. *Trends Neurosci.* **21,** 207–215.
5. McBain, C. J., Traynelis, S. F., and Dingledine, R. (1990) Regional variations of extracellular space in the hippocampus. *Science* **249,** 674–677.
6. D'Ambrosio, R., Wenzel, J., Schwartzkroin, P. A., and Janigro, D. (1998) Functional specialization and topographic segregation of hippocampal astrocytes. *J. Neurosci.* **18,** 1–14.
7. Bradbury, M. W. (1993) The blood-brain barrier. *Exp. Physiol.* **78,** 453–472.
8. Davson, H. and Segal, M. B. (1995) The blood-brain barrier, in *Physiology of the CSF and of the Blood-Brain Barrier* (Davson, H. and Segal, M. B., eds.), CRC, Boca Raton, pp. 49–91.
9. Davson, H. and Segal, M. B. (1995) Blood-brain-CSF interactions, in *Physiology of the CSF and Blood-Brain Barrier*, CRC, Boca Raton.
10. Suchet, S. (1995) The morphology and ultrastructure of oligodendrocytes and their functional implication, in *Neuroglia* (Ransom, B. R. and Kettenmann, H., eds.), Oxford University Press, New York, pp. 23–41.
11. Bunge, R. P. and Fernandez-Valle, C. (1995) Basic biology of the Schwann cell, in *Neuroglia* (Ransom, B. R. and Kettenmann, H., eds.), Oxford University Press, New York, pp. 44–57.
12. Ransom, B. R. and Carlini, W. G. (1986) Electrophysiological properties of astrocytes, in *Astrocytes* (Fedoroff, S. and Vernadakis, A., eds.), vol. 2, Academic, Orlando, pp. 1–49.
13. Sontheimer, H. (1992) Astrocytes, as well as neurons, express a diversity of ion channels. *Can. J. Physiol. Pharmacol.* **70(Suppl.),** S223–238.
14. Bordey, A. and Sontheimer, H. (1997) Postnatal development of ionic currents in rat hippocampal astrocytes in situ. *J. Neurophysiol.* **78,** 461–477.
15. Chesler, M. and Kraig, R. P. (1989) Intracellular pH transients of mammalian astrocytes. *J. Neuroscience* **9,** 2011–2019.
16. Forsyth, R. J. (1996) Astrocytes and the delivery of glucose from plasma to neurons. *Neurochem. Int.* **28,** 231–241.
17. Janzer, R. C. and Raff, M. C. (1987) Astrocytes induce blood-brain barrier properties in endothelial cells. *Nature* **325,** 253–257.
18. Magistretti, P. J. and Pellerin, L. (1995) Cellular basis of brain energy metabolism and their relevance to brain imaging-evidence for a prominent role for astrocytes. *Cereb. Cortex* **5,** 301–306.
19. Newman, E. A. (1986) High potassium conductance in astrocyte endfeet. *Science* **233,** 453–454.
20. Pasti, L., Volterra, A., Pozzan, T., and Carmignoto, G. (1997) Intracellular calcium oscillations in astrocytes: a highly plastic, bidirectional form of communication between neurons and astrocytes in situ. *J. Neurosci.* **17,** 7817–7830.
21. Carmignoto, G., Pasti, L., and Pozzan, T. (1998) On the role of voltage-dependent calcium channels in calcium signaling of astrocytes in situ. *J. Neurosci.* **18,** 4637–4645.
22. Araque, A., Parpura, V., Sanzgiri, R. P., and Haydon, P. G. (1999) Tripartite synapses: glia, the unacknowledged partner. *TINS* **22,** 208–215.
23. Parpura, V., Basarsky, T. A., Liu, F., Jeftinija, K., Jeftinija, S., and Haydon, P. G. (1994) Glutamate-mediated astrocyte-neuron signaling. *Nature* **369,** 744–747.

24. Zhou, W., Cayabyab, F. S., Pennefather, P. S., Schlichter, L. C., and DeCoursey, T. E. (1998) HERG-like channels in microglia. *J. Gen. Physiol.* **111,** 781–794.
25. Emmi, A., Wenzel, H. J., Schwartzkroin, P. A., Taglialatela, M., Castaldo, P., Bianchi, L., et al. (2000) Do glia have heart? Expression and functional role for ether-A-Go-Go currents in hippocampal astrocytes. *J. Neurosci.* **20,** 3915–3925.
26. Schielke, G. P., Moises, H. C., and Betz, A. L. (1991) Blood to brain sodium transport and interstitial fluid potassium concentration during early focal ischemia in the rat. *J. Cereb. Blood. Flow. Metab.* **11,** 466–471.
27. Janigro, D., West, G. A., Gordon, E. L., and Winn, H. R. (1993) ATP-sensitive K+ channels in rat aorta and brain microvascular endothelial cells. *Am. J. Physiol.* **265,** C812–821.
28. Janigro, D., West, G. A., Nguyen, T.-S., and Winn, H. R. (1994) Regulation of blood-brain barrier endothelial cells by nitric oxide. *Circ. Res.* **75,** 528–538.
29. Winn, H. R., Dacey, R. G., and Meyberg, M. R. (1989) Cerebral circulation, in *Textbook of Physiology* (Patton, F. H. S. S., ed.), WB Saunders, Philadelphia, pp. 952–960.
30. Thien-son, N., Winn, H. R., and Janigro, D. (2000) ATP-sensitive K^+ channels may participate in the coupling of neuronal activity and cerebrovascular tone. *Am. Physiol. Soc.* **278,** H878–H885.
31. Winn, H. R., Rubio, R., and Berne, R. M. (1981) The role of adenosine in the regulation of cerebral blood flow. *J. Cerebr. Blood Flow Metab.* **1,** 239–244.
32. Faraci, F. M. and Brian, J. E. J. (1994) Nitric oxide and the cerebral circulation. *Stroke* **25,** 692–703.
33. Iadecola, C. (1993) Regulation of the cerebral microcirculation during neural activity: is nitric oxide the missing link? *Trends Neurosci.* **16,** 206–214.
34. Ishimaru, S., Okada, Y., Mies, G., and Hossmann, K. A. (1993) Relationship between blood flow and blood-brain barrier permeability of sodium and albumin in focal ischaemia of rats: a triple tracer autoradiographic study. *Acta Neurochir. Wien* **120,** 72–80.
35. Kuschinsky, W. (1997) Neuronal-vascular coupling, in *Optical Imaging of Brain Function* (Villringer, A. and Dirnagl, U., eds.), Plenum, New York, pp. 167–176.
36. Pekny, M., Stanness, K. A., Eliasson, C., Betsholtz, C., and Janigro, D. (1998) Impaired induction of blood-brain barrier properties in aortic endothelial cells by astrocytes from GFAP-deficient mice. *Glia* **22,** 390–400.
37. Penkowa, M., Hidalgo, J., and Moos, T. (1997) Increased astrocytic expression of metallothioneins I + II in brainstem of adult rats treated with 6-aminonicotinamide. *Brain Res.* **774,** 256–259.
38. Tsukamoto, H., Hamada, Y., Wu, D., Boado, R. J., and Pardridge, W. M. (1998) GLUT1 glucose transporter: differential gene transcription and mRNA binding to cytosolic and polysome proteins in brain and peripheral tissues. *Brain Res. Mol. Brain Res.* **58,** 170–177.
39. Lauger, P. (1991) Na,K-ATPase, in *Electrogenic Ion Pumps*, Sinauer, Portland, pp. 168–225.
40. Ransom, B. R. and Goldring, S. (1973) Slow hyperpolarization in cells presumed to be glia in cerebral cortex of cat. *J. Neurophysiol.* **36,** 879–892.
41. Janigro, D., Gasparini, S., D'Ambrosio, R., McKhann, G. M., and DiFrancesco, D. (1997) Reduction of K^+ uptake in glia prevents LTD maintenance and causes epileptiform activity. *J. Neurosci.* **17,** 2813–2824.
42. Sypert, G. W. and Ward, A. A. (1971) Unidentified neuroglia potentials during propagated seizures in neocortex. *Exp. Neurol.* **33,** 239–255.
43. O'Connor, E. R., Sontheimer, H., Spencer, D. D., and de-Lanerolle, N. C. (1998) Astrocytes from human hippocampal epileptogenic foci exhibit action potential-like responses. *Epilepsia* **39,** 347–354.
44. Bordey, A., Sontheimer, H., O'Connor, E. R., Sontheimer, H., Spencer, D. D., and de Lanerolle, N. C. (1998) Properties of human glial cells associated with epileptic seizure foci Astrocytes from human hippocampal epileptogenic foci exhibit action potential-like responses. *Epilepsy Res.* **1–2,** 286–303.

45. Lee, S. H., Magge, S., Spencer, D. D., Sontheimer, H., and Cornell-Bell, A. H. (1995) Human epileptic astrocytes exhibit increased gap junction coupling. *Glia* **15,** 195–202.
46. MacFarlane, S. N. and Sontheimer, H. (1997) Electrophysiological changes that accompany reactive gliosis in vitro. *J. Neurosci.* **17,** 7316–7329.
46a. Orkand, R. K. (1966) Effect of nerve impulses on the membrane potential of glial cells in the central nervous system of amphybia. *J. Neurophysiol.* **29,** 788–806.
47. Jan, L. Y. and Jan, Y. N. (1997) Voltage-gated and inwardly rectifying potassium channels. *J. Physiol. (Lond.)* **505,** 267–282.
48. Wible, B. A., Taglialatela, M., Ficker, E., and Brown, A. M. (1994) Gating of inwardly rectifying K channels localized to a single negatively charged residue. *Nature* **371,** 246–249.
49. Robertson, B. (1997) The real life of voltage-gated K^+ channels: more than model behaviour. *Trends Pharmacol. Sci.* **18,** 474–483.
50. Meir, A., Ginsburg, S., Butkevich, A., Kachalsky, S. G., Kaiserman, I., Ahdut, R., Demirgoren, S., and Rahamimoff, R. (1999) Ion channels in presynaptic nerve terminals and control of transmitter release. *Physiol Rev.* **79,** 1019–1088.
51. Levitan, I. B. (1999) Modulation of ion channels by protein phosphorylation. How the brain works. *Adv. Second Messenger Phosphoprotein Res.* **33,** 3–22.
52. Amzica, F. and Steriade, M. (2000) Neuronal and glial membrane potentials during sleep and paroxysmal oscillations in the neocortex. *J. Neurosci.* **20,** 6648–6665.
53. Wadman, W. J., Juta, A. J. A., Kamphuis, W., and Somjen, G. G. (1982) Current source density of sustained potential shifts associated with electrographic seizures and with spreading depression in rat hippocampus. *Brain Res.* **570,** 85–91.
54. Somjen, G. G. (1995) Electrophysiology of mammalian glial cells in situ, in *Neuroglia* (Kettenmann, H. and Ransom, B. R., eds.), Oxford University Press, New York, pp. 319–331.
55. Somjen, G. G. (1979) Extracellular potassium in the mammalian central nervous system. *Annu. Rev. Physiol.* **41,** 159–177.
56. Lux, H. D., Heinemann, U., and Dietzel, I. (1986) Ionic changes and alterations in the size of extracellular space during epileptic activity, in *Advances in Neurology* (Delgado-Escueta, A. V. and Ward, A. A., eds.), vol. 44, Raven, New York, pp. 619–639.
57. Largo, C., Cuevas, P., Somjen, G. G., Martin del Rio, R., and Herreras, O. (1996) The effect of depressing glial function in rat brain *in situ* on ion homeostasis, synaptic transmission, and neuron survival. *J. Neurosci.* **16,** 1219–1229.
58. Lux, H. D. and Neher, E. (1973) The equilibration time course of [K] in rat cortex. *Exp. Brain Res.* **17,** 190–205.
59. Blanco, R. E., Marrero, H., Orkand, P. M., and Orkand, R. K. (1993) Changes in ultrastructure and voltage-dependent currents at the glia limitans of the frog optic nerve following retinal ablation. *Glia* **8,** 97–105.
60. Kuffler, S. W., Nichols, J. G., and Orkand, R. K. (1966) Physiological properties of glial cells in the central nervous system of amphibia. *J. Neurophysiol.* **29,** 768–787.
61. Orkand, R. (1986) Glial-interstitial fluid exchange. *Annu. NY Acad. Sci.* **4,** 269–272.
62. Orkand, R. K. (1966) Effect of nerve impulses on the membrane potential of glial cells in the central nervous system of amphybia. *J. Neurophysiol.* **29,** 788–806.
63. Ransom, B. R. and Orkand, R. K. (1996) Glial-neuronal interactions in non-synaptic areas of the brain: studies in the optic nerve. *Trends Neurosci.* **19,** 352–358.
64. Kuffler, S. W. (1967) Neuroglial cells: physiological properties and a potassium mediated effect of neuronal activity on the glial membrane potential. *Proc. R. Soc. Lond. B. Biol. Sci.* **168,** 1–21.
65. McKhann, G. M., D'Ambrosio, R., and Janigro, D. (1997) Heterogeneity of astrocyte resting membrane potentials revealed by whole cell and gramicidin-perforated patch recordings from cultured neocortical and hippocampal slice astrocytes. *J. Neurosci.* **17,** 6850–6863.

66. Sontheimer, H., Fernandez Marques, E., Ullrich, N., Pappas, C. A., and Waxman, S. G. (1994) Astrocyte Na^+ channels are required for maintenance of Na^+/K^+-ATPase activity. *J. Neurosci.* **14,** 2464–2475.
67. Newman, E. A. (1993) Inward-rectifying potassium channels in retinal glial (Mueller) cells. *J. Neurosci.* **13,** 3333–3345.
68. Ransom, C. B., Sontheimer, H., and Janigro, D. (1996) Astrocytic inwardly-rectifying potassium currents are dependent on extracellular sodium ions. *J. Neurophysiol.* **76,** 626–630.
69. Ransom, C. B. and Sontheimer, H. (1995) Biophysical and pharmacological characterization of inwardly rectifying potassium currents in rat spinal cord astrocytes. *J. Neurophysiol.* **73,** 333–346.
70. Newman, E. A. (1995) Glial cell regulation of extracellular potassium, in *Neuroglia* (Kettenmann, H. and Ransom, B. R., eds.), Oxford University Press, New York, pp. 717–731.
71. Steinhauser, C. (1993) Electrophysiologic characteristics of glial cells. *Hippocampus* **3,** 113–124.
72. Sontheimer, H. (1994) Voltage-dependent ion channels in glial cells. *Glia* **11,** 156–172.
73. Sontheimer, H., Black, J. A., and Waxman, S. G. (1996) Voltage-gated Na^+ channels in glia: properties and possible functions. *Trends Neurosci.* **19,** 325–331.
74. Guatteo, E., Stanness, K. A., and Janigro, D. (1996) Hyperpolarization-activated currents in cultured rat cortical and spinal cord astrocytes. *Glia* **16,** 196–209.
75. Black, J. A., Sontheimer, H., Minturn, J. E., Ransom, B. R., and Waxman, S. G. (1992) The expression of sodium channels in astrocytes in situ and in vitro. *Brain Res.* **94,** 89–107.
76. Hochman, D. W., Baraban, S. C., Owens, J. W., and Schwartzkroin, P. A. (1995) Dissociation of synchronization and excitability in furosemide blockade of epileptiform activity. *Science* **270,** 99–102.
77. Brightman, M., Klatzo, I., Olsson, Y., and Reese, T. S. (1970) The blood-brain barrier to proteins under normal and pathological conditions. *J. Neurol. Sci.* **10,** 215–239.
78. D'Ambrosio, R., Maris, D. O., Grady, M. S., and Janigro, D. (1998) Loss of inward rectification following traumatic brain injury. *Soc. Neurosci. Abstr.* **28,** 673.3.
79. D'Ambrosio, R., Maris, D. O., Grady, M. S., Winn, H. R., and Janigro, D. (1999) Impaired K homeostasis and altered electrophysiological properties of post-traumatic hippocampal glia. *J. Neurosci.* **19,** 8152–8162.
80. Maccaferri, G. and McBain, C. J. (1996) The hyperpolarization-activated current Ih and its contribution to pacemaker activity in rat hippocampal oriens-alveus interneurons. *J. Physiol. (Lond.)* **497,** 119–130.
81. Maccaferri, G., Janigro, D., Lazzari, A., and DiFrancesco, D. (1994) Cesium prevents maintenance of long-term depression in rat hippocampal CA1 neurons. *Neuroreport* **5,** 1813–1816.
82. Maccaferri, G., Mangoni, M., Lazzari, A., and DiFrancesco, D. (1993) Properties of the hyperpolarization-activated current in rat hippocampal CA1 pyramidal cells. *J. Neurophysiol.* **69,** 2129–2136.
83. Hernandez, T. D. and Naritoku, D. K. (1997) Seizures, epilepsy, and functional recovery after traumatic brain injury: a reappraisal. *Neurology* **48,** 803–806.
84. Willmore, L. J. (1993) Post-traumatic seizures. *Neurol. Clin.* **11,** 823–834.
85. Curran, M. E., Splawski, I., Timothy, K. W., Vincent, G. M., Green, E. D., and Keating, M. T. (1995) A molecular basis for cardiac arrhythmia: HERG mutations cause long QT syndrome. *Cell* **80,** 795–803.
86. Kiehn, J., Lacerda, A. E., and Brown, A. M. (1996) Molecular physiology and pharmacology of HERG. *Circulation* **94,** 2572–2579.
87. Felix, R. (2000) Channelopathies: ion channel defects linked to heritable clinical disorders. *J. Med. Genet.* **37,** 729–740.

88. Davis, A. M. and Wilkinson, J. L. (1998) The long QT syndrome and seizures in childhood. *J. Paediatr. Child Health* **34,** 410–411.
89. Yang, W. P., Levesque, P. C., Little, W. A., Conder, M. L., Ramakrishnan, P., Neubauer, M. G., and Blanar, M. A. (1998) Functional expression of two KvLQT1-related potassium channels responsible for an inherited idiopathic epilepsy. *J. Biol. Chem.* **273,** 19,419–19,423.
90. Janigro, D. (1999) Blood-brain barrier ion homeostasis and epilepsy: possible implications towards the understanding of ketogenic diet mechanisms. *Epi. Res.* **37,** 223–232.
91. Grant, G. A., Abbott, N. J., and Janigro, D. (1999) Understanding the physiology of the blood-brain barrier: the role of in vitro models. *NIPS* **13,** 287–293.
92. Cornford, E. M., Hyman, S., Cornford, M. E., and Caron, M. J. (1996) Glut1 glucose transporter activity in human brain injury. *J. Neurotrauma.* **13,** 523–536.
93. Cornford, E. M., Hyman, S., Cornford, M. E., Landaw, E. M., and Delgado-Escueta, A. V. (1998) Interictal seizure resections show two configurations of endothelial Glut1 glucose transporter in the human blood-brain barrier. *J. Cereb. Blood Flow Metab.* **18,** 26–42.
94. Kawai, N., Stummer, W., Ennis, S. R., Betz, A. L., and Keep, R. F. (1999) Blood-brain barrier glutamine transport during normoglycemic and hyperglycemic focal cerebral ischemia. *J. Cereb. Blood Flow Metab.* **19,** 79–86.
95. Klepper, J., Vera, J. C., and De Vivo, D. C. (1998) Deficient transport of dehydroascorbic acid in the glucose transporter protein syndrome. *Ann. Neurol.* **44,** 286–287.
96. Seidner, G., Alvarez, M. G., Yeh, J. I., O'Driscoll, K. R., Klepper, J., Stump, T. S., et al. (1998) GLUT-1 deficiency syndrome caused by haploinsufficiency of the blood-brain barrier hexose carrier. *Nat. Genet.* **18,** 188–191.
97. Bordey, A. and Sontheimer, H. (1998) Electrophysiological properties of human astrocytic tumor cells in situ: enigma of spiking glial cells. *J. Neurophysiol.* **5,** 2782–2793.
98. McFarlane, M. C. and Sontheimer, H. (1997) Electrophysiological changes that accompany injury-induced glial proliferation in vitro. *Soc. Neurosci. Abstr.* **23,** 70.
99. Junger, E. C., Newell, D. W., Grant, G. A., Avellino, A. M., Ghatan, S., Douville, C. M., et al. (1997) Cerebral autoregulation following minor head injury. *J. Neurosurg.* **86,** 425–432.
100. Medlock, M. D., Hanigan, W. C., and Cruse, R. P. (1993) Dissociation of cerebral blood flow, glucose metabolism, and electrical activity in pediatric brain death. Case report. *J. Neurosurg.* **79,** 752–755.

2
Neuronal Energy Requirements

Avital Schurr

1. NEURONAL ENERGY DEMANDS, SUBSTRATES, AND ENERGY GENERATION

Descriptions of cerebral energy requirements found in the literature may confuse many readers. On one hand, Hawkins *(1)* states that, "Although nervous tissue does not participate in processes that require large amounts of energy, such as mechanical work, osmotic work, or extensive biosynthesis, it has almost as high a rate of oxidative metabolism as some tissues that do." On the other, Clarke and Sokoloff *(2)* assert that, "Although it is sometimes stated that the brain is unique among tissues in its high rate of oxidative metabolism, the overall cerebral metabolic rate for O_2 ($CMRO_2$) is of the same order as the unstressed heart and renal cortex." These two contrasting views are not necessarily contradictory. Whether or not the brain has higher energy requirements than other tissues, the brain is unique, both in its energy-demanding functions and the limitations on the types of fuels it uses and their routes of delivery. The above statements are also indicative of the reason brain-energy metabolism has developed into a separate specialty, in which the energy supply and demand of the brain are studied as the basis for many brain dysfunctions and disorders. The past 15 years witnessed several discoveries and new developments in the field of cerebral energy metabolism, which could explain some of the brain's unique energy requirements, and provide a better understanding of various brain disorders.

1.1. Neuronal Energy Requirements, Energy-Demanding Functions, and Energy Substrates

1.1.1. Neuronal Energy Requirements and Energy-Demanding Functions

The majority of the energy-demanding reactions in the brain belong to two categories: biosynthesis and transport. The biosynthesis of macromolecules, such as proteins, polypeptides, and lipids, occurs mostly in cell bodies; that of smaller molecules, such as neurotransmitters, occurs in nerve terminals. Of the multiple transport processes that take place in the brain, ion transport is believed to demand the most energy (as high as 50–60% of all brain-energy-consuming processes) *(1)*. Of these, the maintenance of sodium ions (Na^+) and potassium ions (K^+) gradients is the most demanding. Unlike other tissues, the central nervous system (CNS) stores only minute amounts of endogenous fuel. Brain tissue glucose levels at any given time are much lower than

From: *The Neuronal Environment: Brain Homeostasis in Health and Disease*
Edited by: W. Walz © Humana Press Inc., Totowa, NJ

blood glucose levels (1–2 μmol/g brain wet wt, compared with 5–6 μmol/mL blood) *(1)*. Glycogen stores are not much higher than the glucose stores (2–3 μmol/g wet wt). Cahill and Aoki *(3)* suggested that the required large ratio of water:glycogen (3–4 mL water/1 g glycogen) could cause great fluctuations in volume, which are restricted because of the rigidity of the cranium. Astrocytes are the main source of brain glycogen, a fact that led many *(1)* to suggest that glycogen is not a readily available energy source to neurons. As will be discussed later in this chapter, this view is now changing: New data indicate that shuttle systems exist between astrocytes and neurons for different metabolites. Estimates are that the total brain supplies of both glucose and glycogen are sufficient for no more than 5 min of normal oxygen (O_2) consumption, a period that could get even shorter when excessive energy utilization and blood glucose supplies cannot keep up with the demand. With no O_2 reserves, the brain depends on the blood and its flow, for all its O_2 needs, extracting almost one-third of the total blood O_2 under normal conditions. Consequently, blockade or reduction in blood flow would diminish cerebral energy metabolism, because of O_2 deprivation before any diminution of glucose supply is apparent.

1.1.2. Energy Substrates

As mentioned, the bulk of the energy manufactured by the brain is devoted to nerve excitation and conduction. These two functions are dependent on sustained membrane potential, and, hence, most of the brain's energy is consumed by ion transport. Since the brain's energy stores are limited, an unhindered flow of glucose and O_2 is a prerequisite for continuous, uninterrupted production of adenosine triphosphate (ATP). This concept, i.e., that the normal substrates of cerebral energy metabolism are glucose and O_2 and the products are carbon dioxide (CO_2) and water *(2)*, has not changed in decades, and generally is correct, although, as is explained below, it is an oversimplified one. Consequently, glucose utilization in the brain is regarded as obligatory *(2)*. Thus, the brain is considered to be different from other tissues, which are much more flexible in their ability to utilize alternative fuels to glucose. This conclusion is based on measurements of positive arteriovenous differences only, for glucose and O_2, and consistent negative values only, for CO_2. Normally, neither positive nor negative arteriovenous differences can be demonstrated for lactate or pyruvate. Although the lack of positive differences indicates that the brain does not utilize bloodborne lactate or pyruvate as aerobic energy substrates, the lack of negative differences for at least one of these two products, even during a moderate O_2 shortage, is intriguing, and could bear potentially important implications.

Table 1 encapsulates the stoichiometric relationship between glucose utilization and O_2 consumption. The values in Table 1 are calculated medians of measurements reported in the literature, and glucose equivalent of O_2 consumption is the theoretical one (6 mol O_2/mol glucose) *(2)*.

The normal conscious human brain consumes 156 μmol O_2/100 g tissue/min. A similar CO_2 production yields a respiratory quotient of 1.0. As indicated in Table 1, 156 μmol of O_2 (or CO_2) are equivalent to 26 μmol glucose (based on a ratio of 6 μmol O_2 consumed for 1 μmol glucose utilized). Not withstanding, the measured rate of glucose utilization is 31 μmol/100 g tissue/min, a surplus of 5 μmol glucose, which brings down the O_2:glucose ratio from the theoretical 6.0, to 5.5. The discrepancy between the

Table 1
Stoichiometric Relationship Between Glucose Utilization and O_2 Consumption in Normal Young Adult Man[a]

Function	Rate (μmol/100 g brain tissue/min)
Glucose utilization	31
O_2 consumption	156
CO_2 production	156
Glucose equivalent of O_2 consumption (6 mol O_2/mol glucose for complete oxidation)	26

[a]Adapted from ref. *2*.

calculated and the measured O_2:glucose has remained unexplained, although it has been suggested that the extra, nonoxidized glucose is converted to lactate, pyruvate, and other intermediates of carbohydrate metabolism *(2)*. Moreover, it has been postulated that these intermediates are released from the brain into the blood in such minute amounts that they are not detectable as a significant arteriovenous difference *(2)*.

There are two "facts" supporting the claim that glucose is the only energy substrate that the normal brain utilizes. The first is a respiratory quotient of 1.0 (6 mol O_2 consumed and 6 mol CO_2 formed, a quotient that requires a stoichiometric consumption of 6 mol O_2 for every mol glucose utilized, and thus a ratio of O_2/glucose of 6.0). The second is the inability to detect a positive arteriovenous difference for any other potential energy substrate.

Clarke and Sokoloff *(7)* warn their readers about the discrepancies between in vivo and in vitro results concerning brain tissue, and the great hazard of extrapolating from in vitro data to conclusions about in vivo metabolic function. In vitro systems bypass functions, such as blood flow, but the uniqueness of the brain in vivo stems from the blood–brain barrier (BBB). The assumption that BBB-devoid cerebral tissue, as with in vitro systems, behaves differently from the same tissue in vivo has triggered great skepticism toward many in vitro findings, especially among investigators who use only in vivo systems. Do in vitro systems deserve such skepticism? Are warnings about their pitfalls warranted? The answer to both questions should be affirmative, when crushed tissue preparations, or mitochondrial, synaptosomal, or any other organelle-enriched preparations are concerned. However, many in vitro preparations today maintain whole cells, partial or complete brain regions, and even a whole, perfused brain for hours and days, intact and functional. These in vitro preparations deserve attention, and should be given the chance to prove themselves before anyone blindly criticizes data that have been originated by them. Therefore, this chapter describes data on energy metabolism generated using in vivo systems, along with data produced by in vitro preparations, such as cell cultures, excised CNS nuclei, and brain slices.

If one is to adhere to the premise that the brain is restricted in its choice of energy substrates, because of the BBB, one need measure only glucose and O_2 consumption and CO_2 production to calculate the brain's energy needs. In vivo studies, at least until recently, did just that: measuring what went into the "box" and what came out of the box. Although, ironically, the intricate processes within the box, which are responsible for the consumption of glucose and O_2, and the production of CO_2, were elucidated using in vitro systems, one could ignore them entirely when measurement of energy metabolism in vivo is concerned. It is only in the last 15 yr that in vivo measurements, both in humans and animals, have strongly suggested that brain energy metabolism is not simply $C_6H_{12}O_6 + 6O_2 = 6CO_2 + 6H_2O$. As the methods of measuring brain function improve, insights are gained that provide a more complex and elaborate picture.

Recent studies indicate that, under conditions of cerebral stimulation, intracerebral lactate increases to significantly higher levels than those under resting conditions *(4–9)*. These results confirm what was suspected for many years, i.e., that phasic changes in neural activity are supported by glycolysis *(10)*. As early as 1937 *(11,12)*, it was noted that local tissue O_2 levels increase, rather than decrease, during spontaneous focal seizures in human cerebral cortex. Rapid eye movement sleep caused large increases in blood flow *(13)* and glucose utilization *(14)*, but a decrease in O_2 extraction *(15)*. Figure 1 shows the cerebral metabolic rate (CMR) data in humans *(5)* under resting conditions and upon stimulation. In whole brain under resting conditions, $CMRO_2$ and $CMR_{glucose}$ values were 1.50 and 0.37 µmol/min/100 g, respectively: a molar ratio of 4.1:1. In the visual cortex, under resting conditions, $CMRO_2$ was 1.71 µmol/min/100 g and $CMR_{glucose}$ was 0.42 µmol/min/100 g, or, again, a molar ratio of 4.1:1. Upon visual stimulation the $CMR_{glucose}$ value rose a mean of 0.21 µmol/min/100 g (51%), and $CMRO_2$ value increased by a mean of only 0.08 µmol/min/100 g (5%), to produce a molar ratio, for the increases, of only 0.4:1. These results indicate that the brain, under working (stimulated) conditions, is capable of producing large amounts of lactate (via pyruvate), up to 250% of control. This accumulated lactate remains in the brain, and could later be used aerobically for energy metabolism under resting conditions. Other recent studies show significant increases in lactate levels in human visual cortex *(7,8)* and in rat hippocampus and striatum, on physiological stimulation *(9)*.

Thus, although the impermeability of the BBB to lactate is irrelevant, since the bulk of brain lactate is produced in the brain itself, the claim that the BBB is impermeable to lactate is inaccurate, at best. Lactate entry to the brain via the arterial blood is considered to be negligible, but Rowe et al. *(16)* showed that, immediately after a carbohydrate-rich meal, the higher-than-average lactate blood levels diminished on passing the brain. A relatively high brain uptake index value for lactate has been reported when a low lactate concentration (0.011 m*M*) was injected into the carotid artery of adult rats *(17–19)*. If the normal plasma level of lactate is 1.0 m*M* *(20)*, one could expect brain uptake index for lactate to be even higher. Several studies clearly demonstrate the ability of lactate to quickly permeate the adult BBB *(18,21–24)* via a stereospecific transporter *(18,22)*. A recent study used [3-^{13}C]lactate to demonstrate that the average uptake of lactate during the first 5 min after its injection (iv) to mice was almost 7× higher than the previously calculated V_{max} for uptake of lactate across the BBB *(25a)*.

Since normal lactate concentration in cerebrospinal fluid exceeds (1.6–2.0 m*M*) the plasma lactate level (1.0 m*M*) *(20)*, and the $CMRO_2$:$CMR_{glucose}$ under resting conditions is higher (4.1) than that during cerebral stimulation (2.8), it is obvious that the normal brain produces large amounts of lactate. Moreover, $CMRO_2$:$CMR_{glucose}$ of 4.1 *(5)* is not the expected ratio of 6.0, or even 5.0. $CMRO_2$:$CMR_{glucose}$ values of 5.0 or 4.0 indicate that 17 or 33%, respectively, of the glucose are consumed via nonoxidative metabolism and converted to lactate, and that, under cerebral stimulation, when these

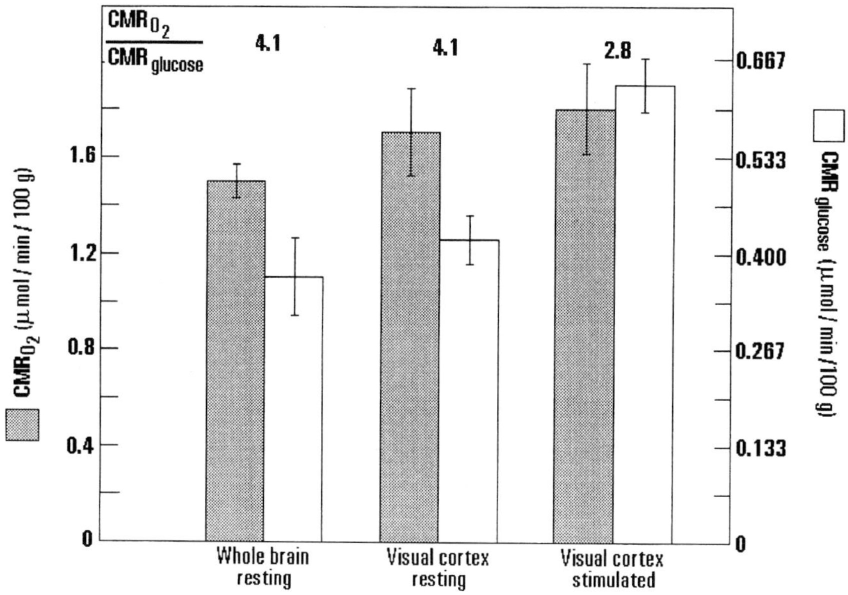

Fig. 1. The ratio of human cerebral metabolic values of oxygen ($CMRO_2$) and glucose ($CMR_{glucose}$) in whole brain and in visual cortex, under resting conditions and during stimulation with a reversing checkerboard stimulus. Under resting conditions, the $CMRO_2$:$CMR_{glucose}$ for the whole brain and the visual cortex were identical (4.1.). However, during stimulation, a significant decrease in this ratio (to 2.8) was observed; the $CMR_{glucose}$ rose by a mean of 0.21 μmol/min/100 g (51%); the $CMRO_2$ rose by a mean of only 0.08 μmol/mi/100 g (5%). This is a molar ratio for the increase in metabolic rate of only 0.4:1. (Modified with permission from ref. *5*.)

values are falling to approx 0.4 *(5)*, 93% of the metabolized glucose is being converted to lactate. Such high production of lactate by the normal brain would explain the higher concentration of lactate in CSF than in plasma *(20)*. Hence, the normal brain continuously produces large amounts of lactate.

If lactate is being produced in the brain regularly as an end product, it should either accumulate there or appear in the venous blood that exits the brain. But in fact neither occurs. A simple explanation for this would be an immediate aerobic cerebral utilization of lactate under resting conditions. Alternatively, lactate could be converted to glycogen, either via gluconeogenesis *(25)* or by direct conversion *(25,26)*. Pyruvate, the precursor of lactate, is as efficient an energy substrate as lactate, it is transported via the same stereospecific transporters that transport lactate, and it is the product that lactate has to be converted to before it can be utilized aerobically. Nevertheless, pyruvate does not accumulate in the brain, and thus the pyruvate:lactate ratio is usually very small.

Before examining in vitro data on lactate as a cerebral energy substrate, two arguments should be reconsidered, that blunt somewhat the warning by Clarke and Sokoloff *(2)* on the limited usefulness of information extracted from in vitro systems resulting from their lack of an intact BBB. First, as has already been pointed out, most, if not all, of the cerebral lactate is produced by the brain itself, and thus the presence or absence of the BBB is irrelevant. Second, as has been shown by the studies mentioned above

(16,18,21–24), the BBB is permeable to lactate, although not to the same degree as glucose. Hence, the relevance and the importance of the in vitro findings described below should not be overlooked.

As early as 1953, McIlwain *(27,28)* was able to demonstrate the ability of guinea pig cortical slices to respire when lactate was the substrate. When such slices, prepared either from guinea pig or rabbit brain, were respiring in glucose-containing media, lactate accumulated in the media *(29)*. This lactate accumulation was initially rapid (50 µmol/g/h), and fell after some minutes, to about 25% of the initial rate of accumulation. This decline results from lactate being served as an oxidizable substrate, when its levels are greater than 3 m*M* *(30)*. Other in vitro results show 2 m*M* to be the minimal lactate concentration that rat hippocampal slices can utilize for energy production *(31)*. The increase in brain lactate levels that accompanies increased cerebral activity, both in humans *(4,5)* and animals *(9)*, has been demonstrated in vitro as a 2–10-fold increase in glycolysis induced by electrical stimulation or incubation with high potassium *(30)*. Thus, rates of lactate formation in vitro, of 100 µmol/g/h during electrical stimulation, are typical, and can be sustained for hours. When the tissue was superfused (3.5 mL/min), lactate formation values rose to 300 µmol/g/h *(30)*. The normal rate of lactate formation by human cerebral tissue in vitro is about 15–20% of the consumed glucose *(30)*.

Evidence published in 1988 *(31)* supports the idea that adult brain tissue is capable of substituting lactate for glucose to fuel its normal function. Later studies *(32–34)* reproduced that finding, demonstrating the ability of adult rat hippocampal slices to utilize lactate (and pyruvate) as the sole energy substrate, upon complete depletion of glucose from the incubation medium. Moreover, in those studies, the inhibition of glycolysis with iodoacetate was ineffective in abolishing synaptic function in lactate-supplemented slices, and completely diminishing such function in glucose-supplemented slices *(31)*.

Cerebellar slices from adult rats exhibited an increase of approx 220% in their ability to oxidize lactate to CO_2, compared to cerebellar slices prepared from early neonates *(34)*. Moreover, at any age, the rate of CO_2 production from lactate is over 300% higher (when slices were supplemented with 10 m*M* lactate) than the rate measured from glucose (when slices were supplemented with 5 m*M* glucose) (Fig. 2). These findings are in agreement with the proposal that, thermodynamically, lactate is a preferred aerobic energy fuel, compared to glucose, since lactate conversion to pyruvate requires no ATP investment; 2 mol ATP are consumed in the conversion of a mole of glucose to pyruvate *(31,35)*. Two recent studies *(25,26)* offer evidence that cultured neonatal mouse astroglial cells are capable of using lactate for gluconeogenesis, and that astroglia-rich primary cultures from neonatal rats are capable of supplying glycogen-derived lactate to neighboring cells.

These in vitro studies strongly suggest a major role for lactate (and pyruvate) as a substrate for cerebral energy metabolism. Based on studies with astrocytic and neuronal cultures, Magistretti et al. *(36–39)* hypothesized that, when the presynaptically released excitatory neurotransmitter, glutamate (Glu) is taken up by astrocytes, it stimulates the production of glycolytic lactate and, consequently, the aerobic utilization of lactate by neurons. The importance of lactate as an aerobic cerebral energy substrate could become even greater under certain conditions, such as hypoxia/ischemia (*see* Sub-

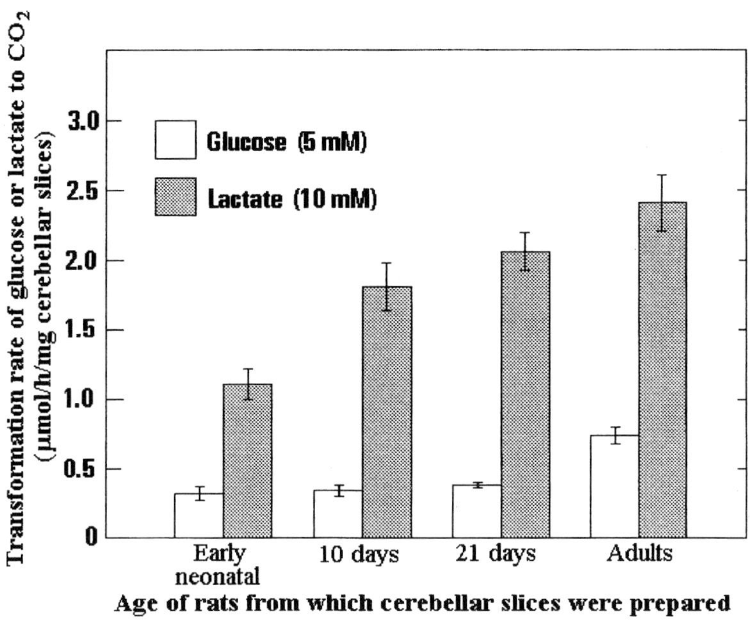

Fig. 2. Conversion of glucose (5 m*M*) or lactate (10 m*M*) to CO_2 by cerebellar slices prepared from rats of different ages. (Modified with permission from ref. *34*.)

heading 4.), as mentioned earlier. Although pyruvate is as useful as lactate for the oxidative production of ATP, it does not accumulate in large quantities, as lactate does. Thus, although pyruvate can support neuronal function in vitro, as a sole energy substrate *(32)*, its levels are too low to account for such support in vivo. Larrabee has shown, using excised chick ganglia, that lactate metabolism competes with glucose metabolism, and vice versa *(40–42)*.

Other substances, such as the ketones, β-hydroxybutyrate and acetoacetate, could, under certain circumstances, serve as energy substrates. Neonates have a great ability to utilize these alternative substrates *(2,43)*. In cases of diabetes or starvation, in which blood levels of ketones are elevated, because of an increase in lipid catabolism, brain tissue shows an adaptive ability to metabolize ketones as energy substrates, using the same monocarboxylate transporters that transport lactate and pyruvate. Ketones are usually converted to acetyl CoA, which directly enters the tricarboxylic acid cycle (TAC).

Although glycogen stores in the brain are low, this polycarbohydrate is the main energy reserve of the brain, and may be utilized during periods of high glucose demands, when glucose supplies cannot keep up with the glycolytic flux. Such high demands, as indicated in the previous subheading, occur regularly, upon increased neuronal activity. It is a long-held dogma that the BBB limits the entry of glucose into astrocytes and neurons during periods of high demands. We have shown, however, using brain slices, in which the BBB is absent, that the rate of neuronal glucose entry is too slow to keep up with its glycolytic consumption *(44)*. If that is true, the role of glycogen, found mainly in astrocytes in the adult brain, could be even more important, especially if some or all of it is converted first to lactate *(26)* before being transported into neurons as an aerobic energy substrate *(40–42)*. Calculations indicate that the glycogen stored in brain tissue (3.3 µmol/g rat brain) could last for less than 5 min as the sole energy source *(2)*. However, in vitro results indicate that astrocytes are capable of synthesizing glycogen from lactate *(25)*, implying that the breakdown and synthesis of glycogen could take place concomitantly.

As the rate of brain glycolysis increases, the level of glycogen's first precursor, glucose-6-phosphate (glucose-6-P) decreases, as the result of an enhancement in the rate of fructose-6-P phosphorylation to fructose-1,6-diP, by the enzyme, phosphofructokinase. This enzyme is the main site of glycolytic regulation. Hence, with the reduction in glucose-6-P levels during periods of energy demands, glycogen synthesis also declines. The uridine diphosphoglucose, formed from glucose-1-P, is the unit transferred and linked, via a α-1,4-glycosidic bond, to the terminal glucose on a nonreducing side of an amylose chain. This reaction, catalyzed by glycogen synthetase, is the rate limiting reaction of glycogen synthesis *(2)*. On the degradation side of glycogen, most of the phosphorylase is in its phosphorylated form, "a," the active form of the enzyme. Nevertheless, it is still unknown if and how glycogenolysis in the brain is regulated. The hydrolysis of glycogen at the α-1,4-glycoside bond leaves a chain of α-1,6-glycoside linkages, upon which glycogen can be resynthesized. The hydrolysis of the α-1,6-glycosidic chain by the debranching enzyme is slower than the hydrolysis of α-1,4-glycosidic bonds, and its product is glucose. This reaction may be the rate-limiting step in glycogenolysis. Approximately 1 mol free glucose is formed for each 11 mol glucose-6-P released, when 1 mol glycogen is completely hydrolyzed. As is discussed in Subheading 2., glycogen metabolism is tightly coupled to neuronal activity. It is rapidly hydrolyzed during shortage in energy supply, and synthesized during adequate supplies of glucose and O_2.

2. COUPLING OF FUNCTION AND ENERGY METABOLISM IN THE BRAIN

The assumption that activation of brain tissue is dependent on energy supplied by oxidative metabolism of glucose was challenged over a decade ago *(4,5)*. As has been seen from the first subheading, magnetic resonance imaging measurements of glucose consumption and concomitant calculations of O_2 consumption from blood flow measurements indicated a possible mismatch between the two, upon activation of the brain. More recent studies *(45,46)* claim that no such mismatch exists, and that the initial increase in glucose consumption, measured upon brain activation, is accompanied by a similar increase in O_2 uptake.

For years, elevated lactate levels have been considered to signal the existence of hypoxia and anaerobic energy metabolism in tissues *(47,48)*. Substantial evidence has been accumulated *(48,49)* to indicate that large amounts of lactate can be produced in many tissues under fully aerobic conditions, but brain tissue has been presumed to be an exception. Lactate production has been promoted as an exclusively anaerobic process, and its accumulation was thought to be a major detrimental factor in ischemic brain damage *(50)*.

Now, however, many studies *(4–9,51)* suggest that the brain is not necessarily different from other tissues, in that it does produce lactate under aerobic conditions. Upon cerebral stimulation, intracerebral lactate increases to significantly higher levels than those under resting conditions, both in humans *(4,5,7,8,37,38)* and rats *(9)*. These results confirmed what had been suspected for some time: Phasic changes in neural activity are supported by glycolysis, a much less efficient ATP biosynthetic pathway than the TCA and oxidative phosphorylation. Moreover, a recent study *(52)* has suggested that, upon activation, a significant pool of lactate is available to the brain as a source of energy. Nevertheless, as has been already mentioned, other studies demonstrate a concomitant increase in O_2 consumption upon activation, to match the increase in glucose utilization. These studies disagree with the idea that increase in energy demands by activated brain is met by glycolysis alone *(45,46)*.

2.1. Glutamate, Neuro–Glial Interaction, and Coupling of Function and Energy Production

Magistretti et al. *(39)*, using primary cultures of mouse cerebral cortical astrocytes, demonstrated the ability of Glu, the main excitatory neurotransmitter in the brain, to stimulate glycolysis, i.e., glucose consumption and lactate production. They hypothesized that Glu, released from active presynaptic neurons and taken up by astrocytes, is the signal that couples neuronal activity to glucose utilization. According to this hypothesis, astrocytic Glu uptake, via Na^+ co-transport, activates the Na^+/K^+-ATPase pump. Glu uptake from the synaptic cleft occurs through specific astroglial Glu transporters, namely, Glu transporter 1 and astrocyte-specific Glu transporter *(52)*. These transporters show a stoichiometry of $3Na^+$ contransported with each Glu molecule. The Na^+-pumping activity, fueled with ATP formed by membrane-bound glycolytic enzymes, increases glycolytic flux and, thus, glucose consumption and lactate production. An additional ATP-requiring reaction takes place, upon Glu entry into astroglia, i.e., Glu conversion to glutamine ($GluNH_2$) by the astroglial enzyme, glutamine synthase *(52)*. Lactate, once released from astrocytes, is taken up by neurons to be utilized aerobically as an energy substrate. Such a mechanism could explain the observed increases in glucose utilization and lactate production, without a concomitant increase in O_2 consumption upon brain stimulation *(4–7,48,49)*. Since neurons consume lactate aerobically *(31–34,53)*, any elevation in its levels should be short-lived.

The coupling of neuronal activity and energy metabolism has also been studied in rat hippocampal slices *(53)*. Although activation of these slices with Glu could result in a significant elevation in the production of lactate, no such increase was detected. Does this lack of increase in lactate tissue level indicate lack of lactate production, or that lactate is consumed as quickly as it is produced? To test whether or not such elevation occurs, lactate utilization must be prevented, or at least slowed down. One way to block lactate use is to inhibit its transport from its site of production to its site of utilization, i.e., from astroglia to neurons. As has been mentioned in Subheading 1., lactate transport occurs via a specific MCT, located in the membrane of both astroglia and neurons *(54,55)*. The use of a specific inhibitor of this transporter, α-cyano-4-hydroxycinnamate (4-CIN), has proved to be an efficient way to block lactate transport into neurons *(53,56)*. Upon activation of neurons with Glu in the presence of 4-CIN, tissue lactate accumulation could be observed (Fig. 3). 4-CIN induced lactate accumulation in brain slices, but it also prevented the recovery of normal neuronal function after the removal of Glu (Fig. 3).

These results lead one to conclude that, under control conditions, any glycolytic lactate formed during exposure to Glu is immediately consumed by neurons as an aero-

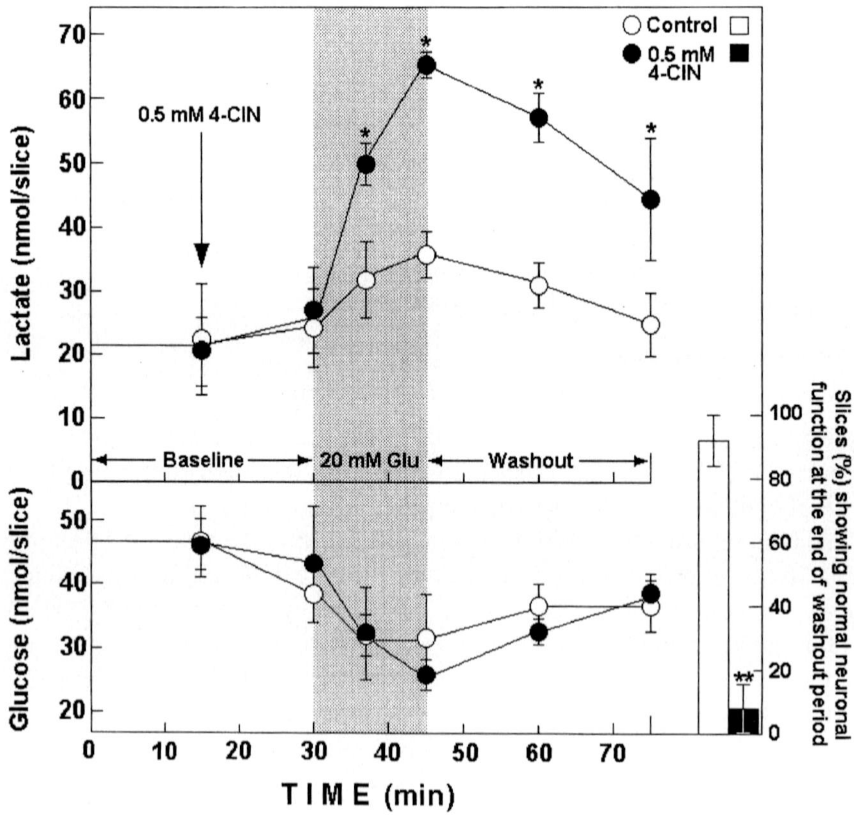

Fig. 3. The content of glucose (**bottom panel**) and lactate (**top panel**), before (baseline) and during exposure to Glu and during Glu washout, in control (open symbols) and 4-CIN-treated (filled symbols) rat hippocampal tissue slices. Slices were supplied with artificial cerebrospinal fluid containing 10 m*M* glucose. The recovery rates of neuronal function at the end of 30 min washout in control (open bar) and 4-CIN-treated slices (filled bar) are shown at the right of the bottom panel. Values are means ± SD. Significantly different from control (*$p < 0.03$; **$p < 0.0001$).

bic energy substrate, and that lactate formation takes place mostly in astroglia. Moreover, lactate appears to be an energy substrate of utmost importance for neuronal recovery after activation. These conclusions are based on three facts: First, in the absence of 4-CIN, no lactate accumulation was observed; second, when 4-CIN was present, inhibition of lactate transport into neurons prevented its utilization, indicating that lactate is produced outside the neuronal compartment; third, 4-CIN prevented the recovery of neuronal function after activation, indicating that lactate could be an obligatory energy substrate during neuronal activation. The blockade by 4-CIN of the neuronal recovery after exposure to Glu ensued, despite the ample supply of glucose (4–10 m*M*) that was present throughout the experimental period.

The importance of glycolysis's role, in the mechanism that couples brain activity with energy metabolism, could be tested by glycolytic inhibition. Such inhibition should weaken the ability of hippocampal slices to maintain neuronal function during exposure to Glu, since glucose utilization would be blocked and lactate would not be

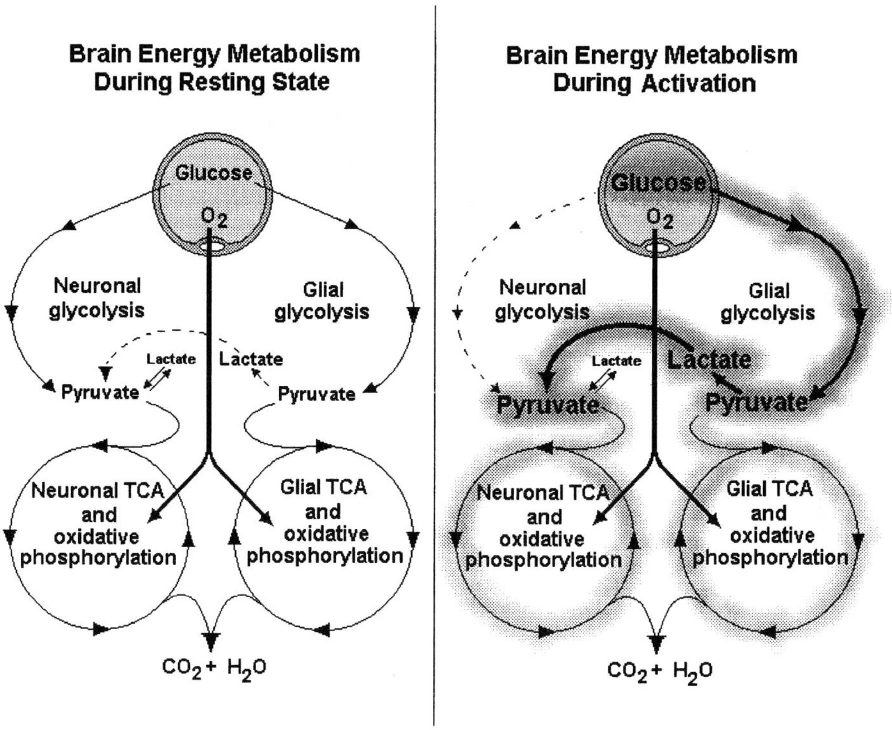

Fig. 4. A schematic diagram of the two main pathways of energy metabolism (glycolysis and oxidative phosphorylation), in neuronal and astroglial compartments, during rest (**left**) and during activation (**right**). For more details, *see* text.

produced. When 2-deoxy-D-glucose (2DG), a nonmetabolizable analog of glucose, was supplied to slices, instead of glucose during exposure to Glu, neuronal function did not recover upon Glu washout. However, when lactate was supplemented along with 2DG, more than 50% of the slices showed neuronal function after Glu washout. This outcome indicates that lactate plays a crucial role in assuring the preservation of neuronal viability during periods of brain tissue activation and/or overactivation. Notwithstanding, a recent study *(57)* claims that lactate is capable of inhibiting glial oxidative glucose metabolism, and consequently O_2 consumption, by up to 40%.

The following scenario (Fig. 4) can be drawn for brain energy metabolism during resting state and during activation (exposure to Glu). During the brain's resting state, most of the glucose taken up from blood supply is metabolized aerobically, both in the astroglial and the neuronal compartments. Under resting conditions, the astroglial production of lactate usually exceeds that of neurons, mostly because of the higher activity of ion pumping via the Na^+,K^+-ATPase pump *(36–39,43,52)*. The majority of this lactate is being transported into neurons, and directly enters the neuronal TCA. Upon activation (with Glu), astroglial Glu uptake immediately ensues, accompanied by Na^+ transport *(36–39,43,52)*. The need to pump out the extra Na^+ brings about a dramatic increase in glial Na^+, K^+-ATPase activity, and thus an equally dramatic increase in glucose consumption, most of which is glycolytic. This increased glycolytic activity is not accompanied by an increase in O_2 consumption, since lactate is its main product.

The large amount of astroglial lactate produced under conditions of activation is handled by two MCT systems. First, lactate is released from glia via the glial MCT system. Second, it is transported into neurons via the neuronal MCT system, where it becomes the main aerobic energy substrate. Thus, the increase in lactate utilization causes a decrease in neuronal glucose utilization *(40–42)*. This scenario could explain the observation made by many investigators, that the stimulation of brain tissue increases glucose uptake and consumption, without a concomitant increase in O_2 consumption.

Are neurons intrinsically programmed to utilize lactate during activation, or is it simply the result of inadequate supply of glucose? To answer this important question, one has to estimate the relative contribution of neurons to the total glucose consumption of the brain. To determine this contribution, the basal (resting) and activation-driven glucose consumption of both neurons and astroglia should be calculated, and the ratio between these two populations of cells in the brain region of study must be taken into account.

Studies *(58–60)* have shown that neurons occupy no more than 50% of the volume of cerebral cortex. In most brain regions, astrocytes outnumber neurons 10:1 *(52,59)*. Taking into account that basal astrocytic glucose consumption is over 3× higher than that of neurons *(61)*, and that, during activation, this consumption increases significantly *(37,52)*, a picture emerges in which neuronal contribution to glucose consumption during activation appears to be minor, if not negligible. Moreover, because of the ratio of astrocytes to neurons (10:1) and the fact that, upon activation, most of the consumed glucose is by astrocytic glycolysis, neurons appear to have no choice but to use lactate as their oxidative energy substrate during activation: This is the only one available to them. The following equations attempt to demonstrate this point.

Astrocytic Energy Substrate Consumption and Metabolite Production

$$\text{glucose} + 5O_2 \longrightarrow 5CO_2 + 5H_2O + 0.33\ \text{lactate} + 32\text{ATP (Basal)}$$

$$\text{glucose} \longrightarrow 2\ \text{lactate} + 2\text{ATP (Activation)}$$

$$2\ \text{glucose} + 5O_2 \longrightarrow 5CO_2 + 5H_2O + 2.33\ \text{lactate} + 34\text{ATP (Basal + activation)}$$

Neuronal Energy Substrate Consumption and Metabolite Production

$$\text{glucose} + 6O_2 \longrightarrow 6CO_2 + 6H_2O + 38\text{ATP (Basal)}$$

$$2\ \text{lactate} + 6O_2 \longrightarrow 6CO_2 + 6H_2O + 34\text{ATP (Activation)}$$

These equations are based on several postulates. First, that $CMR_{glucose}:CMRO_2$ is not the theoretical 6, but the practical 5, meaning that, under resting (basal) conditions, astrocytes use approx 17% of their glucose to support the Na^+,K^+-ATPase pumping activity, through glycolysis, while converting it to lactate. During activation, most if not all of the consumed extra glucose, here assumed to be equal to its basal level, is converted to lactate. Neurons, in contrast, may be capable of a $CMR_{glucose}:CMRO_2$ of 6, whether they utilize glucose or its astrocytic product, lactate, for oxidative energy metabolism. If one relies on Begnami's *(59)* estimate of astrocyte:neuron ratio (10:1), the sum of $CMR_{glucose}$ and $CMRO_2$ for the two cellular compartments during activation is:

$$10 \text{ Astrocytes} \times (2 \text{ glucose} + 5O_2 \longrightarrow 5CO_2 + 5H_2O + 2.33 \text{ lactate} + 34ATP)$$

$$1 \text{ Neuron} \times (2 \text{ lactate} + 6O_2 \longrightarrow 6CO_2 + 6H_2O + 34ATP)$$

$$20 \text{ glucose} + 56O_2 \longrightarrow 56CO_2 + 56H_2O + 21.30 \text{ lactate} + 374ATP)$$

The $CMRO_2$:$CMR_{glucose}$ for the above sum is 2.8, the very value found by Fox et al. *(5)* with activated human visual cortex. It is evident from these equations that 10 astrocytes produce enough lactate to provide for all the oxidative energy needs of one neuron, with enough left over to create a pool of lactate *(62)*.

Until an in vivo quantitative determination can be made of the contribution of each of the two cellular compartments to the total glucose consumption, the question of whether the main neuronal energy substrate is glucose or lactate will remain unanswered. Nevertheless, there is one additional point that must be considered here. With all the lactate that astrocytes produce, does the relatively small number of neurons account for the consumption of this lactate? Considering the above equations, it is clear that neurons would consume only a fraction of the astrocytic lactate produced during activation. However, since no change in the concentration of lactate can be detected in the venous blood exiting the brain during activation (*see also* Subheading 3.), one must assume that all the lactate produced in the brain is consumed by the brain. Hence, if not the activated neurons, then other consumers, such as neurons of surrounding areas, and even white matter and axons, could consume the rest of the lactate.

There is one additional observation that supports Magistretti's hypothesis. Bittar et al. *(63)* studied the distribution of lactate dehydrogenase (LDH) isozymes among neurons (lactate consumers) and astrocytes (lactate producers). They used specific polyclonal antibodies against LDH1, the isozyme found mostly in heart muscle, and LDH5, the one found mainly in skeletal muscle. The antibody raised against LDH1, the isozyme that converts lactate to pyruvate, stained human hippocampal and cortical neurons; hippocampal and cortical astrocytes were immunoreactive with an antibody raised against LDH5. Such selectivity in these two isozymes' distribution indicates that astrocytes in certain brain nuclei, such as the hippocampus and the cortex, contain exclusively the LDH activity that converts pyruvate to lactate, and, as such, are probably very active glycolytically, producing large amounts of lactate. In contrast, the neurons in these regions are enriched with LDH1, the isozyme that converts lactate to pyruvate, typical of tissues that utilize lactate (and pyruvate) for manufacturing energy via the TCA.

Although the greatest emphasis throughout this chapter has been on glucose and lactate (and pyruvate) as the substrates of cerebral energy metabolism, both in rest and during activation, one should not ignore the role played by glycogen. The turnover of this polysaccharide in the brain is very rapid *(43)*, and finely tuned to the energy needs of the activated tissue *(64,65)*. Astrocytic glycogen levels rapidly fall during brain stimulation, but the opposite occurs when brain activity is depressed *(43)*.

3. NEURONAL ENERGY DEPRIVATION

There are three major conditions, or encephalopathies, that could lead to neuronal energy deprivation: Hypoglycemia, or glucose deficiency in the blood; anoxia (hypoxia), or O_2 deficiency in the inspired air; and ischemia, or deficiency in blood supply, which, in essence, combines the first two deficiencies. All three encephalopathies are characterized by an inadequate supply of one of the two major energy substrates, or both. Since glucose and O_2 are the two major energy substrates of the brain, it is easy to understand how their deficiencies will affect the production of cerebral ATP and all the energy-demanding processes. Nevertheless, there are many other encephalopathies of intrinsic origin, such as deficiencies in one enzyme or another of the energy metabolic pathways, and seizures. Others may arise from extrinsic factors, such as vitamin deficiencies, toxicity of heavy metals, and viral infections. This chapter covers only the three major encephalopathies, i.e., hypoglycemia, anoxia (hypoxia), and ischemia.

There are two groups of researchers who study degenerative diseases of the brain. Basic scientists, who attempt to better understand the workings of cerebral energy metabolism and its regulation, and clinical scientists, who delve into the mysteries of these disorders, in an attempt to alleviate their terrible toll. Both groups depend on each other's discoveries to bring them closer to their goals. The basic scientist's approach is the one taken here, keeping the clinical aspects of the major encephalopathies to a necessary minimum. For scientists to study any disorder and its consequences, an experimental model which either mimics or closely resembles this disorder, is a must. An in vivo model usually employs an experimental animal; an in vitro one employs any number of subcellular, cellular, tissue, or organ preparations.

Once established, a good experimental model affords a better understanding of the sequence of events that lead to the final outcome of the disorder under study. The model also opens an opportunity for experimentation with potential treatments or preventive measures that could minimize or eliminate the damaging outcome of the disorder. This is when basic scientists and clinicians closely collaborate, namely, in the discovery, development, and later, the clinical test of these treatments for their potential neuroprotective value.

3.1. *Hypoglycemia and Ischemia/Hypoxia*

Although the phenomena of hypoglycemia, ischemia, and hypoxia have been dealt with separately in many publications, the decision to combine the three under one heading here is based chiefly on information published over the past two decades, which indicated many overlaps and parallels between these three encephalopathies. The main reason they have been traditionally separated has to do more with the distinguished metabolic vehicles that lead to the observed outcome, rather than with the outcome itself.

Hypoglycemia may occur in diabetes through insulin overdose, liver disease, and in many enzyme deficiencies, such as glucose-6-phosphatase, fructose-1,6-biphosphatase, phosphoenol-pyruvate carboxykinase, pyruvate carboxylase, or glycogen synthetase. Each of these deficiencies could affect normal function of the brain, and lead, if not corrected, to neurological deficits.

The most common outcome of the interruption of glucose supply to the brain is hypoglycemia-induced coma, which ensues when glucose blood levels drop to 8–9 mg/100 mL, or about 10% of the normal isoglycemic glucose level of 90 mg/100 mL

(66). In contrast, arterial O_2 content, mean blood pressure, cerebral respiratory quotient, and cerebral blood flow, all remain unchanged. However, cerebral O_2 consumption declines in hypoglycemic coma (56% of normal), along with a dramatic decline in cerebral glucose consumption *(66)*. These declines can be produced by an overdose of insulin, hepatic insufficiency, or by the inhibition of glucose metabolism with sufficiently high dose of 2DG *(67–69)*. Nevertheless, one should be aware that symptomatic hypoglycemia may occur at blood glucose levels that are significantly different from those mentioned above. Thus, in the diabetic, a decrease in blood glucose level from 300 to 100 mg/100 mL, upon insulin administration, may cause symptomatic hypoglycemia *(70)*. Yet, in a better-controlled diabetes, symptomatic hypoglycemia may not occur until the blood glucose levels reach 50 mg/100 mL. In normal, young humans after glucose tolerance test, blood glucose levels may fall to 30–40 mg/100 mL, without the appearance of any symptoms *(70)*.

Intraperitoneal injection of insulin is the most common method employed by investigators for the induction of hypoglycemia. With the right dose of insulin (50–100 U/kg), hypoglycemic coma ensues within 1–2 h following insulin injection in rats or mice *(71)*. In larger animals, such as the dog, higher doses of insulin (200–300 U/kg) are required to achieve coma. For metabolic studies, a small animal is the model of choice. For studies in which cerebral blood flow is involved, a larger animal model is required.

As mentioned earlier, hypoglycemic symptoms can be induced with a pharmacologic dose (400 mg/kg) of 2DG *(67–69)*. When hypoglycemia is induced with 2DG, blood glucose levels actually increase, rather than decrease, which is an indication that symptomatic hypoglycemia is not the result of low levels of blood glucose, but rather the result of a reduction in glucose metabolism or inadequate supply of glucose to its transporters, because of their occupation by 2DG.

Thus, although both methods abolish glucose metabolism, the consequences of these deprivations are markedly different. Insulin enhances glucose metabolism to the point at which the monosaccharide levels are low enough to trigger the utilization of alternative energy substrates, such as glycogen and ketones. In contrast, 2DG competes with glucose for both the glucose transporter and the enzyme, hexokinase, which neither reduces the glucose levels nor triggers the utilization of alternative substrates. Apparently, the effect of 2DG would be more swift and severe than that of insulin injection. Nevertheless, many of the gross consequences of hypoglycemia, such as histologic, electroencephalographic, and behavioral changes induced by both methods, should be similar.

Histologically, cerebral cortex neurons of rats exposed to insulin-induced hypoglycemia exhibit changes dependent on the duration of the hypoglycemic period beyond the onset of isoelectric EEG. A short, 30-min hypoglycemia produces mostly shrinkage and condensation of both cytoplasm and nuclei, but rats that are exposed to a longer, 60-min hypoglycemia exhibit significantly greater number of damaged cells *(71)*. Moreover, hypoglycemia produces cell swelling and the appearance of empty vacuoles. Swollen neurons are more abundant after the longer hypoglycemic period than after the shorter one. Although cell damage observed in other encephlopathies is frequently irreversible, the histologic changes that are recorded in hypoglycemic rats can be reversed by glucose supplement. Within 4 h after the onset of glucose supplementation by iv injection, more than 95% of the neurons have normal appearance *(71)*. The phenomenon of swollen neurons could be explained by the derailment of the energy-dependent ion homeostasis. Shrinkage of cells, as acutely as it occurs in hypoglycemia, remains an unclear phenomenon, especially its reversibility with glucose supplementation.

Slow-wave EEG occurs when glucose blood levels fall to 2 m*M*. When glucose levels fall to 1 m*M* or lower, EEG becomes isoelectric *(71)*. Brain tissue in vitro responds in a similar way when glucose levels in the incubation medium are reduced. Evoked population spikes in the CA1 region of hippocampal slices cannot be sustained when glucose concentration falls below 1–2 m*M* *(31)*. In vivo, cerebral metabolic rate of glucose does not decrease until plasma glucose levels fall below 2.5 m*M* *(72)*. At these levels, the rate of glucose transport into the brain decreases to values 2–3× slower than normal, so that glucose concentration is insufficient to saturate hexokinase *(71,73)*. Rats, like humans, exhibit hypoglycemia-related behavioral changes that range from normal to comatose, in close correlation with the blood glucose levels *(71)*. Coma usually ensues when blood glucose levels fall to 1 m*M* or less. In some cases, however, hypoglycemia may induce seizures and seizure-like EEG, in which event, some of the changes, including the rate of decline in the blood glucose levels, and hence the onset of coma, may occur faster than expected.

Despite all these changes, reports on variations in cerebral blood flow range from an increase to no change during hypoglycemia-induced coma *(71)*. Similarly, depending on the depth of hypoglycemia and accompanying events, such as seizures, O_2 consumption can increase, decrease, or stay unchanged *(71)*.

One of the typical events that accompany hypoglycemia is hypothermia. By all accounts, this phenomenon is directly dependent on the reduction in glucose availability, rather than any of the possible causes of hypoglycemia. The fact that 2DG-induced hypoglycemia is accompanied by hypothermia *(67,69)* is a strong indication that 2DG, although increasing blood glucose levels, also produces functional hypoglycemia *(71)*. Thus, the postulate has been advanced that, when glucose becomes unavailable somehow, through specific temperature-regulating receptors, it induces the observed hypoglycemic hypothermia. The hypothermic effect of hypoglycemia can be reversed by high doses of fructose (5× the dose of 2DG), when administered intraventricularly, along with 2DG *(71,74)*. High doses of glucose should have similar effect to that of fructose, and, based on several recent in vitro studies, lactate, and possibly pyruvate, should also be able to overcome 2DG-induced hypothermia.

Since glucose is the main energy substrate in the brain, it is obvious that any deprivation of this fuel can lead to a shutdown of the very pathways of its metabolism. Thus, both the glycolytic pathway and the mitochondrial TCA are first to be affected by hypoglycemia. The levels of pyruvate, lactate, and glycogen fall significantly, once glucose blood levels fall to 2.5 m*M* or lower *(71)*. Similarly, the levels of the TCA intermediates also decrease in response to hypoglycemia *(71)*. Of the different metabolic effects of hypoglycemia, none is more important than the decline in brain ATP levels and, to a lesser degree, phosphocreatine levels. Notwithstanding, of the different brain regions, the cerebellum does not exhibit such falls in ATP and phosphocreatine levels, even late into the isoelectric period *(71)*. Although this unique ability of the cerebellum to withstand metabolic strain has been documented *(71)* using other models of metabolic stress, the mechanism governing this ability is unknown.

All the hypoglycemic symptoms, functional and behavioral, which are induced by an overdose of insulin, can be completely reversed by an administration of glucose.

Hypoxia is defined as low O_2 content or tension or O_2 deficiency in the inspired air. The extreme case of hypoxia is anoxia, the complete absence of O_2. Ischemia is defined as deficiency in blood supply caused by functional constriction or actual obstruction of a blood vessel *(75)*. Clearly, an organ that suffers from a deficiency in blood supply is also afflicted with hypoxia. Hence, many of the symptoms associated with hypoxia and ischemia are identical. Furthermore, the cellular processes and mechanisms that lead to posthypoxic and postischemic damage are identical. Thus, the term "cerebral ischemia" is being used throughout this subheading, with the understanding that it also refers to cerebral hypoxia.

There are probably more books, review articles, and research papers written about cerebral ischemia than any other encephalopathy. Focal cerebral ischemia (stroke) is the third leading cause of death in the United States and other industrialized countries. Stroke survivors usually suffer permanent and debilitating brain damage. The huge social and economic cost of this encephalopathy, on one hand, and the potential benefits from a future protective modality, on the other, continue to stimulate the research in this field.

The complete dependency of the CNS on glucose and O_2 for the generation of sufficient ATP supplies makes the brain the most vulnerable of all tissues to ischemia. Pure hypoxia is rarely encountered in humans. Pure hypoxia is experienced only in cases of pulmonary insufficiency, such as in pulmonary emphysema, in anemic patients, or in normal individuals who are acutely exposed to altitudes above 2400 m. Obviously, cerebral ischemia of identical duration should be more devastating than hypoxia, or even anoxia, because the ischemic brain suffers not only from oxygen deficiency, but also from shortage of glucose supplies. It is the compounded effect that makes cerebral ischemia so devastating. As is demonstrated below, and has been the case with hypoglycemia, both the brain in vivo, and any of its in vitro preparations, can tolerate either hypoxia or hypoglycemia alone longer than when the two are combined. In other words, the presence of one deficiency sensitizes the brain to the effect of the other. Thus, ischemia of a given duration is always more damaging than either hypoxia or hypoglycemia of a similar duration.

The most common events causing cerebral ischemia are stroke, cardiac arrest, and head injury or trauma. Stroke or focal ischemia occurs when blood supply is interrupted to a specific region of the brain; global ischemia occurs when blood supply to the entire brain is interrupted, as in cardiac arrest. Furthermore, both focal and global ischemia, depending on their severity and duration, may be complete (a total absence of blood flow) or incomplete (a drastic reduction in blood flow). The severity and the duration of these insults determine whether a reversible or irreversible cell injury occurs. An irreversible injury eventually leads to cell death (infarction).

Since there are brain regions that are more sensitive to ischemic injury than others, they are referred to as "selectively vulnerable" regions of neurons. In many of these regions, the neuronal death does not occur during or immediately after the ischemic insult. Frequently, neuronal death is delayed for 3–7 d. In focal ischemia, the vasculature surrounding the ischemic umbra, known as the penumbra, may still provide some collateral, although compromised, circulation, which, if treated early enough, could rescue the penumbric neurons. Of the sensitive brain regions, the most vulnerable neurons are located in the CA1 sector of the hippocampus. Neurons in certain portions of the caudate nucleus, and in layers 3, 5, and 6 of the neocortex, are also selectively vulnerable. Understanding the mechanism that leads to this vulnerability is a major goal of many research laboratories around the world.

Several recently published monographs on cerebral ischemia *(76–78)* describe both in vivo and in vitro models of this encephalopathy. Many of the animal models of yesteryear, such as monkey, dog, cat, rabbit, and pig, are only rarely used today, mostly because of their very high cost. Most of the animal models of the past two decades make use of rodents. These models proved to be both cost-effective and reliable. Moreover, many of the recent advances in this field of research appear to be directly applicable to humans.

The focus of this chapter is on the effect of energy deprivation on neurons, but much can be learned from the enormous number of studies conducted over the last three decades on the cellular mechanisms leading to ischemic neuronal death. The putative roles of lactic acid, Glu, Ca^{2+} influx, and many other factors, in causing the delayed demise of ischemic neurons, have been discussed exhaustively elsewhere and are not the focus of this chapter. However, the hypotheses that were erected to explain this neuronal demise led to many elegant studies that reveal important details about cerebral energy metabolism and the energy-dependent cellular processes that could go awry when ATP is diminished. Of these processes, only those that are directly affected by failure of energy metabolism upon ischemia, hypoxia, and hypoglycemia are of interest here.

Of the many developments characterizing the past two decades, with which cerebral ischemic research is concerned, the employment of in vitro systems, such as the hippocampal slice preparation and cell cultures, has had the greatest impact. These systems can be manipulated, controlled, and analyzed in ways that in vivo systems cannot tolerate. Because most recent knowledge on the effects of energy deprivation on neuronal tissue has emerged from in vitro studies, the last portion of this chapter deals mainly with their findings.

A great advantage of the hippocampal slice preparation, compared to other in vitro systems, such as cell cultures, is the ability of the investigator to use electrophysiologic means to record neuronal function indistinguishable from in vivo records. Moreover, the recording can be made from the most vulnerable sector to ischemia/hypoxia in the hippocampus, the CA1 region. Within 2–3 min of changing the atmosphere that slices are exposed to, from 95% O_2/5% CO_2 to 95% N_2/5% CO_2, the evoked population spike (neuronal function) amplitude falls to 0 mV. The ability of hippocampal slices to recover neuronal function, upon return to normal reoxygenation, depends on the duration of the hypoxic period, and on the concentration of glucose in the artificial cerebrospinal fluid. The higher the glucose concentration, the longer the hypoxic period from which neuronal function can recover (Fig. 5).

Reduction in glucose concentration does not visibly affect neuronal function in hippocampal slices, until it falls below 1.5 m*M*. However, when glucose deprivation is combined with hypoxia (in vitro ischemia), a reduction in glucose concentration, from 10 to 5 m*M*, significantly decreases the ability of hippocampal neuronal function to recover posthypoxia (Fig. 5). In contrast, increasing glucose concentration, from 10 to

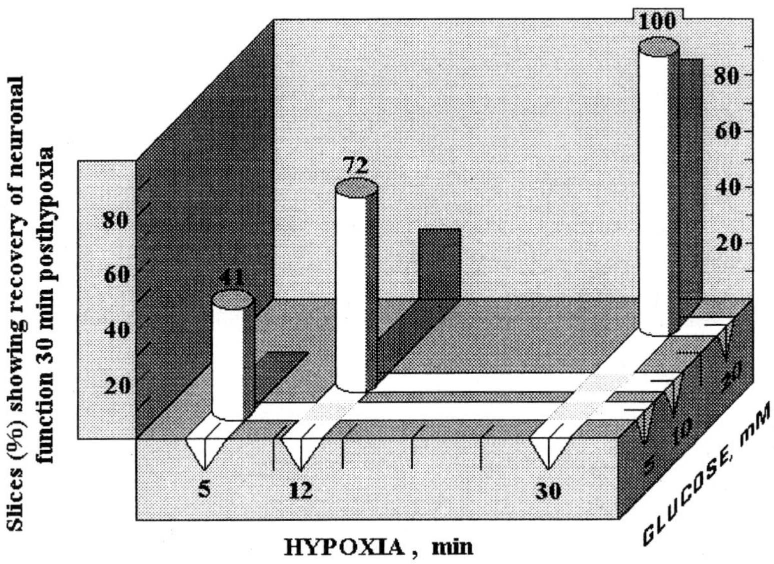

Fig. 5. Rates of recovery of neuronal function (orthodromically evoked CA1 population spike) in rat hippocampal slices after 5, 12, or 30 min of hypoxia and 30 min reoxygenation in the presence of 5, 10, or 20 m*M* glucose, respectively. The higher was the glucose concentration in the perfusion medium, the longer was the duration of hypoxia slices could tolerate.

20 m*M*, significantly prolongs the hypoxic period from which hippocampal neuronal function could recover (Fig. 5). Thus, the higher the concentration of glucose during O_2 deprivation in vitro, the better the recovery of neuronal function posthypoxia. This relationship is in disagreement with in vivo findings, in which hyperglycemia has been shown to aggravate neuronal ischemic damage, a phenomenon known as the "glucose paradox" *(79,80)*. These findings led to the formulation of the "lactic acidosis" hypothesis of cerebral ischemic damage *(50,81)*. The glucose paradox, and other apparent contradictions between results obtained in vitro and in vivo, mostly in the early and mid-1980s, impeded the acceptance of the hippocampal slice preparation as a valuable tool in the study of brain energy metabolism and brain encephalopathies.

A hallmark of ischemic/hypoxic conditions in all tissues, including brain, is the increase in lactate production above the baseline levels under normoxia. When hippocampal slices are exposed to O_2 deprivation, lactate production quickly increased 5–6-fold *(44)*. As mentioned earlier, lactate can support normal neuronal function in hippocampal slices as the sole energy substrate. Moreover, lactate also has been shown to be a preferred energy substrate over glucose in excised sympathetic chick ganglia *(40–42)* and in rat cerebellar slices *(34)*. Lactate is preferable to glucose when recovery from hypoxia is concerned *(44)*. Moreover, lactate has been shown to be a mandatory aerobic energy substrate for recovery of neuronal function posthypoxia in hippocampal slices *(44,56,82)*.

To demonstrate this role of lactate in the recovery of neuronal function posthypoxia, two experimental protocols were used. The first manipulated the ability of the glucose analog, 2DG, to block lactate production during hypoxia via inhibition of glycolysis (Fig. 6). Normally, 100% of slices perfused with 20 m*M* glucose recover their neuronal

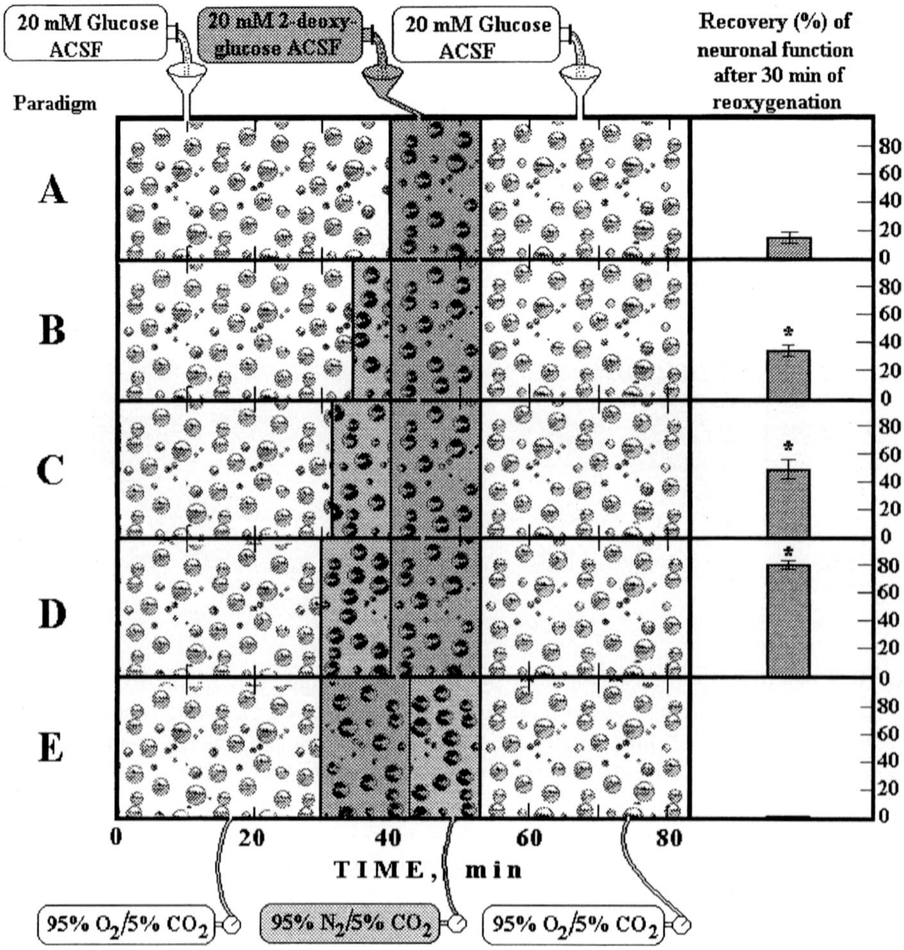

Fig. 6. A schematic illustration of the experimental paradigms used in establishing the role of anaerobically produced lactate in the posthypoxic recovery of neuronal function in rat hippocampal slices. Shown are five different experimental paradigms (**A–E**), the various compositions of artificial cerebrospinal fluid that perfused the slices during each paradigm, and the different gas mixtures (bubbles) that were passed over the slices during these paradigms. In each paradigm, slices were first equilibrated with a medium containing 20 m*M* glucose and an oxygenated atmosphere, for at least 30 min (pretreatment). After a treatment period, slices were allowed to recover for 30 min under pretreatment conditions, before measuring recovery of neuronal function (bars on the right side of each paradigm's panel). In paradigm **A**, slices were exposed to 13-min hypoxia (gray shading), as 20 m*M* glucose-containing medium was replaced with 20 m*M* 2DG-containing medium. In paradigms **B–D**, the hypoxic period started 5, 8, or 10 min prior to the onset of 2DG perfusion, and continued an additional 13 min, for a total of 18, 21, or 23 min of hypoxia, respectively. The longer the hypoxic period prior to the blockade of glycolysis with 2DG, the more lactate produced and the higher the recovery rate of neuronal function posthypoxia. Thus, in what appears as a paradox, slices that were exposed to a total of 23 min hypoxia exhibited a significantly higher rate of neuronal function recovery than slices exposed to only 13 min hypoxia. Paradigm **E** is similar to paradigm **D**, except that, for the first 13 min of hypoxia, slices were perfused with 2DG, and, for the last 10 min, slices were perfused with glucose-containing medium. Slices treated according to paradigm **E** were unable to produce energy and lactate glycolitically during the first 13 min of hypoxia and, hence, the lack of recovery of neuronal function. Histograms on the right are mean values +SD. *Significantly different from paradigm **A** ($p < 0.0005$).

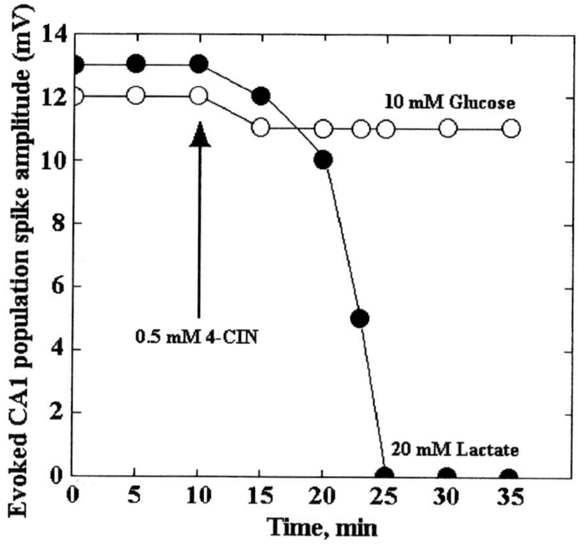

Fig. 7. The effect of 0.5 m*M* 4-CIN, a monocarboxylate transporter inhibitor, on the evoked population spike (neuronal function) amplitude in two slices, over time. One slice (open circles) was perfused with artificial cerebrospinal fluid containing 10 m*M* glucose, while the other was perfused with the same fluid containing 20 m*M* lactate. When 4-CIN was added to the perfusion medium of both slices, only the one that was perfused with the medium containing lactate exhibited diminution of neuronal function, because the inhibitor blocked the entry of lactate into neurons. The inhibitor had no effect on the entry of glucose into neurons.

function after 30-min reoxygenation from 23-min hypoxia. Control slices continued to be perfused with 20 m*M* glucose during the first 10 min of hypoxia, at which time the perfusion was switched to 20 m*M* 2DG for the last 13 min of hypoxia: 80% of these slices exhibited recovery of synaptic function posthypoxia (Fig. 6, paradigm D). Experimental slices were also exposed to 23-min hypoxia; however, 20 m*M* 2DG was perfused during the first 13 min of hypoxia, followed by 20 m*M* glucose for the last 10 min of hypoxia. None of the experimental slices recovered neuronal function posthypoxia (Fig. 6, paradigm E). Although control slices were able to produce lactate during the first 10 min of hypoxia, experimental slices were unable to do so, since glycolysis was inhibited from the onset of hypoxia, thus preventing lactate production. Results shown in Fig. 6 clearly demonstrate the importance of lactate in the posthypoxic recovery of neuronal function. The more time slices were allowed to produce lactate anaerobically before the inhibition of glycolysis with 2DG, the higher the percentage of slices that recovered neuronal function posthypoxia. Hence, more slices that were exposed to a total of 23 min hypoxia recovered their neuronal function than did slices that were exposed to 13-, 18-, or 21-min hypoxia (Fig. 6, paradigms A–D). The most plausible explanation for this phenomenon is the higher levels of lactate found in tissue slices exposed to 23-min hypoxia than the levels in slices exposed to shorter hypoxic periods.

The second experimental protocol employed the specific neuronal lactate transporter inhibitor, 4-CIN *(56,83,84)*. As shown in Fig. 7, this compound is able to block lactate-

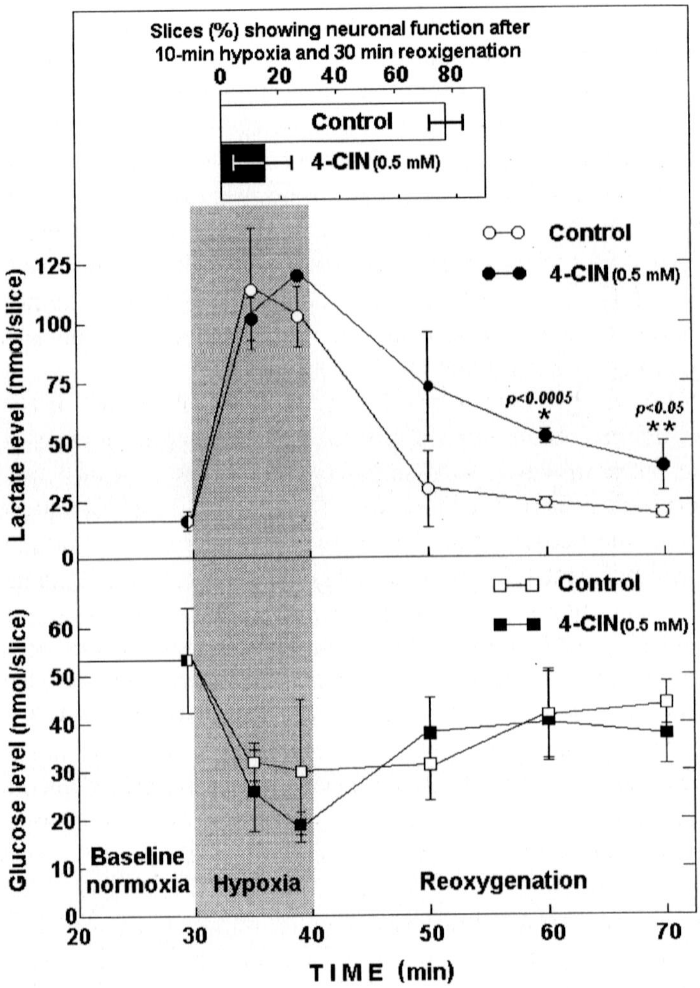

Fig. 8. The effect of 0.5 m*M* 4-CIN on rat hippocampal slices' ability to recover their neuronal function posthypoxia (histogram in **upper panel**), and on their levels of lactate and glucose. A significantly lower percentage of slices recovered neuronal function posthypoxia in the presence of 4-CIN. This poor recovery rate probably resulted from inhibition of neuronal lactate utilization by the monocarboxylate transporter inhibitor. The significantly higher levels of lactate that were present during reoxygenation in 4-CIN-treated slices indicate that neuronal lactate utilization was inhibited. Bars represents means ± SD. *Significantly different from control slices, $p < 0.0005$; **$p < 0.05$.

supported neuronal function in rat hippocampal slices (*see also* ref. *56*). When perfused with 10 m*M* glucose, 78% of control slices recovered neuronal function after 10-min hypoxia; only 15% of experimental slices recovered neuronal function after being exposed to 10-min hypoxia in the presence of 0.5 m*M* 4-CIN (Fig. 8). Both groups of slices produced high levels of lactate during O_2 deprivation, although experimental slices exhibited a significantly slower decline in those levels during reoxygenation. This slower decline could result from blockade of neuronal lactate utilization by 4-CIN (Fig. 8).

The results of the above experiments also explain the ability of elevated glucose concentration, if supplemented before hypoxia, to afford neuroprotection against hypoxic damage in hippocampal slices (Figs. 5 and 6). This neuroprotection is probably the result of two separate processes: First, the increase in glucose availability allows the tissue to maintain glycolytic flux for a longer period of time, for the support of ion homeostasis; second, the longer glycolytic flux produces a significant increase during hypoxia in the lactate level, which in turn is available for aerobic utilization during the initial stages of reoxygenation, when both ATP and glucose levels are very low. The ability of the specific neuronal monocarboxylate transporter inhibitor, 4-CIN, to block posthypoxia recovery of neuronal function, provides further evidence that aerobic lactate utilization is crucial for such recovery.

Although these in vitro results befit the general understanding of both the aerobic and anaerobic energy metabolism pathways and their consequences (including the "more glucose, better tolerance to hypoxia" concept), most investigators in the field of cerebral ischemia find them hard to accept fully. The reason for this skepticism has much to do with the heavily heralded in vivo phenomenon, in which preischemic hyperglycemia significantly exacerbates delayed neuronal damage, as measured 4–7 d postischemia. This glucose paradox phenomenon of cerebral ischemia was first reported almost 25 yr *(79)*, and has been reproduced numerous times in different in vivo models of ischemia. It became the cornerstone of the lactic acidosis hypothesis, which attributes the bulk of the delayed neuronal damage observed 4–7 d postischemia to the increase in lactic acid levels in the brain and the resultant acidosis *(50)*.

The glucose paradox of cerebral ischemia also appears to affirm the main premise of the lactic acidosis hypothesis: more glucose = more lactic acid = more delayed neuronal damage. But does it? If acidosis (pH 6.8–6.5) is detrimental to neurons exposed to O_2 deprivation or to O_2–glucose deprivation, the same trend should be seen in vitro. However, experiments, both in brain slices and in neuronal cultures, showed that acidosis not only did not exacerbate hypoxic or ischemic neuronal damage in vitro, but in essence was actually neuroprotective *(85–87)*. Again, this contradiction between the in vitro and in vivo outcomes was used to criticize the in vitro approach to the study of cerebral ischemia. Nevertheless, the agreement of in vitro results on cerebral ischemia with in vivo findings, over the past two decades, significantly surpasses the disagreement.

If one hypothesizes lactate to be a major energy substrate upon reoxygenation after hypoxia, rather than to be a detrimental factor, perhaps one could uncover supportive in vivo data. Of the vast number of in vivo studies over the past three decades, several included measurements of tissue levels of energy substrates and metabolites, such as glucose, lactate, ATP, and adenylate energy charge, before, during, and after an episode of cerebral ischemia. Such data from three key studies *(88–90)* are depicted in Fig. 9. The trend illustrated by these three studies is typical: Normal preischemic brain levels of glucose (1.7–3.0 μmol/g) and lactate (1.0–1.8 μmol/g) were drastically changed following 5–10 min of anoxia or ischemia. Although glucose levels fell (to 0–0.5 μmol/g), those of lactate rose sharply (to 12–20 μmol/g). After only 15 min of reperfusion/reoxygenation, glucose in the brain climbed to levels significantly higher than those existing preischemia *(88–90)*. Even after 90 min of reperfusion, brain glucose level remained over 200% of preischemic level *(88)*. Despite the dramatic increase

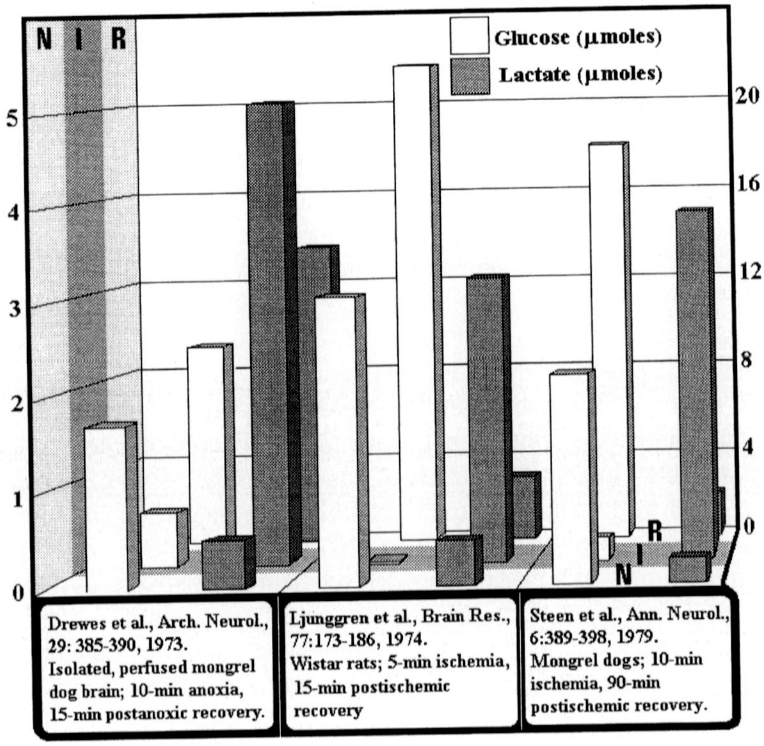

Fig. 9. Brain tissue levels (μmol/g) of glucose (white bars) and lactate (gray bars) at normoxia (N), at the end of ischemia/anoxia (I), and after reperfusion/reoxygenation (R), as reported in three separate and independent studies *(88–90)*.

in lactate concentration after 10 min of anoxia, the rate of its efflux from the brain to the venous blood remained unchanged from the very low control levels (approx 0.2 μmol/g/min) before anoxia *(89)*.

These findings indicate that the brain tightly conserves the lactate it produces, and agree with the results described above for hippocampal slices *(44,56,82)*. The post-ischemic/postanoxic increase in brain glucose was accompanied by a rapid decrease in lactate levels. The investigators themselves termed this decrease "lactate utilization" *(89)*. Energy charge (EC, [ATP + 1/2ADP/{ATP + ADP + AMP}]) was also measured before anoxia (0.89), after 10 min of anoxia (0.39), and after 15 min of reperfusion/reoxygenation (0.92) *(89)*. Similar results were reported in the other two studies *(88,90)*.

A plausible explanation for a postischemic/postanoxic decline in brain lactate levels, and a concomitant increase in glucose levels, could be as follows. Upon reoxygenation, lactate is utilized, aerobically, as an energy substrate, but glucose remains mostly unused, resulting in its accumulation above control levels. This aerobic lactate utilization is sufficient to rapidly restore preanoxic ATP levels and energy charge values.

Although it would be prohibitive to review the vast body of data here, the most revealing information can be extracted from a study *(81)* that was aimed at investigating the role of lactic acidosis in ischemic brain damage under hyperglycemic conditions. But it is the outcome during hypoglycemic conditions that deserves closer attention. Rats fasted for 16–24 h, and exposed to 30 min of severe incomplete ischemia, showed more than a 17-fold increase in brain lactate (from 0.88 to 15.5 μmol/g). Both glucose and glycogen levels fell essentially to zero (from 2.78 and 2.84 μmol/g, respectively) during the same period; ATP levels fell by 95% from the control level. Recirculation for 90 min brought about a significant decrease in brain lactate levels, from 15.5 to 3.19 μmol/g (a decrease of 79%). Glucose levels rose concomitantly to 106% of control, although the rats were not supplemented with glucose during recirculation. ATP rose to 81% of its control level, and 54% of the glycogen pools was restored. Even more striking, however, were the results when another group of fasting rats was infused with glucose (2 mL 50% glucose solution) during the first 10 min of recirculation. After 30 min of recirculation, the lactate level dropped 58% (from 15.5 μmol/g at the end of 30 min of ischemia, to 6.44 μmol/g), declining by 83% after 90 min recirculation. In contrast, glucose levels, which rose only to 3.58 μmol/g in the control, nonischemic animals, rose dramatically in the ischemic rats, to 15.7 μmol/g (439% increase, compared to control) at 30 min recirculation. Even after 90 min recirculation, glucose levels were still 380% of the control level. Again, this 20-yr-old study indicates clearly that, during the first hour or so of postischemia, the brain metabolizes lactate aerobically, while glucose remains unused.

In a recent preliminary study, using a rat model of cardiac-arrest-induced transient global cerebral ischemia (TGI), three important results were obtained *(91)*. First, the glucose paradox, the phenomenon that was heralded as the proof for the detrimental role of lactic acidosis in delayed neuronal damage, can be brought about by hyperglycemia (2 g/kg glucose loading), when induced up to 1 h pre-TGI. Rats that were treated in this way exhibited a significant increase in delayed neuronal damage post-TGI. Second, loading glucose 2–4 h pre-TGI, while rendering the rats hyperglycemic at the time of the ischemic insult, did not exacerbate the delayed neuronal damage post-TGI. Rather, these rats showed significantly less neuronal damage than that measured in control, isoglycemic rats. Third, brain lactate levels were equally high in both hyperglycemic groups, thus refuting the argument that increased lactate level during an ischemic insult is the reason for exacerbation of neuronal damage.

4. CONCLUDING REMARKS

Neuronal energy metabolism has emerged in recent years as an active field of study, with great potential of shedding light on newly explored intrinsic mechanisms supporting special neuronal energy needs. The bulk of the research until now has been done using in vitro systems, laying the foundation on which future in vivo studies will undoubtedly appear. The few in vivo studies already available do not necessarily agree with each other, especially regarding the question of anaerobic vs aerobic glucose utilization during neural activation. Nevertheless, the new information challenges some of the old dogmas of brain cellular energy metabolism both under normal conditions and under conditions of energy deprivation. The next few years promise to be exciting in terms of new knowledge, its understanding, and its practical uses.

REFERENCES

1. Hawkins, R. (1985) Cerebral energy metabolism, in Cerebral Energy Metabolism and Metabolic Encephalopathy (McCandless, D. W., ed.), Plenum, New York, pp. 3–23.
2. Clarke, D. D. and Sokoloff, L. (1994) Circulation and energy metabolism of the brain, in *Basic Neurochemistry* (Siegle, G. J., Agranoff, B. W., Albers, R. W., and Molinoff, P. B., eds.), Raven, New York, pp. 645–680.
3. Cahill, G. F., Jr. and Aoki, T. T. (1988) Alternate fuel utilization by brain, in *Cerebral Metabolism and Neural Function* (Passonneau, J. V., Hawkin, R. A., Lust, D. W., and Welsh, F., eds.), Williams & Wilkins, Baltimore, pp. 234–242.
4. Fox, P. T. and Raichle, M. E. (1986) Focal physiological uncoupling of cerebral blood flow and oxidative metabolism during somatosensory stimulation in human subjects. *Proc. Natl. Acad. Sci. USA* **83,** 1140–1144.
5. Fox, P. T., Raichle, M. E., Mintun, M. A., and Dence, C. (1988) Nonoxidative glucose consumption during focal physiologic neural activity. *Science* **241,** 462–464.
6. Lear, J. L. (1990) Glycolysis: link between PET and proton MR spectroscopic studies of the brain. *Radiology* **174,** 328–330.
7. Prichard, J., Rothman, D., Novotny, E., Petroff, O., Kuwabara, T., Avison, M., et al. (1991) Lactate rise detected by 1H NMR in human visual cortex during physiologic stimulation. *Proc. Natl. Acad. Sci. USA* **88,** 5829–5831.
8. Sappey-Marinier, D., Calabrese, G., Fein, G., Hugg, J. W., Biggins, C., and Weiner, M. W. (1992) Effect of photic stimulation on human visual cortex lactate and phosphates using ^{1}H and ^{31}P magnetic resonance spectroscopy. *J. Cereb. Blood Flow Metab.* **12,** 584–592.
9. Fellows, L. K., Boutelle, M. G., and Fillenze, M. (1993) Physiological stimulation increases nonoxidative glucose metabolism in the brain of the freely moving rat. *J. Neurochem.* **60,** 1258–1263.
10. Raichle, M. E. (1991) The metabolic requirements of functional activity in the human brain: a positron emission tomography study, in *Fuel Homoestasis and the Nervous System. Adv. Exp. Med. Biol.*, vol. 291 (Vranic, M., Efendic, S., and Hollenberg, C. H., eds.), Plenum, New York, pp. 1–4.
11. Penfield, W. (1937) The circulation of the epileptic brain. *Res. Pub. Assoc. Res. Nerv. Ment. Dis.* **18,** 605–637.
12. Plum, F., Posner, J. B., and Troy, B. (1968) Cerebral metabolic and circulatory responses to induced convulsion in animals. *Arch. Neurol.* **18,** 1–13.
13. Reivich, M., Isaccs, G., Evarts, E., and Kety, S. (1968) The effect of slow wave sleep and REM sleep on regional cerebral blood flow in cats. *J. Neurochem.* **15,** 301–306.
14. Heiss, W.-D., Pawlik, G., Herholz, K., and Wienbard, K. (1985) Regional cerebral glucose metabolism in man during wakefulness, sleep and dreaming. *Brain Res.* **327,** 362–366.
15. Santiago, T. V., Guerra, E., Neubauer, J. A., and Edman, N. H. (1984) Correlation between ventilation and brain blood flow during sleep. *J. Clin. Invest.* **73,** 497–506.
16. Rowe, G. G., Maxwell, G. M., Castillo, C. A., Freeman, D. J., and Crumpton, C. W. (1959) A study in man of cerebral blood flow and cerebral glucose, lactate and pyruvate metabolism before and after eating. *J. Clin. Invest.* **38,** 2154–2158.
17. Oldendorf, W. H. (1970) Measurement of brain uptake of radiolabeled substances using a tritiated water internal standard. *Brain Res.* **24,** 372–376.
18. Oldendorf, W. H. (1973) Carrier-mediated blood-brain transport of short-chain monocarboxylic organic acids. *Am. J. Physiol.* **224,** 1450–1453.
19. Oldendorf, W. H. (1980) Clearance of radiolabeled substances by brain after arterial injection using a diffusible internal standard, in *Research Methods in Neurochemistry* (Marks, N. and Rodnight, R., eds.), Plenum, New York, pp. 91–112.
20. Nemoto, E. M., Hoff, J. T., and Severinghaus, J. W. (1974) Lactate uptake and metabolism by brain during hyperlactatemia and hypoglycemia. *Stroke* **5,** 48–53.

21. Nemoto, E. M. and Severinghaus, J. W. (1974) Stereospecific permeability of rat blood Brain barrier to lactic acid. *Stroke* **5,** 81–84.
22. Fishman, R. A. (1992) *Cerebrospinal Fluid in Disease of the Nervous System*, W. B. Saunders, Philadelphia, pp. 183–252.
23. Partridge, W. M. and Oldendorf, W. H. (1977) Transport of metabolic substrates through the blood-brain barrier. *J. Neurochem.* **28,** 5–12.
24. Partridge, W. M. (1983) Brain metabolism: a perspective from the blood-brain barrier. *Physiol. Rev.* **63,** 1481–1535.
25. Dringen, R., Schmoll, D., Cesar, M., and Hamprecht, B. (1993) Incorporation of radioactivity from [^{14}C]lactate into the glycogen of cultured mouse astroglial cells. *Biol. Chem. Hoppe-Seyler* **374,** 343–347.

25a. Hassel, B. and Brathe, A. (2000) Cerebral metabolism of lactate in vivo: evidence for neuronal pyruvate carboxylation. *J. Cereb. Blood Flow Metab.* **20,** 327–336.

26. Dringen, R., Gebhardt, R., and Hamprecht, B. (1993) Glycogen in astrocytes: possible function as lactate supply for neighboring cells. *Brain Res.* **623,** 208–214.
27. McIlwain, H. (1953) Glucose level, metabolism, and response to electrical impulses in cerebral tissues from man and laboratory animals. *Biochem. J.* **55,** 618–624.
28. McIlwain, H. (1953) Substances which support respiration and metabolic response to electrical impulses in human cerebral tissues. *J. Neurol. Neurosurg. Psychiat.* **16,** 257–266.
29. McIlwain, H. (1956) Electrical influences and speed of chemical change in the brain. *Physiol. Rev.* **36,** 355–375.
30. McIlwain, H. and Bachelard, H. S. (1985) *Biochemistry and the Central Nervous System.* Churchill Livingstone, Edinburgh, pp. 7–153.
31. Schurr, A., West, C. A., and Rigor, B. M. (1988) Lactate-supported synaptic function in the rat hippocampal slice preparation. *Science* **240,** 1326–1328.
32. Stittsworth, J. D., Jr. and Lanthorn, T. H. (1993) Lactate mimics only some effects of D-glucose on epileptic depolarization and long-term synaptic failure. *Brain Res.* **630,** 21–27.
33. Izumi, Y., Benz, A. M., Zorumski, C. F., and Olney, J. W. (1994) Effects of lactate and pyruvate on glucose deprivation in rat hippocampal slices. *Neuroreport* **5,** 617–620.
34. Bueno, D., Azzolin, I. R., and Perry, M. L. S. (1994) Ontogenic study of glucose and lactate utilisation by rat cerebellum slices. *Med. Sci. Res.* **22,** 631–632.
35. Schurr, A. and Rigor, B. M. (1992) The mechanism of hypoxic-ischemic damage. *Hippocampus* **2,** 221–228.
36. Magistretti, P. J., Sorg, O., Yu, N., Martin, J.-L., and Pellerin, L. (1993) Neurotransmitters regulates energy metabolism in astrocytes: implications for the metabolic trafficking between neural cells. *Dev. Neurosci.* **15,** 306–312.
37. Pellerin, L. and Magistretti, P. J. (1994) Glutamate uptake into astrocytes stimulates aerobic glycolysis: a mechanism coupling neuronal activity to glucose utilization. *Proc. Natl. Acad. Sci. USA* **91,** 10,625–10,629.
38. Magistretti, P. J., Pellerin, L., Martin, J.-L. (1995) Brain energy metabolism: an integrated cellular perspective, in *Psychopharmacology: The Fourth Generation of Progress* (Bloom, F. E. and Kupfer, D. J., eds.), Raven, New York, pp. 657–670.
39. Pellerin, L. and Magistretti, P. J. (1996) Excitatory amino acids stimulate aerobic glycolysis in astrocytes via activation of the Na^+/K^+ ATPase. *Dev. Neurosci.* **18,** 336–342.
40. Larrabee, M. G. (1983) Lactate uptake and release in the presence of glucose by sympathetic ganglia of chicken embryos and by neuronal and non-neuronal cultures prepared from these ganglia. *J. Neurochem.* **40,** 1237–1250.
41. Larrabee, M. G. (1995) Lactate metabolism and its effects on glucose metabolism in an excised neural tissue. *J. Neuroschem.* **64,** 1734–1741.

42. Larrabee, M. G. (1996) Partitioning of CO_2 production between glucose and lactate in excised sympathetic ganglia, with implications for brain. *J. Neuroschem.* **67,** 1726–1734.
43. Magistretti, P. J. (1999) Brain energy metabolism, in Fundamental Neuroscience (Zigmond, M. J., Bloom, F. E., Landis, S. C., Roberts, J. L., and Squire, L. R., eds.), Academic, New York, pp. 389–413.
44. Schurr, A., Payne, R. S., Miller, J. J., and Rigor, B. M. (1997) Brain lactate, but not glucose, fuels the recovery of synaptic function from hypoxia upon reoxygenation: an in vitro study. *Brain Res.* **744,** 105–111.
45. Malonek, D. and Grinvald, A. (1996) Interaction between electrical activity and cortical microcirculation revealed by imaging spectroscopy: implication for functional brain mapping. *Science* **272,** 551–554.
46. Hyder, F., Chase, J. R., Behar, K. L., Mason, G. F., Siddeek, M., Rothman, D. L., and Shulman, R. G. (1996) Increased tricarboxylic acid cycle flux in rat brain during forepaw stimulation detected with $^1H[^{13}C]$NMR. *Proc. Natl. Acad. Sci. USA* **93,** 7612–7617.
47. Friedmann, T. E. and Barborka, C. (1941) The significance of the ratio of lactic acid to pyruvic acid in blood after exercise. *J. Biol. Chem.* **141,** 993–994.
48. Haljamae, H. (1987) Lactate metabolism. *Intensive Care World* **4,** 118–121.
49. Brooks, G. A. (1986) Lactate production under fully aerobic conditions: the lactate shuttle during rest and exercise. *Fed. Proc.* **45,** 2924–2929.
50. Siesjo, B. K. (1981) Cell damage in the brain: a speculative synthesis. *J. Cereb. Blood Flow Metab.* **1,** 155–185.
51. Magistretti, P. J., Sorg, O., Yu, N., Martin, J.-L., and Pellerin, L. (1993) Neurotransmitters regulate energy metabolism in astrocytes: Implications for the metabolic trafficking between neural cells. *Dev. Neurosci.* **15,** 306–312.
52. Magistretti, P. J. and Pellerin, L. (1999) Cellular mechanisms of brain energy metabolism and their relevance to functional brain imaging. *Phil. Trans. R. Soc. Lond. B.* **354,** 1155–1163.
53. Schurr, A., Miller, J. J., Payne, R. S., and Rigor, B. M. (1999) An increase in lactate output by brain tissue serves to meet the energy needs of glutamate-activated neurons. *J. Neurosci.* **19,** 34–39.
54. Gerhart, D. Z., Enerson, B. E., Zhdankina, O. Y., Leino, R. L., and Drewes, L. R. (1997) Expression of monocarboxylate transporter MCT1 by brain endothelium and glia in adult and suckling rats. *Am. J. Physiol.* **273,** E207–E213.
55. Broer, S., Rahman, B., Pellegri, G., Pellerin, L., Martin, J.-L., Verleysdonk, S., Hamprecht, B., and Magistretti, P. J. (1997) Comparison of lactate transport in astroglial cells an monocarboxylate 1 (MCT1) expressing *Xenopus laevis* oocytes. *J. Biol. Chem.* **272,** 30,096–30,102.
56. Schurr, A., Payne, R. S., Miller, J. J., and Rigor, B. M. (1997) Brain lactate is an obligatory aerobic energy substrate for functional recovery after hypoxia: further in vitro validation. *J. Neurochem.* **69,** 423–426.
57. Griffin, J. L., Rae, C., Radda, G. K., and Matthews, P. M. (1999) Lactate-induced inhibition of glucose catabolism in guinea pig cortical brain slices. *Neurochem. Int.* **35,** 405–409.
58. O'Kusky, J. and Colonnier, M. (1982) A laminar analysis of the number of neurons, glia and synapses in the visual cortex (area 17) of the adult macaque monkey. *J. Comp. Neurol.* **210,** 278–290.
59. Bignami, A. (1991) Glial cells in the central nervous system, in Discussions in Neuroscience, vol. VIII, no. 1 (Magistretti, P. J., ed.), Elsevier, Amsterdam, pp. 1–45.
60. Kimelberg, H. K., Jalonen, T., and Walz, W. (1993) Regulation of brain microenvironment: transmitters and ions, in Astrocytes: Pharmacology and Function (Murphy, S., ed.), Academic, San Diego, pp. 193–228.

61. Magistretti, P. J. and Pellerin, L. (1996) Cellular bases of brain energy metabolism and their relevance to functional brain imaging: evidence for a prominent role of astrocytes. *Cereb. Cortex* **6,** 50–61.
62. Hu, Y. and Wilson, G. S. (1997) A temporary local energy pool coupled to neuronal activity: fluctuations of extracellular lactate levels in rat brain monitored with rapid-response enzyme-based sensor. *J. Neurochem.* **69,** 1484–1490.
63. Bittar, P. G., Charnay, Y., Pellerin, L., Bouras, C., and Magistretti, P. J. (1996) Selective distribution of lactate dehydrogenase isoenzymes in neurons and astrocytes of human brain. *J. Cereb. Blood Flow Metab.* **16,** 1079–1089.
64. Watanabe, H. and Passonneau, J. V. (1973) Factors affecting the turnover of cerebral glycogen and limit dextrin in vivo. *J. Neurochem.* **20,** 1543–1554.
65. Magistretti, P. J., Sorg, O., and Martin, J.-L. (1993) Regulation of glycogen metabolism in astrocytes: physiological, pharmacological, and pathological aspects, in *Astrocytes: Pharmacology and Function* (Murphy, S., ed.), Academic, San Diego, pp. 243–265.
66. Kety, S. S., Woodford, R. B., Harmel, M. H., Freyhan, F. A., Appel, K. F., and Schmidt, C. F. (1948) Cerebral flow and metabolism in schizophrenia. The effect of barbiturate seminarcosis, insulin coma, and electroshock. *Am. J. Psychiatry* **104,** 765–770.
67. Landau, B. R., Laszlo, J., Strengel, J., and Burk, D. (1958) Certain metabolic and pharmacologic effects in cancer patients given infusions of 2-deoxy-D-glucose. *J. Natl. Cancer Inst.* **21,** 485–494.
68. Freinkel, N., Metzger, B. E., Harris, E., Robinson, S., and Mager, M. (1972) The hypothermia of hypoglycemia. *N. Engl. J. Med.* **287,** 841–845.
69. Ritter, S. and Dinh, T. T. (1994) 2-Mercaptoacetate and 2-deoxy-D-glucose induce Fos-like immunoreactivity in rat brain. *Brain Res.* **641,** 111–120.
70. Hass, W. K. (1980) Regional studies of glucose metabolism in coma, in Cerebral Metabolism and Neural Function (Passonneau, J. V., Hawkins, R. A., Lust, W. D., and Welsh, F. A., eds.), Williams & Wilkins, Baltimore, pp. 382–386.
71. McCandless, D. W. and Abel, M. S. (1985) Hypoglycemia and cerebral energy metabolism, in *Cerebral Energy Metabolism and Metabolic Encephalopathy* (McCandless, D. W., ed.), Plenum, New York, pp. 27–41.
72. Ghajar, J. B., Plum, F., and Duffy, T. E. (1982) Cerebral oxidative metabolism and blood flow during acute hypoglycemia and recovery in unanesthetized. *J. Neurochem.* **38,** 397–409.
73. Betz, A. L., Gilboe, D. D., Yudilevich, D. L., and Drewes, L. R. (1973) Kinetics of unidirectional glucose transport into the isolated dog brain. *Am. J. Physiol.* **225,** 586–592.
74. Fiorentini, A. and Muller, E. E. (1972) Sensitivity of central chemoreceptor controlling blood glucose and body temperature during glucose deprivation. *J. Physiol.* **248,** 247–271.
75. Schurr, A. and Rigor, B. M. (1990) *Cerebral Ischemia and Resuscitation*, CRC, Boca Raton, FL.
76. *Dorland's Illustrated Medical Dictionary* (1965), 24th ed., W. B. Saunders, Philadelphia.
77. Marangos, P. J. and Lal, H. (1992) *Emerging Strategies in Neuroprotection*, Birkhauser, Boston.
78. Stamford, J. A. and Strunin, L. (1996) *Bailliare's Clinical Anaesthesiology*, vol. 10/No. 3, *Neuroprotection*, Bailiare Tindall, London.
79. Myers, R. E. and Yamaguchi, S. (1977) Nervous system effects of cardiac arrest in monkeys preservation of vision. *Arch. Neurol.* **34,** 65–74.
80. Ginsberg, M. D., Welsh, F. A., and Budd, W. W. (1980) Deleterious effect of glucose pretreatment on recovery from diffuse cerebral ischemia in the cat. I. Local cerebral blood flow and glucose utilization. *Stroke* **11,** 347–354.
81. Rehncrona, S., Rosen, I., and Siesjo, B. K. (1981) Brain lactic acidosis and ischemic cell damage: 1. Biochemistry and neurophysiology. *J. Cereb. Blood Flow Metab.* **1,** 297–311.
82. Schurr, A., Payne, R. S., Miller, J. J., and Rigor, B. M. (1997) Glia are the main source of lactate utilized by neurons for recovery of function posthypoxia. *Brain Res.* **774,** 221–224.

83. Halestrap, A. P. and Denton, R. M. (1975) The specificity and metabolic implications of the inhibition of pyruvate transport in isolated mitochondria and intact tissue preparations by α-cyano-4-hydroxycinnamate and related compounds. *Biochem. J.* **148,** 97–106.
84. Volk, C., Kempski, B., and Kempski, O. S. (1997) Inhibition of lactate export by quercetin acidifies rat glial cells in vitro. *Neurosci. Lett.* **223,** 121–124.
85. Schurr, A., Dong, W.-Q., Reid, K. H., West, C. A., and Rigor, B. M. (1988) Lactic acidosis and recovery of neuronal function following cerebral hypoxia in vitro. *Brain Res.* **438,** 311–314.
86. Giffard, R. G., Monyer, H., Christine, C. W., and Choi, D. W. (1990) Acidosis reduces NMDA receptor activation, glutamate neurotoxicity, and oxygen-glucose deprivation neuronal injury in cortical cultures. *Brain Res.* **506,** 339–342.
87. Tombaugh, G. C. and Sapolsky, R. M. (1990) Mild acidosis protects hippocampal neurons from injury induced by oxygen and glucose deprivation. *Brain Res.* **506,** 343–345.
88. Ljunggren, B., Norberg, K., and Siesjo, B. K. (1974) Influence of tissue acidosis upon restitution of brain energy metabolism following total ischemia. *Brain Res.* **77,** 173–186.
89. Drewes, L., Gilboe, D. D., and Betz, L. A. (1973) Metabolic alteration in brain during anoxic-anoxia and subsequent recovery. *Arch. Neurol.* **29,** 385–390.
90. Steen, P. A., Michenfelder, J. D., and Milde, J. H. (1979) Incomplete versus complete cerebral ischemia: Improved outcome with a minimal blood flow. *Ann. Neurol.* **6,** 389–398.
91. Schurr, A., Payne, R. S., Tseng, M. T., Miller, J. J., and Rigor, B. M. (1999) The glucose paradox in cerebral ischemia: new insights. *Ann. NY Acad. Sci.* **893,** 386–390.

II

Brain Microenvironment

3
Plasticity of Extracellular Space

Eva Syková

1. INTRODUCTION

The extracellular space (ECS) of the nervous tissue is a microenvironment for nerve cells, and an important communication channel *(1–4)*. It includes ions, transmitters, metabolites, peptides, neurohormones, other neuroactive substances, and molecules of the extracellular matrix (ECM), and directly or indirectly affects neuronal and glial cell functions. Neuroactive substances, released into the ECS by neurons and glia, diffuse via the ECS to their targets, located on nerve as well as glial cells, frequently distant from the release sites. Since glial cells do not have synapses, their communication with neurons is mediated only by the diffusion of ions and neuroactive substances in the ECS.

The structure and physicochemical properties of the ECS vary around each cell, and in different brain regions. Certain synapses, so-called "closed" synapses (Fig. 1), and often even whole neurons, are tightly ensheathed by glial processes or ECM, forming so-called "perineuronal nets" *(5)*; others are more left naked ("open" synapses). Transmitters released from presynaptic terminals can escape from synaptic clefts, particularly during repetitive stimulation, although open synapses are more easily reached by molecules diffusing in the ECS (Fig. 1).

The size and irregular geometry of diffusion channels in the ECS (see further description of tissue tortuosity and anisotropy) substantially differ in various central nervous system (CNS) regions, and thus affect and direct the movement of various neuroactive substances in the ECS (Fig. 2), and thereby modulate neuronal signaling, neuron–glia communication, and extrasynaptic transmission. Dynamic changes in ECS ionic composition, ECM, and ECS volume and geometry accompany neuronal activity, neuronal loss, glial development and proliferation, aging, CNS injury, anoxia/ischemia, spreading depression, tumors, inflammation, dymyelination, and many other pathological (PATH) states in the CNS.

2. ECS

Cellular elements and blood vessels fill ~80% of the total CNS tissue volume, and the remaining portion (15–25%) is the ECS. ECS ionic changes, resulting from transmembrane ionic shifts during neuronal activity, depolarize neighboring neurons and glial cells, enhance or depress their excitability, and affect ion channel permeability

From: *The Neuronal Environment: Brain Homeostasis in Health and Disease*
Edited by: W. Walz © Humana Press Inc., Totowa, NJ

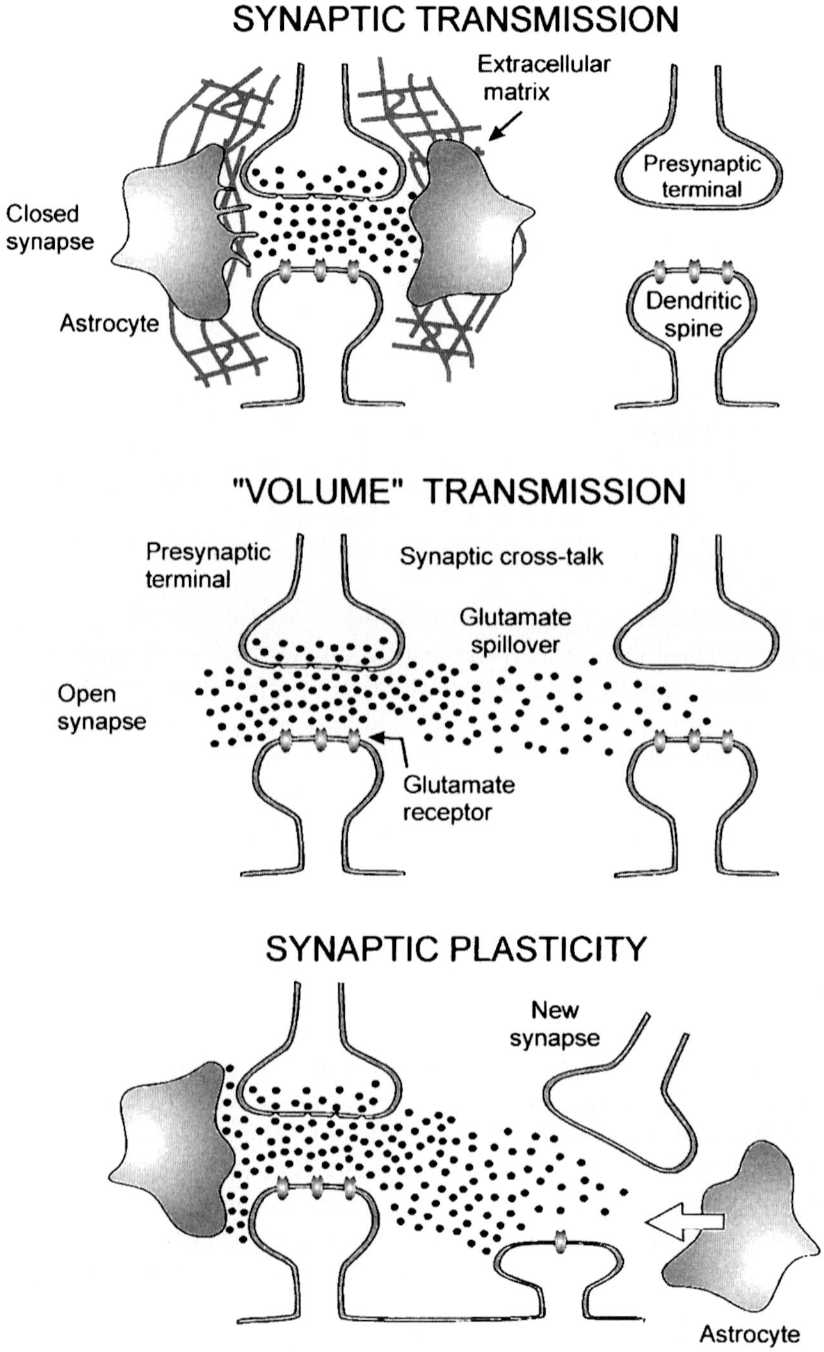

Fig. 1. Concept of synaptic transmission and extrasynaptic VT. Closed synapses are typical of synaptic transmission. This synapse is tightly ensheathed by glial processes and ECM, forming perineuronal or perisynaptic nets. Open synapse is typical of VT. It allows the escape of transmitter (e.g., glutamate, GABA) from the synaptic cleft (spillover), diffusion in ECS, and binding to receptors on nearby synapse. This phenomenon is known as "crosstalk" between synapses. The spillover may also lead to plastic changes, inducing formation of new synapses, or eliciting the rearrangement of astrocytic processes around the synapse.

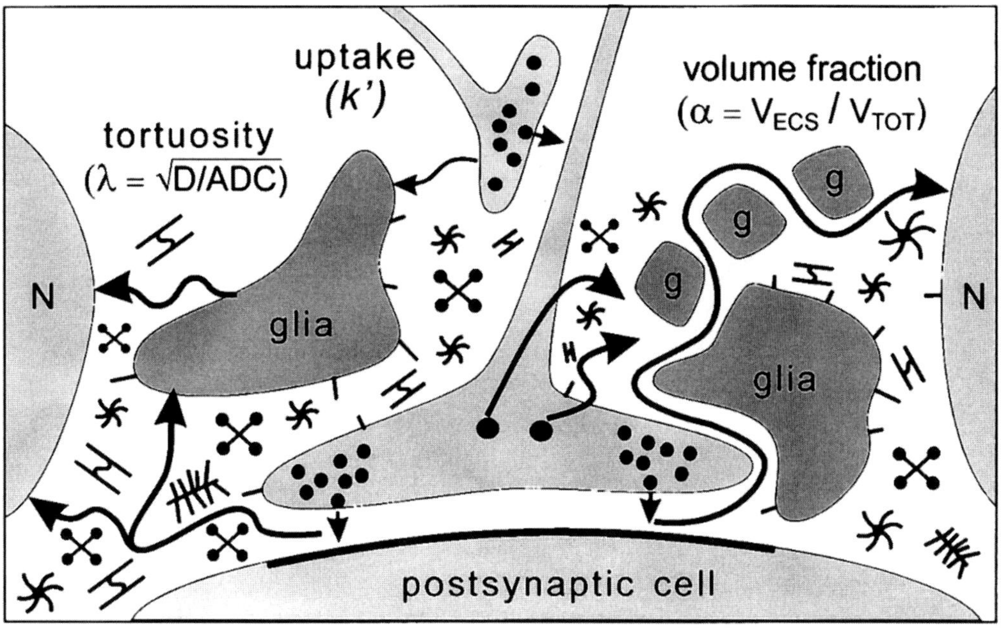

Fig. 2. Schematic of CNS architecture. The CNS architecture is composed of neurons (N), axons, glial cells (Glia), cellular processes (G), molecules of the ECM, and intercellular channels between the cells. This architecture slows down the movement (diffusion) of substances in the brain, which is critically dependent on the ECS diffusion parameters volume fraction (α), tortuosity (λ), and nonspecific uptake (*k*').

(1–3,6–8). These ionic changes may also lead to the synchronization of neuronal activity and stimulate glial cell function *(1–3)*.

Ionic and volume homeostasis in the CNS is maintained by a variety of mechanisms present in neurons, as well as in glial cells. It was shown in a number of studies in vivo and in vitro *(1,2,7)* that changes in EC K^+ concentration ($[K^+]_e$), alkaline and acid shifts in pH_e, and decreases in EC Ca^{2+} concentration ($[Ca^{2+}]_e$) accompany neuronal activity in different brain regions. For example, in the immediate vicinity of individual neurons in the mesencephalic reticular formation of the rat, a structure with a high spontaneous-activity level, dynamic changes in $[K^+]_e$ have been described. During a burst of spontaneous action potentials, K^+ accumulates in the close vicinity of the firing cell, and $[K^+]_e$ steadily increases by as much as 0.1 m*M* *(9)*. Most frequently, neuronal activity results not only in an increase in $[K^+]_e$, but also in a decrease in $[Ca^{2+}]_e$ and in a rapid EC alkaline shift, followed by a slower but longer-lasting acid shift *(2)*. After sustained adequate stimulation of the afferent input, or after repetitive electrical stimulation, the ionic transients reach a certain steady state, the so-called "ceiling" level (Fig. 3). The K^+ ceiling level found in the adult mammalian cortex is about 7 m*M* K^+ *(10)*, and, in the mammalian spinal cord 6–8 m*M* K^+ *(11,12)*. The alkaline shifts in mammalian cortex, cerebellum, or spinal cord do not exceed 0.02 pH units; the acid shifts are about 0.2 pH units (Fig. 3; *2,7,8*). When stimulation is continued, a gradual decrease of both transients, $[K^+]_e$ and pH_e, occurs after the ceiling levels are reached because of homeostatic mechanisms in neurons and glial cells (Fig. 3). The redistribution of activity-

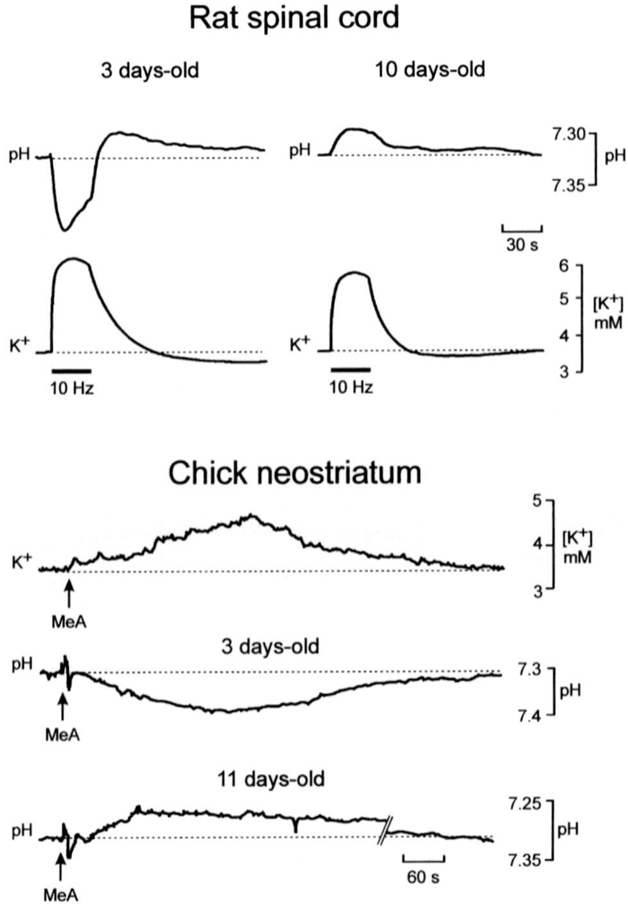

Fig. 3. Stimulation-evoked $[K^+]_e$ and pH_e changes during early postnatal development. Rat spinal cord: Repetitive electrical stimulation of the dorsal root in 3-d-old rats was accompanied by an alkaline shift, and by an increase in $[K^+]_e$. When stimulation was discontinued, a poststimulation acid shift of smaller amplitude appeared, which was accompanied by a K^+-undershoot. In the 10-d-old rat, the $[K^+]_e$ increase was smaller, and was accompanied by an EC acid shift. Chick neostriatum: The effect of taste-aversant methylantranilate (MeA), placed on the tongue of anesthetized chicks on $[K^+]_e$ and pH_e, recorded in the hyperstriatum ventrale at postnatal d 3 and 11. In 3-d-old chicks, MeA evoked an alkaline shift; in 11-d-old chicks, only an acid shift was detected.

related ionic changes is mediated, at least partially, by the Na^+,K^+-ATPase transport mechanism, which is located in neurons and in glial cells. Besides the Na^+/K^+ pump activity, K^+ homeostasis in the ECS is also maintained by three other mechanisms located in glial cells: K^+ spatial buffering, KCl uptake, and Ca^{2+}-activated K^+ channels.

The pH changes in the ECS are caused by neurons, as well as by glial cells. There is now convincing evidence for the neuronal origin of EC alkaline shifts and the glial origin of activity-related acid shifts *(13–15)*. During early postnatal development, when glial cell function is incomplete, activity-related pH_e and $[K^+]_e$ changes are substantially different from those observed in adult animals with completed gliogenesis and glial cell function. In the neonatal rat spinal cord, stimulation-evoked changes in $[K^+]_e$ are much larger than in the adult animal *(13)*. In newborn rats *(13)* or chicks *(16)*, alkaline shifts dominate in the ECS, and the acid shifts generated by glia are small. At postnatal d 10, when gliogenesis in gray matter peaks, the K^+ ceiling level decreases, and stimulation evokes acid shifts in the range of 0.1–0.2 pH unit, which are preceded by scarcely discernible alkaline shifts, also observed in adult rats (Fig. 3). Stimulation-evoked alkaline shifts are, in addition, abolished by blocking synaptic transmission with Mn^{2+} or Mg^{2+}; acid shifts are unaffected *(14)*. Alkaline shifts resulting from neuronal activity are depressed by the application of the aminobutyric acid (GABA) antagonist, picrotoxin, and by glutamate receptor antagonists and channel blockers, such as MK801 (noncompetitive *N*-methyl-D-aspartate receptor antagonist and channel blocker) and CNQX (competitive α-amino-3-hydroxy-5-methyl-4-isoxazolepropionate/kainate receptor antagonist) *(14,17)*.

Glial cells play an important role in buffering activity-related alkaline changes in EC pH. Some of the membrane transport processes regulating intracellular and EC pH, such as Na^+/H^+ exchange and $Na^+/H^+/Cl^-/HCO_3^-$ co-transport, are present in both neurons and glia; others are specific, either for neurons (e.g., H^+ channels, H^+ or HCO_3^- permeability of the ionic channels opened by GABA or glutamate) or for glia (e.g., the voltage-dependent Na^+–HCO_3^- co-transport and lactate extrusion). The glial cell membrane is also readily permeable to CO_2, which reacts with water to form carbonic acid, which in turn quickly dissociates into water and protons. Some of the membrane transport mechanisms result in alkaline shifts in pH_e (acid loaders); others result in acid shifts in pH_e (acid extruders), but it is evident that acid loaders are dominant in neurons, and acid extruders are dominant in glia *(2,7,8)*. EC acid shifts are, therefore, a consequence of activity-related $[K^+]_e$ increases. K^+-induced glial depolarization results in an alkaline shift in glial pH_i, which leads to the stimulation of classical acid extrusion systems in glial cells. Membrane mechanisms responsible for the transport of ions across the cell membrane are always accompanied by the movement of water, thus affecting the cell volume and the size of the ECS. Hence, glial cells play an important role in buffering activity-related alkaline changes in EC pH. This mechanism of negative feedback control suppresses neuronal excitability (Fig. 4).

The solution in the ECS is, however, not a simple salt solution. It has become apparent that long-chain polyelectrolytes, either attached or unattached to cellular membranes, are present in the ECS. The ECS also contains a number of glycosaminoglycans (e.g., hyaluronate), glycoproteins, and proteoglycans that constitute the ECM. The molecular content of the ECM, e.g., chondroitin sulphate proteoglycan, fibronectin, tenascin, laminin, and so on *(5,18)*, dynamically changes during development, aging, wound healing, and many pathological processes. ECM molecules are produced by both neurons and glia. These molecules have been suggested to cordon off distinct functional units in the CNS (groups of neurons, axon tracts, and nuclear groups). As shown in Figs. 2 and 4, these large molecules can slow down the movement (diffusion) of various neuroactive substances through the ECS. More importantly, these molecules can hinder diffusion of molecules, so that they are confined to certain places, and diffusion to other brain regions will be facilitated.

Other important chemical components of the ECS are substances involved in metabolism, particularly glucose and dissolved gasses (O_2 and CO_2). The presence of

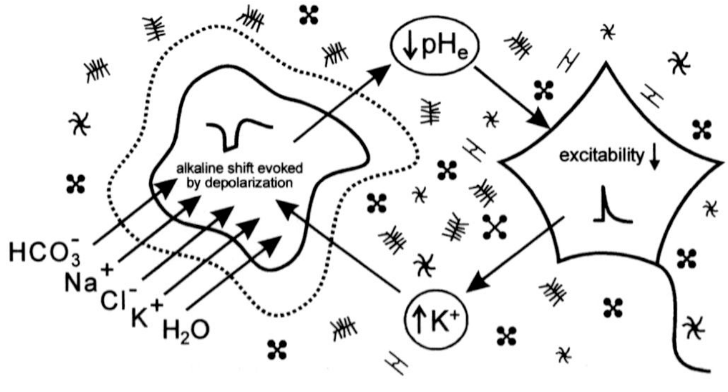

Fig. 4. Schematic of the mechanism of nonspecific feedback suppressing neuronal excitability. Active neurons release K^+, which accumulates in the ECS, depolarizes glial cells, and is taken up by glia. In depolarized glial cells, an alkaline shift is evoked by activation of Na^+/HCO_3^- cotransport. This alkaline shift in glial pH_i causes an acid shift in pH_e. EC acidosis further suppresses neuronal activity. Transmembrane ionic movements are accompanied by water, and will therefore result in glial swelling. ECS volume decreases lead to a greater accumulation of ions and neuroactive substances, and to a crowding of molecules in the ECM of the ECS.

HCO_3^- and CO_2 forms a powerful buffering system that controls EC and intracellular pH. The ECS also contains free-radical scavengers, such as ascorbate and glutathione, which may counteract some potentially lethal products of O_2 metabolism. In addition, the ECS contains amino acids like glutamate and aspartate, catecholamines, indolamines such as dopamine and serotonin, various opioid peptides, nitric oxide, and growth hormones. Transmitters in the ECS bind to extrasynaptically located high-affinity binding sites on neurons and glia.

Fine cellular processes also produce diffusion barriers and anisotropy (Fig. 2). Glial cells, when hypertrophied or proliferating, form diffusion barriers *(3,19,20)*, and, in this way, critically affect the permissiveness of the tissue, synaptic as well as extrasynaptic (volume) transmission, activity-dependent synaptic plasticity, neurogenesis, and regeneration.

3. EXTRASYNAPTIC (VOLUME) TRANSMISSION IN THE CNS

Transmitters that are released nonsynaptically diffuse through the ECS and bind to extrasynaptic, usually high-affinity, binding sites located on neurons, axons, and glial cells. This type of extrasynaptic transmission is also called "volume transmission" (VT) (neuroactive substances move through the volume of the ECS) *(3,4,21–23)*. Populations of neurons can interact both by synapses and by the diffusion of ions and neurotransmitters in the ECS. Diffusion is therefore the underlying mechanism of VT. Diffusion parameters are changing throughout life, e.g., during development and aging, during repeated or prolonged neuronal activity, and in and after pathological states. This mode of communication by diffusion, and without synapses, is therefore very plastic, and provides a mechanism of long-range information processing in functions such as vigilance, sleep, chronic pain, hunger, depression, LTP, LTD, memory formation and other plastic changes in the CNS *(3)*. There are distinct types of VT *(23)*. Besides open synapses, described in Fig. 1, studies using electron microscopy have shown that peptidergic transmission is mediated by extrasynaptic vesicular release, or vesicular release from nonjunctional varicosities, when a transmitter is released from varicosities lacking presynaptic specialization and postsynaptic densities (e.g., catecholaminergic system). Another mechanism for VT is via transmitter uptake carriers. Ions that diffuse through the ECS can function as intercellular signals, as well as gaseous transmitters, e.g., nitric oxide.

3.1. ECS Diffusion Parameters, ECS Inhomogeneity, and Anisotropy

The diffusion of substances in a free medium, such as water or diluted agar, is described by Fick's laws. In contrast to a free medium, diffusion in the ECS of the nervous tissue is hindered by the size of the EC clefts, the presence of membranes, fine neuronal and glial processes, macromolecules of the ECM, and charged molecules, and also by cellular uptake (Fig. 2). To take these factors into account, it was necessary to modify Fick's original diffusion equations *(4,24)*. First, diffusion in the CNS is constrained by the restricted volume of the tissue available for the diffusing particles, i.e., by the ECS volume fraction (α), which is a dimensionless quantity, and is defined as the ratio between the volume of the ECS and the total volume of the tissue ($\alpha = V_{ECS}/V_{TOT}$). It is now evident that the ECS in the adult brain amounts to about 20% of the total brain volume, i.e., $\alpha = 0.2$ (Fig. 5). Second, the free-diffusion coefficient (D) in the brain is reduced by the tortuosity factor (λ). ECS tortuosity is defined as $\lambda = (D/ADC)^{0.5}$, where D is a free-diffusion coefficient and ADC is the apparent diffusion coefficient in the brain (Fig. 2). As a result of tortuosity (in adult brain, λ amounts to 1.55–1.65), D is reduced to an apparent diffusion coefficient, $ADC = D/\lambda^2$. Thus, any substance diffusing in the ECS is hindered by many obstructions. Third, substances released into the ECS are transported across membranes by nonspecific concentration-dependent uptake (k'). In many cases, however, these substances are transported by energy-dependent uptake systems that obey nonlinear kinetics. When these three factors (α, λ, and k') are incorporated into Fick's law, diffusion in the CNS is described satisfactorily *(24)*.

The real-time iontophoretic method is used to determine ECS diffusion parameters and their dynamic changes in nervous tissue in vitro, as well as in vivo *(3,4)*. In principle, ion-sensitive microelectrodes (ISM) are used to measure the diffusion of ions to which the cell membranes are relatively impermeable, e.g., TEA^+, TMA^+, or choline. These substances are injected into the nervous tissue by pressure or by iontophoresis, from an electrode aligned parallel to a double-barreled ISM at a fixed distance (Fig. 5). Such an electrode array is made by gluing together a pressure or iontophoretic pipet and a TMA^+-sensitive ISM, with a tip separation of 60–200 μm. In the case of iontophoretic application, TMA^+ is released into the ECS by applying a current step of +80–100 nA, with a duration of 40–80 s. The released TMA^+ is recorded with the TMA^+–ISM as a diffusion curve (Fig. 5), which is then transferred to a computer. Values of the ECS volume, *ADC*s, tortuosity, and nonspecific cellular uptake are extracted by a nonlinear curve-fitting simplex algorithm applied to the diffusion curves.

The other methods used to study ECS volume and geometry, e.g., intrinsic optical signals (IOS), tissue resistance, integrative optical imaging, and nuclear magnetic reso-

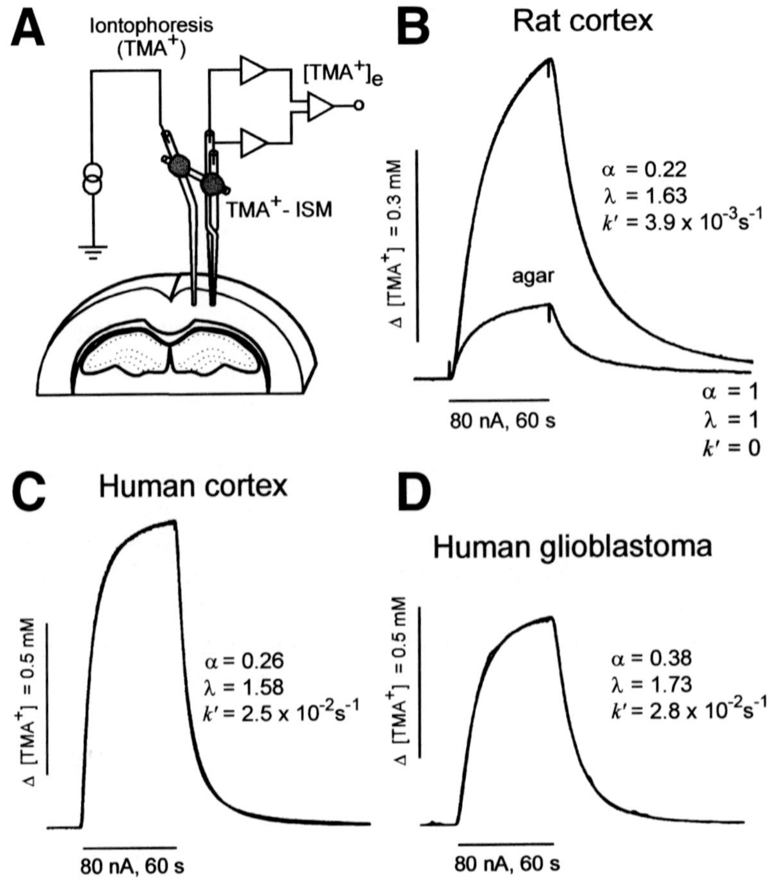

Fig. 5. **(A)** Schema of the experimental arrangement. A TMA^+-selective double-barreled ISM was glued to a bent iontophoresis microelectrode. The separation between electrode tips was 80–200 μm. **(B)** TMA^+ diffusion curves in the rat somatosensory cortex in vivo, and free diffusion measured with the same microelectrode array in agar. **(C)** Typical diffusion curve recorded in a slice (400 μm) from the human temporobasal cortex of an 18-yr-old woman with pharmacoresistant epilepsy. **(D)** Typical diffusion curve in a slice (400 μm) from a human glioblastoma (grade 4, WHO classification). For each curve, the ECS diffusion parameters, α (volume fraction) and λ (tortuosity), were extracted by appropriate nonlinear curve fitting. Experimental and theoretical curves are superimposed in each case. For each figure, the concentration scale is linear.

nance (NMR), are less comprehensive, because they only measure relative changes in the ECS-diffusion parameters, or only some of the three diffusion parameters *(25–29)*. Integrative optical imaging is used to measure the *ADC*s of molecules tagged with fluorescent dye; recordings of IOS, either light transmittance or light reflectance, are believed to reflect changes in the ECS volume; however, direct evidence is missing. On the other hand, diffusion-weighted NMR methods provide information only about the water-diffusion coefficient *(30–33)*. Although a correlation between water-diffusion maps, changes in cell volume, and ECS diffusion parameters has been found, it is still not well-understood. Our recent experiments suggest that tortuosity can also affect the *ADC* of water *(34)*.

Using a light-transmittance method (IOS), we recently found that changes in the ECS volume in brain slices, measured by the TMA method, have a different time-course than those revealed by IOS. Simultaneous measurements, using IOS and the TMA method, were used to determine the absolute values of the ECS volume fraction, α, tortuosity, λ, and nonspecific uptake, k', in the dorsal horns of rat spinal cord slices *(35,36)*. Cell swelling was evoked by a 45-min exposure to hypotonic solutions, elevated potassium or glutamate receptor agonists. Hypotonic solution (160 m*M*/kg), 50 or 80 m*M* K^+, or *N*-methyl-D-aspartate (10^{-4} M) induced a decrease in α of 45–85%, which peaked at 10–15 min; an increase in IOS of 20–35% peaked in the first 5 min. After the initial peak, IOS quickly decreased to control levels; the changes in α and λ persisted throughout the application. There was also no correlation with the changes in ECS tortuosity and uptake. However, the time-course of IOS changes corresponded to increases in neuronal activity, manifested by transient EC potassium increases. These data demonstrate that there is no simple correlation between ECS volume and IOS changes.

Neuroactive substances released constantly into the ECS will accumulate in this limited volume more rapidly than in free solution. Tortuosity (which is absent in a free medium) also causes a greater and more rapid accumulation of released substances. CNS tortuosity reduces the diffusion coefficient for small molecules by a factor of about 2.5, in many CNS regions. Larger molecules (with a relative molecular mass above 10 kDa), have a smaller diffusion coefficient than small molecules, and are significantly more hindered in their diffusion, and therefore exhibit larger tortuosity *(37,38)*. However, even large proteins, e.g., negatively charged globular proteins, such as bovine serum albumin (66 kDa), dextrans of 70 kDa, large copolymers of *N*-(2-hydroxypropyl)methacrylamide (HPMA) polymeric chains with M_r = 1,000,000 and star-like systems, containing either albumin (179,000 M_r) or immunoglobulin G (319,000 M_r) in the center, with HPMA side branches, developed as water-soluble anticancer drug carriers still migrate through the narrow interstices of brain slices *(39,40)*.

ECS diffusion parameters are different in different parts of the CNS. For example, the TMA^+ diffusion parameters in the sensorimotor cortex of the adult rat in vivo are heterogeneous *(14)*. The mean volume fraction gradually increases from $\alpha = 0.19$ in cortical layer II to $\alpha = 0.23$ in cortical layer VI. These typical differences are apparent in each individual animal. In subcortical white matter (corpus callosum) the volume fraction is always lower than in cortical layer VI, often between 0.19 and 0.20 *(42,43)*. There is also a heterogeneity in the spinal cord, the mean values of the volume fraction being highest in the ventral horn and lowest in the white matter *(44–46)*. Similar α values ($\alpha = 0.21$–0.22) have been found throughout the rat brain. In slices from human cortex (temporal and frontal lobe), recently obtained α values were not much different: They ranged between 0.21 and 0.26 (Fig. 5).

By introducing the tortuosity factor into diffusion measurements in the CNS, it soon becomes evident that diffusion in not uniform in all directions, and that it is affected by diffusion barriers. This so-called anisotropic diffusion preferentially channels the movement of neuroactive substances in the ECS in one direction (e.g., along axons), and may, therefore, be responsible for a certain degree of specificity in VT. Significant differences in tortuosity have been found in various brain regions, showing that the

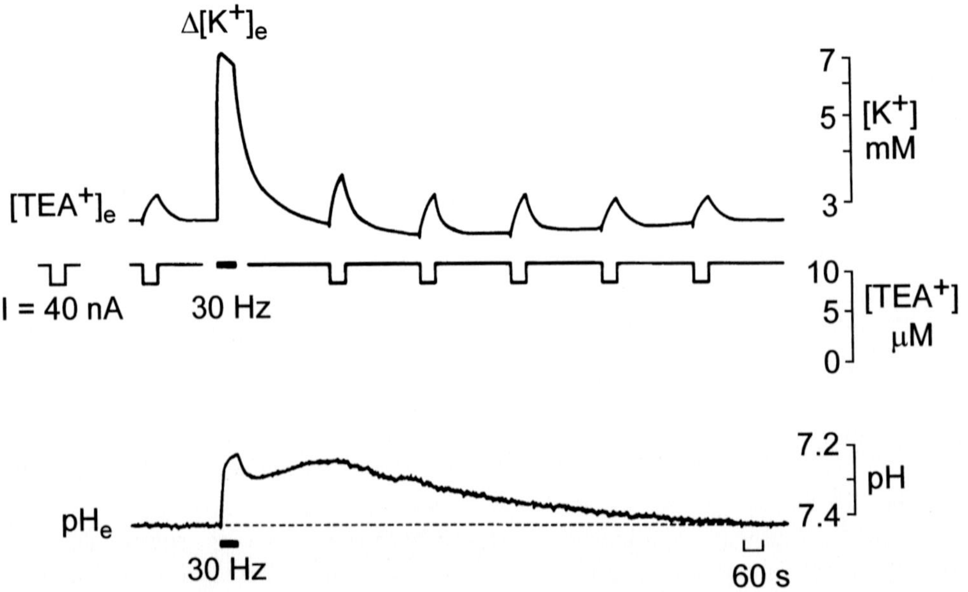

Fig. 6. Effect of repetitive electrical stimulation of dorsal root on the ECS volume, EC potassium increase, and pH_e in the dorsal horn of an isolated frog spinal cord. Current (40 nA) was applied with an iontophoretic pipet filled with TEA^+. Top records show TEA^+ diffusion curves recorded using real-time iontophoretic method, and changes in EC potassium ($\Delta[K^+]_e$) induced by repetitive stimulation (30 Hz, 60 s). pH_e: concomitantly recorded pH_e changes with pH-sensitive microelectrode.

local architecture is significantly different. There is increasing evidence that diffusion in brain tissue is anisotropic. Isotropy is defined as the state of constant λ in any direction from a point source; anisotropy indicates a difference in λ along different axes. To test for anisotropy, the ECS diffusion parameters are measured in three orthogonal axes: x, y, and z. Indeed, anisotropic diffusion was described using the TMA^+ method in the white matter of the corpus callosum and spinal cord *(42,46)*, as well as in the gray matter of the molecular layer of the cerebellum *(47)*, in the hippocampus (Fig. 6; *43*), and in the auditory, but not in the somatosensory, cortex *(20)*. Using MRI, evidence of anisotropic diffusion in white matter was found in cat brain *(48)*, as well as in human brain *(49)*, but so far not in the gray matter. It can, however, be concluded that not only the diffusion of molecules, such as TMA^+ or dextrans, but even the diffusion of water, is hindered by various cellular structures. Because of the distinct diffusion characteristics, the EC molecular traffic will be different in various brain regions.

The anisotropy of white and gray matter could enable different modes of diffusion transmission in these regions. Anisotropy changes during development, aging, and pathological states *(50,51)*. Recently we obtained evidence that changes in anisotropy during development, aging, and pathological states are mediated by various structures, including neurons, dendrites, axons, glial processes, myelin sheaths, and ECM. These variations in molecular structures can affect the migration of substances in the ECS (e.g., preferred diffusion in one direction is lost), and may, therefore, account for an impairment of VT (*see* Subheading 3.2.).

3.2. *Activity-Related Changes in ECS Volume and Geometry*

Transmembrane ionic fluxes during neuronal activity are accompanied by the movement of water and cellular, particularly glial, swelling. Changes in ECS diffusion parameters (ECS volume decrease, tortuosity increase, and *ADC* decrease) are a consequence of activity-related transmembrane ionic shifts and cellular swelling *(52,53)*. In the spinal cord of the rat or frog, adequate repetitive electrical stimulation results in an ECS volume decrease from about 0.24 to 0.12–0.17, i.e., the ECS volume decreases by as much as 30–50% (Fig. 7). The ECS volume in the spinal dorsal horn of the rat also decreases by 20–50%, after injury of the ipsilateral hind paw evoked by subcutaneous injection of turpentine, or after thermal injury. The changes in ECS diffusion parameters therefore persist for many minutes (30 min after electrical stimulation, or even 120 min after peripheral injury) after stimulation has ceased (Fig. 7), suggesting long-term changes in neuronal excitability, neuron–glia communication, and VT.

3.3. *Mechanisms of ECS Volume and Geometry Changes: Role of Glia and Extracellular Matrix*

Astrocyte swelling is an early event in numerous pathological states, accompanied by an elevation of $[K^+]_e$ and a decrease in EC osmolarity *(54)*. It was also shown that, in the isolated turtle cerebellum exposed to hypotonic medium, volume fraction decreased to 0.12; in hypertonic medium, it increased to as much as 0.60 *(38)*. Cell swelling and astrogliosis (manifested as an increase in glial fibrillary acidic protein [GFAP]) were also evoked in isolated rat spinal cords of 4–21-d-old rats, by incubation in either 50 m*M* K^+ (Fig. 6) or hypotonic solution (235 mOsmol kg). Application of K^+ or hypotonic solution resulted, at first, in a decrease in the ECS volume fraction, then in an increase in tortuosity in spinal gray matter *(20)*. These changes resulted from cell swelling, since the total water content in spinal cord was unchanged, and the observed changes were blocked in a Cl^--free solution, and slowed down by furosemide and bumetanide application. During a continuous 45-min application, α and λ often start to return toward control values, apparently as the result of the shrinkage of previously swollen cells, since total water content remains unchanged. This return is blocked by the gliotoxin fluoroacetate, suggesting that most of the changes are caused by the swelling of glia. A 45-min application of 50 m*M* K^+, and, to a lesser degree, of hypotonic solution, evokes astrogliosis, which persists after washing out these solutions with physiological saline. During astrogliosis, λ increases again to values as high as 2.0, while α either returns to, or increases above, control values. This persistent increase in λ after washout is also found in white matter. These data show that glial swelling and astrogliosis are associated with a persistent increase in ECS diffusion barriers.

Many pathological processes in the central nervous system are accompanied by a loss of cells or neuronal processes, astrogliosis, demyelination, and changes in the extracellular matrix, all of which may affect the apparent diffusion coefficients of neuroactive substances. Several animal models have been developed to study changes in ECS diffusion parameters. Brain injury of any kind elicits reactive gliosis, involving both hyperplasia and hypertrophy of astrocytes, which show intense staining for GFAP *(55)*.

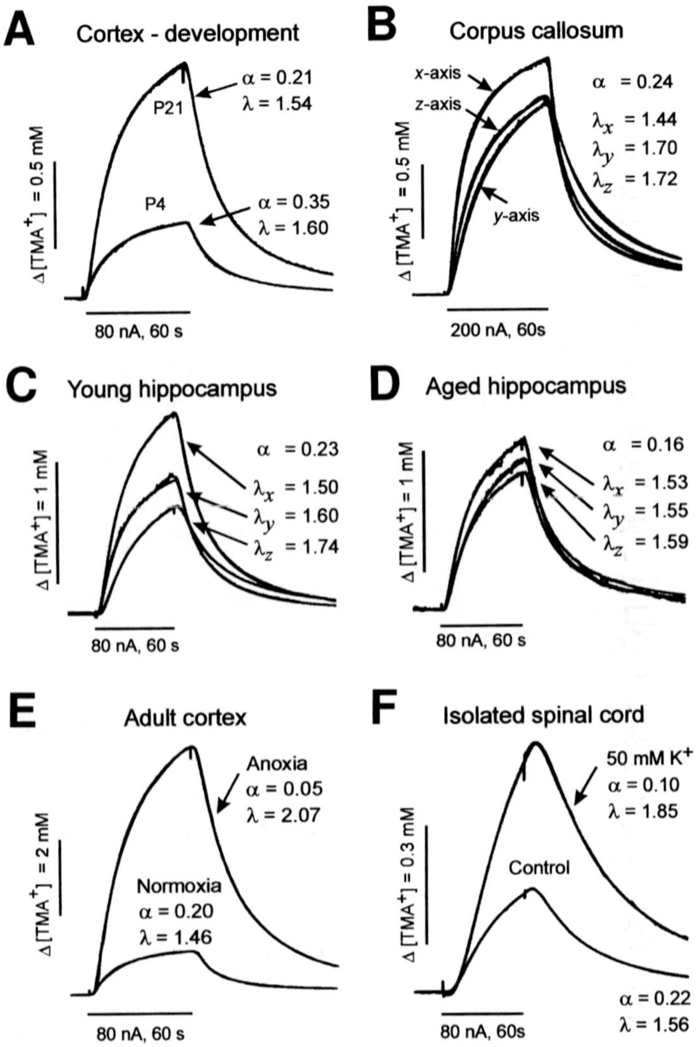

Fig. 7. TMA^+ diffusion curves under different experimental conditions. For each curve, the ECS diffusion parameters, α (volume fraction) and λ (tortuosity), were extracted by appropriate nonlinear curve fitting. Experimental and theoretical curves are superimposed in each case. For each figure, the concentration scale is linear. **(A)** Typical recordings obtained in rat cortex at postnatal d (P) 4 and 21. Values of α and λ are shown with each record. Note that, the larger the curve, the smaller the value of α. **(B)** In the corpus callosum diffusion in the direction perpendicular to the orientation of the axons (*y*- and *z*-axes) is compromised by the number of myelin sheaths. **(C** and **D)** Diffusion parameters in the hippocampus dentate gyrus of a young adult **(C)** and an aged rat **(D)**. Note the anisotropic diffusion in the dentate gyrus of a young adult rat. TMA^+ diffusion curves (concentration-time profiles) were measured along three orthogonal axes (*x*, mediolateral; *y*, rostrocaudal; *z*, dorsoventral). The slower rise in the *z* than in the *y* direction, and in the *y* than in the *x* direction, indicates a higher tortuosity and more-restricted diffusion. The amplitude of the curves shows that TMA^+ concentration is higher along the *x*-axis than along the *y*- and *z*-axes (λx, λy, λz). Actual ECS volume fraction, α, is ~0.2. **(D)** In an aged rat, diffusion curves are higher, showing that α is smaller and anisotropy is almost lost. **(E)** Typical diffusion curves obtained in adult rat cortex in vivo (lamina V) during normoxia, and in the same animal 10 min after cardiac arrest (anoxia). Note the decreases in α and increase in λ during anoxia. **(F)** The effect of 50 m*M* K^+ on ECS diffusion parameters in spinal cord gray matter as measured in 10–13-d-old rats. Note the decrease in α and increase in λ, resulting from K^+-evoked cell swelling.

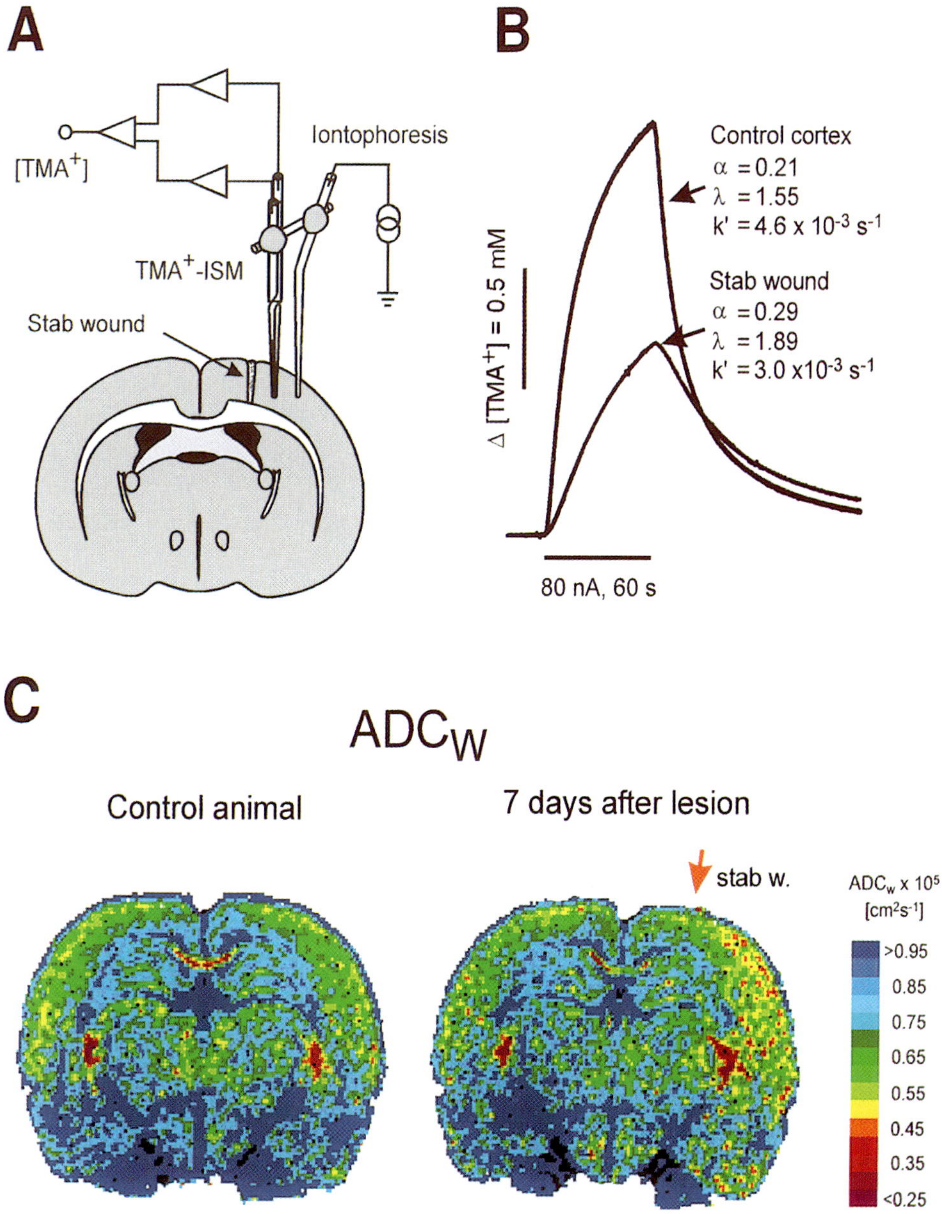

Fig. 8. ECS volume fraction (α), tortuosity (λ), and apparent diffusion coefficient of water (ADC_W), as measured by the TMA method and diffusion-weighted NMR. **(A)** Experimental setup for TMA diffusion measurements. **(B)** The ECS volume fraction and tortuosity, as measured by the TMA method, revealed a significant increase 7 d after the cortical stab wound. α increased only in the vicinity of the wound (up to 1000/μm from the wound); tortuosity was increased in the whole ipsilateral hemisphere. **(C)** Pseudocolor images show ADC_W maps of rat brain in a control animal, and 7 d after a cortical stab wound. Significant decreases in ADC_W were found in the entire ipsilateral hemisphere, including the auditory cortex, which corresponded to an increase in chondroitin sulphate immunostaining (not shown).

Astrogliosis is also a typical characteristic of cortical stab wounds in rodents *(55)*. The lesion is typically accompanied by an ECS volume increase and a substantial tortuosity increase to mean values of α of ~0.26 and λ of ~1.77 (Fig. 8; *19*).

ECM molecules and other large molecules can also affect the tortuosity of the ECS. Their possible effects on changes in TMA^+ diffusion parameters have been studied in rat cortical slices *(37,39,40,56)*, and in isolated rat spinal cord *(57)*. Superfusion of the slice or spinal cord with a solution containing either 40- or 70-kDa dextran or hyaluronic acid (HA) results in a significant increase in λ. In standard physiological solution, λ is ~1.57; in a 1 or 2% solution of 40- or 70-kDa dextran, λ increases to ~1.72–1.77. Application of a 0.1% solution of HA (1.6×10^6 Da) results in an increase in λ to ~2.0. α is either unchanged or decreases by only about 10%, suggesting that these substances have no effect on cell volume and the viability of the preparation. Experiments on tenascin-R knockout mutant mice also revealed remarkable decreases in both α and λ (Syková et al., unpublished results).

Modification of the extracellular matrix can also be achieved by enzymatic treatment. Chondroitin sulphate proteoglycans are essential components of the extracellular matrix, forming so-called perineuronal nets surrounding neurons in cortex and hippocampus. There is increasing evidence that NCAM, the protein backbone of polysialic acid, is involved in synaptic plasticity. Polysialic acid, which is almost exclusively carried by NCAM, is a major modulator of cell adhesion, and is high in areas of continuous neurogenesis, neuronal migration, neurite extension, and synapse formation. It has been found that mice treated with chondroitinase ABC, or with antibodies against NCAM, and transgenic mice lacking the NCAM gene, have impaired LTP *(58)*. It has also been demonstrated that hydrated polysialic acid influences a sufficiently large volume at the cell surface to exert broad steric effects, and that the removal of polysialic acid causes a detectable change in the intercellular space. A single intracortical injection of endoneuroaminidase NE (10 U/L, 5 μU) evoked a significant decrease in tortuosity in the ipsilateral hemisphere 2 h after injection, and this effect persisted at 24 h *(50)*. A decrease in tortuosity and a loss of anisotropy, which may be attributed to changes in the ECM, have also been found during aging (*see* Subheading 3.4.). These results suggest that bigger molecules, such as 40- and 70-kDa dextran, hyaluronic acid and molecules of the ECM may slow the diffusion of small molecules, such as TMA^+ (74 Da), ions, neurotransmitters, metabolites, and so on, in the ECS.

It is therefore possible to assume that regularly found changes in glia and ECM affect VT during physiological, as well as pathological states, and that such ECS plasticity can also contribute to brain plasticity.

3.4. Extracellular Space Diffusion Parameters During Development

Compared to healthy adults, ECS diffusion parameters are significantly different during postnatal development *(41,42,46,59)*. The ECS volume in the cortex is about twice as large (α = 0.36–0.46) in the newborn rat as in the adult rat (α = 0.21–0.23); the tortuosity increases with age (Fig. 7). The reduction in the ECS volume fraction correlates well with the growth of blood vessels. The larger ECS in the first days of postnatal development can be attributed to incomplete neuronal migration, gliogenesis, and angiogenesis, and to the presence of large extracellular matrix proteoglycans, particularly hyaluronic acid, which, because of the mutual repulsion of its highly negatively charged branches, occupies a great deal of space, and holds cells apart. In rat spinal cord gray matter, α decreases with neuronal development and gliogenesis, from postnatal d 4 to 12, by about 15%; while λ significantly increases, showing that the diffusion of molecules becomes more hindered with age. The large ECS channels during development may allow the migration of larger substances (e.g., growth factors), and provide better conditions for cell migration during development. On the other hand, the large ECS in the neonatal brain could significantly dilute ions, metabolites, and neuroactive substances released from cells, relative to release in adults, and may be a factor in the prevention of anoxic injury, seizure, and spreading depression in young individuals. The diffusion parameters could also play an important role in the developmental process itself. Diffusion parameters are substantially different in myelinated and unmyelinated white matter *(42,46)*. Isotropic diffusion is found in the corpus callosum and spinal cord white matter of young rats with incomplete myelination. In myelinated spinal cord and corpus callosum, the tortuosity is higher (the apparent diffusion coefficient is lower), when TMA^+ diffuses across the axons than when it diffuses along the fibers (Fig. 7).

3.5. ECS Diffusion Parameters During Aging

In the mammalian brain, higher cognitive functions such as learning and memory, depend on the circuits that run through the hippocampus. Until recently, learning deficits during aging have been associated with neuronal degeneration and synaptic inefficiency. However, recent observations of a lack of hippocampal cell loss in aged humans, monkeys, and rats *(60–62)* suggest that age-related functional change in the nervous system may not necessarily be a sign of degenerative pathology. The question thus arises whether learning deficits during aging also involve the impairment of extrasynaptic or "volume" transmission, VT, i.e., the diffusion of neuroactive substances in the ECS.

The ECS diffusion parameters, α, $\lambda_{x,y,z}$, and k', were measured in the cortex, corpus callosum and hippocampus (CA1, CA3, and in dentate gyrus). If diffusion in a particular brain region is anisotropic, then the correct value of the ECS volume fraction cannot be calculated from measurements done only in one direction. For anisotropic diffusion, the diagonal components of the tortuosity tensor are not equal, and generally its nondiagonal components need not be zero. Nevertheless, if a suitable referential frame is chosen (i.e., if one measures in three privileged orthogonal directions), neglecting the nondiagonal components becomes possible, and the correct value of the ECS size can thus be determined *(43,47)*. Therefore, TMA^+ diffusion was measured in the ECS, independently, along three orthogonal axes (x, transversal; y, sagital; z, vertical). In all three regions (cortex, corpus callosum, and hippocampus), the mean ECS volume fraction, α, was significantly lower in aged rats (26–32 mo old), ranging from 0.17 to 0.19, than in young adults (3–4 mo old), in which α ranged from 0.21 to 0.22 (Fig. 7). Nonspecific uptake k' was also significantly lower in aged rats. There is a loss of anisotropy in the aging hippocampus, particularly in the CA3 region and the dentate gyrus *(63)*.

Morphological changes during aging include cell loss, loss of dendritic processes, demyelination, astrogliosis, swollen astrocytic processes, and changes in the extracellular matrix. It is reasonable to assume that there is a significant decrease in the *ADC* of many neuroactive substances in the aging brain, which accompanies astrogliosis and changes in the extracellular matrix. One explanation as to why α, in the cortex, corpus callosum and hippocampus of senescent rats, is significantly lower than in young adults could be astrogliosis in the aged brain. Increased GFAP staining and an increase in the size and fibrous character of astrocytes have been found in the cortex, corpus callosum and hippocampus of senescent rats, which may account for changes in the ECS volume fraction *(63)*. Other changes could account for the decreases in λ values, and for the disruption of tissue anisotropy. In the hippocampus in CA1 and CA3, and in the dentate gyrus, we have observed changes in the arrangement of fine astrocytic processes. These are normally organized in parallel in the *x–y* plane, and this organization totally disappears during aging. Moreover, the decreased staining for chondroitin sulfate proteoglycans and for fibronectin, suggests a loss of extracellular matrix macromolecules *(63)*.

Because α is lower in aging rats, we expected some differences in the ECS diffusion parameter changes during ischemia in senescent rats. Indeed, the final values of α, λ, and *k'* induced by cardiac arrest are not significantly different between young and aged rats; however, the time-course of all changes is faster in aged animals *(63)*. Faster changes in ECS volume fraction and tortuosity in nervous tissue during aging can contribute to a faster impairment of signal transmission, e.g., faster accumulation of ions and neuroactive substances released from cells and their slower diffusion away from the hypoxic/ischemic area in the more compacted ECS.

Our recent study also revealed that the degree of learning deficit during aging correlates with the changes in ECS volume, tortuosity, and nonspecific uptake *(64)*. The hippocampus is well-known for its role in memory formation, especially the declarative memory. It is therefore reasonable to assume that diffusion anisotropy, which leads to a certain degree of specificity in extrasynaptic communication, may play an important role in memory formation. There was a significant difference between mildly and severely behaviorally impaired rats (rats were tested in a Morris water maze), which was particularly apparent in the hippocampus. The ECS in the dentate gyrus of severely impaired rats was significantly smaller than in mildly impaired rats. Also, anisotropy in the hippocampus of severely impaired rats, particularly in the dentate gyrus, was greatly reduced; a substantial degree of anisotropy was still present in aged rats with a better learning performance. Anisotropy might be important for extrasynaptic transmission by channeling the flux of substances in a preferential direction. Its loss may severely disrupt volume transmission in the CNS, which has been suggested to play an important role in memory formation *(4,63)*. Chondroitin sulfate proteoglycans participate in multiple cellular processes *(65)*, including axonal outgrowth, axonal branching, and synaptogenesis, which are important for the formation of memory traces.

What is the functional significance of the observed changes in ECS diffusion parameters during aging? We suggest that the alterations in hippocampal diffusion parameters seen in aged animals with severe learning deficits may account for the learning impairment, either because of their effect on VT *(4)* or on crosstalk between synapses, which has been suggested to be involved in LTP and LTD *(66)*, or on both. Anisotropy, which, particularly in the hippocampus and corpus callosum, may help to facilitate the diffusion of neurotransmitters and neuromodulators to regions occupied by their high-affinity extrasynaptical receptors, might have crucial importance for the specificity of signal transmission. The importance of anisotropy for the "spill-over of glutamate," for crosstalk between synapses, and for LTP and LTD has been proposed *(66,67)*. The observed loss of anisotropy in senescent rats could therefore lead to impaired cortical and, particularly, hippocampal function. The decrease in ECS size could be responsible for the greater susceptibility of the aged brain to pathological events, such as ischemia, cell death during anesthesia *(63)*, the poorer outcome of clinical therapy, and the more limited recovery of affected tissue after insult.

3.6. ECS During Pathological States

Pathological states are accompanied by a lack of energy, seizure activity, the excessive release of transmitters and neuroactive substances, neuronal death, glial cell loss or proliferation, glial swelling, the production of damaging metabolites, including free radicals, and the loss of ionic homeostasis. Others are characterized by inflammation, edema, or demyelination. It is therefore evident that they will be accompanied not only by substantial changes in ECS ionic composition *(1,2)*, but also by various changes in ECS diffusion parameters, according to the different functional and anatomical changes.

3.6.1. *Anoxia/Ischemia*

Dramatic K^+ and pH_e changes occur in the brain and spinal cord during anoxia and/or ischemia *(2,44,68)*. Within 2 min of respiratory arrest in adult rats, blood pressure begins to increase and pH_e begins to decrease (by about 0.1 pH U); $[K^+]_e$ remains unchanged. With the subsequent blood pressure decrease, the pH_e decreases by 0.6–0.8 pH U, to pH 6.4–6.6. This pH_e decrease is accompanied by a steep rise in $[K^+]_e$ to about 50–70 m*M*; decreases in $[Na^+]_e$ to 48–59 m*M*, $[Cl^-]_e$ to 70–75 m*M*, $[Ca^{2+}]_e$ to 0.06–0.08 m*M*, and pH_e to 6.1–6.8; an accumulation of excitatory amino acids; a negative DC slow potential shift; and a decrease in ECS volume fraction to 0.04–0.07. The ECS volume starts to decrease when the blood pressure drops below 80 mmHg and $[K^+]_e$ rises above 6 m*M* *(44)*.

Figure 7 shows that during hypoxia and terminal anoxia, the ECS volume fraction in the rat cortex decreases from 0.20 to 0.05; tortuosity increases from about 1.5 to about 2.0 *(44)*. The same ultimate changes have been found in all neonatal, adult, and aged rats, in gray and white matter, in the cortex, corpus callosum and spinal cord. However, the time-course in white matter is significantly slower than in gray matter; and the time-course in neonatal rats is about 10× slower than in adults *(59)*. Linear regression analysis reveals a positive correlation between the normoxic size of the ECS volume and the time-course of the changes. This corresponds to the well-known resistance of the immature CNS and the greater susceptibility of the aged brain to anoxia.

In studies using diffusion-weighted ^{1}H MRS/MRI, the apparent diffusion coefficient of water (ADC_W) was measured during terminal anoxia in rats. Anoxia evokes similar decreases in the apparent diffusion coefficient of ADC_W (measured by the NMR method) and ADC_{TMA} (measured by the TMA method). A comparison of the decreases in ADC_W and ADC_{TMA}, in rats 8–9 d of age, revealed the same time-course, both corresponding to the decrease in ECS volume fraction. Although water moves freely across the cellular membranes, TMA^+ stays predominantly in ECS. Since the total amount of tissue water is not believed to increase *(33)*, this study shows that changes in the *ADC* of brain tissue water measured by diffusion-weighted in vivo MR techniques reflect extra- and intracellular volume changes resulting from a water shift from the extra- to the intracellular compartment.

3.6.2. Brain Injury

A stab wound of the rodent brain is a well-characterized and common model of reactive gliosis, which can impose diffusion barriers in the CNS, because of the hypertrophy of astrocytic processes and an increased production of extracellular matrix components *(55,69,70)*. We compared two different methods for revealing diffusion changes in the rat cortex after injury: the TMA method and diffusion-weighted MR. TMA and MR measurements were performed in the cortex of adult rats from d 3 to d 35 after a unilateral sterile cut through the cortex. Severe astrogliosis is found close to the injury site (up to 1 mm), and mild astrogliosis up to 2 mm from the wound in the ipsilateral cortex, but no astrogliosis is found in the auditory cortex or in the contralateral hemisphere. In contrast to GFAP staining, immunostaining for chondroitin-sulfate preoteoglycan increases in the whole ipsilateral hemisphere *(50,71)*. The mean values of α, ADC_{TMA}, and ADC_W in the contralateral hemisphere were not significantly different from those in nonlesioned, control animals. In the astrogliotic cortex, less than 1 mm distant from the wound, the mean values of α are significantly higher (α = 0.26); the mean values of ADC_{TMA} are lower: 0.42×10^{-5} cm^2/s. The more distant from the wound, the less the values of α and ADC_{TMA} differ from control values. On the other hand, ADC_W is significantly lower in the whole ipsilateral hemisphere, particularly in the auditory cortex: $ADC_W = 0.55 \times 10^{-5}$ cm^2/s. We conclude that an increase in diffusion barriers, manifested by the decrease of both ADC_{TMA} and ADC_W, occurs throughout the entire cortex of the wounded hemisphere, without significant changes in ECS volume. The changes are related to astrogliosis, particularly in and closely around the injured area, and to an increase in the extracellular matrix, which occurs throughout the entire hemisphere.

3.6.3. Demyelination

Changes in the ECS diffusion parameters are found in the spinal cord of rats during myelination, as well as demyelination, e.g., during experimental autoimmune encephalomyelitis (EAE), an experimental model of multiple sclerosis *(45)*, (EAE), which is induced by the injection of guinea-pig myelin basic protein and results in typical morphological changes in the CNS tissue, namely, demyelination, an inflammatory reaction, astrogliosis, BBB damage, and paraparesis, at 14–17 d postinjection. Paraparesis is accompanied by increases in α in the dorsal horn, in the intermediate region, in the ventral horn, and in white matter from ~0.18 to ~0.30. The λ in the dorsal horn and the intermediate region significantly decreases, and k' decreases in the intermediate region and the ventral horn *(45)*. There is a close correlation between the changes in ECS diffusion parameters and the manifestation of neurological abnormalities.

These results suggest that the expansion of the ECS alters diffusion parameters in inflammatory and demyelinating diseases, and may affect the accumulation and movement of ions, neurotransmitters, neuromodulators, and metabolites in the CNS, in these disorders, possibly by interfering with axonal conduction.

3.6.4. Brain Tumors

Cancer is the second leading cause of death in many industrialized countries, and malignant brain tumors, particularly the gliomas, are among the deadliest of tumors: Many patients, including children, die within 2 yr. Only recently have new basic findings been made about tumor cell division, differentiation, and migration, the relationship between glial cells and gliomas, the existence of multiple glial precursor cell populations, and new insights into the developmental biology of glial cells. One of the recently discussed issues is the existence of CNS-specific extracellular matrix proteins, e.g., brain-enriched hyaluronan binding proteins and brevican, which are expressed at high levels during initial gliogenesis, and also in all types and grades of human gliomas *(72)*. Hyaluronan binding proteins (BEHAB) help cells to move through tissue during development, and have also been associated with invasive cancers. It has also been suggested that the migration of cells could be critically dependent on their shape and size, their binding to various proteins in the ECS, such as hyaluronan-binding protein, which can boost the invasiveness of tumors, and on the size and geometry of the ECS. The delivery of drugs to tumors is affected by the permeability of the blood–brain barrier, their diffusion through the ECS in normal and malignant tissue, and their side effects on healthy cells surrounding the tumor. It is therefore crucial to quantify the size, composition, and geometry of the ECS, because these factors critically affect cell migration and the diffusion of substances in the brain *(3,4)*.

Older studies, particularly using sucrose space and electron microscopic methods, showed that the visible EC compartment is larger in brain tumors, particularly in gliomas, than in normal brain tissue *(73,74)*. However, these methods resulted either in small values for the size of the ECS, because of tissue shrinkage during fixation and embedding, or did not allow one to measure the absolute values of ECS size and geometry, or to follow the diffusion of molecules of different sizes and shapes. Increased ECS, along with the above mentioned ECM proteins, could allow cells to migrate more easily in tumors and into surrounding tissue. Shrinkage of the ECS and ECM proteins could slow or substantially limit the diffusion and migration of cells.

To quantify ECS size, geometry, and diffusion properties in tumors (particularly malignant gliomas), TMA measurements were performed in slices from surgically removed pieces of patients' brains. We found that, the more malignant the glioma, the more dramatic the increase in ECS volume. In many brain slices from glioblastomas (grade 4, WHO classification system), the ECS volume is as large as 37–46% of total tissue volume (Fig. 5). There is also a significant increase in tortuosity, which may be caused either by the frequently observed astrogliosis or by changes in the ECM. It is therefore reasonable to assume that ECS composition, volume, and geometry play an important role in cancer malignancy and invasiveness. The size of the ECS in malignant tumors, and their geometry and structure, could be taken into account during therapeutic drug delivery.

3.6.5. Grafted Tissue and Regeneration

Both α and λ are increased in cortical grafts of fetal tissue transplanted to the midbrain, where severe astrogliosis, compared to host cortex, is found, but not in fetal grafts placed into the cavity of the cortex, where only mild astrogliosis occurs *(75)*. Another characteristic feature of cortical grafts into the midbrain is the variability of α and λ. The different values found at various depths of the graft correlate with the morphological heterogeneity of the graft neuropil. These measurements show that, even when the ECS in gliotic tissue or in cortical grafts is larger than in normal cortex, the tortuosity is still higher, and the diffusion of chemical signals in such tissue may be hindered. Limited diffusion may also have a negative impact on the viability of grafts in host brains. Compared to host cortex, immunohistochemistry shows myelinated patches and a larger number of hypertrophic astrocytes in areas of high λ values, suggesting that more numerous and/or thicker glial cell processes may be the cause of increased tortuosity.

Extrasynaptic VT of dopamine plays an important role in Parkinson's disease. The 6-hydroxydopamine rat model of Parkinson's was used to study how the grafting of fetal dopaminergic cells affects diffusion of dopamine from a graft into the striatum in vivo *(76)*. Lesioned rats were divided into three groups: They either received grafts of E14 cell suspensions of ventral midbrain, shamgrafts (saline injection), or remained untreated. Age-matched nonlesioned rats served as controls. The cells for intrastriatal transplantation (100,000 in 4 μL) were unilaterally distributed by either two deposits by a metal cannula (macrografts) or eight deposits by a glass capillary (micrografts). The degree of lesion, the survival, function of transplants, and astrogliosis were studied by amphetamine-induced rotational behavior, and by immunohistochemistry of tyrosine hydroxylase and GFAP. We found a functional recovery, good survival of tyrosine hydroxylase-positive cells, and astrogliosis in grafted rats 3–5 mo after grafting. Prior to diffusion studies, grafts were localized by T2 or diffusion-weighted MR, rats were anesthetized, and TMA diffusion parameters investigated in the striatum. In nongrafted rats, α and λ were unchanged, compared to controls, and k' decreased. In both grafted and shamgrafted rats, α increased, and k' remained decreased, suggesting a lower density of cells in injured tissue. λ increased in the grafts and in the tissue adjacent to them. The increase in ECS diffusion barriers corresponded to astrogliosis in and around the grafts. The increase of EC tortuosity therefore suggests an impediment of dopamine VT from the grafts (particularly for macrografts) into the lesioned striatum.

4. CONCLUSION

Brain plasticity is defined as the adaptive capacity of the brain: its ability to modify its own structural organization and function. We suggest that brain plasticity also includes changes in ECS volume and geometry, up- and downregulation of extrasynaptic receptors, and reorganization of areas without hard-wired circuitry changes. There is increasing evidence that long-term changes in the physical and chemical parameters of the ECS accompany many physiological and pathological states. The acute, or relatively fast, changes in the size of the intercellular channels are apparently a consequence of cellular (particularly glial) swelling. An abrupt ECS volume decrease may cause cellular or molecular crowding, which can lead to an acute increase in tortuosity. Long-term, or even permanent, changes in ECS diffusion parameters would require either permanent changes in the size of the intercellular channels, changes in ECM molecules, or changes in the number and thickness of cellular processes. Available data suggest that, in some pathophysiological states, α and λ behave as independent variables. A persistent increase in λ (without a decrease in ECS volume) is always found during astrogliosis and in myelinated tissue, suggesting that glial cells can form diffusion barriers, render the nervous tissue less permissive, and play an important role in signal transmission, tissue regeneration, and pathologic states. This observation has important implications for understanding the function of glial cells. The ECM apparently also contributes to diffusion barriers and to diffusional anisotropy, particularly since its loss, e.g., during aging, correlates with a tortuosity decrease and a loss of anisotropy. Various changes, transient or permanent, in glial cell morphology and in the ECM (which accompany physiological and pathological states), alter the diffusion properties of the ECS, which in turn affects the communication efficacy between the nerve cells. One critical plasticity-related question, which is important for rehabilitation and tissue repair, is the following: What factors affect the adaptation and reorganization capacity of the adult CNS? In answering this question, extrasynaptic VT plasticity should be taken into account, because new functional connections can be achieved by facilitation of VT (i.e., diffusion of neuroactive substances), and by the development of new extrasynaptic receptors.

ACKNOWLEDGMENT

Supported by grants GACR 307/96/K226, GACR 309/97/K048, GACR 309/99/0657, GACR 309/00/1430.

REFERENCES

1. Syková, E. (1983) Extracellular K^+ accumulation in the central nervous system. *Prog. BioPhysiol. Mol. Biol.* **42,** 135–189.
2. Syková, E. (1992) *Ionic and Volume Changes in the Microenvironment of Nerve and Receptor Cells*, Springer-Verlag, Heidelberg.
3. Syková, E. (1997) The extracellular space in the CNS: its regulation, volume and geometry in normal and pathological neuronal function. *Neuroscientist* **3,** 28–41.
4. Nicholson, C. and Syková, E. (1998) Extracellular space structure revealed by diffusion analysis. *Trends Neurosci.* **21,** 207–215.
5. Celio, M. R., Spreafico, R., De Biasi, S., and Vitellaro-Zuccarello, L. (1998) Perineuronal nets: past and present. *Trends Neurosci.* **21,** 510–515.
6. Walz, W. (1989) Role of glial cells in the regulation of the brain ion microenvironment. *Progr. Neurobiol.* **33,** 309–333.
7. Chesler, M. (1990) The regulation and modulation of pH in the nervous system. *Progr. Neurobiol.* **34,** 401–427.
8. Deitmer, J. W. and Rose, C. R. (1996) pH regulation and proton signalling by glial cells. *Progr. Neurobiol.* **48,** 73–103.
9. Syková, E., Rothenberg, S., and Krekule, I. (1974) Changes of extracellular potassium concentration during spontaneous activity in the mesencephalic reticular formation of the rat. *Brain Res.* **79,** 333–337.
10. Heinemann, V. and Lux, H. D. (1977) Ceiling of stimulus induced rises in extracellular potassium concentration in the cerebral cortex of the cat. *Brain Res.* **120,** 231–249.
11. Kříž, N., Syková, E., Ujec, E., and Vyklický, L. (1974) Changes of extracellular potassium concentration induced by neuronal activity in the spinal cord of the cat. *J. Physiol.* **238,** 1–15.
12. Syková, E. and Svoboda, J. (1990) Extracellular alkaline-acid-alkaline transients in the rat spinal cord evoked by peripheral stimulation. *Brain Res.* **512,** 181–189.
13. Jendelová, P. and Syková, E. (1991) Role of glia in K^+ and pH homeostasis in the neonatal rat spinal cord. *Glia* **4,** 56–63.
14. Syková, E., Jendelová, P., Šimonová, Z., and Chvátal, A. (1992a) K^+ and pH homeostasis in the developing rat spinal cord is impaired by postnatal X-irradiation. *Brain Res.* **594,** 19–30.
15. Syková, E., Jendelová, P., Svoboda J., and Chvátal, A. (1992b) Extracellular K^+, pH, and

volume changes in spinal cord of adult rats during postnatal development. *Can. J. Physiol. Pharmacol.* **70(Suppl.),** S301–309.
16. Ng, K. T., Gibbs, M. E., Crowe, S. F., Sedman, G. L., Hua, F., Zhao, W., et al. (1991) Molecular mechanisms of memory formation. *Mol. Neurobiol.* **5,** 333–350.
17. Jendelová, P., Chvátal, A., Šimonová, Z., and Syková, E. (1994) Effect of excitory amino acids on extracellular pH in isolated rat spinal cord, in *Abstracts of the 17th Annual Meeting of ENA*. Vienna 199.
18. Thomas, L. B. and Steindler, D. A. (1995) Glial boundaries and scars: programs for normal development and wound healing in the brain. *Neuroscientist* **1,** 142–154.
19. Roitbak, T. and Syková, E. (1999) Diffusion barriers evoked in the rat cortex by reactive astrogliosis. *Glia* **28,** 40–48.
20. Syková, E., Mazel, T., Roitbak, T., and Šimonová, Z. (1999a) Morphological changes and diffusion barriers in auditory cortex and hippocampus of aged rats. *Assoc. Res. Otolaryngol. Abs.* **22,** 17.
21. Fuxe, K. and Agnati, L. F. (1991) *Volume Transmission in the Brain: Novel Mechanisms for Neural Transmission*, Raven, New York.
22. Agnati, L. F., Zoli, M., Stromberg, I., and Fuxe, K. (1995) Intercellular communication in the brain: wiring versus volume transmission. *Neuroscience* **69,** 711–726.
23. Zoli, M., Jansson, A., Syková, E., Agnati, L. F., and Fuxe, K. (1999) Intercellular communication in the central nervous system. The emergence of the volume transmission concept and its relevance for neuropsychopharmacology. *Trends Pharmacol. Sci.* **20,** 142–150.
24. Nicholson, C. and Phillips, J. M. (1981) Ion diffusion modified by tortuosity and volume fraction in the extracellular microenvironment of the rat cerebellum. *J. Physiol. (Lond.)* **321,** 225–257.
25. Van Harreveld, A., Dafny, N., and Khattab, F. I. (1971) Effects of calcium on electrical resistance and the extracellular space of cerebral cortex. *Exp. Neurol.* **31,** 358–367.
26. Matsuoka, Y. and Hossmann, K. A. (1982) Cortical impedance and extracellular volume changes following middle cerebral artery occlusion in cats. *J. Cereb. Blood Flow Metab.* **2,** 466–474.
27. Korf, J., Klein, H. C., and Postrema, F. (1988) Increases in striatal and hippocampal impedance and extracellular levels of amino acids by cardiac arrest in freely moving rats. *J. Neurochem.* **50,** 1087–1096.
28. Nicholson, C. and Tao, L. (1993) Hindered diffusion of high molecular weight compounds in brain extracellular microenvironment measured with integrative optical imaging. *Biophys. J.* **65,** 2277–2290.
29. Andrew, R. D. and MacVicar, B. A. (1994) Imaging cell volume changes and neuronal excitation in the hippocampal slice. *Neuroscience* **62,** 371–383.
30. Benveniste, H., Hedlund, L. W., and Johnson, G. A. (1992) Mechanism of detection of acute cerebral ischemia in rats by diffusion-weighted magnetic resonance microscopy. *Stroke* **23,** 746–754.
31. Latour, L. L., Svoboda, K., Mitra, P. P., and Sotak, C. H. (1994) Time-dependent diffusion of water in a biological model system. *Proc. Natl. Acad. Sci. USA* **91,** 1229–1233.
32. Norris, D. G., Niendorf, T., and Leibfritz, D. (1994) Healthy and infarcted brain tissues studied at short diffusion times: the origins of apparent restriction and the reduction in apparent diffusion coefficient. *NMR Biomed.* **7,** 304–310.
33. Van der Toorn, A., Syková, E., Dijkhuizen, R. M., Vořišek, I., Vargová, L., Škobisová, E., et al. (1996) Dynamic changes in water ADC, energy metabolism, extracellular space volume, and tortuosity in neonatal rat brain during global ischemia. *Magn. Reson. Med.* **36,** 52–60.
34. Syková, E., Vořišek, I., Fiala, J., and del Bigio, M. R. (2000a) Altered diffusion parameters

in adult rats with kaolin-induced hydrocephalus. *Magn. Reson. Mat. Phys. Biol. Med. Abs.* **11,** 36.
35. Vargová, L., Prokopová, Š., Chvátal, A., and Syková, E. (1999) Are the changes in intrinsic optical signals a tool to measure changes in extracellular space volume? *Soc. Neurosci.* **25,** 741.
36. Syková, E., Vargová, L., Jendelová, P., and Chvátal, A. (2000) Intrinsic optical signals evoked in spinal cord slices and their relation to neural activity and cell swelling. *Soc. Neurosci. Abstr.* **26,** 1619.
37. Tao, L., Voříšek, I., Lehmenkühler, A., Syková, E., and Nicholson, C. (1995) Comparison of extracellular tortuosity derived from diffusion of 3 kDa dextran and TMA^+ in rat cortical slices. *Soc. Neurosci. Abstr.* **21,** 604.
38. Križaj, D., Rice, M. E., Wardle, R. A., and Nicholson, C. (1996) Water compartmentalization and extracellular tortuosity after osmotic changes in cerebellum of Trachemys scripta. *J. Physiol. (Lond.)* **492,** 887–896.
39. Tao, L. and Nicholson, C. (1996) Diffusion of albumins in rat cortical slices and relevance to volume transmission. *Neuroscience* **75,** 839–847.
40. Prokopová-Kubinová, Š., Vargová, L., Tao, K., Šubr, E., Syková, E., and Nicholson, C. Diffusion characteristics of PHPMA polymers in brain extracellular space. *Biophys. J.* **80,** 542–548.
41. Lehmenkühler, A., Syková, E., Svoboda, J., Zilles, K., and Nicholson, C. (1993) Extracellular space parameters in the rat neocortex and subcortical white matter during postnatal development determined by diffusion analysis. *Neuroscience* **55,** 339–351.
42. Voříšek, I. and Syková, E. (1997a) Evolution of anisotropic diffusion in the developing rat corpus callosum. *J. Neurophysiol.* **78,** 912–919.
43. Mazel, T., Šimonová, Z., and Syková, E. (1998) Diffusion heterogeneity and anisotropy in rat hippocampus. *Neuroreport* **9,** 1299–1304.
44. Syková, E., Svoboda, J., Polák, J., and Chvátal, A. (1994) Extracellular volume fraction and diffusion characteristics during progressive ischemia and terminal anoxia in the spinal cord of the rat. *J. Cereb. Blood Flow Metab.* **14,** 301–311.
45. Šimonová, Z., Svoboda, J., Orkand, R., Bernard, C. C. A., Lassmann, H., and Syková, E. (1996) Changes of extracellular space volume and tortuosity in the spinal cord of Lewis rats with experimental autoimmune encephalomyelitis. *Physiol. Res.* **45,** 11–22.
46. Prokopová, Š., Vargová, L., and Syková, E. (1997) Heterogeneous and anisotropic diffusion in the developing rat spinal cord. *Neuroreport* **8,** 3527–3532.
47. Rice, M. E., Okada, Y. C., and Nicholson, C. (1993) Anisotropic and heterogeneous diffusion in the turtle cerebellum: implications for volume transmission. *J. Neurophysiol.* **70,** 2035–2044.
48. Moseley, M. E., Cohen, Y., Mintorovitch, J., Chileuitt, L., Shimizu, H., Kucharczyk, J., Wendland, M. F., and Weinstein, P. R. (1990) Early detection of regional cerebral ischemia in cats: comparison of diffusion- and T2-weighted MRI and spectroscopy. *Magn. Reson. Med.* **14,** 330–346.
49. Le Bihan, D., Turner, R., and Douek, P. (1993) Is water diffusion restricted in human brain white matter? An echoplanar NMR imaging study. *Neuroreport* **4,** 887–890.
50. Syková, E., Mazel, T., Vargová, L., Voříšek, I., and Prokopová-Kubinová, S. (2000) Extracellular space diffusion and pathological states, in *Progress in Brain Research: Volume Transmission Revisited* (Fuxe, K., Agnati, L. F., Nicholson, C., and Syková, E., eds.), Elsevier, Amsterdam, in press.
51. Syková, E. Glial diffusion barriers during aging and pathological states, in *Progress in Brain Research: Glial Cells Function in Health and Disease* (Niento-Sampedro, M. and Castellano, B., eds.), Elsevier, Amsterdam, in press.
52. Syková, E. (1987) Modulation of spinal cord transmission by changes in extracellular K^+ activity and extracellular volume. *Can. J. Physiol. Pharmacol.* **65,** 1058–1066.

53. Svoboda, J. and Syková, E. (1991) Extracellular space volume changes in the rat spinal cord produced by nerve stimulation and peripheral injury. *Brain Res.* **560,** 216–224.
54. Kimelberg, H. K., Sankar, P., O'Connor, E. R., Jalonen, T., and Goderie, S. K. (1992) Functional consequences of astrocyte swelling. *Progr. Brain Res.* **94,** 57–68.
55. Norton, W. T., Aquino, D. A., Hosumi, I., Chiu, F. C., and Brosnan, C. F. (1992) Quantitative aspects of reactive gliosis: a review. *Neurochem. Res.* **17,** 877–885.
56. Vargová, L., Tao, L., Syková, E., Ulbrich, K., Šubr, V., and Nicholson, C. (1998) Diffusion of large polymers in rat cortical slices measured by integrative optical imaging. *J. Physiol. (Lond.)* **511,** 16P.
57. Prokopová, Š., Nicholson, C., and Syková, E. (1996) The effect of 40-kDa or 70-kDa dextran and hyaluronic acid solution on extracellular space tortuosity in isolated rat spinal cord. *Physiol. Res.* **45,** P28.
58. Becker, C. G., Artola, A., Gerardy-Schahn, R., Becker, T., Welzl, H., and Schachner, M. (1996) The polysialic acid modification of the neural cell adhesion molecule is involved in spatial learning and hippocampal long-term potentiation. *J. Neurosci. Res.* **45,** 143–152.
59. Voříšek, I. and Syková, E. (1997b) Ischemia-induced changes in the extracellular space diffusion parameters, K^+ and pH in the developing rat cortex and corpus callosum. *J. Cereb. Blood Flow Metab.* **17,** 191–203.
60. West, M. J. (1993) Regionally specific loss of neurons in the aging human hippocampus. *Neurobiol. Aging* **14,** 287–293.
61. Rapp, P. R. and Gallagher, M. (1996) Preserved neuron number in the hippocampus of aged rats with spatial learning deficits. *Proc. Natl. Acad. Sci. USA* **93,** 9926–9930.
62. Rasmussen, T., Schliemann, T., Sorensen, J. C., Zimmer, J., and West, M. J. (1996) Memory impaired aged rats: No loss of principal hippocampal and subicular neurons. *Neurobiol. Aging* **17,** 143–147.
63. Syková, E., Mazel, T., and Šimonová, Z. (1998b) Diffusion constraints and neuron-glia interaction during aging. *Exp. Gerontol.* **33,** 837–851.
64. Syková, E., Mazel, T., Frisch, C., Šimonová, Z., Hasenöhrl, R. U., and Huston, J. P. (1998a) Spatial memory and diffusion parameters in aged rat cortex, corpus callosum and hippocampus. *Soc. Neurosci. Abstr.* **24,** 1420.
65. Hardington, T. E. and Fosang, A. J. (1992) Proteoglycans: many forms and many functions. *FASEB J.* **6,** 861–870.
66. Kullmann, D. M., Erdemli, G., and Asztely, F. (1996) LTP of AMPA and NMDA receptor-mediated signals: evidence for presynaptic expression and extrasynaptic glutamate spillover. *Neuron* **17,** 461–474.
67. Asztely, F., Erdemli, G., and Kullmann, D. M. (1997) Extrasynaptic glutamate spillover in the hippocampus: dependence on temperature and the role of active glutamate uptake. *Neuron* **18,** 281–293.
68. Xie, Y., Zacharias, E., Hoff, P., and Tegtmeier, F. (1995) Ion channel involvement in anoxic depolarisation induced by cardiac arrest in rat brain. *J. Cereb. Blood Flow Metab.* **15,** 587–594.
69. Hatten, M. E., Liem, R. K. H., Shelanski, M. L., and Mason, C. A. (1991) Astroglia in CNS injury. *Glia* **4,** 233–243.
70. Ridet, J. I., Malhotra, S. K., Privat, A., and Gage, F. H. (1997) Reactive astrocytes: cellular and molecular cues to biological function. *Trends Neurosci.* **20,** 570–577.
71. Voříšek, I., Roitbak, T., Nicolay, K., and Syková, E. (1999) Water ADC, extracellular space volume and tortuosity in the rat cortex during astrogliosis. *Physiol. Res.* **48,** S136.
72. Jaworski, D. M., Kelly, G. M., and Hockfield, S. (1996) The CNS-specific hyaluronan binding protein, BEHAB, is expressed during periods of glial cell generation and motility. *Semin. Neurosci.* **8,** 391–396.
73. Bakay, L. (1970a) The extracellular space in brain tumor. I. Morphological considerations. *Brain* **93,** 693–698.

74. Bakay, L. (1970b) The extracellular space in brain tumor. II. The sucrose space. *Brain* **93,** 699–707.
75. Syková, E., Roitbak, T., Mazel, T., Šimonová, Z., and Harvey, A. R. (1999b) Astrocytes, oligodendroglia, extracellular space volume and geometry in rat fetal brain grafts. *Neuroscience* **91,** 783–798.
76. Olshausen, F., Reum, T., Mazel, T., Voříšek, I., Morgenstern, R., and Syková, E. (2000) Assessment of extracellular diffusion in the striatum of 6-OHDA-lesioned and grafted rats. *Soc. Neurosci. Abstr.* **26,** 870.

4

Transmitter–Receptor Mismatches in Central Dopamine, Serotonin, and Neuropeptide Systems

Further Evidence for Volume Transmission

Anders Jansson, Laurent Descarries, Virginia Cornea-Hébert, Mustapha Riad, Daniel Vergé, Mircea Bancila, Luigi Francesco Agnati, and Kjell Fuxe

1. INTRODUCTION

In the central nervous system (CNS), the existence of neurotransmitter-defined neurons, with projections showing a regional distribution distinctly separate from that of the corresponding receptors on target cells, represents a compelling argument in favor of volume transmission (VT). Such mismatches imply that the transmitter reaches its functional targets by diffusion into the extracellular space (ECS) *(1–4)*, as initially postulated to be the case for cortical monoamines released from axon terminals (varicosities) often lacking synaptic membrane specializations (*5*; for review, *see* refs. *6* and *7*). In the past 10 yr, it has been further realized that the VT modality may also apply to entirely synaptic neuronal systems, such as those using γ-aminobutyric acid and glutamate, in which spillover of transmitter beyond synaptic clefts has been described *(8–10)*. This chapter focuses on recent immunofluorescence, confocal laser, and immunoelectron microscopic (IEM) observations, which illustrate the diversity of topological relationships between dopamine (DA)-, serotonin (5-hydroxytryptamine [5-HT])-, and neuropeptide (NP)-containing neurons and their receptors in various regions of the brain. The clinical relevance of some of these data is also emphasized *(11)*. Noradrenaline and acetylcholine systems are not considered here, since they have been recently reviewed elsewhere *(7,12,13)*. In each of these transmitter-defined systems, and presumably many others, the bulk of the data supports the view that VT plays a major role throughout the CNS.

2. DOPAMINE (DA) SYSTEM

In several DA-innervated regions, which have not all been systematically examined for the synaptic frequency of their DA varicosities, the evidence for VT comes mostly from some of the striking mismatches observed by light microscopic immunocytochemistry between these axon terminals and DA receptors, and from IEM observations on

From: *The Neuronal Environment: Brain Homeostasis in Health and Disease*
Edited by: W. Walz © Humana Press Inc., Totowa, NJ

the subcellular distribution of the latter, combined or not with a simultaneous visualization of the DA terminals.

A detailed immunofluorescence analysis of the distribution of D_1 and D_2 receptors, in relation to tyrosine hydroxylase (TH)- and DA transporter-labeled elements has recently been carried out in the rat nucleus accumbens *(14)*. In the dorsomedial part of the shell of this nucleus, the existence of patches with a high density of D_1 receptors, but a low density of D_2 receptors, TH, and DA transporter were demonstrated (Figs. 1 and 2). In contrast, these D_1-rich, but poorly DA-innervated, patches were surrounded by a high density of TH-containing nerve terminals, coincident with the distribution of D_2 receptors. These D_1-rich patches formed a continuous tubular-shaped network through the shell of the nucleus, suggesting that its population of D_1-bearing neurons may be modulated by DA released and having diffused from surrounding DA nerve terminals.

Extensive observations were also made in the islands of Calleja complex (ICC), which consists of dense aggregations of small granule cells (7–9 μm), closely grouped together and showing somatic appositions compatible with ephaptic (electrical) coupling *(15)*. A recent study *(16)* has shown that activation of D_3 receptors in the ICC results in the formation of gap junctions between its granule cells. The DA afferents from the substantia nigra and ventral tegmental area do not reach into the ICC, but mostly form a dense network around its islets *(17,18)*. Only a few fine DA fibers enter the core of the islets (*see* ref. *19*), as also described previously for cholecystokinin and substance P (SP) fibers *(20)*. Abundant D_3-receptor mRNA and a high density of D_3 receptor binding sites are present in most ICC granule cells *(21–25)*. D_1 receptors are also expressed in these cells, as indicated by high D_1 mRNA levels, a high density of D_1 receptor binding sites, and functional responses to D_1 receptor activation *(26–29)*. The presence of these two DA receptor subtypes, away from the corresponding afferents, strongly argues in favor of VT/modulation by DA in the ICC.

In the neostriatum of rat, the asynaptic nature of 60–70% of DA terminals, identified electron microscopically after [^{3}H]DA uptake autoradiography or DA immunocytochemistry *(30)*, has provided a strong indication of a dual mode of communication, diffuse as well as synaptic. A major extrasynatic localization of D_1 and D_2 receptors has been described in several parts of the basal ganglia of rat, monkey, and human *(28,31,32)*. In rat neostriatum, it has been emphasized that a majority of membrane-associated D_1 or D_2 receptors were located along portions of dendrites lacking a synaptic input. Extracellular (EC) DA concentrations are presumably sufficient to activate high-affinity D_1 and D_2 receptors in the basal ganglia *(33)*, especially in view of the increases in EC efflux of DA associated with burst firing *(34)*. It has been calculated that the distance traveled by DA molecules in the ECS might be on the order of 10 μm during one half-life *(33,35)*. This could allow DA, released from nonjunctional varicosities and spilling over from synaptic varicosities, to exert its influence on several thousand neighboring axon terminals or dendritic spines, provided that these potential targets are endowed with high-affinity DA receptors (for bases of this estimate, *see* ref. *12*).

The major mechanism for terminating EC DA signaling is the plasma membrane transporter for DA *(36)*, located along all parts of DA neurons, but outside synapses *(37,38)*. Intraregional differences in the release of DA have already been documented in the neostriatum of adult rat *(39,40)*, presumably corresponding to the islands and

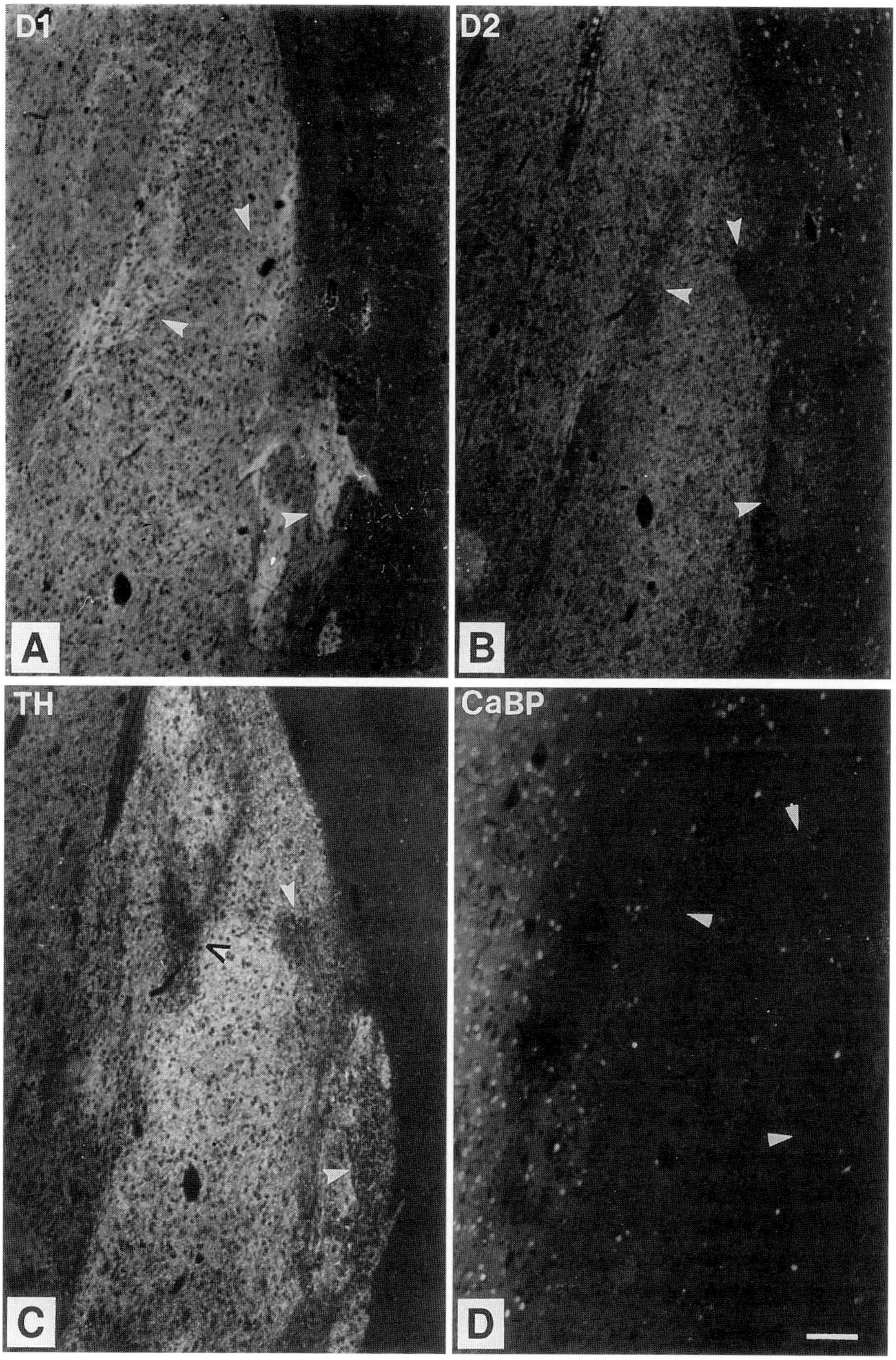

Fig. 1. Patches of D_1 receptor-immunoreactivity in the shell of the nucleus accumbens (bregma 1.5 mm; adult rat). As shown in **(A)**, these form rostrocaudally running bands on consecutive sections. Note the lack of immunoreactivity to D_2 receptor **(B)** and TH **(C)** in the corresponding areas of adjacent 10-µm-thick sections. Calbindin immunoreactivity is weak in this region **(D)**. Immunofluorescence microphotographs. Scale bar = 100 µm.

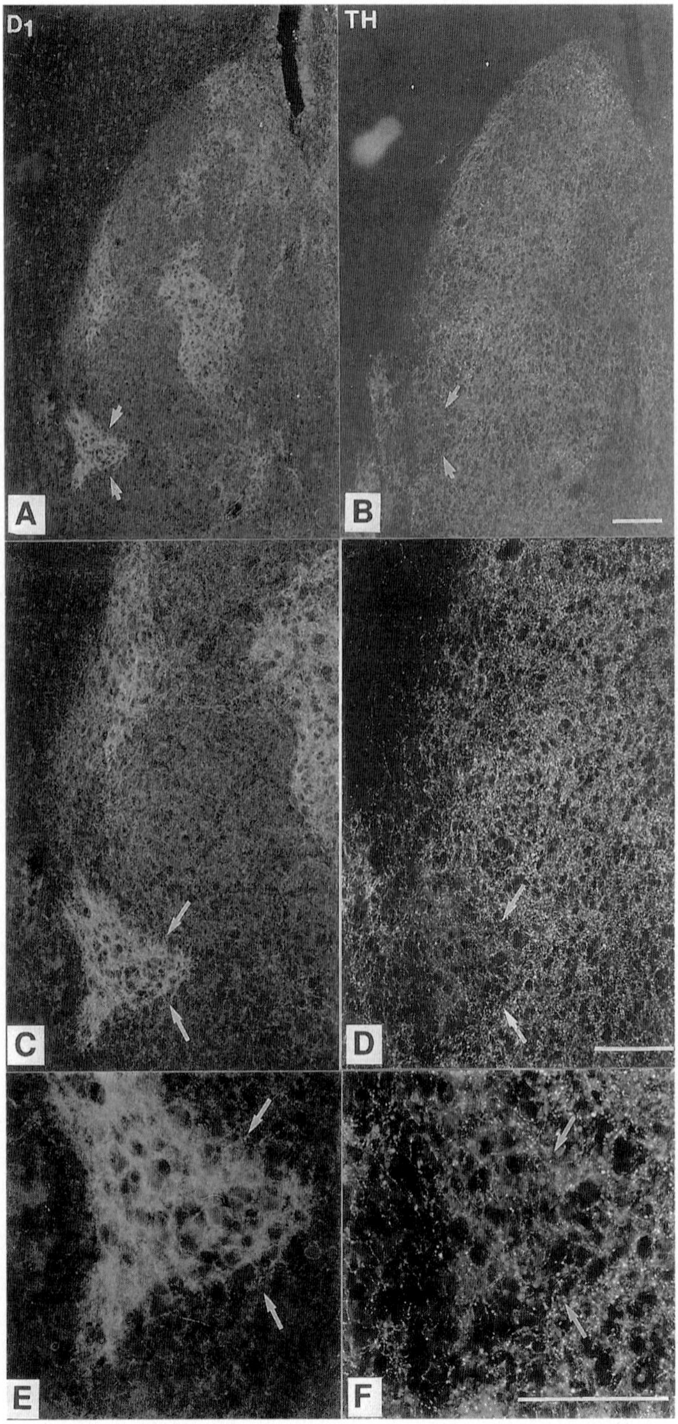

Fig. 2. Double immunofluorescence labeling with a polyclonal D_1 receptor Ab and a mouse monoclonal TH Ab in the shell of nucleus accumbens (11-day-old rat). Arrows designate the patches of strong D_1-immunoreactivity **(A,C,E)** and low TH-immunoreactivity **(B,D,F)**. (For details, *see* ref. *14*). Scale bar = 100 μm.

matrix of hyperdense and dense DA innervation, respectively *(41)*. A similar heterogeneity might be expected in the nucleus accumbens, in view of its patchy DA innervation. Moreover, a higher density of plasmalemmal transporter immunoreactivity has been reported on DA axons in the core vs the shell of the nucleus accumbens *(38)*. Recent IEM observations *(42)* on DA (TH-immunostained) terminals in the prelimbic division of rat prefrontal cortex have shown that the membrane transporter for DA predominates on preterminal or intervaricose axon segments, compared to axon varicosities. Such a location of the DA transporter away from release sites may be in keeping with the even greater diffusion of DA measured in the prefrontal cortex, compared with the striatum *(43)*. It could also account for the markedly higher release rate and lower uptake of DA in this brain region *(44)*.

The issue of VT is clearly of relevance in the pathophysiology and treatment of Parkinson's disease. Considerable experimental evidence indicates that the loss of nigrostriatal DA innervation underlying this disease may be, at least initially, compensated by DA spreading from residual DA terminals *(30,45–50)*. In a recent study showing spontaneous recovery in 1-methyl-4-phenyl-1,2,3,6-tetrahydropyridine-treated cats *(50)*, decrease DA uptake by surviving DA neurons has also been suggested as a possible compensatory mechanism. Of similar interest was the finding of a regional downregulation of TH in mice lacking DA transporter *(51)*, a reduction more pronounced in the dorsal striatum than in nucleus accumbens, where the uptake of DA is purported to be less efficient *(52)*. It may then be expected that the treatment of patients with Parkinson's disease would benefit from an efficient DA transporter inhibition, increasing the residual levels of DA, and thus compensating for the loss of DA terminals via VT mechanisms *(49,53)*.

3. SEROTONIN SYSTEM

5-HT neurons *(54–56)* have long been known to innervate the brain with axon varicosities (terminals) that often lack morphologically defined synaptic specializations *(5)*. For example, in rat neocortex, neostriatum, and hippocampus, the synaptic incidence of 5-HT terminals has been shown to be in the order of 20–45%, depending on the region and/or the layer considered *(57)*. In the cerebral cortex, both junctional and nonjunctional 5-HT varicosities have been observed along the same axons *(58)*, dispelling the suggestion of two categories of terminals issued from different cell groups of origin *(59)*. The existence of asynaptic cortical 5-HT varicosities has now been confirmed in all mammalian species examined, including primates, strengthening the view of a widely distributed system using both volume and wiring modalities of transmission *(7)*.

The cyclic voltammetry studies of Bunin et al. *(60–62)* on rat brain slices, comprising the mesencephalic raphe nuclei and substantia nigra, have provided functional evidence for VT from synaptic, as well as asynaptic, 5-HT release sites. Those authors demonstrated comparable electrically evoked diffusion of 5-HT in the substantia nigra, pars reticulata, in which an abundant 5-HT innervation is known to be entirely junctional *(63)*, and in the dorsal raphe nucleus, where most of the rare 5-HT varicosities appear to be nonjunctional *(64)* and 5-HT release occurs mostly from somata/dendrites *(65–68)*. Such results clearly indicate that rises in EC 5-HT, upon electrical stimulation, may be derived from leaky synapses, as well as 5-HT somata/dendrites and/or asynaptic varicosities. Moreover, in the substantia nigra, they have led to estimating the half-life of 5-HT in the order of 200 ms, during which diffusion could take place over 20 µm *(61)*, which is a huge distance at the subcellular level.

Recent IEM studies *(69,70)* on the ultrastructural distribution of the plasma membrane transporter for 5-HT have also yielded results consistent with a diffusion of 5-HT away from synaptic and/or asynaptic release sites. A predominant localization of the transporter along nonsynaptic portions of the membrane of 5-HT axons was then demonstrated in several regions of adult rat brain: Uptake experiments confirmed that the transporter was indeed functional in such a location *(69)*. Additional studies *(71)* have shown the 5-HT transporter to be scant on the membrane of 5-HT neuron somata/dendrites, and mostly concentrated along extrasynaptic portions of the axolemma of their axons and varicosities.

As in the case of DA, double immunolabeling experiments at the light microscopic level have been carried out to investigate the spatial relationships between 5 HT neurons and some of their receptor targets. For example, using anti-5-HT and anti-5-HT_{2A} receptor antibodies (Abs) in nucleus raphe pallidus of the medulla oblongata, 5-HT_{2A}-immunoreactive axon-like processes were found within the dendritic arbors of 5-HT-immunoreactive neurons *(72,73)*. In this instance, both types of processes remained a short (10–50 µm) distance away from each another, suggesting that these 5-HT_{2A} receptors may be activated by somatodendritically released 5-HT that reaches them by diffusion. In the ICC, a striking 5-HT–5-HT_{2A} receptor mismatch was observed, resembling that of the DA system (Fig. 3). A large number of 5-HT_{2A}-immunoreactive somata/dendrites contrasted with a paucity of 5-HT-immunoreactive fibers in the islets proper and Calleja magna; these were surrounded by a strong 5-HT innervation *(73)*.

In contrast, in the facial and trigeminal motor nuclei (Fig. 4), point-to-point relationships were apparent between 5-HT axon varicosities and the dendritic branches of large motor neurons showing 5-HT_{2A} receptors, in keeping with previous electron microscopic demonstrations of the existence of axosomatic and axodendritic synapses formed by 5-HT varicosities in these two nuclei *(74,75)*. Similarly, 5-HT-immunopositive axons, abutting on 5-HT_{2A} immunopositive neurons, were visible in the nucleus ambiguus, where the existence of direct appositions between both types of elements has in fact been established by dual IEM labeling (Fig. 5). In the ventral tegmental nucleus, there was no relation of proximity between 5-HT-immunoreactive terminals and 5-HT_{2A}-immunoreactive cell bodies and dendrites. Instead, these 5-HT_{2A}-immunoreactive neurons are presumably reached by diffusion of 5-HT released from cell bodies and dendrites in the neighboring dorsal raphe nucleus or 5-HT axons passing through this nucleus *(72)*.

Such a diversity of topological relationships between 5-HT elements and one of their receptor targets should be sufficient evidence in itself that 5-HT transmission/modulation may involve both VT and wiring transmission. In many regions of adult rat brain and the spinal cord, however, it has also been possible to investigate the distribution of the 5-HT_{2A} receptors at the ultrastructural level. After immunogold labeling, it was then found that these 5-HT receptors are mainly cytoplasmic, rather than membrane-bound, whether located on the somata, dendrites, or myelinated axons of central neurons *(76)*. Of particular interest was their prominent localization in pyramidal neurons of adult rat cerebral cortex *(77,78)*, in which quantitative analysis after

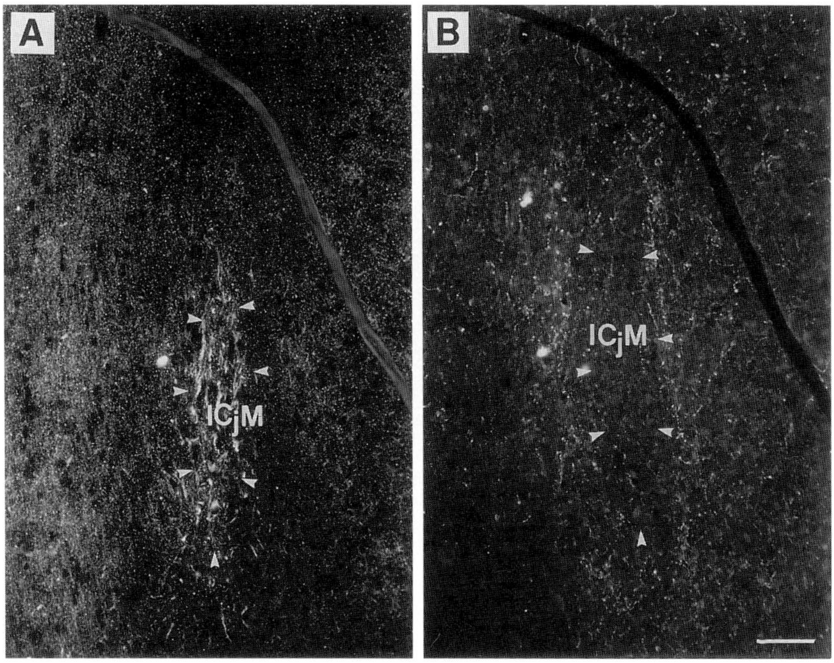

Fig. 3. Double immunofluorescence labeling for 5-HT2A receptor **(A)** and 5-HT itself **(B)** in the basal forebrain (bregma 1.2 mm; adult rat). Although strongly 5-HT_{2A}-immunoreactive processes, probably dendrites, are visualized in the island of Calleja magna (ICjM) (A), a dense network of 5-HT-immunoreactive nerve terminals surrounds this island (B). Arrowheads indicate the border of the ICjM (*see also* ref. *73*). Scale bar = 50 µm.

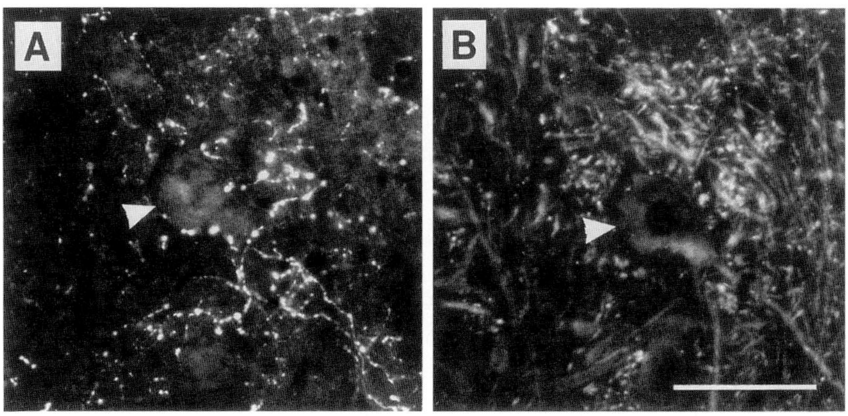

Fig. 4. Double immunofluorescence labeling in the facial nucleus (adult rat), showing 5 HT-immunoreactive terminals in close relationships with a 5-HT2A-immunoreactive neuronal cell body. Scale bar = 50 µm.

immunogold labeling excluded any enrichment of the plasma membrane (*76*; Fig. 6). Moreover, in these and other spiny dendrites, the 5-HT_{2A} receptors were always absent from dendritic spines, even though these represent a major site of synaptic afferents

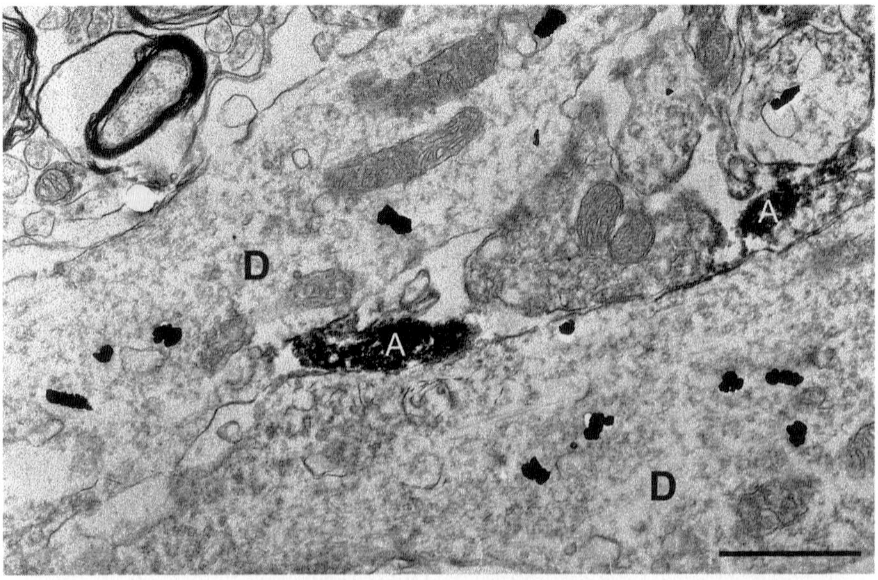

Fig. 5. Double immuno-electron microscopic labeling for 5-HT (rabbit polyclonal anti-5-HT Ab/DAB labeling) and 5-HT_{2A} receptor (mouse monoclonal Ab/immunogold labeling) in the nucleus ambiguus (adult rat). A DAB-labeled 5-HT axon terminal (A) is visualized in direct apposition with at least two dendrites (D) displaying internal 5-HT_{2A} receptor. Scale bar = 10 μm.

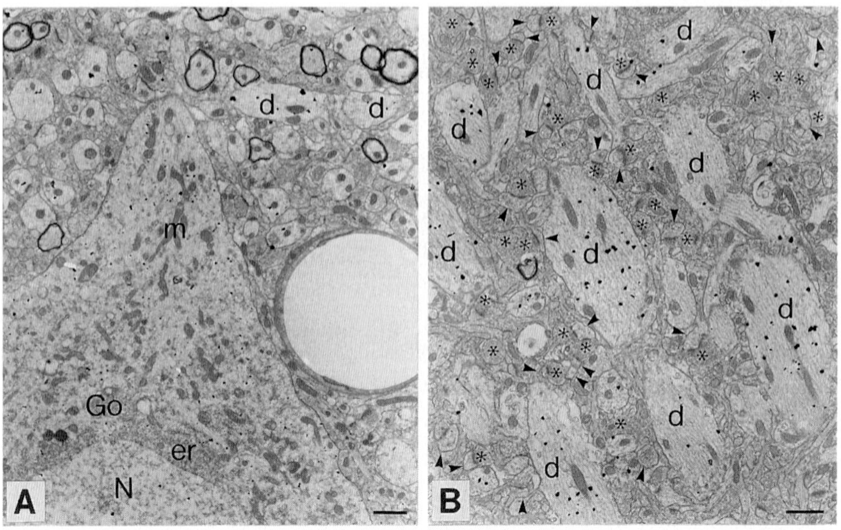

Fig. 6. Electron micrographs from layer V of the parietal cortex (adult rat), illustrating the subcellular localization of immunogold-labeled 5-HT_{2A} receptors. In both, the cell body of a pyramidal neuron **(A)** N in nucleus, and relatively large dendrites (d) in the neuropil **(B)**, most of the immunogold particles are found in the cytoplasm, and very few on the plasma membrane. In (A), note the absence of labeling on the endoplasmic reticulum (er), Golgi apparatus (Go), and mitochondria (m). In (B), note that dendritic spines (arrowheads), as well as axon terminals (asterisks) remain immunonegative. Scale bar = 1 μm. (Adapted with permission from ref. *78a*.)

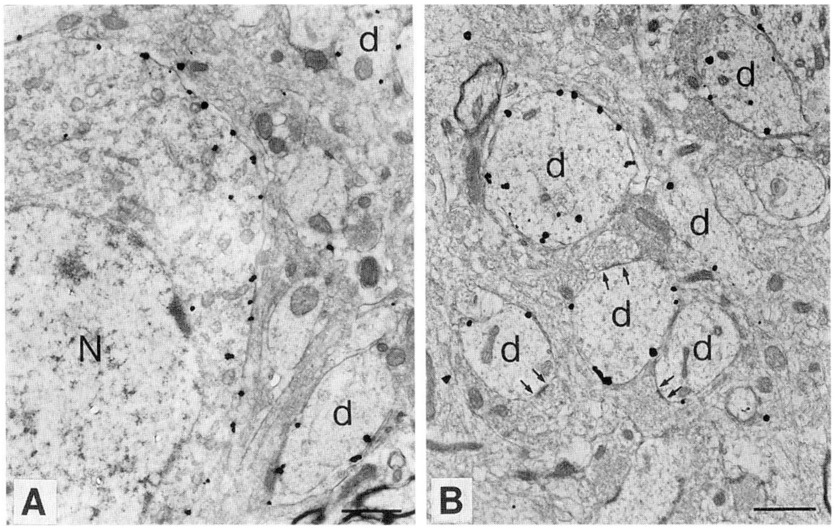

Fig. 7. Immunogold labeling of 5-HT_{1A} receptors on part of a neuronal cell body **(A)** N in nucleus, and dendrites (A and B) in the nucleus raphe dorsalis (adult rat). In both the cell body and dendrites, most of the immunogold particles are associated with the plasma membrane. In dendrites, however, such particles are never seen on postsynaptic specializations (between arrows in **[B]**). Scale bar = 1 μm. (Adapted with permission from ref. *78b* and *82*).

(Fig. 6B). The suggestion was therefore made that 5-HT_{2A} receptors are permanently internalized under physiological conditions, and that one of their main functions is to participate in retrograde signaling processes, activated also by other 5-HT receptors located on the same cells. In IF experiments, a redistribution of 5-HT_{2A} receptors from apical dendrites into the cell bodies of pyramidal cortical neurons, has now been reported in vivo and in vitro, following treatment with the atypical antipsychotic, clozapine *(79,80)*. This observation lends further support to a retrograde signaling function of 5-HT_{2A} receptors.

Electron microscopic observations after immunogold labeling have also been made on both 5-HT_{1A} and 5-HT_{1B} receptors in adult rat brain *(81–84)*. In nucleus raphe dorsalis, where 5-HT_{1A} receptors are known to behave as autoreceptors, their immunogold labeling was exclusively on the somata/dendrites of 5-HT neurons, and exhibited a clearcut predilection for the plasma membrane, but to the exclusion of its synaptically differentiated portions (Fig. 7). Because this 5-HT nucleus receives only a scant, mostly asynaptic, 5-HT innervation *(64)*, presumably from other raphe nuclei, it could be assumed that these high-affinity autoreceptors are activated by somatodendritically released 5-HT. In the hippocampal formation, both the somata and dendrites of pyramidal and granule cells displayed extrasynaptic, membrane-associated 5-HT_{1A} heteroreceptors, which appeared to be particularly abundant on the dendritic spines of pyramidal cells, at least in CA3 *(82)*. In this hippocampal sector, not only is the 5-HT innervation known to be mostly asynaptic, but it has also been shown to synaptically contact only dendritic branches, and never dendritic spines *(85)*. Thus, in every respect, the topological relationships between 5-HT somata/dendrites or axon terminals and 5-HT_{1A} auto- and heteroreceptors favor the diffuse transmission paradigm.

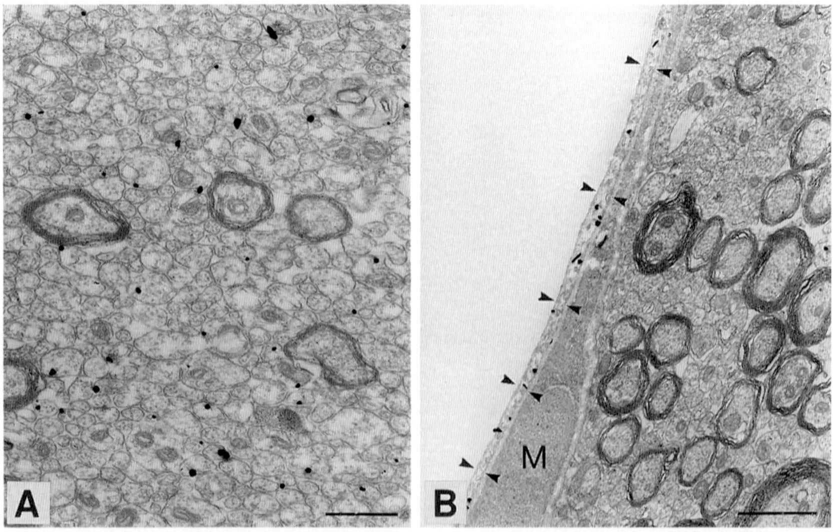

Fig. 8. Immunogold labeling of 5-HT_{1B} receptors in the globus pallidus (adult rat). In this field of neuropil, **(A)**, most of the immunogold particles are associated with the membrane of fine, unmyelinated, preterminal axons. On a nearby microvessel, **(B)**, the immunogold labeling is confined to the endothelial lining (between arrowheads), and spares an underlying myocyte (M in nucleus). Scale bar = 1 μm. (Adapted with permission from ref. *78b* and *82*).

5-HT_{1B} receptors have been visualized in both the substantia nigra and globus pallidus of adult rat (*82–84*; Fig. 8). These receptors were then shown to be preferentially located on the membrane of fine, unmyelinated, preterminal axons, presumably belonging to striatonigral and striatopallidal neurons (Fig. 8A). 5-HT_{1B} receptors on nerve terminals in the substantia nigra and superior colliculus were consistently found in extrasynaptic locations *(83,84)*. A selective localization to the cytoplasm of endothelial cells of cerebral microvessels was also observed (*81*; Fig. 8B). The presence of 5-HT_{1B} receptors on the membrane of preterminal axons suggested that they may mediate a 5-HT control of transmitter release through effects on axonal conduction. Together with their microvascular localization, it represents a further aspect of 5-HT function that appears incompatible with wiring transmission.

It seems likely that novel antidepressants, such as fluoxetine, which selectively block the 5-HT transporter, exert their therapeutic effects through increased EC levels of 5-HT. These will then activate extrasynaptic 5-HT receptor subtypes that may be otherwise silent or merely mediating long-term influences of an ambient, resting level of 5-HT. The 5-HT_{2A} receptors, and notably their downregulation, have been implicated in the etiology of depression *(86–90)*. Their involvement is further evidenced by clinical and genetic observations of altered 5-HT_{2A} receptor function in depressed patients *(91–93)*, and by their mediation of estrogen- or androgen-induced mood changes in animals *(94,95)*.

5-HT_{1A} receptors are a major target for psychotherapeutic drugs *(96)*. The implication of these receptors in the therapeutic efficacy of antidepressants is well-established *(97)*. In particular, their desensitization as autoreceptors on the somata/dendrites of 5-HT neurons in the dorsal raphe nucleus accounts for the therapeutic delay following treatment with monoamine oxidase inhibitors or selective 5-HT reuptake inhibitors *(98)*. A recent study in adult rat has provided an explanation for this desensitization. Using quantitative IEM, after treatment with the selective 5-HT_{1A} agonist, 8-hydroxy-2-(n-dipropylamine)tetralin hydrobromide, a significant decrease in the amount of receptor located on the membrane of dorsal raphe somata/dendrites, and increase in their cytoplasmic labeling, was demonstrated 1 h after intravenous injection of the drug *(99)*.

These findings are in keeping with previous electrophysiological results having shown a desensitization of 5-HT_{1A} autoreceptors after treatment with selective 5-HT reuptake inhibitors *(100)*. They strongly suggest that this desensitization is the result of an internalization. Internalization of 5-HT_{1A} autoreceptors would thus prevent inhibition of the firing of 5-HT neurons and their release of 5-HT, and hence increase their functional activity upon stimulation.

4. PEPTIDERGIC SYSTEMS

During the past decade, considerable attention has been devoted to the diverse properties of NPs, several of which are now recognized as neuromodulators with potent physiological effects. There have been no systematic studies to determine the synaptic or asynaptic character of these innervations, but mismatches between presumed sites of NP release and their receptors were recognized early, during investigations of the distribution of β-endorphin- and enkephalin-immunoreactive terminals and opiate receptor binding sites *(2,3)*. Furthermore, it was subsequently documented that exogenous neuropeptide tyrosine (NPY) *(4)*, β-endorphin and prolactin *(101)* may diffuse for long distances in brain, and that endogenously released β-endorphin seems capable also of globally modulating brain function via the cerebrospinal fluid (CSF) *(102)*. Topographical analysis of these systems has also greatly benefited from the development of specific Abs against various NP receptors, allowing analysis of the relationships between release and target sites, by double immunolabeling at both the light and electron microscopic levels.

4.1. *Enkephalin*

The light microscopic observations of a mismatch between putative opioid binding and release sites have now been confirmed by immunofluorescence observations on the distribution of opioid receptors and axon terminals. In rat entopeduncular nucleus, for example, enkephalin-immunoreactive terminals were found in a medial subregion; the neuronal cell bodies and dendrites bearing delta-opioid receptor were located laterally, suggesting VT signaling *(103)*. Whether such remote delta-opioid receptors are fully functional, or are spare receptors *(104)*, remains to be established, but they may be reached by EC β-endorphin released from surrounding plexa of β-endorphin terminals or coming from the CSF *(102)*. An IEM study of rat dentate gyrus *(105)* has demonstrated delta-opioid receptors in granule and nongranule cell bodies, dendrites, dendritic spines, axon terminals, and astroglia. It concluded that the distribution of these profiles overlapped with, but was not restricted to, regions known to contain enkephalin. Again, such delta-opioid receptors could be reached by β-endorphin from the CSF.

In early electron microscope autoradiographic studies of rat neostriatum with radioligands, μ-opioid receptors were observed in para- and extrasynaptic locations, along the plasma membrane of somata/dendrites and axons *(106)*. More recently, a double-immunolabeling study of μ-opioid receptors and leu^5-enkephalin, carried out in the patch compartment of neostriatum *(107)*, also demonstrated an extrasynaptic plasma membrane distribution of these receptors on the soma-dendrites of spiny neurons, including some containing leu^5-enkephalin. μ-opioid receptors were also found in axon terminals, some of which contained leu^5-enkephalin then did not form recognizable synaptic junctions. Opioids, acting in an autocrine or paracrine fashion, might thus be in a position to exert a variety of effects. They could modulate their own release from either somata/dendrites or axon terminals through autoreceptors, but also influence neuronal excitability and/or the release of other transmitters through somatodendritic and axon terminal heteroreceptors. Moreover, they are likely to exert these effects as coexistent transmitters, as already shown for other peptides, such as galanin in the noradrenaline neurons of rat locus coeruleus *(108)*.

4.2. Neuropeptide Y

Early immunocytochemical studies of NPY and its receptors also revealed significant mismatches *(4,109)*. This work has now been extended by detailed immunocytochemical observations of NPY terminals and Y1 receptor bearing profiles in the same rat brain sections *(110)*. In the hippocampal formation, there were no close associations between Y1-immunopositive somata and dendrites and a sparse plexus of weakly NPY-immunoreactive nerve terminals. This contrasted with areas of co-distribution of the two markers in the hypothalamus (paraventricular nucleus and dorsomedial nucleus). In the arcuate nucleus, some of the cell bodies and dendrites showing Y1 receptors and also immunoreactive for β-endorphin *(111)*, were closely surrounded by NPY terminals, suggesting appositional or synaptic contacts in addition to short-distance relationships (μm range). In the rostral and dorsal part of the spinal tract nucleus of the Vth nerve (Fig. 9), the area displaying Y1-immunoreactive neurons was considerably larger than that pervaded by NPY terminals, suggesting activation of the constituent neurons by diffusing NPY, in addition to point-to-point contacts. More caudally, the distance between the terminals and receptors was even greater, the former being confined to the superficial layer and the Y1-immunoreactive processes to the deeper magnocellular layer.

As with DA and 5-HT and some of their receptors, the ICC also displayed an obvious mismatch between NPY-immunoreactive terminals and Y1 receptors. The islands and Calleja magna were surrounded by a dense network of NPY terminals (Fig. 10); there were only a few within the ICC, in which numerous dendritic profiles exhibited a high density of Y1 receptors. Binding of a peptide YY derivative, selective for the Y1 receptor has previously been shown to be highly concentrated in the islands of Calleja *(112)*. These findings imply that, in order to activate Y1 receptors in the islands, NPY released in their surroundings must diffuse for distances of 100–200 μm *(110)*.

Somewhat divergent observations were, however, made in a double-labeling IEM study *(113)* on the localization of Y1 receptors and NPY, carried out on rat nucleus accumbens. It was then shown that, in the core and shell of the nucleus, Y1 receptors were present within neuronal cell bodies, dendrites, and axon terminals that contained or were apposed to NPY-immunoreactive neurons, and also in numerous other spiny or

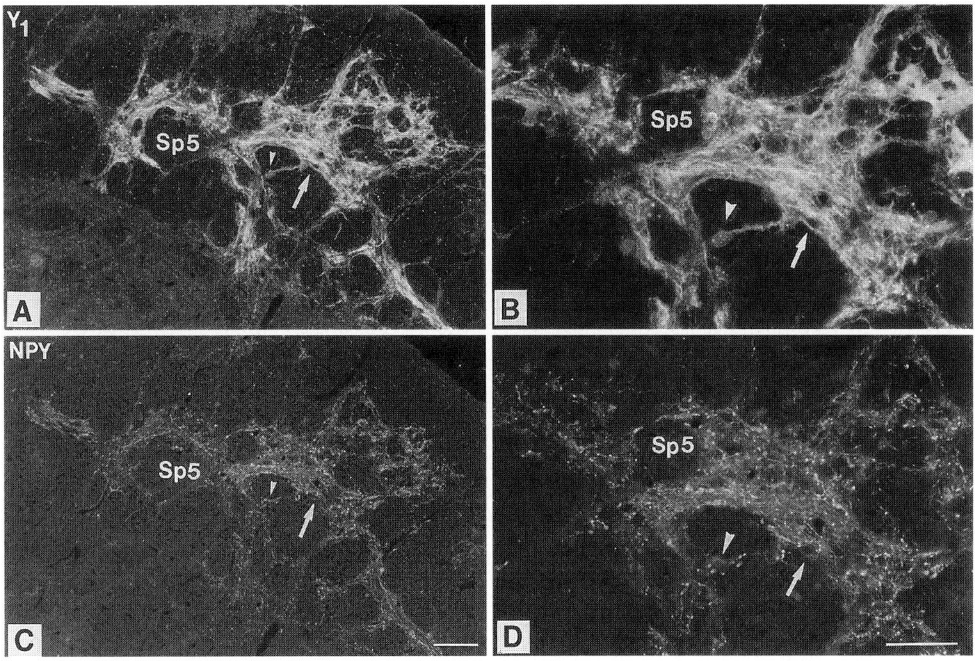

Fig. 9. Double immunofluorescence labeling for NPY Y1 receptor **(A,B)** and NPY **(C,D)** in transverse sections from the dorso-rostral part of the spinal trigeminal nucleus (Sp5) (adult rat; bregma 13.5 mm). Note the high degree of overlap between the regions, showing Y1 receptor-IR dendritic profiles (A and B) and NPY-IR terminals (C and D). As indicated by corresponding arrows and arrowheads, many NPY-IR terminals are closely associated with Y1 receptor-IR somata/dendrites, suggestive of point-to-point connections. Scale bar = 50 μm.

aspiny neurons, axon terminals, and astrocytes. It was concluded from those observations that Y1 receptors may play a role in the autoregulation of NPY neurons, in which they would be internalized, along with endogenous NPY, but may also be implicated in effects of NPY on other transmitters and glial function.

4.3. Substance P

Ultrastructural data are also available in the case of SP and the neurokinin receptors. In cerebral cortex, striatum, and spinal cord, a subcellular mismatch has been described at the EM level, after double immunolabeling for SP and SP receptors *(114)*. It was then estimated that ~70% of the plasma membrane of receptor-immunoreactive somata/dendrites was laden with receptor, but that less than 15% was contacted by synaptic boutons, only some of which were SP-immunoreactive.

In contrast, in the superficial laminae of the dorsal horn, another double-immunolabeling study *(115)* has demonstrated that most dendrites endowed with NK-1 receptors were apposed by SP-containing boutons, and this more frequently than their receptor-immunonegative counterparts. A preferential relationship between SP-immunoreactive boutons and neurokinin-1 receptor sites has also been described by light microscope in the so-called "striocapsules" encircling striosomes *(116)*, within which SP-immunoreactive puncta were reported to densely innervate SP-immunoreactive

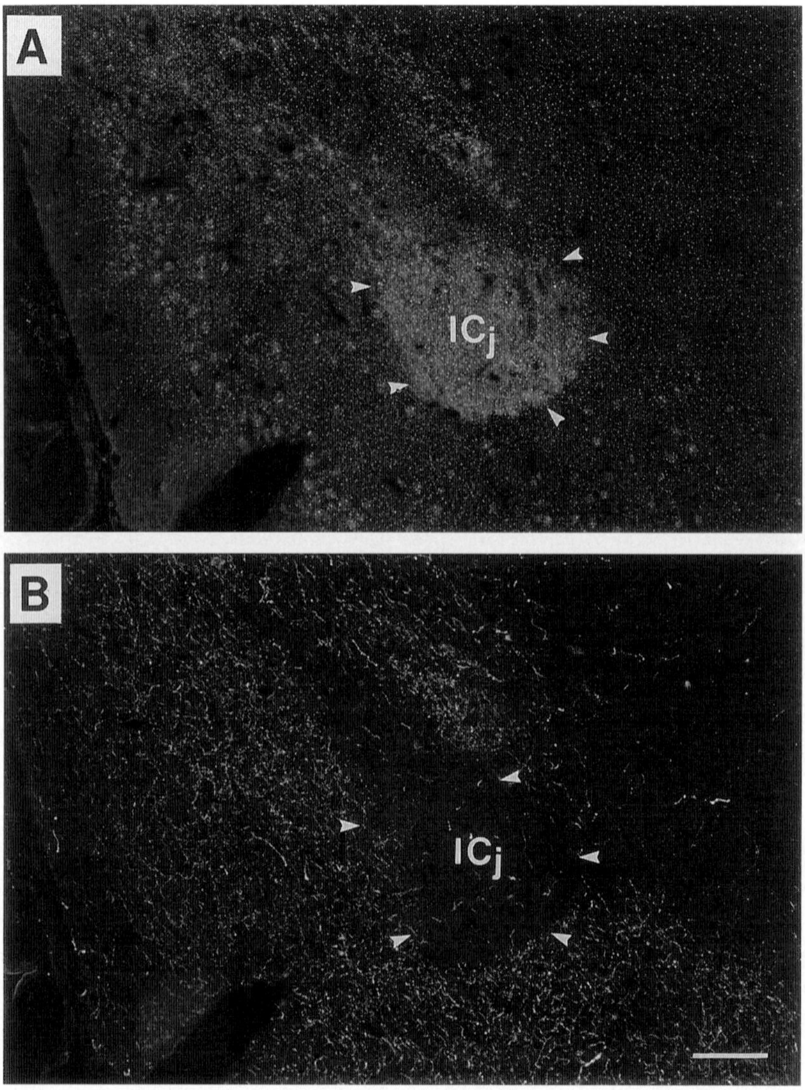

Fig. 10. IF labeling for NPY Y1 receptor **(A)** and NPY **(B)** in adjacent transverse sections from the ventral forebrain (adult rat; bregma 1.7 mm). As shown in (A), Y1 receptor-immunoreactivity is dense inside the islands of Calleja (ICj); in (B), NPY terminals surround the ICj (arrowheads) and very few penetrate it. (For details, *see* ref. *110*). Scale bar = 50 μm. Bregma: 1.7 mm.

dendritic processes. In the dorsal vagal complex of rat and cat, however, NK-1 receptors were found mostly in an extrasynaptic, somatodendritic plasma membrane localization, even if many SP terminals make synaptic contact with these neurons *(117)*. This was taken as an indication that these SP receptors would be activated by VT, but from synaptic release sites. In the rat spinal cord, immunoreactivity to NK-2 receptors (receptor for neurokinin A) was found to be primarily associated with astroglia *(118)*; that for NK-3 receptors (receptor for neurokinin B) decorated the plasma membrane of spines and axon terminals, but outside of synaptic differentiations *(119)*. There is indeed good evidence *(120,121)* that SP has the ability to spread over considerable distances in the ECS, at least under conditions of endopeptidase inhibition, or of endogenous release of calcitonin gene-related peptide, competitive substrate for these enzymes. Another tachykinin, neurokinin A, has been detected in CSF 30 min after its release *(122)*, and, in controlled superfusion experiments, displays substantial spread and clearance within the spinal cord, in line with the astroglial location of NK-2 receptors *(121)*.

4.4. *Neurotensin*

New and interesting data have also emerged from the study of neurotensin (NT). Previous immunocytochemical mapping studies had shown extensive axon networks immunoreactive for this peptide in rat basal forebrain, hypothalamus, and some brainstem regions, including those containing the cell bodies of origin and terminal arborizations of nigrostriatal and mesolimbic DA *(123)*. Autoradiographic studies later demonstrated NT-binding sites in the same regions, including the mesostriatal and mesolimbic systems, although transmitter/receptor mismatches appeared to be present in other regions, e.g., the cingulate cortex *(124)*.

NT receptor-immunoreactivity has now been studied in great detail, at both the light and electron microscopic levels *(125,126)*, disclosing unexpected findings. First, a number of regions endowed with NT-binding sites showed no or divergent patterns of NT receptor immunostaining, suggesting that they may express a molecularly distinct form of the receptor *(127–129)*. Second, in rat substantia nigra, NT receptor-immunoreactivity was highly correlated with $[^{125}I]$NT binding, and both markers were located extrasynaptically, along the plasma membrane of presumed DA neurons, and less heavily in myelinated and unmyelinated axons and astrocytic leaflets, suggesting functional sites activated by VT *(126)*. A similar conclusion had already been reached from an electron microscope autoradiographic study *(130)* of the ultrastructural distribution of NT-binding sites in the ventral tegmental area and interfascicular nucleus of rat. Moreover, a subsequent double-immunolabeling study had shown that NT terminals directly abutted, but only infrequently established, differentiated synaptic contacts with TH-containing perikarya and dendrites, furthering the view of a widespread, paracrine type of action of NT in the midbrain tegmentum *(131)*. A similar ultrastructural localization of somatostatin sst2A receptors, mostly involving extrasynaptic portions of somatodendritic plasma membrane, was recently described by Dournaud et al. *(132)*, in rat brain regions displaying diffuse, rather than dense, sst2A receptor labeling and only moderate or low densities of somatostatin-immunoreactive fibers and varicosities.

4.5. *Extracellular (EC) Enzymes*

The study of EC brain enzymes has also thrown new light on VT by NPs. For example, endoprotease 3.4.24.16, which participates in the degradation of NT *(133)*, is present in nerve cells and glial cells, but also exists in a soluble form. The neuronal form is membrane-bound: the glial form is soluble and secreted into the ECS *(134)*. Such a localization suggests a role in restricting the diffusion of released NT.

Another example is the prohormone convertase 5 (PC5), an endoprotease responsible for the formation of peptides and peptide fragments. PC5 is associated with nerve cell bodies and dendrites, but not with astrocytes *(135)*. It has an extensive distribution in the forebrain, and is only one of at least six prohormone convertases thus far identified in rat brain. The isoform, PC5-A, has been shown to process NT, nerve growth factor, brain-derived neurotrophic factor, and other signaling proteins *(135,136)*. It is probable that secreted endopeptidases or endoproteases, bound to the cell membranes bordering the ECS, act in the EC environment, to enhance or terminate NP signaling, through the formation of active or inactive fragments, respectively, as already suggested for galanin and galanin fragments *(137,138)*. EC enzymes, attached to the plasma membrane and/or present in the ECS, are themselves regionally distributed in brain in a differential manner, thus creating, in the case of NPs, neurochemical networks for VT molecules. In the clinic, increased activity of prolylendopeptidase has been measured in serum from patients with posttraumatic stress disorder *(139)*, and it has been suggested *(140)* that peptidase inhibitors could be used to modulate nociceptive pain. Future research will tell how components of the ECS, notably various enzymes, may actually influence the chemical pathways and networks for VT signaling.

4.6. *Clinical Relevance*

The involvement of NPs in the etiology and treatment of psychiatric disorders may depend directly or indirectly on VT. Indeed, NPs are now recognized as possible regulators of DA and 5-HT transmission/modulation, as shown, for example, by the effects of NT or cholecystokinin-8 on the binding of D_2 DA receptor antagonists *(141)*. Clinical investigations have indicated that schizophrenic symptoms may be associated with low levels of NT in CSF *(142–144)*; reduced amounts of NT receptor binding have been measured postmortem in several cortical regions of the brain of schizophrenic subjects *(145,146)*. These findings have drawn attention to the possible treatment of psychotic disorders with NT receptor agonists *(147)*. In patients with major depression, recent studies have suggested an association between NPY and mood disorders, despite earlier conflicting reports regarding the CSF levels of NPY in such patients *(148,149)*. Low plasma levels of NPY were measured in patients with a history of suicidal attempts *(150)*, and reduced NPY mRNA levels in the prefrontal cortex of subjects diagnosed with bipolar disorder *(151)*.

The implication of NPY and Y1 receptors in depression is also supported by observations on a genetic animal model, in which NPY-immunoreactivity was found to be decreased and Y1 binding increased in the hippocampus *(152)*. Because of increased CSF levels in vivo *(153)*, and an altered morphology of hippocampal NPY terminals postmortem *(154)*, NPY has also been implicated in schizophrenia. Specific changes in CSF concentrations of NPY, somatostatin, and CRF, parallel to clinical recovery, have indeed been measured after effective electroconvulsive therapy in both depressed and schizophrenic patients *(155)*. Since NPY, given systemically to healthy subjects, promotes and improves sleep *(156)*, it will be of interest to determine whether it improves the sleep disturbances so commonly found in depression. New leads for the development of antidepressant drugs are also provided by recent clinical trials demonstrating an improvement of major depression after treatment with SP receptor antagonists *(157,158)*.

5. CONCLUSIONS

Clearly, on the basis of increasing light, confocal and electron microscopic immunocytochemical evidence, and depending on the region and chemical messenger considered, monoamine and NP communication in brain may be both synaptic and/or diffuse, whether in the form of short- or long-distance VT. The two modes of transmission presumably operate in parallel, complementary to each other. When it will become possible to determine discrete sites of transmitter release, as well as effect, in vivo, their respective roles and relative importance, should be amenable to investigation. The fact that a majority of the monoamine and NP receptors display an extrasynaptic location favors a predominance of VT in the case of these neuromodulators. In some NP systems, such as the β-endorphin, long-distance communication, including CSF convection, has already been demonstrated *(102)*. In others, such as the NPY and neurokinins, both short- and long-distance VT may be involved, in addition to wiring transmission. Long-distance communication by active peptide fragments must also be taken into account, as initially shown for vasopressin *(159)*, some fragments of which are highly active and improve performance in behavioral and memory experiments *(160,161)*. Such a mechanism is probably important for many NPs, as exemplified by the existence of EC enzymes degrading and modifying the activity of, e.g., NT *(153,154)*. Additional evidence should also be forthcoming from the study of prohormone convertases, e.g., PC5A, which have the ability to form active fragments or terminate the action of several NPs.

REFERENCES

1. Kuhar, M. J., Unnerstall, J. R., and DeSouza, E. B. (1985) Receptor mapping in neuropharmacology by autoradiography: some technical problems. *Nida Res. Monogr.* **62,** 1–12.
2. Agnati, L. F., Fuxe, K., Zoli, M., Ozini, I., Toffano, G., and Ferraguti, F. (1986) A correlation analysis of the regional distribution of central enkephalin and beta-endorphin immunoreactive terminals and of opiate receptors in adult and old male rats. Evidence for the existence of two main types of communication in the central nervous system: the volume transmission and the wiring transmission. *Acta Physiol. Scand.* **128,** 201–207.
3. Herkenham, M. (1987) Mismatches between neurotransmitter and receptor localizations in brain: observations and implications. *Neuroscience* **23,** 1–38.
4. Fuxe, K. and Agnati, L. F. (1991) Two principal modes of electrochemical communication in the brain: volume versus wiring transmission, in *Volume Transmission in the Brain. Novel Mechanisms for Neural Transmission* (Fuxe, K. and Agnati, L. F., eds.), Raven, New York, pp. 1–9.
5. Descarries, L., Beaudet, A., and Watkins, K. C. (1975) Serotonin nerve terminals in adult rat neocortex. *Brain Res.* **100,** 563–588.
6. Descarries, L., Séguéla, P., and Watkins, K. C. (1991) Nonjunctional relationships of monoamine axon terminals in the cerebral cortex of adult rat, in *Volume Transmission in the Brain: Novel Mechanisms for Neural Transmission* (Fuxe, K. and Agnati, L. F., eds.), Raven, New York, pp. 53–62.
7. Descarries, L. and Mechawar, N. (2000) Ultrastructural evidence for diffuse transmission by monoamine and acetylcholine neurons of the central nervous system, in *Progress Brain Research* (Agnati, L. F., Fuxe, K., Nicholson, C., and Syková, E., eds.), Volume Transmission Revisited. 125, Elsevier, Amsterdam.
8. Isaacson, J. S. and Nicoll, R. A. (1993) The uptake inhibitor L-trans-PDC enhances responses to glutamate but fails to alter the kinetics of excitatory synaptic currents in the hippocampus. *J. Neurophysiol.* **70,** 2187–2191.

9. Kullman, D. M. (1999) Synaptic and extrasynaptic roles of glutamate in the mammalian hippocampus. *Acta Physiol. Scand.* **166,** 79–83.
10. Kullman, D. M. (2000) Spillover and synaptic cross talk mediated by glutamate and GABA in the mammalian brain, in *Progress Brain Research* (Agnati, L. F., Fuxe, K., Nicholson, C., and Syková, E., eds.), 125, Elsevier, Amsterdam.
11. Zoli, M., Jansson, A., Syková, E., Agnati, L. F., and Fuxe, K. (1999) Volume transmission in the CNS and its relevance for neuropsychopharmacology. *Trends Pharmacol. Sci.* **20,** 142–150.
12. Descarries, L., Gisiger, V., and Steriade, M. (1997) Diffuse transmission by acetylcholine in the CNS. *Progr. Neurobiol.* **53,** 603–625.
13. Descarries, L. (1998) The hypothesis of an ambient level of acetylcholine in the central nervous system. *J. Physiol. Paris.* **92,** 215–220.
14. Jansson, A., Goldstein, M., Tinner, B., Zoli, M., Meador, W. J., Lew, J. Y., et al. (1999) On the distribution patterns of D1, D2, tyrosine hydroxylase and dopamine transporter immunoreactivities in the ventral striatum of the rat. *Neuroscience* **89,** 473–489.
15. Ribak, C. E. and Fallon, J. H. (1982) The island of Calleja complex of rat basal forebrain. I. Light and electron microscopic observations. *J. Comp. Neurol.* **205,** 207–218.
16. Halliwell, J. V. and Horne, A. L. (1998) Evidence for enhancement of gap junctional coupling between rat island of Calleja granule cells in vitro by the activation of dopamine D3 receptors. *J. Physiol. (Lond.)* **506,** 175–194.
17. Hökfelt, T., Johansson, O., Fuxe, K., Goldstein, M., and Park, D. (1977) Immunocytochemical studies on the localization and distribution of monoamine neuron systems in the rat brain II. Tyrosine hydroxylase in the telencephalon. *Med. Biol.* **55,** 21–40.
18. Voorn, P., Jorritsma-Byham, B., Van Dijk, C., and Buijs, R. M. (1986) The dopaminergic innervation of the ventral striatum in the rat: a light- and electron-microscopical study with antibodies against dopamine. *J. Comp. Neurol.* **251,** 84–99.
19. Seifert, U., Hartig, W., Grosche, J., Bruckner, G., Riedel, A., and Brauer, K. (1998) Axonal expression sites of tyrosine hydroxylase, calretinin- and calbindin-immunoreactivity in striato-pallidal and septal nuclei of the rat brain: a double-immunolabelling study. *Brain Res.* **795,** 227–246.
20. Fallon, J. H., Loughlin, S. E., and Ribak, C. E. (1983) The islands of Calleja complex of rat basal forebrain. III. Histochemical evidence for a striatopallidal system. *J. Comp. Neurol.* **218,** 91–120.
21. Diaz, J., Levesque, D., Lammers, C. H., Griffon, N., Martres, M. P., Schwartz, J. C., and Sokoloff, P. (1995) Phenotypical characterization of neurons expressing the dopamine D3 receptor in the rat brain. *Neuroscience* **65,** 731–745.
22. Gurevich, E. V., Himes, J. W., and Joyce, J. N. (1999) Developmental regulation of expression of the D3 dopamine receptor in rat nucleus accumbens and islands of Calleja. *J. Pharmacol. Exp. Ther.* **289,** 587–598.
23. Ridray, S., Griffon, N., Mignon, V., Souil, E., Carboni, S., Diaz, J., Schwartz, J. C., and Sokoloff, P. (1998) Coexpression of dopamine D1 and D3 receptors in islands of Calleja and shell of nucleus accumbens of the rat: opposite and synergistic functional interactions. *Eur. J. Neurosci.* **10,** 1676–1686.
24. Schwartz, J. C., Diaz, J., Bordet, R., Griffon, N., Perachon, S., Pilon, C., Ridray, S., and Sokoloff, P. (1998) Functional implications of multiple dopamine receptor subtypes: the D1/D3 receptor coexistence. *Brain Res. Brain Res. Rev.* **26,** 236–242.
25. Hillefors, M., von Euler, M., Hedlund, P. B., and von Euler, G. (1999) Prominent binding of the dopamine D3 agonist [^{3}H]PD 128907 in the caudate-putamen of the adult rat. *Brain Res.* **822,** 126–131.
26. Wamsley, J. K., Gehlert, D. R., Filloux, F. M., and Dawson, T. M. (1989) Comparison of the distribution of D-1 and D-2 dopamine receptors in the rat brain. *J. Chem. Neuroanat.* **2,** 119–137.

27. Mengod, G., Vilaro, M. T., Niznik, H. B., Sunahara, R. K., Seeman, P., O'Dowd, B. F., and Palacios, J. M. (1991) Visualization of a dopamine D1 receptor mRNA in human and rat brain. *Brain Res. Mol. Brain Res.* **10,** 185–191.
28. Caillé, I., Dumartin, B., and Bloch, B. (1996) Ultrastructural localization of D1 dopamine receptor immunoreactivity in the rat striatonigral neurons and its relation with dopaminergic innervation. *Brain Res.* **730,** 17–31.
29. Wirtshafter, D. (1998) D1 dopamine receptors mediate neuroleptic-induced Fos expression in the islands of Calleja. *Synapse* **28,** 154–159.
30. Descarries, L., Watkins, K. C., Garcia, S., Bosler, O., and Doucet, G. (1996) Dual character, asynaptic and synaptic, of the dopamine innervation in adult rat neostriatum: a quantitative autoradiographic and immunocytochemical analysis. *J. Comp. Neurol.* **375,** 167–186.
31. Levey, A. I., Herch, S. M., Rye, D. B., Sunahara, R., Niznik, H. B., Kitt, C. A., et al. (1993) Localization of D1 and D2 dopamine receptors in rat, monkey and human brain with subtype-specific antibodies. *Proc. Natl. Acad. Sci. USA* **90,** 8861–8865.
32. Yung, K. K. L., Bolam, J. P., Smith, A. D., Hersch, S. M., Ciliax, B. J., and Levey, A. I. (1995) Immunocytochemical localization of D_1 and D_2 dopamine receptors in the basal ganglia of the rat: light and electron microscopy. *Neuroscience* **65,** 709–730.
33. Garris, P. A., Ciolkowski, E. L., Pastore, P., and Wightman, R. M. (1994) Efflux of dopamine from the synaptic cleft in the nucleus accumbens of the rat brain. *J. Neurosci.* **14,** 6084–6093.
34. Chergui, K., Suaud-Chagny, M. F., and Gonon, F. (1994) Nonlinear relationship between impulse flow, dopamine release and dopamine elimination in the rat brain in vivo. *Neuroscience* **62,** 641–645.
35. Gonon, F. (1997) Prolonged and extrasynaptic excitatory action of dopamine mediated by D1 receptors in the rat striatum in vivo. *J. Neurosci.* **17,** 5972–5978.
36. Jones, S. R., Garris, P. A., Kilts, C. D., and Wightman, R. M. (1995) Comparison of dopamine uptake in the basolateral amygdaloid nucleus, caudate-putamen, and nucleus accumbens of the rat. *J. Neurochem.* **64,** 2581–2589.
37. Nirenberg, M. J., Vaughan, R. A., Uhl, G. R., Kuhar, M. J., and Pickel, V. M. (1996) The dopamine transporter is localized to dendritic and axonal plasma membranes of the nigrostriatal dopaminergic neurons. *J. Neurosci.* **16,** 436–447.
38. Nirenberg, M. J., Chan, J., Pohorille, A., Vaughan, R. A., Uhl, G. R., Kuhar, M. J., and Pickel, V. M. (1997) The dopamine transporter: comparative ultrastructure of dopaminergic axons in limbic and motor compartments of the nucleus accumbens. *J. Neurosci.* **17,** 6899–6907.
39. Garris, P. A., Ciolkowski, E. L., and Wightman, R. M. (1994) Heterogeneity of evoked dopamine overflow within the striatal and striatoamygdaloid regions. *Neuroscience* **59,** 417–427.
40. Peters, J. L. and Michael, A. C. (2000) Changes in the kinetics of dopamine release and uptake have differential effects on the spatial distribution of extracellular dopamine concentration in rat striatum. *J. Neurochem.* **74,** 1563–1573.
41. Doucet, G., Descarries, L., and Garcia, S. (1986) Quantification of the dopamine innervation in adult rat neostriatum. *Neuroscience* **19,** 427–445.
42. Sesack, S. R., Hawrylak, V. A., Matus, C., Guido, M. A., and Levey, A. I. (1998) Dopamine axon varicosities in the prelimbic division of the rat prefrontal cortex exhibit sparse immunoreactivity for the dopamine transporter. *J. Neurosci.* **18,** 2697–2708.
43. Cass, W. A. and Gerhardt, G. A. (1995) In vivo assessment of dopamine uptake in rat medial prefrontal cortex: comparison with dorsal striatum and nucleus accumbens. *J. Neurochem.* **65,** 201–207.
44. Garris, P. A. and Wightman, R. M. (1994) Different kinetics govern dopaminergic transmission in the amygdala, prefrontal cortex, and striatum: an in vivo voltammetric study. *J. Neurosci.* **14,** 442–450.

45. Zigmond, M. J., Abercrombie, E. D., Berger, T. W., Grace, A. A., and Stricker, E. M. (1990) Compensations after lesions of central dopaminergic neurons: some clinical and basic implications. *Trends Neurosci.* **13,** 290–296.
46. Rothblat, D. S. and Schneider, J. S. (1994) Spontaneous functional recovery from parkinsonism is not due to reinnervation of the dorsal striatum by residual dopaminergic neurons. *Brain Res. Bull.* **34,** 309–312.
47. Schneider, J. S., Rothblat, D. S., and DiStefano, L. (1994) Volume transmission of dopamine over large distances may contribute to recovery from experimental parkinsonism. *Brain Res.* **643,** 86–91.
48. Chritin, M., Blanchard, V., Raisman, V. R., Feuerstein, C., Agid, Y., Javoy, A. F., and Savasta, M. (1996) DA uptake sites, D1 and D2 receptors, D2 and preproenkephalin mRNAs and Fos immunoreactivity in rat striatal subregions after partial dopaminergic degeneration. *Eur. J. Neurosci.* **8,** 2511–2520.
49. Zoli, M., Torri, C., Ferrari, R., Jansson, A., Zini, I., Fuxe, K., and Agnati, L. F. (1998) The emergence of the volume transmission concept. *Brain Res. Rev.* **29,** 136–147.
50. Rothblat, D. S. and Schneider, J. S. (1999) Regional differences in striatal dopamine uptake and release associated with recovery from MPTP-induced parkinsonism: an in vivo electrochemical study. *J. Neurochem.* **72,** 724–733.
51. Jaber, M., Dumartin, B., Sagne, C., Haycock, J. W., Roubert, C., Giros, B., Bloch, B., and Caron, M. G. (1999) Differential regulation of tyrosine hydroxylase in the basal ganglia of mice lacking the dopamine transporter. *Eur. J. Neurosci.* **11,** 3499–3511.
52. Cass, W. A., Zahniser, N. R., Flach, K. A., and Gerhardt, G. A. (1993) Clearance of exogenous dopamine in rat dorsal striatum and nucleus accumbens: role of metabolism and effects of locally applied uptake inhibitors. *J. Neurochem.* **61,** 2269–2278.
53. Garris, P. A., Walker, Q. D., and Wightman, R. M. (1997) Dopamine release and uptake rates both decrease in the partially denervated striatum in proportion to the loss of dopamine terminals. *Brain Res.* **753,** 225–234.
54. Fuxe, K. (1965) Evidence for the existence of monoamine neurons in the central nervous system. III. The monoamine terminal. *Z. Zellforsch.* **65,** 573–596.
55. Fuxe, K. (1965) Evidence for the existence of monoamine containing neurons in the central nervous system. IV. The distribution of monoamine terminals in the central nervous system. *Acta Physiol. Scand.* **64,** 39–85.
56. Steinbusch, H. W. M. (1981) Distribution of serotonin-immunoreactivity in the central nervous system of the rat: cell bodies and terminals. *Neuroscience* **6,** 557–618.
57. Descarries, L. and Umbriaco, D. (1995) Ultrastructural basis of monoamine and acetylcholine function in the CNS. *Semin. Neurosci.* **7,** 309–318.
58. Séguéla, P., Watkins, K. C., and Descarries, L. (1989) Ultrastructural relationships of serotonin axon terminals in the cerebral cortex of the adult rat. *J. Comp. Neurol.* **289,** 129–142.
59. Kosofsky, B. E. and Molliver, M. E. (1987) The serotoninergic innervation of cerebral cortex: different classes of axon terminals arise from dorsal and median raphe nuclei. *Synapse* **1,** 153–168.
60. Bunin, M. A. and Wightman, R. M. (1998) Quantitative evaluation of 5-hydroxytryptamine (serotonin) neuronal release and uptake: an investigation of extrasynaptic transmission. *J. Neurosci.* **18,** 4854–4860.
61. Bunin, M. A., Prioleau, C., Mailman, R. B., and Wightman, R. M. (1998) Release and uptake rates of 5-hydroxytryptamine in the dorsal raphe and substantia nigra reticulata of the rat brain. *J. Neurochem.* **70,** 1077–1087.
62. Bunin, M. A. and Wightman, R. M. (1999) Paracrine neurotransmission in the CNS: involvement of 5-HT. *Trends Neurosci.* **22,** 377–382.
63. Moukhles, H., Bosler, O., Bolam, J. P., Vallée, A., Umbriaco, D., Geffard, M., and Doucet, G. (1997) Quantitative and morphometric data indicate precise cellular inter-

actions between serotonin terminals and postsynaptic targets in rat substantia nigra. *Neuroscience* **76,** 1159–1171.
64. Descarries, L., Watkins, K. C., Garcia, S., and Beaudet, A. (1982) The serotonin neurons in nucleus raphe dorsalis of adult rat: a light and electron microscope radioautographic study. *J. Comp. Neurol.* **207,** 239–254.
65. Adell, A., Carceller, A., and Artigas, F. (1993) In vivo brain dialysis study of the somatodendritic release of serotonin in the Raphe nuclei of the rat: effects of 8-hydroxy-2-(di-n-propylamino)tetralin. *J. Neurochem.* **60,** 1673–1681.
66. Adell, A. and Artigas, F. (1998) A microdialysis study of the in vivo release of 5-HT in the median raphe nucleus of the rat. *Br. J. Pharmacol.* **125,** 1361–1367.
67. Héry, F., Faudon, M., and Ternaux, J. P. (1982) In vivo release of serotonin in two raphe nuclei (raphe dorsalis and magnus) of the cat. *Brain Res. Bull.* **8,** 123–129.
68. Matos, F. F., Urban, C., and Yocca, F. D. (1996) Serotonin (5-HT) release in the dorsal raphe and ventral hippocampus: raphe control of somatodendritic and terminal 5-HT release. *J. Neural. Transm.* **103,** 173–190.
69. Zhou, F. C., Tao, C. J., Segu, L., Patel, T., and Wang, Y. (1998) Serotonin transporters are located on the axons beyond the synaptic junctions: anatomical and functional evidence. *Brain Res.* **805,** 241–254.
70. Pickel, V. M. and Chan, J. (1999) Ultrastructural localization of the serotonin transporter in limbic and motor compartments of the nucleus accumbens. *J. Neurosci.* **19,** 7356–7366.
71. Tao-Cheng, J. H. and Zhou, F. C. (1999) Differential polarization of serotonin transporters in axons versus soma-dendrites: an immunogold electron microscopy study. *Neuroscience* **94,** 821–830.
72. Jansson, A., Tinner, B., Steinbusch, H. W. M., Agnati, L. F., and Fuxe, K. (1998) On the relationship of 5-hydroxytryptamine neurons to 5- hydroxytryptamine 2A receptor-immunoreactive neuronal processes in the brain stem of rats. A double immunolabelling analysis. *Neuroreport* **9,** 2505–2511.
73. Jansson, A., Tinner, B., Steinbusch, H. W. M., Agnati, L. F., and Fuxe, K. (1999) Relationships of the 5-hydroxytryptamine-immunoreactive nerve terminal network to 5-hydroxytryptamine 2A receptor-immunoreactive neuronal processes in the rat forebrain. *Abstr. Soc. Neurosci.* **25,** 1204.
74. Aghajanian, G. K. and McCall, R. B. (1980) Serotonergic synaptic input to facial motoneurons: localization by electron-microscopic autoradiography. *Neuroscience* **5,** 2155–2162.
75. Schaffar, N., Jean, A., and Calas, A. (1984) Radioautographic study of serotoninergic axon terminals in the rat trigeminal motor nucleus. *Neurosci. Lett.* **44,** 31–36.
76. Cornea-Hébert, V., Riad, M., Wu, C., Singh, S., and Descarries, L. (1999) Cellular and subcellular distribution of the serotonin 5-HT2A receptor in the central nervous system of adult rat. *J. Comp. Neurol.* **409,** 187–209.
77. Willins, D. L., Deutch, A. Y., and Roth, B. L. (1997) Serotonin 5-HT2A receptors are expressed on pyramidal cells and interneurons in the rat cortex. *Synapse* **27,** 79–82.
78. Wu, C., Yoder, E. J., Shih, J., Chen, K., Dias, P., Shi, L., et al. (1998) Development and characterization of monoclonal antibodies specific to the serotonin 5-HT2A receptor. *J. Histochem. Cytochem.* **46,** 811–824.
78a. Cornea-Hébert, V., Riad, M., Wu, C., Singh, S., and Descarries, L. Copyright © (1999 Wiley-Liss, Inc.) Cellular and subcellular distribution of the serotonin 5-HT2A receptor in the central nervous system of adult rat. *J. Comparat. Neurol.*
78b. Riad, M., Garcia, S., Watkins, K. C., et al. Copyright © (2000 Wiley-Liss, Inc.) Somatodendritic localization of 5-HT1A and preterminal axonal localization of 5-HT1B serotonin receptors in adult rat brain. *J. Comparat. Neurol.*
79. Willins, D. L., Alsayegh, L., Berry, S. A., Backstrom, J. R., Sanders-Bush, E., Friedman, L., Khan, N., and Roth, B. L. (1998) Serotonergic antagonist effects on trafficking of serotonin 5-HT2A receptors in vitro and in vivo. *Ann. NY Acad. Sci.* **861,** 121–127.

80. Willins, D. L., Berry, S. A., Alsayegh, L., Backstrom, J. R., Sanders-Bush, E., Friedman, L., and Roth, B. L. (1999) Clozapine and other 5-hydroxytryptamine-2A receptor antagonists alter the subcellular distribution of 5-hydroxytryptamine-2A receptors in vitro and in vivo. *Neuroscience* **91,** 599–606.
81. Riad, M., Tong, X. K., El Mestikawy, S., Hamon, M., Hamel, E., and Descarries, L. (1998) Endothelial expression of the 5-hydroxytryptamine1B antimigraine drug receptor in rat and human brain microvessels. *Neuroscience* **86,** 1031–1035.
82. Riad, M., Garcia, S., Watkins, K. C., Jodoin, N., Doucet, E., Langlois, X., et al. Copyright © (2000 Wiley-Liss, Inc.) Somatodendritic localization of 5-HT1A and preterminal axonal localization of 5-HT1B serotonin receptors in adult rat brain. *J. Comp. Neurol.* **417,** 181–194.
83. Sari, Y., Lefèvre, K., Bancila, M., Quignon, M., Miquel, M. C., Langlois, X., Hamon, M., and Vergé, D. (1997) Light and electron microscopic immunocytochemical visualization of 5-HT1B receptors in the rat brain. *Brain Res.* **760,** 281–286.
84. Sari, Y., Miquel, M. C., Brisorgueil, M. J., Ruiz, G., Doucet, E., Hamon, M., and Vergé, D. (1999) Cellular and subcellular localization of 5-hydroxytryptamine1B receptors in the rat central nervous system: immunocytochemical, autoradiographic and lesion studies. *Neuroscience* **88,** 899–915.
85. Oleskevich, S., Descarries, L., Watkins, K. C., Séguéla, P., and Daszuta, A. (1991) Ultrastructural features of the serotonin innervation in adult rat hippocampus: an immunocytochemical description in single and serial thin sections. *Neuroscience* **42,** 777–791.
86. Burnet, P. W., Mead, A., Eastwood, S. L., Lacey, K., Harrison, P. J., and Sharp, T. (1995) Repeated ECS differentially affects rat brain 5-HT1A and 5-HT2A receptor expression. *Neuroreport* **6,** 901–904.
87. Fernandes, C., McKittrick, C. R., File, S. E., and McEwen, B. S. (1997) Decreased 5-HT1A and increased 5-HT2A receptor binding after chronic corticosterone associated with a behavioural indication of depression but not anxiety. *Psychoneuroendocrinology* **22,** 477–491.
88. Ichikawa, J., Kuroki, T., Dai, J., and Meltzer, H. Y. (1998) Effect of antipsychotic drugs on extracellular serotonin levels in rat medial prefrontal cortex and nucleus accumbens. *Eur. J. Pharmacol.* **351,** 163–171.
89. Jagadeesh, S. R. and Subhash, M. N. (1998) Effect of antidepressants on intracellular Ca^{++} mobilization in human frontal cortex. *Biol. Psychiatry* **44,** 617–621.
90. Maes, M., Libbrecht, I., van Hunsel, F., Campens, D., and Meltzer, H. Y. (1999) Pindolol and mianserin augment the antidepressant activity of fluoxetine in hospitalized major depressed patients, including those with treatment resistance. *J. Clin. Psychopharmacol.* **19,** 177–182.
91. Staley, J. K., Malison, R. T., and Innis, R. B. (1998) Imaging of the serotonergic system: interactions of neuroanatomical and functional abnormalities of depression. *Biol. Psychiatry* **44,** 534–549.
92. Du, L., Bakish, D., Lapierre, Y. D., Ravindran, A. V., and Hrdina, P. D. (2000) Association of polymorphism of serotonin 2A receptor gene with suicidal ideation in major depressive disorder. *Am. J. Med. Genet.* **96,** 56–60.
93. Plein, H. and Berk, M. (2000) Changes in the platelet intracellular calcium response to serotonin in patients with major depression treated with electroconvulsive therapy: state or trait marker status. *Int. Clin. Psychopharmacol.* **15,** 93–98.
94. Fink, G., Sumner, B. E., McQueen, J. K., Wilson, H., and Rosie, R. (1998) Sex steroid control of mood, mental state and memory. *Clin. Exp. Pharmacol. Physiol.* **25,** 764–775.
95. Fink, G., Sumner, B., Rosie, R., Wilson, H., and McQueen, J. (1999) Androgen actions on central serotonin neurotransmission: relevance for mood, mental state and memory. *Behav. Brain Res.* **105,** 53–68.

96. De Vry, J. (1995) 5-HT1A receptor agonists: recent developments and controversial issues. *Psychopharmacology (Berl.)* **121,** 1–26.
97. Hamon, M. (1997) The main features of central 5-HT1A receptors, in *Serotonergic Neurons and 5-HT Receptors in the CNS* (Baumbarten, H. G. and Göthert, M., eds.), *Handbook Experimental Pharmacology* 129, Springer-Verlag, Berlin, pp. 239–268.
98. Blier, P. and de Montigny, C. (1999) Serotonin and drug-induced therapeutic responses in major depression, obsessive-compulsive and panic disorders. *Neuropsychopharmacology* **21,** 91S–98S.
99. Riad, M., Doucet, E., Hamon, M., and Descarries, L. (1999) 8-OH-DPAT internalizes 5-HT1A autoreceptors in serotonin neurons of nucleus raphe dorsalis. *Soc. Neurosci. Abstr.* **25,** 1466.
100. Le Poul, E., Laaris, N., Doucet, E., Laporte, A. M., Hamon, M., and Lanfumey, L. (1995) Early desensitization of somato-dendritic 5-HT1A autoreceptors in rats treated with fluoxetine or paroxetine. *Naunyn Schmiedebergs Arch. Pharmacol.* **352,** 141–148.
101. Bjelke, B. (1994) Volume transmission: experimental evidence for chemical communication via extracellular fluid pathways in the dopamine and β-endorphin neurons of the brain, *Academic Thesis*, Department of Neuroscience, Karolinska Institutet, Stockholm.
102. MacMillan, S. A. J., Mark, M. A., and Duggan, A. W. (1998) The release of β-endorphin and the neuropeptide-receptor mismatch in the brain. *Brain Res.* **794,** 127–136.
103. Elde, R., Arvidsson, U., Riedl, M., Vulchanova, L., Lee, J. H., Dado, R., et al. (1995) Distribution of neuropeptide receptors. New views of peptidergic neurotransmission made possible by antibodies to opioid receptors. *Ann. NY Acad. Sci.* **757,** 390–404.
104. Bowers, C. W. (1994) Superfluous neurotransmitters? *Trends Neurosci.* **17,** 315–320.
105. Commons, K. G. and Milner, T. A. (1996) Cellular and subcellular localization of delta opioid receptor immunoreactivity in the rat dentate gyrus. *Brain Res.* **738,** 181–195.
106. Hamel, E. and Beaudet, A. (1987) Opioid receptors in rat neostriatum: radioautographic distribution at the electron microscopic level. *Brain Res.* **401,** 239–257.
107. Wang, H., Moriwaki, A., Wang, J. B., Uhl, G. R., and Pickel, V. M. (1996) Ultrastructural immunocytochemical localization of μ opioid receptors and Leu^5-enkephalin in the patch compartment of the rat caudate-putamen nucleus. *J. Comp. Neurol.* **375,** 659–674.
108. Pieribone, V. A., Xu, Z. Q., Zhang, X., Grillner, S., Bartfai, T., and Hökfelt, T. (1995) Galanin induces a hyperpolarization of norepinephrine-containing locus coeruleus neurons in the brainstem slice. *Neuroscience* **64,** 861–874.
109. Zoli, M., Agnati, L. F., Fuxe, K., and Bjelke, B. (1989) Demonstration of NPY transmitter receptor mismatches in the central nervous system of the male rat. *Acta Physiol. Scand.* **135,** 201–202.
110. Caberlotto, L., Tinner, B., Bunnemann, B., Agnati, L. F., and Fuxe, K. (1998) On the relationship of neuropeptide Y Y1 receptor-immunoreactive neuronal structures to the neuropeptide Y-immunoreactive nerve terminal networks. A double immunolabelling analysis in the rat brain. *Neuroscience* **86,** 827–845.
111. Fuxe, K., Tinner, B., Caberlotto, L., Bunnemann, B., and Agnati, L. F. (1997) NPY Y1 receptor like immunoreactivity exists in a subpopulation of beta- endorphin immunoreactive nerve cells in the arcuate nucleus: a double immunolabelling analysis in the rat. *Neurosci. Lett.* **225,** 49–52.
112. Dumont, Y., Fournier, A., Quirion, R., et al. (1996) Autoradiographic distribution of [^{125}I]Leu31,Pro34]PYY and [^{125}I]PYY3-36 binding sites in the rat brain evaluated with two newly developed Y1 and Y2 receptor radioligands. *Synapse* **22,** 139–158.
113. Pickel, V. M., Beck-Sickinger, A. G., Chan, J., and Weiland, H. A. (1998) Y1 receptors in the nucleus accumbens: ultrastructural localization and association with neuropeptide Y. *J. Neurosci. Res.* **52,** 54–68.
114. Liu, H., Brown, J. L., Jasmin, L., Maggio, J. E., Vigna, S. R., Mantyh, P. W., and Basbaum, A. I. (1994) Synaptic relationship between substance P and the substance P

receptor: light and electron microscopic characterization of the mismatch between neuropeptides and their receptors. *Proc. Natl. Acad. Sci. USA* **91,** 1009–1013.
115. McLeod, A. L., Krause, J. E., Cuello, A. C., and Ribeiro-da-Silva, A. (1998) Preferential synaptic relationships between substance P-immunoreactive boutons and neurokinin 1 receptor sites in the rat spinal cord. *Proc. Natl. Acad. Sci. USA* **95,** 15,775–15,780.
116. Jakab, R. L., Hazrati, L. N., and Goldman, R. P. (1996) Distribution and neurochemical character of substance P receptor (SPR)-immunoreactive striatal neurons of the macaque monkey: accumulation of SP fibers and SPR neurons and dendrites in "striocapsules" encircling striosomes. *J. Comp. Neurol.* **369,** 137–149.
117. Baude, A. and Shigemoto, R. (1998) Cellular and subcellular distribution of substance P receptor immunoreactivity in the dorsal vagal complex of the rat and cat: a light and electron microscope study. *J. Comp. Neurol.* **402,** 181–196.
118. Zerari, F., Karpitskiy, V., Krause, J., Descarries, L., and Couture, R. (1998) Astroglial distribution of neurokinin-2 receptor immunoreactivity in the rat spinal cord. *Neuroscience* **84,** 1233–1246.
119. Zerari, F., Karpitskiy, V., Krause, J., Descarries, L., and Couture, R. (1997) Immunoelectron microscopic localization of NK-3 receptor in the rat spinal cord. *Neuroreport* **8,** 2661–2664.
120. Duggan, A. W., Schaible, H. G., Hope, P. J., and Lang, C. W. (1992) Effect of peptidase inhibition on the pattern of intraspinally released immunoreactive substance P detected with antibody microprobes. *Brain Res.* **579,** 261–269.
121. Beck, H., Schrock, H., and Sandkuhler, J. (1995) Controlled superfusion of the rat spinal cord for studying non-synaptic transmission: an autoradiographic analysis. *J. Neurosci. Methods* **58,** 193–202.
122. Hope, P. J., Jarrott, B., Schaible, H. G., Clarke, R. W., and Duggan, A. W. (1990) Release and spread of immunoreactive neurokinin A in the cat spinal cord in a model of acute arthritis. *Brain Res.* **533,** 292–299.
123. Jennes, L., Stumpf, W. E., and Kalivas, P. W. (1982) Neurotensin: topographical distribution in rat brain by immunohistochemistry. *J. Comp. Neurol.* **210,** 211–224.
124. Moyse, E., Rostene, W., Vial, M., Leonard, K., Mazella, J., Kitabgi, P., Vincent, J. P., and Beaudet, A. (1987) Distribution of neurotensin binding sites in rat brain: a light microscopic radioautographic study using monoiodo [^{125}I]Tyr3-neurotensin. *Neuroscience* **22,** 525–536.
125. Boudin, H., Pelaprat, D., Rostene, W., and Beaudet, A. (1996) Cellular distribution of neurotensin receptors in rat brain: immunohistochemical study using an antipeptide antibody against the cloned high affinity receptor. *J. Comp. Neurol.* **373,** 76–89.
126. Boudin, H., Pelaprat, D., Rostene, W., Pickel, V. M., and Beaudet, A. (1998) Correlative ultrastructural distribution of neurotensin receptor proteins and binding sites in the rat substantia nigra. *J. Neurosci.* **18,** 8473–8484.
127. Botto, J. M., Sarret, P., Vincent, J. P., and Mazella, J. (1997) Identification and expression of a variant isoform of the levocabastine-sensitive neurotensin receptor in the mouse central nervous system. *FEBS Lett.* **400,** 211–214.
128. Nouel, D., Sarret, P., Vincent, J. P., Mazella, J., and Beaudet, A. (1999) Pharmacological, molecular and functional characterization of glial neurotensin receptors. *Neuroscience* **94,** 1189–1197.
129. Vincent, J. P., Mazella, J., and Kitabgi, P. (1999) Neurotensin and neurotensin receptors. *Trends Pharmacol. Sci.* **20,** 302–309.
130. Dana, C., Vial, M., Leonard, K., Beauregard, A., Kitabgi, P., Vincent, J. P., Rostene, W., and Beaudet, A. (1989) Electron microscopic localization of neurotensin binding sites in the midbrain tegmentum of the rat. I. Ventral tegmental area and the interfascicular nucleus. *J. Neurosci.* **9,** 2247–2257.

131. Woulfe, J. and Beaudet, A. (1992) Neurotensin terminals form synapses primarily with neurons lacking detectable tyrosine hydroxylase immunoreactivity in the rat substantia nigra and ventral tegmental area. *J. Comp. Neurol.* **321,** 163–176.
132. Dournaud, P., Boudin, H., Schonbrunn, A., Tannenbaum, G. S., and Beaudet, A. (1998) Interrelationships between somatostatin sst2A receptors and somatostatin-containing axons in rat brain: evidence for regulation of cell surface receptors by endogenous somatostatin. *J. Neurosci.* **18,** 1056–1071.
133. Checler, F., Barelli, H., Kitabgi, P., and Vincent, J. P. (1988) Neurotensin metabolism in various tissues of central and peripheral origins: ubiquitous involvement of a novel neurotensin degrading metalloendopeptidase. *Biochimie* **70,** 75–82.
134. Vincent, B., Beaudet, A., Dauch, P., Vincent, J. P., and Checler, F. (1996) Distinct properties of neuronal and astrocytic endopeptidase 3.4.24.16: a study on differentiation, subcellular distribution, and secretion processes. *J. Neurosci.* **16,** 5049–5059.
135. Villeneuve, P., Seidah, N. G., and Beaudet, A. (1999) Immunohistochemical distribution of the prohormone convertase PC5-A in rat brain. *Neuroscience* **92,** 641–654.
136. Barbero, P., Rovere, C., De Bie, I., Seidah, N., Beaudet, A., and Kitabgi, P. (1998) PC5-A-mediated processing of pro-neurotensin in early compartments of the regulated secretory pathway of PC5-transfected PC12 cells. *J. Biol. Chem.* **273,** 25,339–25,346.
137. Hedlund, P. B., Yanaihara, N., and Fuxe, K. (1992) Evidence for specific N-terminal galanin fragment binding sites in the rat brain. *Eur. J. Pharmacol.* **224,** 203–205.
138. Hedlund, P. B., Finnman, U. B., Yanaihara, N., and Fuxe, K. (1994) Galanin-(1-15), but not galanin-(1-29), modulates 5-HT1A receptors in the dorsal hippocampus of the rat brain: possible existence of galanin receptor subtypes. *Brain Res.* **634,** 163–167.
139. Maes, M., Lin, A. H., Bonaccorso, S., Goossens, F., Van Gastel, A., Pioli, R., Delmeire, L., and Scharpe, S. (1999) Higher serum prolyl endopeptidase activity in patients with post-traumatic stress disorder. *J. Affect. Disord.* **53,** 27–34.
140. Noble, F., Smadja, C., Valverde, O., Maldonado, R., Coric, P., Turcaud, S., Fournie-Zaluski, M. C., and Roques, B. P. (1997) Pain-suppressive effects on various nociceptive stimuli (thermal, chemical, electrical and inflammatory) of the first orally active enkephalin-metabolizing enzyme inhibitor RB 120. *Pain* **73,** 383–391.
141. Fuxe, K., Li, X. M., Tanganelli, S., Hedlund, P., O'Connor, W. T., Ferraro, L., Ungerstedt, U., and Agnati, L. F. (1995) Receptor-receptor interactions and their relevance for receptor diversity. Focus on neuropeptide/dopamine interactions. *Ann. NY Acad. Sci.* **757,** 365–376.
142. Breslin, N. A., Suddath, R. L., Bissette, G., Nemeroff, C. B., Lowrimore, P., and Weinberger, D. R. (1994) CSF concentrations of neurotensin in schizophrenia: an investigation of clinical and biochemical correlates. *Schizophr. Res.* **12,** 35–41.
143. Lindström, L. H., Widerlöv, E., Bisette, G., and Nemeroff, C. (1988) Reduced CSF neurotensin concentration in drug-free schizophrenic patients. *Schizophr. Res.* **1,** 55–59.
144. Sharma, R. P., Janicak, P. G., Bissette, G., and Nemeroff, C. B. (1997) CSF neurotensin concentrations and antipsychotic treatment in schizophrenia and schizoaffective disorder. *Am. J. Psychiatry* **154,** 1019–1021.
145. Lahti, R. A., Cochrane, E. V., Roberts, R. C., Conley, R. R., and Tamminga, C. A. (1998) [^{3}H]Neurotensin receptor densities in human postmortem brain tissue obtained from normal and schizophrenic persons. An autoradiographic study. *J. Neural. Transm.* **105,** 507–516.
146. Wolf, S. S., Hyde, T. M., Saunders, R. C., Herman, M. M., Weinberger, D. R., and Kleinman, J. E. (1995) Autoradiographic characterization of neurotensin receptors in the entorhinal cortex of schizophrenic patients and control subjects. *J. Neural. Transm. General Section* **102,** 55–65.
147. Kinkead, B., Binder, E. B., and Nemeroff, C. B. (1999) Does neurotensin mediate the effects of antipsychotic drugs? *Biol. Psychiatry* **46,** 340–351.

148. Berrettini, W. H., Doran, A. R., Kelsoe, J., Roy, A., and Pickar, D. (1987) Cerebrospinal fluid neuropeptide Y in depression and schizophrenia. *Neuropsychopharmacology* **1,** 81–83.
149. Widerlöv, E., Lindström, L. H., Wahlestedt, C., and Ekman, R. (1988) Neuropeptide Y and peptide YY as possible cerebrospinal fluid markers for major depression and schizophrenia, respectively. *J. Psychiatr. Res.* **22,** 69–79.
150. Westrin, A., Ekman, R., and Traskman-Bendz, L. (1999) Alterations of corticotropin releasing hormone (CRH) and neuropeptide Y (NPY) plasma levels in mood disorder patients with a recent suicide attempt. *Eur. Neuropsychopharmacol.* **9,** 205–211.
151. Caberlotto, L. and Hurd, Y. L. (1999) Reduced neuropeptide Y mRNA expression in the prefrontal cortex of subjects with bipolar disorder. *Neuroreport* **10,** 1747–1750.
152. Caberlotto, L., Jimenez, P., Overstreet, D. H., Hurd, Y. L., Mathé, A. A., and Fuxe, K. (1999) Alterations in neuropeptide Y levels and Y1 binding sites in the Flinders Sensitive Line rats, a genetic animal model of depression. *Neurosci. Lett.* **265,** 191–194.
153. Peters, J., Van Kammen, D. P., Gelernter, J., Yao, J., and Shaw, D. (1990) Neuropeptide Y-like immunoreactivity in schizophrenia. Relationships with clinical measures. *Schizophr. Res.* **3,** 287–294.
154. Iritani, S., Kuroki, N., Niizato, K., and Ikeda, K. (2000) Morphological changes in neuropeptide Y-positive fiber in the hippocampal formation of schizophrenics. *Progr. Neuropsychopharmacol. Biol. Psychiatry* **24,** 241–249.
155. Mathé, A. A. (1999) Neuropeptides and electroconvulsive treatment. *J. Ect.* **15,** 60–75.
156. Antonijevic, I. A., Murck, H., Bohlhalter, S., Frieboes, R. M., Holsboer, F., and Steiger, A. (2000) Neuropeptide Y promotes sleep and inhibits ACTH and cortisol release in young men. *Neuropharmacology* **39,** 1474–1481.
157. Kramer, M. S., Cutler, N., Feighner, J., Shrivastava, R., Carman, J., Sramek, J. J., et al. (1998) Distinct mechanism for antidepressant activity by blockade of central substance P receptors. *Science* **281,** 1640–1645.
158. Rupniak, N. M. and Kramer, M. S. (1999) Discovery of the antidepressant and anti-emetic efficacy of substance P receptor (NK1) antagonists. *Trends Pharmacol. Sci.* **20,** 485–490.
159. DeWied, D. and Jolles, J. (1982) Neuropeptides derived from proopiocortin: behavioural, physiological and neurochemical aspects. *Physiol. Rev.* **62,** 976–1059.
160. Burbach, J. P., Wang, X. C., Ten Haaf, J. A., and De Wied, D. (1984) Substances resembling C-terminal vasopressin fragments are present in the brain but not in the pituitary gland. *Brain Res.* **306,** 384–387.
161. Vawter, M. P., De Wied, D., and Van Ree, J. M. (1997) Vasopressin fragment, AVP-(4-8), improves long-term and short-term memory in the hole board search task. *Neuropeptides* **31,** 489–494.

5

The Extracellular Matrix in Neural Development, Plasticity, and Regeneration

Jeremy Garwood, Nicolas Heck, Franck Rigato, and Andreas Faissner

1. INTRODUCTION: GENETIC MECHANISMS VS ENVIRONMENT

Differentiation and morphogenesis of neural tissues involve a diversity of interactions between neural cells and their environment. Many potentially important interactions occur with the extracellular matrix (ECM), a complex association of extracellular molecules organized into aggregates and polymers *(1–3)*.

In general terms, a distinction can be made in the development of the organism between events that are genetically determined and those that are more subject to localized environmental factors. The former apparently underlie the early stages of proliferation and stem cell determination and the initial migration out of the neuroepithelial zones of proliferation; the further the neuronal cells pass along the route of morphogenetic development, the greater is the role of the latter environmental cues in determining the cell's behavior.

Developing cells of the nervous system are embedded in complex fields of mechanical tension and diffusible biochemical signals. This constantly changing pattern of spatial and temporal information for each cell is primarily generated by the cells of the nervous system themselves, and represents the major environmental forces that drive the sequence of developmental processes. The dynamic interactions between these environmental influences and the cells' response to them account for the considerable transformations that the cells undergo during periods of growth and survival, migration and sorting, and morphological and biochemical differentiation.

This chapter presents an overview of the different proteins expressed in the ECM of the nervous system, together with some of the membrane-bound cell surface receptors with which they interact. The mechanisms underlying signaling between cells and the ECM are then discussed, concentrating on those mediated by the integrins, and by receptors that can modify intracellular protein tyrosine phosphorylation. Subsequent changes in response to such signals are illustrated by showing how these can result in adjustments to the cell's cytoskeleton. The molecular interactions between cells and the ECM are then related, through the underlying processes of cell proliferation, migration, and differentiation, to changes in the development and maintenance of

From: *The Neuronal Environment: Brain Homeostasis in Health and Disease*
Edited by: W. Walz © Humana Press Inc., Totowa, NJ

the morphology of nervous tissue. Finally, possible roles of the ECM in the pathology and regeneration of the nervous system are presented.

2. COMPOSITION AND STRUCTURE OF ECM

2.1. ECM Organization in General

ECM is the term generally given to any material produced by cells and secreted into the surrounding medium. The ECM can be divided into three component classes: fibrous elements, particularly collagen and elastin; link proteins, such as fibronectin and laminin, and space-filling molecules, principally glycosaminoglycan polymers (GAGs).

The ECM of connective tissue is particularly extensive, and determines the properties of the tissue: It may be mineralized to resist compression, as in bone, or predominantly composed of tension-resistant fibers, as in tendons.

Basement membrane is the ECM characteristically found under epithelial cells, and consists of two distinct layers: The basal lamina, immediately adjacent to the cells, is a product of the epithelial cells themselves, and contains collagen type IV; the reticular lamina is produced by fibroblasts of the underlying connective tissue, and contains fibrillar collagen.

ECM can be considered as both a structural support for the overall shape of a tissue, as a means of delimiting different tissue subtypes, e.g., cortical layers, and as a medium of cell–cell communication, filled with many kinds of signaling molecules. Hence, the ECM of tissue serves not only to hold the cells together in a tissue, and to give the tissue its characteristic shape and texture, but it also plays an important role in the development and functional maintenance of tissue, influencing the proliferation, survival, migration, and differentiation of cells *(4–9)*.

2.2. Neuronal ECM

Histology studies reveal two different general patterns for the formation and maintenance of the ECM. On the one hand, there are cells forming an epithelium, which are linked by both tight junctions and gap junctions, thereby excluding an extracellular space (ECS) between them, but which, as epithelial layers, are attached to basement membranes. On the other hand, cells that do not form an epithelium have an ECS between them, through which the development of the tissue and communication between cells is achieved.

Although some populations of astrocytes, beneath the meninges and at certain other locations in the central nervous system (CNS), are linked by gap junctions, with a subsequently narrow ECS between them *(10)*, the neurons and glial cells forming the CNS are mostly separated by ECS, filled with a dynamic ECM composed of a wide variety of proteins and carbohydrate structures *(11,12)*. In quantitative terms, it is estimated that this ECM could represent up to 20% of the total volume of the adult brain, with all of the potential roles in development and functional maintenance of the tissue that this implies *(13)*.

Although most of the ECM of the CNS seems to have a fairly loose structure, there are several regions where relatively organized connective tissue and basement membranes are formed, most notably the meningeal coverings: the dura mater, the arachnoid mater, and the pia mater, which are composed of interlacing collagenous bundles surrounded by fine elastic networks *(14)*.

The dura mater is the outermost layer, lining the skull. The arachnoid mater adheres to the inner surface of the dura mater, and constitutes the outer layer of the subarachnoid spaces, where the cerebrospinal fluid circulates. The cerebrospinal fluid is secreted by the choroid plexus, which is a highly irrigated epithelium, and there is connective tissue between the choroid endothelium of blood capillaries and the choroid epithelium. The nervous tissue and cerebrospinal fluid are separated by a layer of ependymal cells, forming an epithelium on a basement membrane *(15)*.

The pia mater contains the blood vessels, and directly covers the brain and the spinal cord. From here, the blood capillaries enter the nervous tissue. The endothelial cells comprising the capillary walls are linked by tight junctions, to form a restrictive blood–brain barrier *(16)*. A basement membrane is formed between the vascular endothelial cells and the endfeet of the surrounding astrocytes. The marginal layer of astrocytes also sends processes to the external surface of the CNS, where there is a basement membrane with the fibroblasts of the pia mater, forming the external glia limitans.

In compositional terms, the ECM of the CNS is relatively rich in chondroitin sulfate proteoglycans and hyaluronan (HA) but poor in fibrous elements. It is also possible to extract many of the elements of the ECM from tissue homogenates, using physiological buffers suggestive of a much more open structure than that found in other tissues. Indeed, fixation techniques employed in electron microscopy have often removed the ECM from the CNS, a problem which at one time led to doubts as to the very existence of ECM there *(17–19)*.

In contrast to the CNS, the peripheral nervous system (PNS) has a highly organized ECM, because of the need to protect peripheral nerves with sleeves of fibrous connective tissue as they traverse the rest of the body. Nerves are made up of many peripheral axons, embedded in many layers of connective tissue. Each axon is covered by Schwann cells and a basal membrane that separates each axonal fiber. These nerve fibers are in turn surrounded by capillaries and collagen fibers, which together form a very loose connective tissue called "endonerve." Bundles of endonerve and nerve fibers are surrounded by perinerve, composed of epithelial cells and fibroblasts. The perinerve is finally enveloped by the epinerve, a connective tissue containing fibroblasts, blood vessels, lymphatic vessels, and fatty tissue. Ganglia contain the cell bodies of the peripheral nerves, covered by satellite Schwann cells and a basal membrane. This is covered by the endoganglional connective tissue, composed of fibroblasts, and an extension of the perinerve.

2.3. Type of Molecules Present in the Extracellular Space (ECS)

The ECM in neural tissues is a complex and dynamic entity composed of many types of molecules that have distinct patterns of spatial and temporal expression. Many ECM components, originally discovered in nonneural tissues, are also present in developing neural tissues, including fibronectin, laminin, vitronectin, collagens, proteoglycans (PGs), and thrombospondin *(20–28)*.

The ECS is filled with an assortment of molecules, ranging from small ionic species to huge polymers attaining sizes of several million Daltons. The ECM is composed of proteins and carbohydrate polymers, which are secreted into the space between the cells, and which are likely to contribute to the higher organization structure of the ECM, by interacting with each other and with the surfaces of the cells. The interactions of the molecules of the ECM with the membrane-bound receptors on the cell surface, provides structural coherence to the organization of the tissue, and a means of transmitting extracellular signals into the cells, or, alternatively, of sending signals from cells via molecules in the ECM.

As a general feature, the proteins described tend to be of a modular nature, that is, they are composed of varying combinations of conserved protein domains that occur in other extracellular proteins, the number and juxtaposition of these domains tending to be characteristic of particular protein families *(29–33)*.

In addition, there is extensive glycosylation of many of these proteins, the most extreme example being some of the PGs, in which the sugar modifications account for the majority of the molecule.

2.3.1. *Proteoglycans and GAGs*

PGs are proteins that have been posttranslationally modified by the covalent addition of at least one GAG chain. Depending on the type of GAG they carry, PGs can be divided into three general classes: the CSPGs, heparan sulfate PGs (HSPGs), and keratan sulfate PGs (KSPGs) *(34–37)*.

GAG chains are polymers of disaccharides that are covalently linked to the core protein: chondroitin sulfate (CS) has a disaccharide unit composed of D-glucuronic acid and *N*-acetyl-galactosamine (Fig. 1; dermatan sulfate is a form of CS, in which there is epimerization of the uronic acid to iduronic); heparan sulfate (HS) has a disaccharide unit composed of D-glucuronic acid and *N*-acetyl-glucosamine; and keratan sulfate (KS) has a disaccharide of D-galactose and *N*-acetylglucosamine. There is in fact another class of glycosaminoglycan, HA, with a disaccharide composed of glucuronic acid linked to *N*-acetylglucosamine, although, unlike the first three GAG types, HA is not bound covalently to proteins, and it is not sulfated. However, the sugars of most GAGs are further chemically modified by O-sulfation, *N*-deacetylation followed by *N*-sulfation, and/or epimerization (isomerization) of glucuronic acid to iduronic acid. Because each disaccharide in a GAG chain may be modified to a different degree, the large-scale structures of GAGs can be exceedingly complex, e.g., an HS disaccharide can be modified in up to five ways: hence, a hexasaccharide could have over 30,000 possible structures. Each disaccharide is repeated 40–100× in a single GAG chain. Because one disaccharide is 1 nm in length (with a mol wt of 440 Daltons), and the chains are fairly rigid linear structures, one can imagine the huge size of the chains and the resulting PGs. In addition, the abundance of negative charges from the acidic and sulfate groups on the GAG makes them very hydrophilic, hence the hydrodynamic properties of PGs.

CS, HS, and KS GAG chains are usually attached to the core protein on serine residues, although a clear consensus sequence has not yet been established *(38)*. The GAG chain begins with the addition of a tetrasaccharide composed of a xylose, two galactose sugars, and a glucuronic acid. However, *O*-glycosylation of threonine residues via a *N*-acetylglucosamine has also been shown for some KS, and corneal KS is *N*-linked to asparagine.

The number of GAG chains carried by a protein can vary from a single chain to many dozens (e.g., aggrecan). Some PGs can bear more than one type of GAG chain at the same time (e.g., syndecan has HS and CS; and phosphacan can be found with both

CS and KS). Although most PGs are always expressed as PGs, there are also some part-time PGs, that is, some proteins can be expressed either with or without the addition of GAG chains (e.g., amyloid precursor protein [APP]/appican *[39]*). In addition, there may be cell-type and stage-specific variations in the number of GAG chains present on a PG, e.g., the addition of KS to phosphacan *(40)*. Also, there is often further *N*- and *O*-linked glycosylation of the PG core protein with other classes of oligosaccharide. Hence, PGs can be very complex proteins, presenting a wide variety of sites for interaction with other molecules in the ECM *(41–44)*.

At least 25 different PG core proteins have been detected in preparations of developing and adult rat brain plasma membrane *(45)*, primarily substituted with HS. Many PGs, primarily substituted with CS, can be extracted from brain, using physiological buffers without detergent, indicating a loose association with the ECM.

2.3.1.1. Hyaluronic Acid/Hyaluronan

HA is the only GAG chain not covalently bound to a protein core. It is synthesized on the inner surface of the plasma membrane and secreted during its elongation; the synthesis of other GAG chain types is achieved during exocytosis through the Golgi. HA is evolutionarily a well-conserved molecule, being found even in bacteria. As a linear GAG polymer, it can attain sizes of several million Daltons, presenting many sites of interaction for HA-binding proteins *(17,46)*.

HA is bound noncovalently by a range of EC proteins, including the CSPGs from the aggrecan/lectican family, link protein, and brain enriched HA-binding protein (BEHAB). The C-lectins, including the lecticans, bind HA via the PG tandem repeat binding site. Link protein, first characterized in cartilage, but also apparently expressed in mouse brain *(47)*, possesses separate binding sites for HA and aggrecan, and hence can link HA and aggrecan together. Brevican/BEHAB is specific to the CNS *(48–50)*. Other HA-binding proteins include hyaluronectin and the glial HA-binding protein *(17)*. Hyaluronectin (68 kDa) is found in the Ranvier's node of adult rat nerves, in myelinated peripheral nerves, and is expressed by oligodendrocytes in vitro. Glial HA-binding protein (60–70 kDa, 45 kDa after deglycosylation) is specific to the CNS, and is found mainly in the white matter. It is identical to the HA-binding region of Versican.

The first receptor identified for HA was CD44, a transmembrane protein involved in immunological functions *(51)*. CD44 acts on small G proteins to regulate lamellipodial outgrowth in epithelial cells *(52)*. More recently, another receptor for HA has been identified and cloned: murine cell surface receptor for hyaluronan-mediated motility (RHAMM), which is a glycosyl-phosphatidyl-inositol (GPI)-anchored protein. RHAMM is implicated in the motility of astrocytes and microglia, and in axon growth *(53–55)*. RHAMM could be an oncogene *(56)*. HA has been recently found *(46)* in the cytosol, and putative functions for this intracellular localization have been suggested. Intracellular hyaluronicacid-binding protein (IHABP), hsp90, cdc37, and an intracellular form of RHAMM are intracellular receptors for HA, but these results still need to be confirmed in the context of the nervous system.

2.3.1.2. Small Leucine-Rich PGs

These PGs constitute a family of structurally related PGs arising from distinct genes. More than 70% of their core protein is occupied by leucine-rich repeats. Decorin and biglycan are two members of this family that are associated with the CNS *(41,57)*.

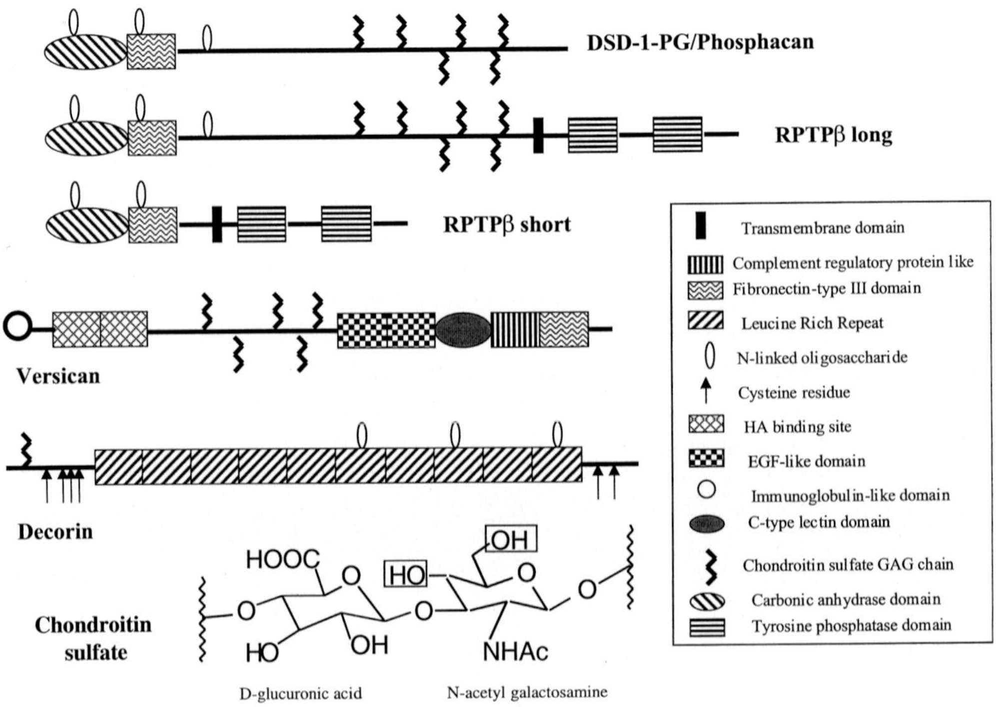

Fig. 1. Modular structure of PGs. Represented are three CSPGs and the disaccharide monomer of CS. The GAG chain is formed from the association of the disaccharide, and can be subsequently modified by selective sulfation of N-acetyl galactosamine at the boxed positions shown, or by epimerization of the GA. The GAG chain is covalently attached to the protein core at certain serine residues. Decorin has a single CS GAG chain; versican and phosphacan have several. DSD-1-PG/phosphacan is a secreted form of a transmembrane receptor tyrosine phosphatase, RPTPβ, which occurs in a long and a short form, as indicated. The protein cores of PGs have modular domain structures resembling those found in other EC proteins, e.g., FN III and Ig-like domains.

Biglycan has a 38-kDa core protein that carries two CS GAG chains, is expressed by astrocytes, and is associated with the cell surface or pericellular matrix. Biglycan has been shown to bind transforming growth factor β (TGF-β) and the epidermal growth factor (EGF) receptor, thereby playing a role in cell cycle regulation. It can also bind fibronectin and thrombospondin.

Decorin has a 36-kDa core protein that carries a single CS GAG chain (Fig. 1). It is expressed by astrocytes, and upregulated in the context of lesions. In connective tissue, decorin can modify the kinetics of collagen fibril formation, and affects the morphology of collagen fibrils, although its role in the CNS is not known *(58)*.

2.3.1.3. Lectican/Aggrecan Family

These CSPGs are characterized by HA-binding domains at their N-termini and C-type lectin-like domains at their C-terminal ends. The central regions of these proteins have no apparent homology with each other, and are considered to represent those parts of all proteins of this family where glycosaminoglycan chains are attached *(41–44,59–61)*.

This family includes aggrecan, the archetypal PG. Aggrecan is a very large CSPG found primarily in cartilage and connective tissue, but which also occurs in the CNS. It is heavily modified with GAG chains, and has a bottlebrush-like appearance, when observed by electron microscopy using rotary shadowing.

Neurocan is a CSPG specific to the CNS, which has been found to co-localize in brain with the neural cell adhesion molecules, NCAM and neuron-glia (Ng)-CAM, and with tenascin during certain developmental stages, and to interact with these morphoregulatory proteins in binding and aggregation inhibition assays. It bears 2–3 chondroitin sulfate chains and 30–40 sialylated *O*-linked oligosaccharides. During the first postnatal month, neurocan is increasingly proteolytically processed in the central region of the molecule, resulting in the generation of a distinct C-terminal fragment, neurocan-C. This fragment has been shown to retain the ability to interact with the neural cell adhesion molecules and to inhibit their homophilic interaction *(44,62)*.

Brevican (BEHAB) is a part-time CSPG (145 kDa), of which a truncated form (80 kDa) has also been detected (in bovine brain). In rat brain, two isoforms are known, sharing 82% homology with bovine brevican. One of these isoforms may be modified with a GPI-anchor *(63)*. Brevican is brain-specific, and is expressed by astrocytes *(48,50)*.

Versican is a CSPG (Fig. 1; *64*), which can also be expressed as a brain-specific splice variant named "V2" *(65)*. It is associated with myelin fibers, and it has been found to be inhibitory for neurite growth in vitro *(66,67)*. In addition, versican is implicated in the regulation of cell adhesion and proliferation *(68,69)*.

2.3.1.4. DSD-1-PG/Phosphacan/RPTP-beta

Phosphacan (also known as DSD-1-PG *[70]* and 6B4-PG *[71]*) is a large CSPG (180-kDa protein core, but up to 800 kDa with glycosylation), which represents the secreted isoform of a receptor protein tyrosine phosphatase, RPTP-β *(72,73)* (also called PTP-ζ; Fig. 1). At its N-terminus, there is a carbonic anhydrase-like domain, followed by a single fibronectin type III domain, although the remainder of the protein has no apparent strong homology with other known proteins. It is equivalent to the entire EC portion of the long form of RPTP-β up to its transmembrane domain. There is a shorter variant of the receptor, in which the second half of the EC region is absent. Both forms of RPTP-β possess two IC tyrosine phosphatase domains, only the first of which seems to be active *(74)*.

In addition to CS GAG chains, of which there are at least four with a mol wt of ~25 kDa each, phosphacan can possess keratan sulfate GAG chains (of ~10 kDa), and a variety of smaller N-linked sugars, including the HNK-1 epitope and Lewis-X *(40,75)*.

Studies of the distribution of both the phosphacan mRNA *(76)* and of the expressed protein *(77)* show that, for example, at E13-16, the phosphacan mRNA is mostly confined to areas of active cell proliferation, such as the ventricular zone of the brain and the ependymal layer surrounding the central canal of the spinal cord. Also, although the mRNA is mostly in the neuroepithelium of the embryonic brain and spinal cord, the protein is widely distributed in these tissues, presumably as a consequence of transport in or along glial processes, local secretion, and/or redistribution as a consequence of cell migration *(77)*. Based on the punctual expression pattern of phosphacan throughout the developing nervous system, it has been proposed *(71,77)* that it may play a role in neuronal migration, differentiation, and circuit formation.

Phosphacan can bind specifically to cell adhesion molecules, including NCAM, NrCAM, NgCAM *(72)*, F3/contactin, and TAG-1/axonin-1 *(78)*, as well as many other cell surface and ECM proteins, including tenascin-C, tenascin-R, heparin-binding growth-associated molecule (HB-GAM/pleiotrophin), and tissue plasminogen activator (tPA/plasmin). Some of these interactions are mediated through sugar residues, e.g., sialylated complex-type oligosaccharides in the CA and FN III domains are necessary for binding to NgCAM/L1, NCAM, and tenascin-C *(79,80)*.

The distribution of DSD-1-PG/phosphacan during development has been found to correspond to regions related to the formation of axonal trajectories. In this respect, it may play either a neurite-promoting role, as in the interrhomberic boundaries in chick *(81)*, or an inhibitory role, which would correspond to its presence in glial barrel field boundaries in the developing somatosensory cortex of mouse *(82)*.

Neurite-outgrowth studies with cell culture systems have demonstrated that phosphacan can play both outgrowth-promoting and inhibitory roles. Outgrowth-promotion from embryonic hippocampal neurons was shown to be dependent on a CS GAG chain motif, recognized by the monoclonal antibody, 473HD; inhibition of laminin-promoted outgrowth from dorsal root ganglion explants was associated with the core glycoprotein *(70)*.

Several studies have reported the enhanced expression of growth-inhibiting CSPGs in the context of CNS lesions (*83–86*; *see* Subheading 7.1.), and DSD-1-PG is also strongly upregulated upon wounding in the CNS *(87–89)*.

2.3.1.5. Syndecans and Glypicans

Syndecans and glypicans are cell surface HSPGs bearing a variable number of HS GAGs, and sometimes CS chains (Fig. 3).

The syndecan family contains four members (syndecan-1/syndecan, syndecan-2/fibroglycan, syndecan-3/N-syndecan, syndecan-4/ryudocan), possessing homologous transmembrane and short cytoplasmic domains with four well-conserved tyrosine residues *(90–92)*. The cytoplasmic tail of syndecan-1 interacts with intracellular microfilaments, and that of syndecan-4 with focal adhesion molecules, e.g., through the PDZ domains of syntenin, and of the serine kinase, CASK-LIN-2 *(93–95)*.

The glypican family is composed of five members (glypican-1/glypican, glypican-2/cerebroglypican, glypican-3/OCI-5, glypican-4/K-glypican, glypican-5). Glypicans are linked to the cell surface by a GPI anchor at the carboxyl-terminus *(96–98)*. They possess an EC region, with GAG attachment sites and 14 invariant cysteine residues, which stabilize a compact tertiary structure. They are mainly targeted to apical surfaces, a process partially dependent on glycosylation. In neural tissues, syndecan-3 is the most expressed HSPG receptor, but syndecan-4 and glypican-4 are also expressed.

2.3.1.6. NG2

NG2 is a CSPG that exists in three forms: one transmembrane, full-length form of 300 kDa; and two shorter forms, one of 275 kDa, which remains attached to the membrane, and one of 290 kDa, which is secreted into the ECM *(35,41,42,99,100)*. However, these different forms arise from EC proteolytic cleavage, and not from mRNA splicing. In vivo, NG2 is expressed in oligodendrocyte precursor cells/platelet-derived growth factor (PDGFα) receptor-positive. In adult rodents, some NG2-positive cells can be observed, which express the PDGFα receptor but not glial fibrillary acidic protein; hence, some authors (101) argue that these cells are either a pool of precursors of oligodendrocytes still present in the adult brain, or a new type of astrocyte. The functions of NG2 have been studied in vitro: the PG inhibits neurite outgrowth of cerebellar neurons and dorsal root ganglions in the presence of laminin *(102)*. NG2 could be involved in the inhibition of axonal regeneration, because it is upregulated in sites of lesion after brain injury *(103,409)*. The molecular mechanisms of NG2 action are unknown; no clear evidence for the existence of any receptor has been given *(104)*, but the range of interactions with other molecules from the ECM is better known: NG2 can bind to different types of collagen, to tenascin, and laminin *(105,106)*.

2.3.1.7. Appican

Appican is the CSPG form of amyloid precursor protein (APP) *(39,41,42,107,108)*, and is expressed by certain transformed cell lines of neural origin, namely, C6 cells and N2a neuroblastomas, and is detected in both human and rat brain, and, in primary cultures, is expressed by astrocytes, but not neurons. The core protein of appican has been shown to be an alternatively spliced isoform of APP, lacking exon 15 of the APP gene; splicing out of exon 15 results in the joining of exons 14 and 16, and formation of a CS GAG chain attachment site at serine 619, which lies 16 amino acids upstream of the Abeta peptide sequence, implicated in the pathology of Alzheimer's disease (AD). Levels of appican expression could be regulated by growth conditions independent of APP, suggesting that these molecules may serve distinct physiological roles within the cell. Morphological changes were also observed in both astrocytic and transformed cell cultures, suggesting that appican may be a CAM. Appican-enriched ECM was also observed to serve as a better substrate for attachment of N2a neuroblastomas, pheochromocytoma PC12 cells, and primary astrocytes, compared to APP-enriched ECM.

2.3.2. *Glycoproteins*

2.3.2.1. Fibronectin

Fibronectin is a glycoprotein that can mediate or modulate cell adhesion, morphology, and migration *(109–111)*. Fibronectin is composed of three types of repeating unit, and these structural modules are termed types I, II, and III (Fig. 2). These modular binding units can bind to cells, as well as to other key EC molecules; hence FN type I binds to heparin (HSPGs), fibrin, and collagen; FN II binds to collagen, and FN type III can bind to heparin and cell surfaces, but not to collagen. Various pairings of these binding domains can account for the ability of FN to form crosslinks between different ECM proteins, as well as its ability to mediate cell adhesion to collagen or fibrin.

Alternative splicing of the single FN gene results in the insertion or absence of three FN III domains, ED-A (EIIIA), ED-B (EIIIB), and IIICS, processing of the latter domain being more complex, occurring at three sites.

There are six sites in the FN molecule capable of mediating cell adhesion, all of which occur in the type III domains. These can be subdivided into three regions. First, there is the central binding domain, which contains the amino acid sequence, Arg-Gly-Asp (RGD), a key recognition unit for integrin receptor binding, found in a number of EC adhesion proteins. However, full receptor binding specificity and strength of binding to the major FN receptor requires another region containing the sequence, Pro-His-Ser-Arg-Asn, which functions in synergy with this basic adhesion sequence. Second, there are two sites in the alternatively spliced IIICS region,

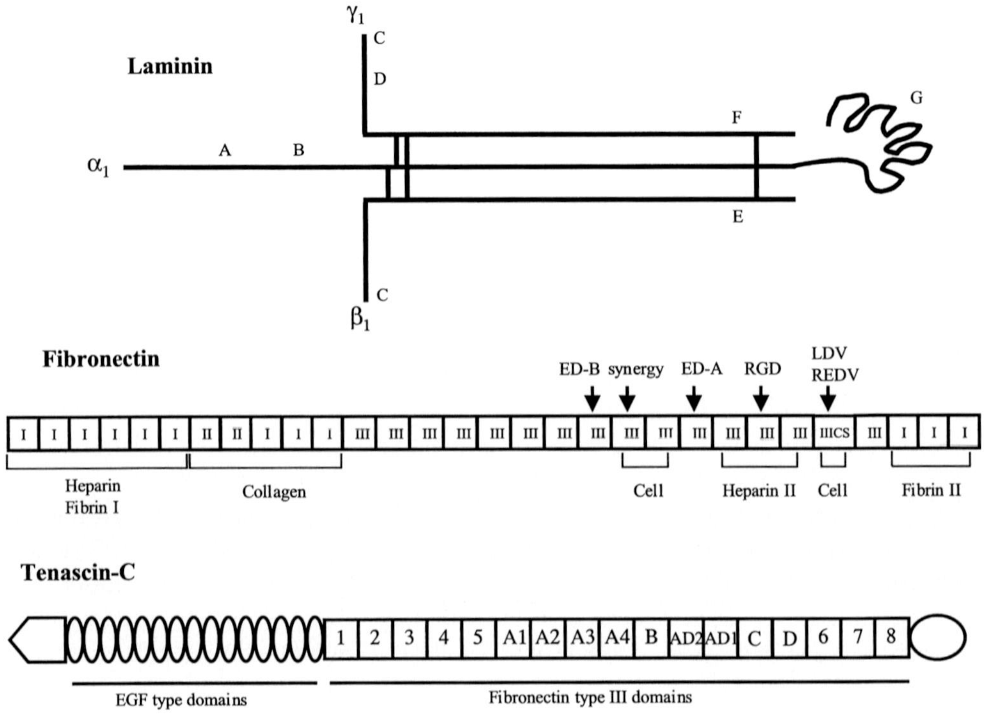

Fig. 2. Modular structure of GPs in the ECM. LAM is a heterotrimer composed of α, β, and γ chains. Lines connecting the monomers represent disulfide bridges. A = RGD sequence; B = cell interaction site; C = Collagen type IV interaction site; D = Nidogen interaction site; E = α6β1 integrin cell interaction site; F = P20 sequence involved in neuronal adhesion; G = Heparin interaction site, which promotes aggregation of laminin into a noncovalently associated polymer. FN is composed of multiple repeats of three types of structural module, termed I, II, III. Interaction sites with other molecules and the sites of cell interaction are indicated. Arrows show specific peptide sequences. FN forms a dimer of nonidentical subunits linked by disulfide bridges. Human TN-C is composed of repeats of EGF-type and FN III sequences. The FN III repeats are numbered, and alternative splicing occurs between numbers 5 and 6. The N-terminus is composed of a Cys-rich region; the C-terminus has a fibrinogen-like sequence.

with the consensus peptides, Leu-Asp-Val (LDV) and Arg-Glu-Asp-Val (REDV). These can confer cell-type specificity to the FN recognition; hence, although most cells adhere to the RGD and synergy sequences, adhesive recognition of the latter two IIICS sites is more restricted. For example, cells derived from the neural crest, such as those of the PNS (neurons from the sympathetic and sensory ganglia), bind to LDV, as do lymphocytes. Finally, there is the heparin-binding domain, containing two peptide sequences recognized by melanoma cells.

FN protein can be modified by glycosylation (mostly by N-linked sugars), phosphorylation, and sulfation. The FN monomer can form multimers by disulphide bonding at the carboxyl-terminus. From these, it can form detergent-insoluble fibrils. This is achieved by disulfide bridges or by transglutaminase crosslinking. For matrix assembly, the central binding domain and the N-terminal part are involved, because fibrillogenesis is inhibited by specific antibodies directed against these regions.

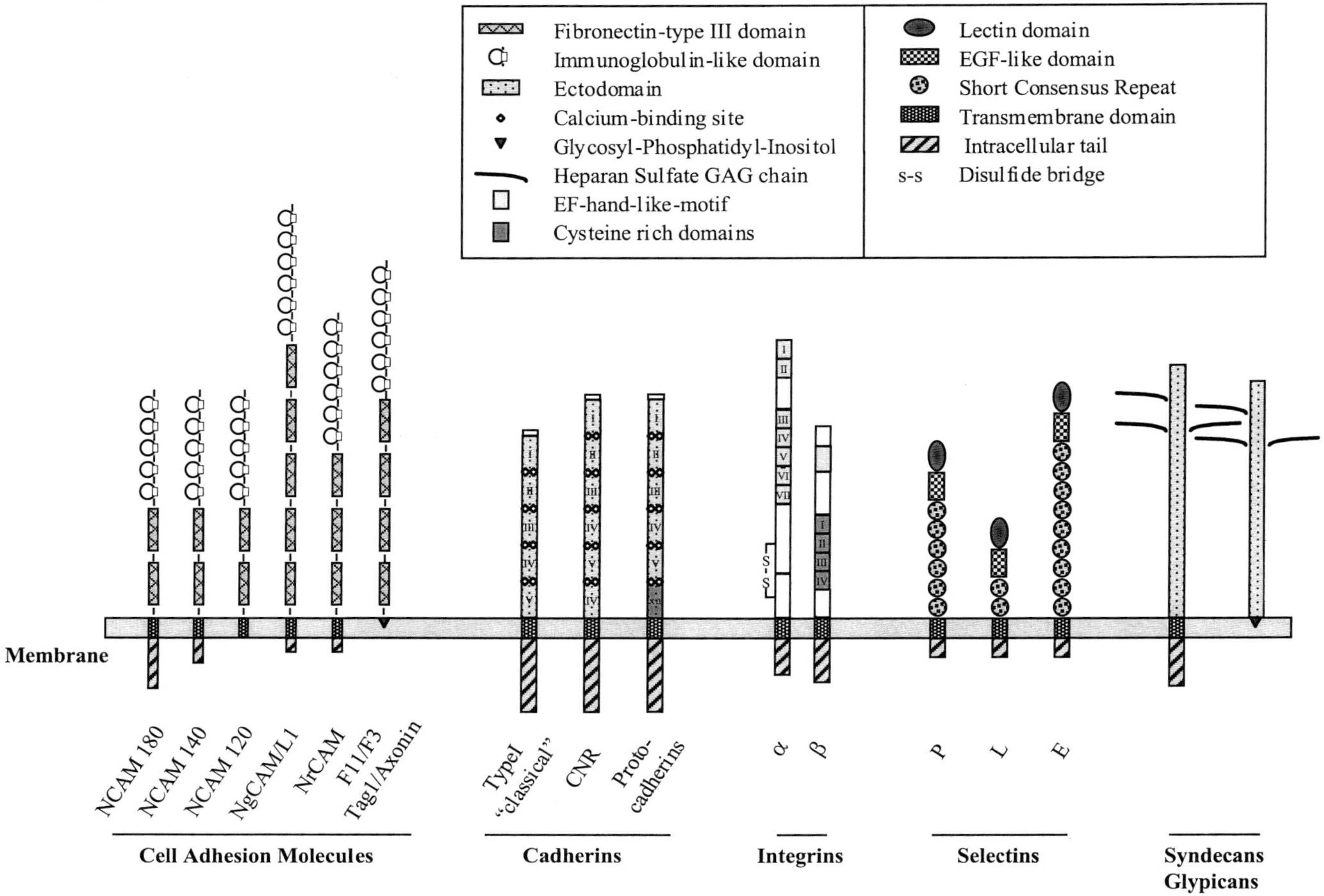

Fig. 3. Schematic presentation of the modular structure of some of the membrane-bound proteins in contact with the ECM.

Fibrillogenesis is inhibited by an antibody directed against the α5β1 integrin, leading to the hypothesis that some receptor-like system is involved between cells and secreted FN during the formation of an organized ECM.

2.3.2.2. LAMININ

There are at least seven structurally distinct forms of laminin, of which the most abundant form is classic LAM, and they are mostly found as prominent constituents of basement membranes *(112–115)*. LAM is a 850-kDa heterotrimeric glycoprotein, containing three noncovalently associated chains, termed α (400 kDa), β (210 kDa) and γ (200 kDa) (Fig. 2). Classic LAM (LAM-1) comprises α1, β1, and γ1 (formerly called A, B1, and B2). Each monomer is expressed by a single gene giving rise to different isoforms by alternative splicing. Each of the chains forms a short arm; the long arm consists of all three subunits intertwined in a rigid coiled-coil structure terminated by a globular terminal domain. 25% of the total weight of LAM results from carbohydrate moieties. The short arms of all monomers contain an EGF repeat.

Like fibronectin, LAM plays important roles in mediating the adhesion and migration of cells, although LAM seems to be more specialized for neuronal and epithelial cells. LAM is principally expressed in basement membranes where it is organized in a complex polymerized structure associated with other ECM molecules, such as nidogen, perlecan, and collagen IV *(116)*, but can appear transiently elsewhere to play a role in other processes, such as neurite outgrowth. LAM-1 could be implied in the promotion of axon growth, and the α6β1 integrin, known as a receptor for LAM, plays critical roles in mammalian neural development *(117,118)*.

LAM contains differents domains associated with different functions. There are a number of cell-binding sites, including a RGD sequence on the α chain, and a neurite interaction site, which involve all monomers. There is also a central cell binding site, less active for binding neurons, which could be a cryptic domain. Indeed, several reported peptide-adhesive sites appear to be cryptic, and probably require proteolysis or denaturation to be exposed or released. Unlike FN, LAM can also bind to cells through sulfated glycolipids or cell-surface glycosyl transferases.

The HS (or HSPG)-binding domain of the multiglobular part of the A chain is important for basement membrane integrity. LAM can also bind type IV collagen and entactin, other major structural components of basement membranes. Entactin (also named nidogen) binds the short arm of LAM in a noncovalent manner, with a stoichiometry of 1:1, forming a highly stable complex. This binding is such that purified LAM is in fact usually a LAM–entactin complex *(119)*.

The other forms of LAM include merosin (LAM-2) with the structure α2, β1, γ1, which is particularly present in the brain and PNS, synaptic- or S-LAM (LAM-3), with the structure α1, β2, γ1, which is enriched in synaptic sites, and which binds to ciliary ganglion neurons, using a Leu-Arg-Glu (LRE) sequence present in the neurite interaction site, and S-merosin (LAM-4), with the structure α2, β2, γ1, which is expressed by Schwann cells.

Cell-surface receptors for LAM include the integrins *(120)*, although there are many nonintegrin-binding proteins, such as lectins, elastin, dystroglycan *(121)*, cranin (a 120-kDa form of dystroglycan, whose LAM binding is inhibited by heparin: Cranin is localized in synapses, and contains carbohydrates, including HNK-1) *(122,123)*, and galactosyl transferase, an enzyme localized on growth cones, which transfers galactose residues from a donor (UDP-galactose) to an acceptor (N-acetyl-glucosamine) located on LAM, especially in the E8 domain. This enzyme plays a role in PC12 neurite outgrowth, and it may also change LAM-binding cell properties by changing the number of lectin binding sites on the molecule.

LAM α2 is expressed in the brain, where it seems to play critical roles, because the mutation affecting its gene cause severe neurological defects. Abnormal myelination is observed in these mutants, but the mechanisms responsible for such defects are unclear *(124–126)*.

2.3.2.3. Thrombospondin and F-Spondin

The thrombospondin (TSP) family comprises five members: TSP-1–4 and COMP (now renamed TSP-5) *(28,127)*. All TSPs contain EGF repeats (named "TSP type 2 repeat") and seven calcium-binding domains (TSP type 3 repeat) at their carboxyl-terminus. TSP-1 and TSP-2 contain, in addition, a procollagen homology domain and three TSP type 1 repeats. TSP-1 and 2 are assembled as homotrimers; TSP-3–5 are pentamers. All TSPs but TSP-5 are expressed in the developing nervous system.

The interaction of TSP-1 with cellular and EC binding molecules has been extensively studied. The N-terminus of TSP contains a heparin binding domain, which in addition to membrane-inserted HSPGs, such as syndecan, can bind sulfated lipids on cell surfaces. TSP inhibits adhesion of FN to cells. It also contains an RGD sequence in its last calcium binding domain, and this probably explains the binding of TSP to integrins αVβ3 and α2β3 *(128,129)*. The α3β1 integrin is another receptor for TSP, through which it can promote neurite growth *(130)*. TSP is expressed by microglia *(131)*, and neurons *(132)*, and is a strong neurite growth promoter. It is also found in basement membranes during embryonic development, and is associated with peripheral nerves. In addition, TSP-1 regulates plasminogen activity, and can bind and activate TGF-β *(28)*.

F-Spondin and mindin form a family of molecules containing a TSP type I repeat at the C-terminus and two spondin domains, called FS1 and FS2, at the N-terminus *(28)*. F-Spondin is required for the appropriate pathfinding of commissural axons of the floor plate *(133,134)*, and is expressed in the hippocampus, where it promotes outgrowth of embryonic neurons *(135)*.

2.3.2.4. Tenascins

The tenascin (TN) family comprises five distinct members: TN-C, TN-R, TN-W, TN-X, and TN-Y *(136)*. The TNs display highly restricted and dynamic patterns of expression in the embryo, particularly during neural development, skeletogenesis, and vasculogenesis. These molecules are re-expressed in the adult during normal processes, such as wound healing, nerve regeneration, and tissue involution, and in pathological states, including vascular disease, tumorigenesis, and metastasis *(137,410)*.

TNs are characterized by a serial arrangement of a cysteine-rich N-terminus, followed by varying numbers of EGF-type repeats, succeeded by multiple FN III repeats, and a carboxy-terminus, with homologies to fibrinogen-β and -γ (Fig. 2; *136*).

The basic TN-C protein in mouse comprises 14.5 EGF-type repeats, followed by eight FN III repeats, although there is considerable splicing of up to six additional FN III domains between the fifth and sixth FN III units of the basic structure, potentially resulting in up to 64 different isoforms, of which 27 have been shown to be expressed as mRNAs *(138)*. The N-terminal region contributes to the central knob, a cysteine-rich structure that assembles six monomers via disulfide bridges, to form a hexameric polymer. In the CNS, TN-C is principally expressed by astrocytes and radial glia at early stages of development. In the adult, the protein is restricted to regions such as the molecular layer of the cerebellum, the olfactory bulb, and the optic nerve head *(139–141)*. TN-C has been shown, in different in vitro assays, to exert both inhibitory and stimulatory actions on neuronal cells *(89,136,142–144,411,412)*. These diverse functions have been localized to distinct functional domains, using proteolytic fragments and fusion proteins. Hence, for example, an analysis of the alternatively spliced FN III domains shows that, together, they support short-term, but not long-term, adhesion of embryonic and early postnatal neurons; FN III domains A1, A2, and A4 together exhibit repulsive properties on HC neurons; FN III B and D promote neurite outgrowth *(143)*. These contrary cellular functions also seem to depend on their mode of presentation, i.e., whether the proteins are soluble or substrate-bound, and also upon the cell types and differentiation states of the target tissues *(144)*.

TN-R is shorter than TN-C, with 4.5 EGF-type repeats and 8–9 FN III repeats. It is predominantly expressed by oligodendrocytes during the onset and early phases of myelin formation, and remains expressed by some oligodendrocytes in the adult, although it is also expressed by some neurons in the spinal cord, retina, cerebellum, olfactory bulb, and hippocampus *(145–147)*. TN-R, like TN-C, seems to be a multifunctional molecule that promotes neurite outgrowth when presented as a uniform substrate, inhibits growth cone advance when offered as a sharp substrate boundary, and induces axonal defasciculation in vitro *(146,148,149)*.

Expression of TNs is regulated by a variety of growth factors (GFs), cytokines, vasoactive peptides, ECM proteins, and biomechanical factors *(150–152)*. The signals generated by these factors converge on particular combinations of *cis*-regulatory elements within the recently identified TN gene promoters, via specific transcriptional activators or repressors *(145,153–159)*, e.g., the transcription factor, Sp1, activates the expression of the mouse *TN-X* gene *(154)*, Sp1 and Ets proteins act as potent activators of TN-C expression *(155)*, and OTX2 binds and represses the human TN-C promoter *(157)*.

2.3.2.5. Bifunctional Axon Guidance Molecules

Over the past few years, significant progress has been made in understanding the molecular basis of axonal guidance, by the identification of several families of proteins that can act as substratum-bound, short-range, or long-range diffusible axon guidance signals *(160,161)*.

The netrin protein family was first discovered in *C. elegans* with unc-6, but two mammalian homologs are known: netrin-1 and netrin-2 *(162)*. The netrins contain globular and EGF repeat domains highly related to domains V and VI of the LAM β-chain, and a unique basic C-terminal amino acid sequence. Netrins are secreted proteins exerting both repulsive and attractive effects on axonal growth cones *(163)*. They are considered to be diffusible factors acting as cues for growing axons, e.g., of the commissural neurons of the spinal cord *(164)*. Two transmembrane netrin receptors are known, possessing EC domains related to the immunoglobulin (Ig) family of CAMs: Both unc-5 in *C. elegans*, and its mammalian homolog, contain two Ig domains and two TSP type I domains; unc-40 in *C. elegans*, and its mammalian homologs, DCC and neogenin, contain four Ig domains and six FN III domains *(165,166)*.

The semaphorins are the largest family of signaling molecules implicated in axonal guidance. They can be divided into eight classes (1–7, and V), and can be both secreted and transmembrane molecules *(167–169)*. They are characterized by the presence of a conserved semaphorin domain of 500 residues found in both invertebrate and vertebrate proteins. The secreted forms contain an Ig domain, in addition to the Sema domain; the transmembrane forms contain an IC domain and the Sema domain, but can occur with a TSP domain, instead of the Ig domain. Semaphorin was first termed as "fasciclin" in the grasshopper, and as "collapsin" in the chick *(165,170)*. The Semaphorins play critical roles in axon guidance, by acting as repulsive cues, inducing growth cone collapse in in vitro tests; e.g., exposure of sensory neurons to the secreted protein, SemD, or its chick homolog, collapsin-1, induces a rapid but reversible collapse of their growth cones *(171,172)*, and a limited contact of growth cones with immobilized Sema 3A (collapsin) can redirect the trajectories of growing axons, an effect mediated by the neuropilin membrane receptors. Semaphorins can also act as attractive cues during neuronal morphogenesis, e.g., controlling dendritic arborization *(173)*.

First characterized in *Drosophila*, Slit is a secreted protein containing four LRR and seven EGF repeats expressed by midline glia, which binds to the Roundabout (Robo) receptor *(174–176)*. Although *Drosophila* and *C. elegans* have a single gene coding for Slit, three genes occur in mammals, coding for Slit proteins, which are known to be involved in the repulsion of motor and olfactory bulb axons *(177)*. In addition, Slit is the first known protein to act as a promoter of axon branching, as demonstrated on sensory neurons *(178)*. Slit can bind to netrin and LAM, and can act as a ligand for glypican-1 *(177)*.

2.3.2.6. Astrotactin

Astrotactin is a GP containing three EGF domains and two FN III domains. In the developing cerebellum, it is expressed by postmitotic immature neurons during their migration along the radial glia. Astrotactin therefore appears to be a ligand for neuronal migration in the CNS along astroglial fibers. It is also expressed in the developing hippocampal formation and olfactory bulb *(179–181)*.

2.3.2.7. Reelin

The Reeler mouse is a mutant that shows a disturbance in the formation of its cortical layers. Characterization of this mutant identified the reelin protein. Reelin is a GP expressed by Cajal-Retzius cells in the developing cortex, and by neurons from the external and internal granule layer of the developing cerebellum. It contains EGF repeats, and its N-terminus reveals 25% identity with F-spondin *(182,183)*. It is a ligand for lipoprotein receptors *(184,185)*, for $\alpha3\beta1$ integrin *(186)*, and for proteins of the cadherin-related neuronal receptors (CNR) family: These are proteins containing six EC cadherin ectodomains and a cytoplasmic domain interacting with the tyrosine kinase (TK) Fyn *(187)*.

2.3.3. Collagen

Collagen accounts for up to 30% of the total protein mass in mammals. The collagen family is a heterogeneous group of proteins, consisting of at least 19 different types.

They are macromolecular proteins of approximate mol wt 300,000, composed of three helical polypeptide chains wound round each other to form a triple helix. The amino acid composition of these chains is characterized by a high content of glycine, proline, alanine, absence of Cys and tryptophan, low content of tyrosine, and the presence of unique hydroxylated amino acids (hydroxyproline and hydroxylysine) *(188–191)*.

Collagen types I, II, III, V, and XI, the fibrillar collagens, are composed of long, uninterrupted triple helices, flanked by globular domains that are cleaved off on maturation: These molecules associate into long fibrils. Collagen fibers provide mechanical support to the connective tissue, e.g., in the meninges *(192,193)*. They exhibit structural variation from tissue to tissue, with reference to the orientation of fibrils and in the diameters and the density of packing of fibrils in the tissue. This variation is continued at the molecular level, where there are tissue-specific differences in the lateral packing of individual molecules into fibrils. This fibrillar diversity in structure is related to the specific mechanical and structural properties characteristic of each tissue. In addition, the collagens also form an essential substrate for cellular adhesion and migration.

Collagen type IV differs considerably from the fibrillar collagens, with many interruptions in its triple helix, and several noncollagenous domains: a highly disulphide-bonded triple-helical N-terminal domain, 7S, and a C-terminal globular domain, NC1, which persists in its mature form. Carbohydrate accounts for 10% of its mass. Collagen type IV is the major component of basement membranes. It is highly expressed by the endothelium of blood vessels *(194)*. The globular terminal extensions of the molecule associate to form a network; hence, association of four molecules at their N-termini gives rise to tetramers, which can further aggregate into a regular tetragonal or irregular polygonal meshwork. Variations in the size and the nature of these pores may account for the variations in the properties of the basement membranes of different tissues *(190)*.

2.3.4. *Proteases*

The matrix metalloproteinases (MMPs) are a family of more than 20 secreted and cell-surface zinc-endopeptidases, which can degrade the proteins of the ECM, and are thereby involved in the remodeling of the ECM in a variety of physiological and pathological processes *(195–201)*.

All of the MMPs are initially expressed as inactive pro-MMP zymogens, with a common 80-amino acid propeptide, followed by the N-terminus catalytic domain and a C-terminus domain (except in MMP-7). A Zn atom in the catalytic domain is bound to a Cys in the propeptide, and the enzyme is activated (notably by plasmin) by disrupting this interaction, thus exposing the active site, which cleaves the propeptide by autocatalysis. The MMPs bind to their ECM substrates via the C-terminus domain, and there is an additional C-terminal transmembrane domain in the membrane-type matrix metalloproteinases (MT-MMPs). Although there is some substrate specificity, many of the MMPs can degrade several different classes of ECM protein, e.g., MMP-2 (gelatinase A) and MMP-9 (gelatinase B) degrade collagens I, IV, V, VII, and X, gelatin, elastin, FN, and PGs; MMP-3 (stromelysin-1) and MMP-10 (stromelysin-2) degrade collagens IV, IX, and X, LAM, elastin, FN, and PGs.

Given their potential capacity to catalyze the degradation of all of the protein components of the ECM, their activities are kept under tight control, to prevent tissue destruction. In addition to the control of proenzyme activation, there are numerous transcription factors that regulate their gene expression, and there is a family of inhibitory molecules (tissue inhibitors of MMPs [TIMPs]), which can regulate the activity of the enzyme, once it is activated.

A number of different signals, including cytokines, growth factors, integrins and viruses, can activate transcription of the genes encoding MMPs; hence, there is an upregulation of MMPs associated with wound healing and angiogenesis, but also with various pathologies, such as tumor invasion and metastasis, and multiple sclerosis *(198)*.

An important activator of pro-MMPs is plasmin, a serine proteinase generated from plasminogen by the action of the tissue- or urokinase-plasminogen activator (tPA/uPA), which is anchored to the cell surface by its receptor (tPA-R/uPA-R). Pro-MT-MMPs are also activated by furin, another Ser proteinase. The localization of this activation mechanism on the cell surface implies a concentration of proteolysis in the immediate pericellular environment, where it can influence cell–cell and cell–ECM interactions *(199)*.

There are currently four members of the family of TIMPs, of which three are expressed in the brain (TIMPs 2–4) *(202)*. These bind with high affinity to the C-terminus domain of MMPs, but can also bind to the proenzymes by other mechanisms. It seems that a balance between the MMPs and the TIMPs can regulate a number of cell–matrix interactions, and that their relative stoichiometry is critical. For example, TIMP-2 can bind MMP-2 and MT-MMPs, but it can also serve as an adaptor molecule, allowing pro-MMP-2 to associate with, and be activated by, MT-MMPs *(199,202)*.

2.3.5. Lectins

Lectins are proteins that recognize specific carbohydrate structures *(203,204)*. They bind through their carbohydrate recognition domains, and play an important role in the ECM, where there are many carbohydrate structures to which they can bind, including GAG chains and N-linked sugar residues, such as sialic acid and mannose *(205)*. Lectins can belong to other protein families, according to the nature of the other protein domains that make up their structure.

Lectins can divided into two kinds: I-lectins and C-lectins. I-lectins belong to the Ig-like superfamily. A number of I-lectins are present in the nervous system, including P0, which can bind to HNK-1 (a sulphated glucuronic structure), ICAM 1 (I for intercellular), which binds HA, NCAM, and the fibroblast growth factor (FGF)-receptor, which is the only non-CAM I-lectin that can bind to HS. In addition, there are the sialoadhesins, a family of sialic-acid-binding I-lectin CAMs (also denoted as Siglec-1–4), which includes myelin-associated glycoprotein, Schwann myelin protein, and sialoadhesin. Sialoadhesin is a 185-kDa protein expressed in murine macrophages in the PNS, and at specific sites of the blood–brain barrier, such as the choroid plexus *(206)*. I-CAM1 is a protein that can bind HA by its Ig domain. It can mediate intercellular interactions in the immune and nervous system, and is expressed by microglia and astrocytes in vitro and under pathological conditions in vivo.

C-lectins are calcium-dependent and include L-selectin on lymphocytes, as well as the lectican family of CSPGs (which includes aggrecan, neurocan, versican, and brevican). In addition to a C-type lectin domain, a C-lectin must contain a HA-binding site and a Ig-like domain toward its N-terminus *(203)*.

2.3.6. Growth Factors

An important function of FN and other extracellular molecules is to serve as carriers of growth and differentiation factors, such as TGF-β *(207,208)*. Growth factors can bind to the ECM via protein domains or sugar groups, particularly GAG chains. ECM binding of growth factors limits diffusion, providing localized storage of factors, which may persist after growth factors production has ceased; e.g., matrix-bound FGF is degraded more slowly than free FGF, prolonging its activity *(209)*. Such matrix-bound growth factors can be released following focalized protease destruction of ECM molecules *(36)*, e.g., PDGF *(210)*.

HS and heparin can bind to and modulate the activity of various growth factors, including FGFs 1–9, vascular endothelial growth factor, and heparin-binding epithelial growth factor *(96,211)*. As such, it has been shown that HSPGs can regulate the mitogenicity of FGFs, and that these interactions are prerequisites for the binding of the growth factors to their high-affinity transmembrane receptors. However, the critical roles of HSPGs in developmental processes and specific signaling pathways have recently been illustrated by the identification of mutations involved in the biosynthesis of HS in *Drosophila* and mice.

In *Drosophila*, mutations in the UDP-D-glucose dehydrogenase, which produces the UDPGlcA HS precursor sugar (*sugarless*), and in the N-sulphotransferase, which sulphates the HS GAG chain (*sulfateless*), result in a disruption of the FGF, *Wingless*, and *Hedgehog* signaling pathways. The loss of activity of these pathways results in severe phenotypes, indicating that the HSPGs are absolutely required for signaling of these growth factors and morphogens *(212,213)*.

Some ECM molecules also possess intrinsic growth factor activity that might be localized to growth factor-like sequences; e.g., EGF-like regions are present in LAM *(214)*, TN *(215)*, and TSP *(216)*, all of which have been reported to possess mitogenic activity *(217)*.

3. MEMBRANE BOUND PROTEINS: CELL CONTACT TO THE ECM

EC molecules interact with each other to form and organize the ECM, but they interact also with receptors on cell surfaces. These receptors can be divided into five major groups: CAMs, integrins, selectins, cadherins, and syndecans/glypicans (discussed in Subheading 2.3.1.5.). All are represented in the CNS, and induce signal transduction, allowing the cells to adapt their behavior to their microenvironment. Although cadherins are only involved in direct cell–cell interactions, the other receptor groups can all mediate both cell–cell contact and cellular interactions with the ECM. The modular structure of these cell–surface molecules is illustrated in Fig. 3.

3.1. Cell Adhesion Molecules (CAMs)

CAMs are cell-surface macromolecules from the Ig superfamily *(218,219)*. They contain a variable number of Ig-like domains and FN III repeats, and CAMs bind their ligands in a Ca^{2+}-independent manner. In this family of adhesion molecules, molecules can be distinguished that have a transmembrane domain followed by a cytoplasmic C-terminus, such as NCAM180/140, NrCAM, NgCAM/L1, or neurofascin, and CAMs that are linked to the membrane by a GPI anchor, such as NCAM120, F11/F3/contactin, or TAG-1/axonin *(220–224)*. GPI-attached molecules are generally associated with others in the membrane.

In the CNS, Ng(neuron-glia)CAM/L1, N(neuronal)CAM, F11/F3/contactin, and TAG-1/axonin have all been shown to mediate fundamental processes in brain development and neuronal plasticity, such as process outgrowth, guidance, and fasciculation *(225–227)*.

CAMs can also possess important posttranslational modifications, such as glycosylation; e.g., a form of NCAM exists with a polysialic acid (PSA) modification (known as PSA-NCAM), and it has been demonstrated that the functional activity of NCAM in neuronal plasticity can be regulated by the presence of this carbohydrate moiety *(228,229)*.

3.2. Integrins

Integrins are the major group of receptors for EC molecules. They are heterodimeric receptors that result from the association of two subunits, α and β (Fig. 3). At least 20 α (α 1–9, α V, α IIb, α L, α M, α X, α PS2, α IEL) and nine β (β 1–9) subunits have been identified. One α-subunit can bind to one of several β-subunits, and vice versa, giving rise to a large number of possible combinations: e.g., β1 can bind 10 α units (α 1–9, respectively, and α V) *(230)*. In this way, the ligand specificity of the integrins can be modulated *(231)*.

Each of the integrin subunits is composed of a large N-terminal EC domain, a transmembrane segment, and a short C-terminal cytoplasmic region. The EC region of α-subunits consists of seven homologous repeats, the repeats IV–VII, containing motifs similar to divalent-cation-binding EF-hands found in calmodulin, and their fixation of Ca^{2+} or Mg^{2+} not only stabilizes the tertiary structure of the molecule, but is also necessary for the interaction of the dimeric receptor with ligands. The N-terminal region of the β-subunits is characterized by a highly conserved segment that also contains an EF-hand-like motif and an EC region consisting of cysteine-rich repeats *(230)*.

In the brain, integrins are found in all regions, and seem to participate in all developmental processes, plasticity, and repair mechanisms *(232)*.

3.3. Selectins

Selectins are a group of glycoproteins composed of an N-terminal C-type lectin domain, an EGF-like domain, and a variable number of short consensus repeat units similar to those found in complement regulatory proteins of the immune system, followed by a transmembrane and a short cytoplasmic domain (Fig. 3). The selectin family consists of L-selectin (lymphocytes), E-selectin (endothelium), and P-selectin (platelets) *(233)*. These molecules are expressed on blood cells and the endothelium of vascular vessels, where they play an important role in inflammatory processes *(234)*, although there is no evidence that selectins interact directly with neural cells in brain inflammatory mechanisms.

3.4. Cadherins

Cadherins form a large superfamily of transmembrane glycoproteins comprising at least 50 different members sharing similar structures *(235–237)*. Cadherins have been subclassified into several categories: the type I includes N-, E-, P-, R-, and M- "classical" cadherins; the type II comprises cadherins 5–12, and finally, there are the T-cadherin, protocadherins, cadherins from desmosomes, and the CNR cadherins.

Typically, they all possess an extracellular region composed of several homologous β-sandwich EC domains (five for the classical cadherins, and 6–7 for protocadherins), which are topologically similar, but not identical, to Ig domains (*238,239*; Fig. 3).

Cadherins are primarily responsible for specific adhesion between cells, during which the binding will be homophilic, between like cadherin molecules on adjacent cells, although there are some known cases of interaction between different cadherins, and cadherin molecules can also interact on the same cell surface. All of these binding interactions are calcium-dependent, cadherins possessing two calcium-binding sites at the junction between each of the EC domains *(240)*. The binding of calcium stabilizes the structure of the molecule in a functional conformation *(241)*.

Most of the cadherins are expressed in the CNS. E- and N-cadherins are involved in neural connections, and in regulating synaptic mechanisms *(242)*.

4. SIGNALING MECHANISMS AND THE ECM

Interactions of cells with each other, and with the ECM during development and remodeling, depend on a multifactorial, complex array of molecular interactions and signaling events that are constantly being integrated by the cell to produce behavior appropriate for the spatial and temporal environment (Fig. 4).

The development of different cell types in the brain depends on interactions between the molecules of the ECM and their receptors on the cell surfaces. Signals from the cell's microenvironment can be transmitted through these receptors, allowing the cells to respond and to adapt appropriately *(243)*.

Different types of interaction can be distinguished. First, receptors can pass on extracellular signals directly via secondary messengers: This is the case for integrins and changes to the cytoskeleton. But receptors can also be associated to other cell-surface molecules when transmitting information from the ECM; e.g., ECM interactions with the GPI protein, F3/F11/contactin, in turn depend on its interaction with other transmembrane CAMs, such as NCAM or Caspr/Paranodin, which can modulate the activity of Fyn kinase on the cytoplasmic surface of the cell membrane *(244,245)*.

4.1. *Integrins and IC Signal Cascades*

The interaction of cells with components of the ECM is vital for cell growth, cell differentiation, and cell migration. The family of cell-surface receptors chiefly responsible for these interactions is the integrin receptor family, a group of highly versatile, heterodimeric, transmembrane glycoproteins. Signaling via integrins is achieved through interactions with IC effectors that couple integrins and growth factor receptors to downstream components (*246–249*; Fig. 4).

4.1.1. Outside-In Signaling

The ligand-binding or antibody-mediated clustering of integrins results in the activation of a variety of intracellular signaling processes, such as stimulation of the activities of cytoplasmic tyrosine and serine/threonine protein kinases, induction of calcium and hydrogen transients via activation of calcium channels and the Na^+/H^+ antiporter, stimulation of lipid metabolism and generation of lipid second messengers, and organization of the actin cytoskeleton *(250–253)*. Although capable of direct signaling, integrin-mediated adhesion can also regulate signaling events emanating from other

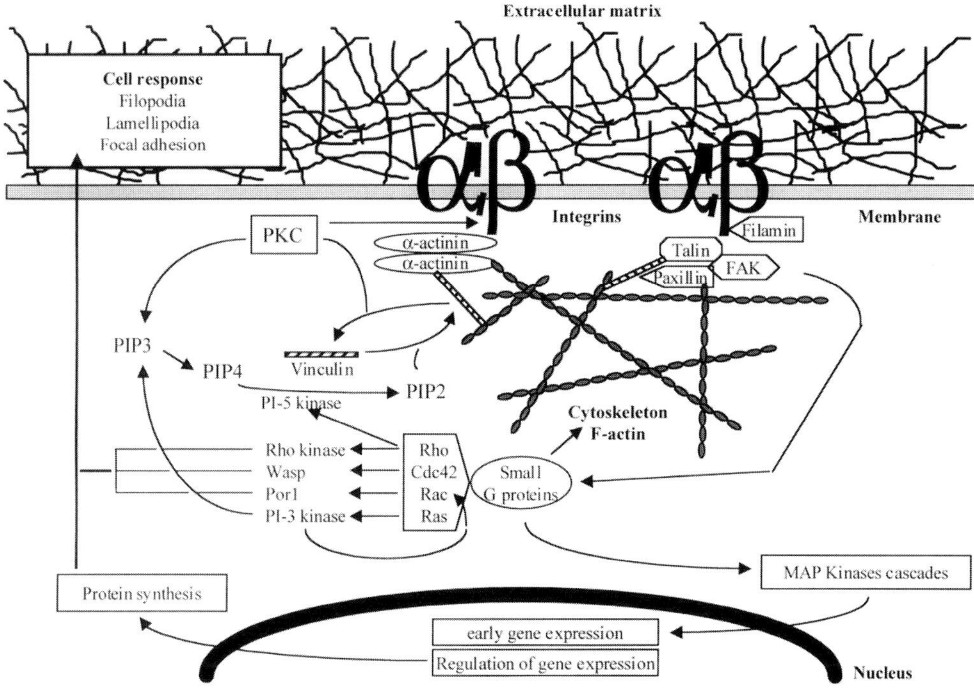

Fig. 4. Integrin-mediated signaling pathways linking the ECM to cytoskeletal adjustments. αβ = integrins make up a large family of heterodimeric proteins composed of α- and β-subunits that mediate cell–ECM adhesive connections. The EC domains interact with ECM molecules, such as FN, LAM, and TN, and the IC domain interacts with the actin cytoskeleton through several intermediate proteins. In addition to ligand binding, integrin signaling also requires integrin clustering, which is dependent on the GTPase, Rho. Through such clustering, a focal adhesion plaque is formed, which is a large complex of IC proteins, composed of cytoskeletal and signaling molecules, such as talin, α-actinin, filamin, tensin, vinculin, and FAK. Here, FAK becomes hyperphosphorylated. The major signaling pathways involving the Rho family of GTPases (small G proteins), Rho, Cdc42, Rac, and Ras are also illustrated. Integrins can activate Ras via the phosphorylated FAK, and subsequently Rac via Ras, and Rho. Cdc42, Rac, and Rho are interconnected in a hierarchical fashion, and interact with a number of target effectors. Some of these, such as WASP (Wiskott-Aldrich syndrome protein), Por1, Rho kinase, and phosphatidylinositol phosphate 5-kinase (PI-5 kinase), are responsible for organizing cytoskeletal structures, such as filopodia, lamellipodia, and focal adhesions. Other effectors, such as the MAPKs, activate elements that influence gene expression, resulting in protein synthesis. Another pathway involves the conversion in the membrane of PIPs by the action of enzymes, such as phospholipase C and protein kinase C (PKC). For example, vinculin regulates the anchorage of the cytoskeleton to the membrane: It is recruited to the membrane by Rho-dependent synthesis of PIP2, a process that can be inhibited through phosphorylation of vinculin by PKC. PKC also plays a role in activation of integrins, probably via its substrate Rack-1, which binds directly to integrin β tails.

receptors, e.g., receptor tyrosine kinases (RTKs). Engagement of integrins results in tyrosine phosphorylation of focal adhesion kinase (FAK), a protein tyrosine kinase present in focal adhesions. FAK associates with several different signaling proteins, such as Src-family protein tyrosine kinases, p130Cas, Shc, *Grb2*, PI-3 kinase, and paxillin *(254)*. This enables FAK to function within a network of integrin-stimulated signaling pathways, leading to the activation of targets, such as the extracellular-signal-regulated kinase (ERK) and mitogen-activated protein kinase (MAPK) pathways *(253)*.

The net effect of integrin engagement is cell-type-specific, and depends on the context in which the cell finds itself. Integrin-mediated adhesion to the ECM can regulate apoptosis and the cell-cycle machinery, mitogenesis, cell differentiation, cell migration, and cell-to-cell adhesion.

4.1.2. *Inside-Out Signaling*

The adhesive property of integrins is of fundamental importance during various stages of embryonic development, as shown by the targeted elimination of many of the integrin subunits by gene-knockout strategies, which results in a spectrum of embryonic lethality *(255)*. The binding of integrin to its ligands is a dynamic process that can be regulated. Hence, integrins can either be active and capable of binding to their ligands, then of forming productive cytoskeletal linkages via their cytoplasmic tails, or they can remain inert and incapable of forming such connections. The switch from inert to active integrins can be influenced by factors inside the cell, and this has been termed "inside-out" signaling *(256,257)*. Such integrin activation can promote the assembly of ECMs around and between cells, indicating a true signaling process from the inside to the outside of the cell. The resulting increase in cell adhesion varies according to the stimulus and the cell type.

Many factors have been implicated in the activation of integrins, including phorbol esters, lipid mediators, calcium ionophores, thrombin, chemokines and chemoattractants acting through serpentine receptors, aggregation of various coreceptors, including CD2, CD28, and CD31, protein kinase C, and TSP binding *(257)*.

4.2. *Protein Tyrosine Phosphorylation*

Protein phosphorylation and dephosphorylation are fundamental mechanisms used to control biological processes. Perhaps one-third of cellular proteins are phosphorylated, of which the vast majority are phosphorylated on serine or threonine residues. However, although tyrosine phosphorylation only accounts for less than 0.1% of total protein phosphorylation, it has become clear that tyrosine phosphorylation is involved in the regulation of numerous cell functions, including passage through the cell cycle, proliferation and differentiation, motility, and cytoskeletal organization *(263)*. Hence, in neurons, signal transduction based on protein tyrosine phosphorylation of cytoplasmic proteins seems to regulate cell fate choices, cell differentiation, survival, and axogenesis, and is also implicated in synapse formation.

The level of tyrosine phosphorylation is regulated by the opposing actions of protein tyrosine kinases and protein tyrosine phosphatases (PTPs). Both protein families are very large, with ~100 PTPs so far identified, and many more kinases known. Both PTKs and phosphatases can be either cytoplasmic or single-pass transmembrane receptors. Through their extracellular domains, the receptors can interact with extracellular ligands, triggering the enzymatic activity of their cytoplasmic domains.

4.2.1. *Receptor Tyrosine Kinases (RTKs)*

EC signal molecules, such as EGF, PDGF, and insulin bind to RTKs *(258)*. These receptors possess one or two cytoplasmic tyrosine kinase regions. Upon ligand binding, RTKs autophosphorylate. The resulting phosphotyrosine residues act as highly selective binding sites for so-called Src homology domain 2 (SH2)-containing proteins, which transduce the signal by changing their enzymatic activity or recruiting other proteins *(259,260)*. Among these SH2-containing proteins are Ras-GTPase activating protein and phospholipase C. Phospholipase C hydrolyzes phosphatidylinositol 4,5-bisphosphate (PIP2) into inositol 1,4,5-trisphosphate (IP3) and diacylglycerol (DAG). IP3 releases Ca^{2+} from intracellular stores, and diacylglycerol activates protein kinase C. The SH2-containing adaptor molecules, and growth factor receptor-bound protein 2 (GRB2), also bind to phosphorylated RTKs *(261)*. The latter recruit G-nucleotide exchange factors toward the plasma membrane, where they activate the small G-protein Ras by stimulating the exchange of GDP for GTP on it. Activated Ras-GTP in turn stimulates a protein kinase cascade, resulting in MAPK/ERK activation *(261)*. Stimulation of MAPK/ERKs results in transcriptional activation and gene expression *(253)*.

4.2.2. Receptor Protein Tyrosine Phosphatases (RPTPs)

RPTPs contain a variable extracellular domain, a transmembrane domain, and 1–2 intracellular Tyr phosphatase domains. Depending on the RPTP subtype, the EC domain exhibits Ig- and FN-like regions and other sequence motifs involved in cell–cell adhesion *(262,263)*.

RPTP β contains in its EC domain a region with sequence homology to the enzyme, carbonic anhydrase, followed by a FN (FN III) and a long, unique sequence termed the "spacer domain," which is spliced out in a shorter RPTP β *(74,264)*. The entire EC part of RPTP β is expressed as a PG, DSD-1-PG/phosphacan (Fig. 1; described in Subheading 2.3.1.4.). The different forms of RPTP β bind to multiple ligands *(43,265,266)*, including the neuronal recognition molecules, NCAM, NgCAM *(72)*, and TAG-1 *(78)*, the ECM molecules, TN-C and TN-R *(79,148,267,268)*, and with several heparin-binding growth factors, including HB-GAM *(71)*, amphoterin *(269)*, and bFGF *(270)*. The multifunctional interaction of phosphacan and RPTP β with different ligands involves different domains of the receptor and, at least for phosphacan, depends on the presence of CS, or is mediated by N-linked oligosaccharides *(266)*. RPTP β is predominantly expressed in the nervous system, where it is found mostly in glial precursors, radial glia, and astrocytes *(271)*, as well as in certain neurons *(272)*. The expression of the different forms of RPTP β is regulated during development of the glial lineage. Early in development, high levels of the receptor forms are found in proliferating precursor cells at the subventricular zones. As development progresses and cells mature, these receptor forms are replaced by the secreted nonphosphatase form *(273,274)*.

The RPTP, CD45, is involved in the activation of B- and T-lymphocytes. It stimulates the Src-like kinases, Lck and Fyn, by dephosphorylating them *(263)*. Similarly, RPTP α is thought to activate Src, and binds to the adapter protein, GRB2, implying signaling links through SH2- and SH3-domain binding. Both Src and Fyn kinases are required for CAM-mediated axon outgrowth *(262)*.

The RPTPs, μ and κ, have been shown to modulate cell–cell interaction, and it has been proposed that there is an association of the cytoplasmic domain of RPTP μ with a protein complex containing cadherin transmembrane adhesion molecules and catenins (which are linked to the underlying actin cytoskeleton) *(263)*.

RPTP δ expression generally correlates with populations of differentiating neurons. DLAR, a *Drosophila* homolog of RPTP δ, is required for motor axon pathfinding.

RPTP σ is expressed in both proliferating and differentiated cells in the CNS. However, there is still relatively little known about the cytoplasmic targets of the RPTPs *(262)*.

4.3. Cytoskeletal Adjustments in Response to ECM

The IC signal cascades induced by the ECM in different developmental processes and mechanisms of plasticity, such as process outgrowth and guidance, cell migration, synaptogenesis, or regeneration, result predominantly in cytoskeletal adjustments *(254)*. Cell matrix adhesion regulates actin cytoskeleton organization, through distinct steps, from formation of filopodia and lamellipodia in the early phases of cell adhesion, to organization of focal adhesions and stress fibers in fully adherent cells. Focal adhesions are formed by the clustering of integrins that are associated with a number of cytoplasmic proteins, which include α-actinin, talin, tensin, paxillin, vinculin, and many others. Focal adhesions are also anchoring points for actin-myosin microfilaments bundled in stress fibers.

The polymerization of the cytoskeletal components, such as globular actin to filamentous actin (F-actin), or α- and β-tubulin, can be regulated by a number of different factors. These regulators, which are targets of IC cascades from receptors, essentially integrins, can modulate both the polymerization of the cytoskeleton and its anchorage to the membrane *(249,251,252,275,276)*. Many cytoplasmic proteins are involved in cytoskeletal adjustments, and include talin, vinculin, α-actinin, filamin, ankyrin, spectrin, profilin, myosins, and zyxin (Fig. 4). Their activities are modulated by small G proteins and kinases. Small G proteins are monomeric GTPases, and include Rho, Rac, and Cdc42 *(249)*. They regulate the actin cytoskeleton, and influence gene expression by interacting with multiple effectors. The small GTPases do this by controlling the activity of different types of kinases, such as those that are involved in the PIP cascades, but they can also directly modulate proteins associated with the cytoskeleton *(246)*. *De novo* protein synthesis is a key element to the cell's response to extracellular signals; e.g., many new components are necessary to modify the structure of synapses in synaptogenesis and memory.

Talin, vinculin, and filamin regulate anchorage of cytoskeletal molecules to the cytoplasmic membrane. These proteins transmit traction forces through the cytoplasmic membrane from the extracellular space or from cytoplasmic compartments. Talin is a protein of 270 kDa, and possesses binding sites for membrane proteins, such as β integrins *(277–279)*, F-actin *(280)*, vinculin *(281)*, and focal adhesion kinase (FAK) *(282)*. Two forms of talin have been described (talin 1 and 2). They play an important role in linking integrins to F-actin. In fact, talin, forming dimers, acts as a hinge between the ECM and the cytoskeleton, and is also involved in regulation of integrin activation *(283)*. As such, talins have a key role in adhesion to the ECM and the connection of the cytoskeleton to EC components.

Vinculin, a cytoplasmic protein, participates in intramolecular interactions with talin, α-actinin and F-actin. Its role consists in the regulation of anchorage of the cytoskeleton to the membrane. Vinculin is amphitropic, i.e., it binds weakly (reversibly) to membrane lipids, and this process regulates its function. Recent studies *(284)* have shown that Rho-dependent synthesis of PIP2 promotes the recruitment of vinculin to the membrane, and that PIP2 inhibits interaction of vinculin with F-actin *(285)*. Moreover, vinculin can be phosphorylated by protein kinase C, inhibiting its association to the membrane, since this depends on acidic phospholipids *(286,287)*. Filamin is involved in focal adhesion of the cytoskeleton to the cytoplasmic leaflet of the membrane. Three members have been characterized, which interact with β-integrin subunits *(288)* and small G proteins *(289)*. Several studies have shown that these proteins participate in cell migration mechanisms *(290)*.

5. DEVELOPMENTAL PROCESSES AND THE ECM

In general terms, major functions of the ECM include providing structural support, tensile strength or conditioning, providing substrates and pathways for cell migration and regulating cellular differentiation and metabolic functions.

5.1. Cell Proliferation and Cell Survival

In the nervous system, apoptosis of developing neurons occurs if insufficient target-derived growth factor is available *(291)*. It is now appreciated that developmental apoptosis (programmed cell death) often results from an interaction between responsive cells that are competing for trophic support: The cells that compete successfully survive; the remainder undergo apoptosis. In addition, many cell types require integrin-mediated adhesion to ECM proteins, in order to survive *(292)*. In the absence of appropriate ECM contacts, cells also undergo apoptosis. The signaling pathways from growth factor receptors and integrins converge at relatively early points; hence, control of cell survival by growth factors and adhesion may be interdependent and involve similar mechanisms *(293–295)*. Signaling intermediates for integrins and cell survival include FAK, MAPK, and phospho-inositide 3-kinase (PI-3 kinase); p53 and Bcl-2 appear to be mediators of the subsequent apoptosis. However, by binding to them, the molecules of the ECM may also influence the diffusion of growth factors through the ECM. Hence, by regulating the accessibility of the factors for their cell-surface receptors, the ECM can regulate the transmission of the signal; e.g., reservoirs of factors stocked in the ECM may be released by a localized activation of proteases *(210)*.

5.2. Cell Migration

Migration of cells plays a significant role in brain morphogenesis *(181,296–298)*. Most neurons travel long distances through the complex extracellular terrain of the developing embryo to reach their final position. The most common mechanism for neuron movement appears to be a combination of the extension of a cell process and its attachment to the substratum, followed by the pulling of the entire cell toward the point of attachment by means of contractile proteins associated with an intracellular network of microfilaments. Directional control of cell movement appears to be of two types, with cells either moving along a scaffold made up of guide cells, or following a concentration gradient of a trophic factor laid down in a multicellular terrain. In both cases, small molecules diffusing through, or attached to, the ECM can alter a cell's behavior. Reelin, for example, appears to function as an instructive signal in the regulation of cell patterning during cortical development *(183)*. Other molecules, such as astrotactin *(179)*, are implied in the formation of cortical and cerebellar layers, and FN, LAM, and TN are found in the neuroepithelium *(23–25,144)*. Finally, integrins have been shown to be necessary keys for the establishment of the laminar organization of the cerebral cortex *(299)*.

In the complex neuropil of the cerebral cortex, molecular signals undoubtedly permit neurons to distinguish radial glial processes, along which they travel from the other neuronal, glial, and endothelial surfaces. Neural crest cell movement is another model of molecular control of cell migration, and contrasts with the work from the cortical systems. Since there are no glial cells to create highways for the crest cells, the migration of these cells into the spaces surrounding the neural tube depends on the nature of the ECM *(300)*. In order to migrate to specific destinations, a number of signals must be regulated in a coordinated fashion, both spatially and temporally. Nearly all of the major components of the matrix (collagen, FN, LAM, F-spondin *[301]*), TSP-1 *(302)*, PGs such as versican *(303)*, and HA appear to play a role in regulating neural crest cell migration *(297,304)*.

5.3. Morphological Differentiation and Neuronal Polarity

A variety of substrate- and cell-attached factors influence neural development, by regulating adhesion properties of cells. Interactions occur directly between cells or between a cell and the ECM of the microenvironment *(305)*. Growth cones are located at the leading edge of a neurite, and display two types of motile structure: long, spike-like structures called "filopodia," and thin, broad sheets of membrane called "lamellipodia." These delicate structures are important in pathfinding and outgrowth branching. Although growth cone movements can be influenced by soluble factors, such as NGF and FGF, adhesion molecules, either on the cell surface or in the ECM, also seem to play an important role in neurite outgrowth, since these molecules mediate neurite–neurite and neurite–glial interactions. The initiation of process outgrowth, rate of elongation, and degree of branching of neurites are all strongly influenced by interactions with the ECM. The molecules mediating these interactions have been implicated in regulating the specificity and timing of cell–cell adhesion and the consequences on cell morphology and physiology *(306)*. Hence, they influence the ability of the cells not only to migrate, but also to identify themselves, stabilizing the spatial relationships considered important for the process of differentiation. The CAMs, integrins, and cadherins all belong in this category *(299,307–313)*. In general, many of the same soluble and matrix components that regulate cell proliferation, survival, and migration have been shown to also mediate process outgrowth, both in vivo and in vitro *(314,315)*. Most PGs act as inhibitors or promoters of neurite growth, although the respective roles in these effects of the core protein vs the GAG chains are still a matter of debate *(60,70,102,265,316–319)*. It seems that most of the glycoproteins of the CNS ECM can play a role in the regulation of neurite growth *(320)*. Some are strong promoters of axon growth, such as LAM *(113)*; others can act as promoters or inhibitors, depending on the context, such as the type of neurons tested. This is the case, e.g., with TN-C *(143)*. Finally, some proteins of the ECM act as guidance cues, instructing the growth cone, so that it can reach its target: netrin and semaphorin are such proteins *(321)*.

5.4. Glial Boundaries and Axon Highways

Axon pathfinding during development, synaptogenesis and regeneration is supported by glial cells and by ECM molecules produced by glia *(322–326)*. Numerous molecules, produced by astrocytes in the neuronal environment such as LAM, are known to encourage axon growth. However, another important role of glia in the CNS is the production of molecules that negatively modulate axon growth. These adverse effects on axon pathfinding may be important for cell distribution and the establishment of the interneuronal circuitry, suggesting the existence of tissue segmentation and boundary formation in the CNS.

Many experiments have shown that ECM molecules are associated with boundaries *(327,328)*. TN, for example, is associated with glial boundaries. It is found with developing axon pathways in the spinal cord *(85,329)*, in the optic nerve *(141)*, in the optic tectum *(139)*, the olfactory bulb *(330)*, and the somatosensory cortex *(140)*. Similarly, PGs, particularly CSPGs, have been implicated in the formation of glial boundaries at the roof plate and the midline dorsal tectum *(331)*, the dorsal root entry zone and dorsal columns in the spinal cord *(85)*, cortical barrels, thalamic nuclei, geniculate nucleus, and in the cerebellum *(332–336)*.

The transitory expression of these molecules in different areas of the CNS suggests a contribution of such boundaries to the regulation of axonal growth and the establishment of axonal highways. Indeed, it has been demonstrated that molecules associated with boundaries can regulate outgrowth. This is the case for different forms of TN, which have positive or negative effects on outgrowth *(137,327)*, and PGs, which can inhibit neurite growth via their GAG sugar chains *(331,337–339)*. But other types of mechanisms could be involved, such as cell–cell interactions mediated by Ephrin molecules *(340)*, and the idea of glial boundaries playing an instructive role for growing axons still remains controversial *(341)*.

5.5. *Synaptic Cleft and Neuromuscular Junction*

In the CNS, the composition of the ECM in the highly specialized extracellular space, localized between the pre- and postsynaptic elements, is still relatively ill-defined, but some of the proteins that have been localized in synapses include the PGs, TN *(342)*, TAP-1 *(343,344)*, and syndecan-2 *(94)*, and GPs, such as agrin *(345,346)* or TSP-4 *(132)*, and the cell–cell contact protein, cadherin *(347)*.

There is, however, much more known about the nature of synapses in the PNS, particularly, the neuromuscular junction. Here, the synaptic cleft (50 nm wide) is a continuum of the basal lamina ensheathing the muscle, composed of type IV collagen, perlecan, and S-LAM (LAM-4). Agrin is also a major component, which is synthesized and secreted by both the nerve ending and muscle cells. During the development of the neuromuscular junction, the 225-kDa core protein becomes glycosylated, notably by the addition of HS GAG chains, giving it an apparent mol wt of 400–600 kDa. Its structure includes EGF-type repeats and LAM-type, follistatin-like domains. Agrin can interact with α-dystroglycan and LAM. The main function attributed to agrin is the clustering of acetylcholine receptors on the muscle cell surface beneath the presynaptic element. Agrin is also necessary for the maintenance of the synapse *(348)*.

The best understanding of the structure of the synapse has come from studies of the interaction of agrin with other molecules, such as dystroglycan. It appears that, as in other basement membranes, collagen IV and LAM-4 form a coherent matrix structure, and that the postsynaptic element (i.e., the muscle cell) can be linked to this structure via agrin and dystroglycan. The actin cytoskeleton of the muscle cell is involved in this interaction. Hence, it seems that, in the neuromuscular junction, there is a link between the structure of the ECM, cell surface receptors, and the actin cytoskeleton *(349)*.

The synapse of the neuromuscular junction is flanked by Schwann cells, and experiments on knockout mice have demonstrated a crucial function for LAM-4 in excluding the invasion of the synaptic cleft by the surrounding Schwann cells, illustrating the role this protein plays in the construction of the cleft itself *(350)*.

Astrocytes flank the synapses of the CNS, and contribute to the regulation of synapse function, e.g., in glutamatergic synapses, where the astrocytes remove the glutamate neurotransmitter, preventing overstimulation and apoptosis of the neurons. Other functions have emerged in recent years, but little information is available concerning the regulation by astrocytes of the composition and structure of the synapse during its development and maintenance *(351)*.

6. PLASTICITY IN THE MATURE BRAIN AND THE ECM

6.1. *Long-Term Potentiation and Memory: Synaptic Plasticity*

Changes in extracellular composition are a consequence of neuronal activity and glial regulation. These events allow the adaptation of neurons as a function of their activity and of their connections with other cells, modulating their synaptic transmission.

Contacts between neurons, the synapses, are regulated, and can change, enduring from minutes to hours to days. The presence of adhesion molecules in the synaptic cleft suggests that these molecules can participate in initiating and maintaining synaptic remodeling *(352,353)*. Long-term potentiation (LTP) is the predominant event underlying these changes, and most studies of LTP have been made in the hippocampus, a brain structure implicated in memory formation *(354)*. Several studies have shown that different types of adhesion molecules are involved in LTP and learning processes, and that expression of these molecules is modulated by LTP, indicating a role for adhesion molecules in plasticity processes *(355,356)*.

This is the case for TN-C, which is overexpressed during LTP, in correlation with changes in synaptic plasticity *(357)*. Cadherins are also clearly implicated in synaptic cell contacts: They can modulate the structure of the synaptic cleft or the maintenance of the cell junction, affecting the quality of the synaptic transmission between two neurons *(358)*. LAM is another regulator of LTP: Its degradation by plasmin regulates LTP, which can be induced, but not maintained, in the presence of plasmin *(359)*.

The most interesting and well-known molecule implicated in LTP is NCAM *(360)*. Hippocampal tissues from NCAM-deficient mice exhibit a strongly reduced capacity to generate LTP *(361)*. Moreover, several studies provide evidence for the role of the polysialated form of NCAM (PSA-NCAM) in LTP, and support results suggesting an important role for CAMs in learning and memory *(227,362)*. PSA-NCAM expressed in the synaptic cleft induces a structural remodeling, which affects transduction mechanisms to potentiate LTP.

6.2. *Song Birds and Rodent Olfactory Bulbs*

Although it was once thought that the anatomy of the adult vertebrate brain remained relatively static throughout the animal's life-span, it is increasingly evident that there are populations of progenitor cells present in the adult CNS, and that a degree of anatomical plasticity persists. Two well-studied examples of this are the seasonal brain changes in the songbird, and the olfactory bulb in mammals. The role of the ECM in these processes is most evident in the migration pathways of the neuronal progenitor cells, from the proliferative ventricular and subventricular layers to their final destinations *(363–365)*.

Songbirds use song to attract mates and to defend breeding territories. There are seasonal changes in song behavior and the brain regions that control it: The entire volumes of several song nuclei, including the neostriatal higher vocal center and the robust nucleus of the archistriatum, are considerably larger during the spring breeding season than in autumn and winter. In spotted towhees, the volume of the higher vocal center almost triples, and the adult canary higher vocal center can generate over 1.4% of its neurons daily. Neurogenesis in the ependymal/subependymal zone is followed by daughter cell migration into the parenchyma, along radial guide cells, as in forebrain development. It seems that, as in development, the radial guide cells and the new neurons derive from a common progenitor. It has been suggested that release of neurons from the subventricullar zone (SVZ) may require a downregulation of N-cadherin, and that the expression of CAMs by the neurons, notably NgCAM, is necessary for the subsequent migration, as indicated by antibody-blocking studies *(363)*.

Neuronal migration in the postnatal rodent brain is a robust, ongoing process, with constant neuronal recruitment to the olfactory bulb from progenitor cells distributed throughout the rostral forebrain subependyma. However, in contrast to the widespread parenchymal migration in adult birds, in which radial cells guide neurons to their destinations, the migration in the adult mammal seems to be a persistent manifestation of the rostral migratory stream, in which cells migrate over great distances within the plane of the ventricular subependyma. Migrating neurons are restricted to the olfactory stream, with little or no trangression of its borders. The guidance cues required for this tight spatial restriction are not well understood, but the migrating chains of neurons are surrounded and demarcated by glial sheaths that form topologically contiguous tubes. The stream is also demarcated by the presence of heterophilic ECM ligands for CAMs such as NCAM and NgCAM, including TN, and CSPGs, such as DSD-1-PG/phosphacan *(363)*.

6.3. Perineuronal Nets

Described for the first time by Golgi in 1893, who defined them as “a finely reticular covering” adhering to the surface of the cell, perineuronal nets (PNs) are present on the cell body and proximal dendrites of some neurons. PNs are seen on Purkinje cells in the cerebellum, in the spinal cord, the deep cerebellar nuclei, the hippocampus, and the cerebral cortex, independent of the class of neurotransmitter secreted by the neurons. They are also seen on interneurons expressing parvalbumin, a calcium binding protein, these cells being nonpyramidal and GABAergic. In functional terms, it is not clear what role PNs play in the functioning of the nervous system, although their selective distribution, associated with glial endfeet and the surfaces of only certain subsets of neurons, have led to a number of suggestions, notably regarding the formation and stabilization of synaptic contacts, and the maintenance of cellular relationships in the adult brain *(366–368)*.

In developmental terms, PNs appear postnatally, during the second week, but astrocytes can assemble PNs in vitro from cells prepared from earlier stages. PNs have not been extensively studied, because their visualization relies on relatively difficult methods, i.e., Golgi staining, periodic acid Schiff reaction, or using the particle exclusion assay *(369)*. Nevertheless, PNs can be observed using certain biotinylated lectins that recognize sugar epitopes such as N-acetyl-galactosamine *(18,370,371)*.

The presence of such carbohydrates in PNs suggests a role for PGs, and it has been shown, by use of specific antibodies, that the PN contains a number of typical neuronal ECM proteins. A relative abundance of HA in PN is revealed by staining for hyaluronectin, a HA-binding protein. CSPGs are also present (notably DSD-1-PG/phosphacan) *(147,372–374)*, and TN-C and -R are among the glycoproteins present. Some subsets of PN have also been characterized by the presence of parvalbumins, small (12-kDa) Ca^{2+}-binding proteins, and this has led to the suggestion that they create polyionic microenvironments. The PN is probably expressed by both the neuron and the astrocytic processes, and it also seems clear that PNs are very heterogeneous in their make-up, their association with different sets of neurons being reflected in distinct molecular compositions.

Attempts at visualizing a higher order of PN organization have recently led to the proposal that some nerve cells are surrounded by three almost-concentric nets: two extracellular, the first composed of interlaced astrocytic processes, the other of ECM molecules interposed between these astrocytic endfeet and the neuronal surface; the third net is intracellular and consists of a membrane-cytoskeleton composed of spectrin and ankyrin *(375,376)*.

7. NEURAL ECM IN REGENERATION AND PATHOLOGY

7.1. Glial Scar and Lesions

When the CNS is damaged, a glial reaction (gliosis) leads to the formation of a glial scar. This response results in a recruitment of microglia, oligodendrocyte precursors, meningeal cells, astrocytes, and stem cells *(377–382)*. Most of these cells produce EC molecules that essentially inhibit axon regeneration *(84,383,384)*. In this glial environment, different molecules of the ECM might act as inhibitors for regeneration. This is the case for PGs and glycoproteins, such as TNs and CD44. TN-R, which is present in white matter produced by oligodendrocytes, is upregulated after injury to the CNS. Likewise, TN-C is overexpressed in glial scars by reactive astrocytes. Both TN-R and some forms of TN-C inhibit axon growth in vitro *(136,142)*.

A second important class of inhibitory molecule in glial scars are CSPGs *(385)*, essentially expressed by astrocytes, but also produced by oligodendrocyte precursors and meningeal cells. There is a considerable increase of CSPG expression in CNS injuries after a few days, which can last for several weeks. Phosphacan, neurocan, NG2 decorin, and biglycan are upregulated in CNS injuries *(386–392)*, and other molecules, such as aggrecan, versican, and brevican *(393,394)*, which might also be overexpressed in nervous system lesions, are known to inhibit axon growth in vitro, interacting with various neuronal receptors, such as the CAMs, L1, F3/F11, and TAG-1/axonin.

7.2. Extracellular Aggregates and Alzheimer's

AD is a progressive neurodegenerative disorder that accounts for 70% of all cases of dementia. It is characterized by two diagnostic pathological lesions: EC senile plaques, and intraneuronal neurofibrillary tangles (bundles of abnormal filaments within neurons) *(395,396)*. The major components of the senile plaques are an irregular, loosely arranged aggregate of neuronal and glial processes, and the β-amyloid peptide (Abeta).

The discovery of familial AD mutations, which cause increases in Abeta production, has illuminated the importance of Abeta deposition in AD.

Abeta is a 39–42-amino-acid peptide that arises from proteolytic processing of the amyloid precursor protein (APP). The Abeta peptide, that is found in senile plaques and cerebrovascular deposits, exists as a multimeric aggregate with a fibrillar appearance *(397)*. The amyloid hypothesis proposes that the accumulation of Abeta aggregates, in brain regions important for memory, reasoning, and cognition, are the principal cause of the neurotoxicity and inflammatory processes observed in AD *(398,399)*. Therefore, the processes that lead to fibrillar Abeta accumulation are a major focus in AD research.

Several other extracellular molecules have also been shown to be associated with amyloid deposits, and in vitro studies of fibrillogenesis suggest that they may be important in the aggregation and persistence of the Abeta fibrils in vivo. These include apolipoprotein E *(400,401)*, LAM *(402,403)*, acetylcholinesterase *(404)*, and PGs, notably the HSPGs, perlecan *(405)*, and agrin *(406)*.

7.3. *Tumor Invasiveness*

Gliomas, a type of devastating primary brain tumor, are distinct from other solid, nonneural neoplasms, in that they display extensive infiltrative invasive behavior, but seldom metastasize outside the CNS. They are thought to be derived from astrocytes, oligodendrocytes, or ependymal cells, and display a correspondingly broad spectrum of histopathological features. Malignant gliomas are the most common type of malignant brain tumor in adults, and are second only to stroke as the leading cause of death from neurological disease *(407)*.

Although they rarely metastasize outside the CNS, gliomas are extremely invasive tumors in the surrounding normal brain, and this local invasiveness that contributes substantially to the inability to achieve total resection by surgery, and often results in recurrences at the primary site and at locations on the opposite side of the brain.

Glial cell invasion is a multistep process that is inevitably accompanied by proteolytic remodeling of the ECM. Initially, the cancerous cells change their affinity for each other or for the ECM, in order to be released from the primary tumor. To allow for cell migration, the surrounding ECM must then be remodeled by the local production of proteolytic enzymes. Numerous studies have demonstrated a close association between gliomal cell migration and the expression of various proteases, such as cysteine proteases (cathepsin B), serine proteases (including the plasminogen activation/plasmin system), and MMPs. However, MMPs seem to be responsible for much of this activity, and a strong correlation can be found between the invasiveness of glioma cells in vitro or in vivo and their production of MMP-2, MMP-9, or MT-MMPs *(408)*. The levels of the TIMPs can also be lower in malignant gliomas, suggesting that a decreased inhibition of MMPs could contribute to their disregulation by these cells *(198,300)*.

8. CONCLUDING REMARKS

The purpose of this chapter has been to present an overview of the complex mixture of proteins and carbohydrates that constitute the neuronal ECM, and to relate these to the cellular receptor and signaling processes through which the cells of the nervous system orient themselves and respond. This is a rapidly expanding domain, which is providing many new insights into the mechanisms underlying the cell's interaction with its environment, and the roles that these play in the development and functional maintenance of the nervous tissue. In addition, from an applied viewpoint, it is increasingly clear that the ECM is involved in many pathological conditions, and that a greater understanding of its dynamic nature should contribute to the treatment of such diseases.

REFERENCES

1. Adams, J. C. and Watt, F. M. (1993) Regulation of development and differentiation by the extracellular matrix. *Development* **117,** 1183–1198.
2. Gumbiner, B. M. (1996) Cell adhesion: the molecular basis of tissue architecture and morphogenesis. *Cell* **84,** 345–57.
3. Hutter, H., et al. (2000), Conservation and novelty in the evolution of cell adhesion and extracellular matrix genes. *Science* **287,** 989–994.
4. Hay, E. D. (1993) Extracellular matrix alters epithelial differentiation. *Curr. Opin. Cell Biol.* **5,** 1029–1035.
5. Boudreau, N., Werb, Z., and Bissell, M. J. (1996) Suppression of apoptosis by basement membrane requires three-dimensional tissue organization and withdrawal from the cell cycle. *Proc. Natl. Acad. Sci. USA* **93,** 3509–3513.
6. Lukashev, M. E. and Werb, Z. (1998) ECM signalling: orchestrating cell behaviour and misbehaviour. *Trends Cell. Biol.* **8,** 437–441.
7. Roskelley, C. D., Srebrow, A., and Bissell, M. J. (1995) A hierarchy of ECM-mediated signalling regulates tissue-specific gene expression. *Curr. Opin. Cell Biol.* **7,** 736–747.
8. Streuli, C. (1999) Extracellular matrix remodelling and cellular differentiation. *Curr. Opin. Cell Biol.* **11,** 634–640.
9. Kappler, J., et al. (1997) Chondroitin/dermatan sulphate promotes the survival of neurons from rat embryonic neocortex. *Eur. J. Neurosci.* **9,** 306–318.
10. Brightmann, M. W. and Reese, M. B. T. (1975) Membrane specializations of ependymal cells and astrocytes, in *The Nervous System: The Basic Neurosciences* (Tower, R. B. D., ed.), Raven, New York, pp. 267–277.
11. Nicholson, C. (1995) Extracellular space as the pathway for neuron-glial cell interaction, in *Neuroglia* (Ransom, H. K. B., ed.), Oxford University Press, New York.
12. McKeon, R. J. and Silver, R. M. J. (1995) Functional significance of glial-derived matrix during development and regeneration, in *Neuroglia* (Ransom, H. K. B., ed.), Oxford University Press, New York, pp. 398–410.
13. Nicholson, C. and Sykova, E. (1998) Extracellular space structure revealed by diffusion analysis [see comments]. *Trends Neurosci.* **21,** 207–215.
14. Angevine, J. (1975) The nervous tissue, in *A Textbook of Histology* (Bloom, W. and Fawcett, W. B. D., eds.), WB Saunders, Philadelphia, pp. 333–385.
15. Del Bigio, M. R. (1995) The ependyma: a protective barrier between brain and cerebrospinal fluid. *Glia* **14,** 1–13.
16. Allt, G. and Lawrenson, J. G. (2000) The blood-nerve barrier: enzymes, transporters and receptors: a comparison with the blood-brain barrier. *Brain Res. Bull.* **52,** 1–12.
17. Bignami, A., Hosley, M., and Dahl, D. (1993) Hyaluronic acid and hyaluronic acid-binding proteins in brain extracellular matrix. *Anat. Embryol. (Berl.)* **188,** 419–433.
18. Bruckner, G., et al. (1996) Extracellular matrix organization in various regions of rat brain grey matter. *J. Neurocytol.* **25,** 333–346.
19. Van Harreveld, A. and Steiner, J. (1970) The magnitude of the extracellular space in electron micrographs of superficial and deep regions of the cerebral cortex. *J. Cell Sci.* **6,** 793–805.
20. Letourneau, P. C., Condic, M. L., and Snow, D. M. (1994) Interactions of developing neurons with the extracellular matrix. *J. Neurosci.* **14,** 915–928.

21. Hagg, T., et al. (1989) Laminin-like antigen in rat CNS neurons: distribution and changes upon brain injury and nerve growth factor treatment. *Neuron* **3,** 721–732.
22. Zhou, F. C. (1990) Four patterns of laminin-immunoreactive structure in developing rat brain. *Brain Res. Dev. Brain Res.* **55,** 191–201.
23. Pearlman, A. L. and Sheppard, A. M. (1996) Extracellular matrix in early cortical development. *Progr. Brain Res.* **108,** 117–34.
24. Sheppard, A. M., Hamilton, S. K., and Pearlman, A. L. (1991) Changes in the distribution of extracellular matrix components accompany early morphogenetic events of mammalian cortical development. *J. Neurosci.* **11,** 3928–3942.
25. Sheppard, A. M., et al. (1995) Neuronal production of fibronectin in the cerebral cortex during migration and layer formation is unique to specific cortical domains. *Dev. Biol.* **172,** 504–518.
26. Thomas, T. and Dziadek, M. (1993) Genes coding for basement membrane glycoproteins laminin, nidogen, and collagen IV are differentially expressed in the nervous system and by epithelial, endothelial, and mesenchymal cells of the mouse embryo. *Exp. Cell Res.* **208,** 54–67.
27. Julliard, A. K. and Hartmann, D. J. (1998) Spatiotemporal patterns of expression of extracellular matrix molecules in the developing and adult rat olfactory system. *Neuroscience* **84,** 1135–1150.
28. Adams, J. C. and Tucker, R. P. (2000) The thrombospondin type 1 repeat (TSR) superfamily: diverse proteins with related roles in neuronal development. *Dev. Dyn.* **218,** 280–299.
29. Schachner, M. and Martini, R. (1995) Glycans and the modulation of neural-recognition molecule function. *Trends Neurosci.* **18,** 183–191.
30. Fryer, H. J. and Hockfield, S. (1996) The role of polysialic acid and other carbohydrate polymers in neural structural plasticity. *Curr. Opin. Neurobiol.* **6,** 113–118.
31. Imperiali, B. and O'Connor, S. E. (1999) Effect of N-linked glycosylation on glycopeptide and glycoprotein structure. *Curr. Opin. Chem. Biol.* **3,** 643–649.
32. Varki, A. (1998) Factors controlling the glycosylation potential of the Golgi apparatus. *Trends Cell Biol.* **8,** 34–40.
33. Yamada, Y. and Kleinman, H. K. (1992) Functional domains of cell adhesion molecules. *Curr. Opin. Cell Biol.* **4,** 819–823.
34. Kjellen, L. and Lindahl, U. (1991) Proteoglycans: structures and interactions. *Annu. Rev. Biochem.* **60,** 443–475.
35. Lander, A. D. (1993) Proteoglycans in the nervous system. *Curr. Opin. Neurobiol.* **3,** 716–723.
36. Lander, A. D. (1998) Proteoglycans: master regulators of molecular encounter? *Matrix Biol.* **17,** 465–472.
37. Ruoslahti, E. (1988) Structure and biology of proteoglycans. *Annu. Rev. Cell Biol.* **4,** 229–255.
38. Hileman, R. E., et al. (1998) Glycosaminoglycan-protein interactions: definition of consensus sites in glycosaminoglycan binding proteins. *Bioessays* **20,** 156–167.
39. Pangalos, M. N., et al. (1996) Characterization of appican, the chondroitin sulfate proteoglycan form of the Alzheimer amyloid precursor protein. *Neurodegeneration* **5,** 445–451.
40. Meyer-Puttlitz, B., et al. (1995) Chondroitin sulfate and chondroitin/keratan sulfate proteoglycans of nervous tissue: developmental changes of neurocan and phosphacan. *J. Neurochem.* **65,** 2327–2337.
41. Dow, K. E. and Wang, W. (1998) Cell biology of astrocyte proteoglycans. *Cell Mol. Life Sci.* **54,** 567–581.
42. Small, D. H., et al. (1996) Role of proteoglycans in neural development, regeneration, and the aging brain. *J. Neurochem.* **67,** 889–899.

43. Oohira, A., et al. (2000) Molecular interactions of neural chondroitin sulfate proteoglycans in the brain development. *Arch. Biochem. Biophys.* **374,** 24–34.
44. Grumet, M., Friedlander, D. R., and Sakurai, T. (1996) Functions of brain chondroitin sulfate proteoglycans during developments: interactions with adhesion molecules. *Perspect. Dev. Neurobiol.* **3,** 319–330.
45. Herndon, M. E. and Lander, A. D. (1990) A diverse set of developmentally regulated proteoglycans is expressed in the rat central nervous system. *Neuron* **4,** 949–961.
46. Lee, J. Y. and Spicer, A. P. (2000) Hyaluronan: a multifunctional, megaDalton, stealth molecule. *Curr. Opin. Cell Biol.* **12,** 581–586.
47. Marks, M. S., Chi-Rosso, G., and Toole, B. P. (1990) Hyaluronate-binding proteins of murine brain. *J. Neurochem.* **54,** 171–180.
48. Gary, S. C., Kelly, G. M., and Hockfield, S. (1998) BEHAB/brevican: a brain-specific lectican implicated in gliomas and glial cell motility. *Curr. Opin. Neurobiol.* **8,** 576–581.
49. Jaworski, D. M., Kelly, G. M., and Hockfield, S. BEHAB, a new member of the proteoglycan tandem repeat family of hyaluronan-binding proteins that is restricted to the brain. *J. Cell Biol.* **125,** 495–509.
50. Yamaguchi, Y. (1996) Brevican: a major proteoglycan in adult brain. *Perspect. Dev. Neurobiol.* **3,** 307–317.
51. Takahashi, K., et al. (1996) Keratan sulfate modification of CD44 modulates adhesion to hyaluronate. *J. Biol. Chem.* **271,** 9490–9496.
52. Oliferenko, S., et al. (2000) Hyaluronic acid (HA) binding to CD44 activates Rac1 and induces lamellipodia outgrowth. *J. Cell Biol.* **148,** 1159–1164.
53. Nagy, J. I., et al. (1995) Requirement of the hyaluronan receptor RHAMM in neurite extension and motility as demonstrated in primary neurons and neuronal cell lines. *J. Neurosci.* **15(1 Pt 1),** 241–252.
54. Nagy, J. I., et al. (1998) The hyaluronan receptor RHAMM in noradrenergic fibers contributes to axon growth capacity of locus coeruleus neurons in an intraocular transplant model. *Neuroscience* **86,** 241–255.
55. Turley, E. A., et al. (1994) Astrocyte and microglial motility in vitro is functionally dependent on the hyaluronan receptor RHAMM. *Glia* **12,** 68–80.
56. Hofmann, M., et al. (1998) Problems with RHAMM: a new link between surface adhesion and oncogenesis? *Cell* **95,** 591–592; discussion 592–593.
57. Iozzo, R. V. (1999) The biology of the small leucine-rich proteoglycans. Functional network of interactive proteins. *J. Biol. Chem.* **274,** 18,843–18,846.
58. Kappler, J., et al. (1998) Developmental regulation of decorin expression in postnatal rat brain. *Brain Res.* **793,** 328–332.
59. Ruoslahti, E. (1996) Brain extracellular matrix. *Glycobiology* **6,** 489–492.
60. Bovolenta, P. and Fernaud-Espinosa, I. (2000) Nervous system proteoglycans as modulators of neurite outgrowth. *Progr. Neurobiol.* **61,** 113–132.
61. Yamaguchi, Y. (2000) Lecticans: organizers of the brain extracellular matrix. *Cell Mol. Life Sci.* **57,** 276–289.
62. Rauch, U., et al. (1992) Cloning and primary structure of neurocan, a developmentally regulated, aggregating chondroitin sulfate proteoglycan of brain. *J. Biol. Chem.* **267,** 19,536–19,547.
63. Seidenbecher, C. I., et al. (1998) Transcripts for secreted and GPI-anchored brevican are differentially distributed in rat brain. *Eur. J. Neurosci.* **10,** 1621–1630.
64. Milev, P., et al. (1998) Differential regulation of expression of hyaluronan-binding proteoglycans in developing brain: aggrecan, versican, neurocan, and brevican. *Biochem. Biophys. Res. Commun.* **247,** 207–212.
65. Schmalfeldt, M., et al. (1998) Versican V2 is a major extracellular matrix component of the mature bovine brain. *J. Biol. Chem.* **273,** 15,758–15,764.

66. Schmalfeldt, M., et al. (2000) Brain derived versican V2 is a potent inhibitor of axonal growth. *J. Cell Sci.* **113(Pt 5),** 807–816.
67. Niederost, B. P., et al. (1999) Bovine CNS myelin contains neurite growth-inhibitory activity associated with chondroitin sulfate proteoglycans. *J. Neurosci.* **19,** 8979–8989.
68. Yang, B. L., et al. (1999) Cell adhesion and proliferation mediated through the G1 domain of versican. *J. Cell Biochem.* **72,** 210–220.
69. Zhang, Y., et al. (1998) The G3 domain of versican enhances cell proliferation via epidermial growth factor-like motifs. *J. Biol. Chem.* **273,** 21,342–21,351.
70. Garwood, J., et al. (1999) DSD-1-proteoglycan is the mouse homolog of phosphacan and displays opposing effects on neurite outgrowth dependent on neuronal lineage. *J. Neurosci.* **19,** 3888–3899.
71. Maeda, N., et al. (1996) 6B4 proteoglycan/phosphacan, an extracellular variant of receptor-like protein-tyrosine phosphatase zeta/RPTPbeta, binds pleiotrophin/heparin-binding growth-associated molecule (HB-GAM). *J. Biol. Chem.* **271,** 21,446–21,452.
72. Milev, P., et al. (1994) Interactions of the chondroitin sulfate proteoglycan phosphacan, the extracellular domain of a receptor-type protein tyrosine phosphatase, with neurons, glia, and neural cell adhesion molecules. *J. Cell Biol.* **127(6 Pt 1),** 1703–1715.
73. Maurel, P., et al. (1995) Nucleotide sequence and molecular variants of rat receptor-type protein tyrosine phosphatase-zeta/beta. *DNA Seq.* **5,** 323–328.
74. Krueger, N. X. and Saito, H. (1992) A human transmembrane protein-tyrosine-phosphatase, PTP zeta, is expressed in brain and has an N-terminal receptor domain homologous to carbonic anhydrases. *Proc. Natl. Acad. Sci. USA* **89,** 7417–7421.
75. Allendoerfer, K. L., Magnani, J. L., and Patterson, P. H. (1995) FORSE-1, an antibody that labels regionally restricted subpopulations of progenitor cells in the embryonic central nervous system, recognizes the Le(x) carbohydrate on a proteoglycan and two glycolipid antigens. *Mol. Cell Neurosci.* **6,** 381–395.
76. Engel, M., et al. (1996) Chondroitin sulfate proteoglycans in the developing central nervous system. I. cellular sites of synthesis of neurocan and phosphacan. *J. Comp. Neurol.* **366,** 34–43.
77. Meyer-Puttlitz, B., et al. (1996) Chondroitin sulfate proteoglycans in the developing central nervous system. II. Immunocytochemical localization of neurocan and phosphacan. *J. Comp. Neurol.* **366,** 44–54.
78. Milev, P., et al. (1996) TAG-1/axonin-1 is a high-affinity ligand of neurocan, phosphacan/protein-tyrosine phosphatase-zeta/beta, and N-CAM. *J. Biol. Chem.* **271,** 15,716–15,723.
79. Milev, P., et al. (1995) Complex-type asparagine-linked oligosaccharides on phosphacan and protein-tyrosine phosphatase-zeta/beta mediate their binding to neural cell adhesion molecules and tenascin. *J. Biol. Chem.* **270,** 24,650–24,653.
80. Milev, P., et al. (1997) The fibrinogen-like globe of tenascin-C mediates its interactions with neurocan and phosphacan/protein-tyrosine phosphatase-zeta/beta. *J. Biol. Chem.* **272,** 15,501–15,509.
81. Heyman, I., Faissner, A., and Lumsden, A. (1995) Cell and matrix specialisations of rhombomere boundaries. *Dev. Dyn.* **204,** 301–315.
82. Steindler, D. A., et al. (1995) Tenascin knockout mice: barrels, boundary molecules, and glial scars. *J. Neurosci.* **15,** 1971–1983.
83. Bovolenta, P., Wandosell, F., and Nieto-Sampedro, M. (1993) Neurite outgrowth inhibitors associated with glial cells and glial cell lines. *Neuroreport* **5,** 345–348.
84. McKeon, R. J., et al. (1991) Reduction of neurite outgrowth in a model of glial scarring following CNS injury is correlated with the expression of inhibitory molecules on reactive astrocytes. *J. Neurosci.* **11,** 3398–3411.

85. Pindzola, R. R., Doller, C., and Silver, J. (1993) Putative inhibitory extracellular matrix molecules at the dorsal root entry zone of the spinal cord during development and after root and sciatic nerve lesions. *Dev. Biol.* **156,** 34–48.
86. Lips, K., Stichel, C. C., and Muller, H. W. (1995) Restricted appearance of tenascin and chondroitin sulphate proteoglycans after transection and sprouting of adult rat post-commissural fornix. *J. Neurocytol.* **24,** 449–464.
87. Barker, R. A., et al. (1996) The time course of loss of dopaminergic neurons and the gliotic reaction surrounding grafts of embryonic mesencephalon to the striatum. *Exp. Neurol.* **141,** 79–93.
88. Laywell, E. D. and Steindler, D. A. (1991) Boundaries and wounds, glia and glyco-conjugates. Cellular and molecular analyses of developmental partitions and adult brain lesions. *Ann. NY Acad. Sci.* **633,** 122–141.
89. Deller, T., et al. (1997) Up-regulation of astrocyte-derived tenascin-C correlates with neurite outgrowth in the rat dentate gyrus after unilateral entorhinal cortex lesion. *Neuroscience* **81,** 829–846.
90. Bernfield, M., et al. (1992) Biology of the syndecans: a family of transmembrane heparan sulfate proteoglycans. *Annu. Rev. Cell Biol.* **8,** 365–393.
91. Carey, D. J. (1997) Syndecans: multifunctional cell-surface co-receptors. *Biochem. J.* **327,** 1–16.
92. Couchman, J. R. and Woods, A. (1996) Syndecans, signaling, and cell adhesion. *J. Cell Biochem.* **61,** 578–584.
93. Grootjans, J. J., et al. (1997) Syntenin, a PDZ protein that binds syndecan cytoplasmic domains. *Proc. Natl. Acad. Sci. USA* **94,** 13,683–13,688.
94. Hsueh, Y. P. and Sheng, M. (1999) Regulated expression and subcellular localization of syndecan heparan sulfate proteoglycans and the syndecan-binding protein CASK/LIN-2 during rat brain development. *J. Neurosci.* **19,** 7415–7425.
95. Hsueh, Y. P., et al. (2000) Nuclear translocation and transcription regulation by the membrane-associated guanylate kinase CASK/LIN-2. *Nature* **404,** 298–302.
96. Lander, A. D., Stipp, C. S., and Ivins, J. K. (1996) The glypican family of heparan sulfate proteoglycans: major cell-surface proteoglycans of the developing nervous system. *Perspect. Dev. Neurobiol.* **3,** 347–358.
97. Veugelers, M., et al. (1997) Characterization of glypican-5 and chromosomal localization of human GPC5, a new member of the glypican gene family. *Genomics* **40,** 24–30.
98. Rosenberg, R. D., et al. (1997) Heparan sulfate proteoglycans of the cardiovascular system. Specific structures emerge but how is synthesis regulated? *J. Clin. Invest.* **100(11 Suppl),** S67–75.
99. Levine, J. M. and Nishiyama, A. (1996) The NG2 chondroitin sulfate proteoglycan: a multifunctional proteoglycan associated with immature cells. *Perspect. Dev. Neurobiol.* **3,** 245–259.
100. Lin, X. H., Dahlin-Huppe, K., and Stallcup, W. B. (1996) Interaction of the NG2 proteoglycan with the actin cytoskeleton. *J. Cell Biochem.* **63,** 463–477.
101. Nishiyama, A., Chang, A., and Trapp, B. D. (1999) NG2+ glial cells: a novel glial cell population in the adult brain. *J. Neuropathol. Exp. Neurol.* **58,** 1113–1124.
102. Dou, C. L. and Levine, J. M. (1994) Inhibition of neurite growth by the NG2 chondroitin sulfate proteoglycan. *J. Neurosci.* **14,** 7616–7628.
103. Levine, J. M. (1994) Increased expression of the NG2 chondroitin-sulfate proteoglycan after brain injury. *J. Neurosci.* **14,** 4716–4730.
104. Dou, C. L. and Levine, J. M. (1997) Identification of a neuronal cell surface receptor for a growth inhibitory chondroitin sulfate proteoglycan (NG2). *J. Neurochem.* **68,** 1021–1030.
105. Tillet, E., et al. (1997) The membrane-spanning proteoglycan NG2 binds to collagens V and VI through the central nonglobular domain of its core protein. *J. Biol. Chem.* **272,** 10,769–10,776.

106. Burg, M. A., et al. (1996) Binding of the NG2 proteoglycan to type VI collagen and other extracellular matrix molecules. *J. Biol. Chem.* **271,** 26,110–26,116.
107. Wu, A., et al. (1997) Appican expression induces morphological changes in C6 glioma cells and promotes adhesion of neural cells to the extracellular matrix. *J. Neurosci.* **17,** 4987–4993.
108. Salinero, O., Moreno-Flores, M. T., and Wandosell, F. (2000) Increasing neurite outgrowth capacity of beta-amyloid precursor protein proteoglycan in Alzheimer's disease. *J. Neurosci. Res.* **60,** 87–97.
109. Rogers, S. L., Letourneau, P. C., and Pech, I. V. (1989) The role of fibronectin in neural development. *Dev. Neurosci.* **11,** 248–265.
110. Schwarzbauer, J. E. and Sechler, J. L. (1999) Fibronectin fibrillogenesis: a paradigm for extracellular matrix assembly. *Curr. Opin. Cell Biol.* **11,** 622–627.
111. Pankov, R., et al. (2000) Integrin dynamics and matrix assembly: tensin-dependent translocation of alpha(5)beta(1) integrins promotes early fibronectin fibrillogenesis. *J. Cell Biol.* **148,** 1075–1090.
112. Sephel, G. C., Burrous, B. A., and Kleinman, H. K. (1989) Laminin neural activity and binding proteins. *Dev. Neurosci.* **11,** 313–331.
113. Luckenbill-Edds, L. (1997) Laminin and the mechanism of neuronal outgrowth. *Brain Res. Brain Res. Rev.* **23,** 1–27.
114. Colognato, H. and Yurchenco, P. D. (2000) Form and function: the laminin family of heterotrimers. *Dev. Dyn.* **218,** 213–234.
115. Burgeson, R. E., et al. (1994) A new nomenclature for the laminins. *Matrix Biol.* **14,** 209–211.
116. Colognato, H., Winkelmann, D. A., and Yurchenco, P. D. (1999) Laminin polymerization induces a receptor-cytoskeleton network. *J. Cell Biol.* **145,** 619–631.
117. Georges-Labouesse, E., et al. (1998) Essential role of alpha 6 integrins in cortical and retinal lamination. *Curr. Biol.* **8,** 983–986.
118. Jacques, T. S., et al. (1998) Neural precursor cell chain migration and division are regulated through different beta1 integrins. *Development* **125,** 3167–3177.
119. Durkin, M. E., et al. (1988) Amino acid sequence and domain structure of entactin. Homology with epidermal growth factor precursor and low density lipoprotein receptor. *J. Cell Biol.* **107(6 Pt 2),** 2749–2756.
120. Colognato, H., et al. (1997) The laminin alpha2-chain short arm mediates cell adhesion through both the alpha1beta1 and alpha2beta1 integrins. *J. Biol. Chem.* **272,** 29,330–29,336.
121. Durbeej, M., Henry, M. D., and Campbell, K. P. (1998) Dystroglycan in development and disease. *Curr. Opin. Cell Biol.* **10,** 594–601.
122. Smalheiser, N. R. and Schwartz, N. B. (1987) Cranin: a laminin-binding protein of cell membranes. *Proc. Natl. Acad. Sci. USA* **84,** 6457–6461.
123. Smalheiser, N. R. and Kim, E. (1995) Purification of cranin, a laminin binding membrane protein. Identity with dystroglycan and reassessment of its carbohydrate moieties. *J. Biol. Chem.* **270,** 15,425–15,433.
124. Sunada, Y., et al. (1995) Identification of a novel mutant transcript of laminin alpha 2 chain gene responsible for muscular dystrophy and dysmyelination in dy2J mice. *Hum. Mol. Genet.* **4,** 1055–1061.
125. van der Knaap, M. S., et al. (1997) Magnetic resonance imaging in classification of congenital muscular dystrophies with brain abnormalities. *Ann. Neurol.* **42,** 50–59.
126. Buttery, P. C. and ffrech-Constant, C. (1999) Laminin-2/integrin interactions enhance myelin membrane formation by oligodendrocytes. *Mol. Cell Neurosci.* **14,** 199–212.
127. Lawler, J. (2000) The functions of thrombospondin-1 and -2. *Curr. Opin. Cell Biol.* **12,** 634–640.
128. Lawler, J., Weinstein, R., and Hynes, R. O. (1988) Cell attachment to thrombospondin: the role of ARG-GLY-ASP, calcium, and integrin receptors. *J. Cell Biol.* **107(6 Pt 1),** 2351–2361.

129. Gao, A. G., et al. (1996) Thrombospondin modulates alpha v beta 3 function through integrin-associated protein. *J. Cell Biol.* **135,** 533–544.
130. DeFreitas, M. F., et al. (1995) Identification of integrin alpha 3 beta 1 as a neuronal thrombospondin receptor mediating neurite outgrowth. *Neuron* **15,** 333–343.
131. Chamak, B., Morandi, V., and Mallat, M. (1994) Brain macrophages stimulate neurite growth and regeneration by secreting thrombospondin. *J. Neurosci. Res.* **38,** 221–233.
132. Arber, S. and Caroni, P. (1995) Thrombospondin-4, an extracellular matrix protein expressed in the developing and adult nervous system promotes neurite outgrowth. *J. Cell Biol.* **131,** 1083–1094.
133. Klar, A., Baldassare, M., and Jessell, T. M. (1992) F-spondin: a gene expressed at high levels in the floor plate encodes a secreted protein that promotes neural cell adhesion and neurite extension. *Cell* **69,** 95–110.
134. Burstyn-Cohen, T., et al. (1999) F-Spondin is required for accurate pathfinding of commissural axons at the floor plate. *Neuron* **23,** 233–246.
135. Feinstein, Y., et al. (1999) F-spondin and mindin: two structurally and functionally related genes expressed in the hippocampus that promote outgrowth of embryonic hippocampal neurons. *Development* **126,** 3637–3648.
136. Faissner, A. (1997) The tenascin gene family in axon growth and guidance. *Cell Tissue Res.* **290,** 331–341.
137. Chiquet-Ehrismann, R., Hagios, C., and Schenk, S. (1995) The complexity in regulating the expression of tenascins. *Bioessays* **17,** 873–878.
138. Joester, A. and Faissner, A. (1999) Evidence for combinatorial variability of tenascin-C isoforms and developmental regulation in the mouse central nervous system. *J. Biol. Chem.* **274,** 17,144–17,151.
139. Perez, R. G. and Halfter, W. (1993) Tenascin in the developing chick visual system: distribution and potential role as a modulator of retinal axon growth. *Dev. Biol.* **156,** 278–292.
140. Mitrovic, N., Dorries, U., and Schachner, M. (1994) Expression of the extracellular matrix glycoprotein tenascin in the somatosensory cortex of the mouse during postnatal development: an immunocytochemical and in situ hybridization analysis. *J. Neurocytol.* **23,** 364–378.
141. Bartsch, U., et al. (1994) Tenascin demarcates the boundary between the myelinated and nonmyelinated part of retinal ganglion cell axons in the developing and adult mouse. *J. Neurosci.* **14,** 4756–4768.
142. Meiners, S., Powell, E. M., and Geller, H. M. (1995) A distinct subset of tenascin/CS-6-PG-rich astrocytes restricts neuronal growth in vitro. *J. Neurosci.* **15,** 8096–8108.
143. Gotz, B., et al. (1996) Tenascin-C contains distinct adhesive, anti-adhesive, and neurite outgrowth promoting sites for neurons. *J. Cell Biol.* **132,** 681–699.
144. Gotz, M., et al. (1997) Tenascin-C synthesis and influence on axonal growth during rat cortical development. *Eur. J. Neurosci.* **9,** 496–506.
145. Leprini, A., et al. (1998) Rat tenascin-R gene: structure, chromosome location and transcriptional activity of promoter and exon 1. *Cytogenet. Cell Genet.* **83,** 115–123.
146. Weber, P., et al. (1999) Mice deficient for tenascin-R display alterations of the extracellular matrix and decreased axonal conduction velocities in the CNS. *J. Neurosci.* **19,** 4245–4262.
147. Haunso, A., et al. (2000) Morphology of perineuronal nets in tenascin-R and parvalbumin single and double knockout mice. *Brain Res.* **864,** 142–145.
148. Xiao, Z. C., et al. (1997) Isolation of a tenascin-R binding protein from mouse brain membranes. A phosphacan-related chondroitin sulfate proteoglycan. *J. Biol. Chem.* **272,** 32,092–32,101.
149. Xiao, Z. C., et al. (1999) Tenascin-R is a functional modulator of sodium channel beta subunits. *J. Biol. Chem.* **274,** 26,511–26,517.

150. Burkhardt-Holm, P., et al. (1996) Testosterone elevates expression of tenascin-R and oligomannosidic carbohydrates in developing male zebra finches. *J. Neurosci. Res.* **46,** 385–392.
151. Jones, P. L., et al. (1999) Induction of vascular smooth muscle cell tenascin-C gene expression by denatured type I collagen is dependent upon a beta3 integrin-mediated mitogen-activated protein kinase pathway and a 122-base pair promoter element. *J. Cell Sci.* **112(Pt 4),** 435–445.
152. Gonzalez-Sancho, J. M., Alvarez-Dolado, M., and Munoz, A. (1998) 1,25-Dihydroxy-vitamin D3 inhibits tenascin-C expression in mammary epithelial cells. *FEBS Lett.* **426,** 225–228.
153. Gherzi, R., et al. (1998) Structure of 5' region of human tenascin-R gene and characterization of its promoter. *DNA Cell Biol.* **17,** 275–282.
154. Minamitani, T., Ariga, H., and Matsumoto, K. (2000) Transcription factor Sp1 activates the expression of the mouse tenascin-X gene. *Biochem. Biophys. Res. Commun.* **267,** 626–631.
155. Shirasaki, F., et al. (1999) Ets transcription factors cooperate with Sp1 to activate the human tenascin-C promoter. *Oncogene* **18,** 7755–7764.
156. Alvarez-Dolado, M., et al. (1999) Retinoic acid and 1,25-dihydroxyvitamin D3 inhibit tenascin-C expression in rat glioma C6 cells. *J. Neurosci. Res.* **58,** 293–300.
157. Gherzi, R., et al. (1997) The human homeodomain protein OTX2 binds to the human tenascin-C promoter and trans-represses its activity in transfected cells. *DNA Cell Biol.* **16,** 559–567.
158. Copertino, D. W., Edelman, G. M., and Jones, F. S. (1997) Multiple promoter elements differentially regulate the expression of the mouse tenascin gene. *Proc. Natl. Acad. Sci. USA* **94,** 1846–1851.
159. Speek, M., Barry, F., and Miller, W. L. (1996) Alternate promoters and alternate splicing of human tenascin-X, a gene with 5' and 3' ends buried in other genes. *Hum. Mol. Genet.* **5,** 1749–1758.
160. Nieto, M. A. (1996) Molecular biology of axon guidance. *Neuron* **17,** 1039–1048.
161. Tessier-Lavigne, M. and Goodman, C. S. (1996) The molecular biology of axon guidance. *Science* **274,** 1123–1133.
162. Serafini, T., et al. (1994) The netrins define a family of axon outgrowth-promoting proteins homologous to C. elegans UNC-6. *Cell* **78,** 409–424.
163. Colamarino, S. A. and Tessier-Lavigne, M. (1995) The axonal chemoattractant netrin-1 is also a chemorepellent for trochlear motor axons. *Cell* **81,** 621–629.
164. Kennedy, T. E., et al. (1994) Netrins are diffusible chemotropic factors for commissural axons in the embryonic spinal cord. *Cell* **78,** 425–435.
165. Culotti, J. G. and Kolodkin, A. L. (1996) Functions of netrins and semaphorins in axon guidance. *Curr. Opin. Neurobiol.* **6,** 81–88.
166. Culotti, J. G. and Merz, D. C. (1998) DCC and netrins. *Curr. Opin. Cell Biol.* **10,** 609–613.
167. Bagnard, D., et al. (1998) Semaphorins act as attractive and repulsive guidance signals during the development of cortical projections. *Development* **125,** 5043–5053.
168. Bagnard, D., et al. (2000) Spatial distributions of guidance molecules regulate chemorepulsion and chemoattraction of growth cones. *J. Neurosci.* **20,** 1030–1035.
169. Raper, J. A. (2000) Semaphorins and their receptors in vertebrates and invertebrates. *Curr. Opin. Neurobiol.* **10,** 88–94.
170. Mark, M. D., Lohrum, M., and Puschel, A. W. (1997) Patterning neuronal connections by chemorepulsion: the semaphorins. *Cell Tissue Res.* **290,** 299–306.
171. Kapfhammer, J. P. and Raper, J. A. (1987) Collapse of growth cone structure on contact with specific neurites in culture. *J. Neurosci.* **7,** 201–212.
172. Luo, Y., Raible, D., and Raper, J. A. (1993) Collapsin: a protein in brain that induces the collapse and paralysis of neuronal growth cones. *Cell* **75,** 217–227.

173. Polleux, F., Morrow, T., and Ghosh, A. (2000) Semaphorin 3A is a chemoattractant for cortical apical dendrites. *Nature* **404,** 567–573.
174. Van Vactor, D. and Flanagan, J. G. (1999) The middle and the end: slit brings guidance and branching together in axon pathway selection. *Neuron* **22,** 649–652.
175. Zinn, K. and Sun, Q. (1999) Slit branches out: a secreted protein mediates both attractive and repulsive axon guidance. *Cell* **97,** 1–4.
176. Brose, K. and Tessier-Lavigne, M. (2000) Slit proteins: key regulators of axon guidance, axonal branching, and cell migration. *Curr. Opin. Neurobiol.* **10,** 95–102.
177. Liang, Y., et al. (1999) Mammalian homologues of the Drosophila slit protein are ligands of the heparan sulfate proteoglycan glypican-1 in brain. *J. Biol. Chem.* **274,** 17,885–17,892.
178. Wang, K. H., et al. (1999) Biochemical purification of a mammalian slit protein as a positive regulator of sensory axon elongation and branching. *Cell* **96,** 771–784.
179. Zheng, C., Heintz, N., and Hatten, M. E. (1996) CNS gene encoding astrotactin, which supports neuronal migration along glial fibers. *Science* **272,** 417–419.
180. Fink, J. M., et al. (1997) Astrotactin (ASTN), a gene for glial-guided neuronal migration, maps to human chromosome 1q25. 2. *Genomics* **40,** 202–205.
181. Hatten, M. E. (1999) Central nervous system neuronal migration. *Annu. Rev. Neurosci.* **22,** 511–539.
182. Gilmore, E. C. and Herrup, K. (2000) Cortical development: receiving reelin. *Curr. Biol.* **10,** R162–166.
183. Curran, T. and D'Arcangelo, G. (1998) Role of reelin in the control of brain development. *Brain Res. Brain Res. Rev.* **26,** 285–294.
184. Hiesberger, T., et al. (1999) Direct binding of Reelin to VLDL receptor and ApoE receptor 2 induces tyrosine phosphorylation of disabled-1 and modulates tau phosphorylation. *Neuron* **24,** 481–489.
185. D'Arcangelo, G., et al. (1999) Reelin is a ligand for lipoprotein receptors. *Neuron* **24,** 471–479.
186. Dulabon, L., et al. (2000) Reelin binds alpha3beta1 integrin and inhibits neuronal migration. *Neuron* **27,** 33–44.
187. Senzaki, K., Ogawa, M., and Yagi, T. (1999) Proteins of the CNR family are multiple receptors for Reelin. *Cell* **99(6),** 635–647.
188. Bornstein, P. and Sage, H. (1980) Structurally distinct collagen types. *Annu. Rev. Biochem.* **49,** 957–1003.
189. Vuorio, E. and de Crombrugghe, B. (1990) The family of collagen genes. *Annu. Rev. Biochem.* **59,** 837–872.
190. Linsenmayer, T. F. (1991) Collagen, in *Cell Biology of the Extracellular Matrix* (Hay, E. D., ed.), Plenum, New York.
191. Olsen, B. R. (1995) New insights into the function of collagens from genetic analysis. *Curr. Opin. Cell Biol.* **7,** 720–727.
192. Sievers, J., et al. (1994) Meningeal cells organize the superficial glia limitans of the cerebellum and produce components of both the interstitial matrix and the basement membrane. *J. Neurocytol.* **23,** 135–149.
193. Goldberg, H., et al. (1992) Tissue-specific expression of the mouse alpha 2(I) collagen promoter. Studies in transgenic mice and in tissue culture cells. *J. Biol. Chem.* **267,** 19,622–19,630.
194. Shellswell, G. B., et al. (1979) Identification and differential distribution of collagen types in the central and peripheral nervous systems. *FEBS Lett.* **106,** 305–308.
195. Pittman, R. N. and Buettner, H. M. (1989) Degradation of extracellular matrix by neuronal proteases. *Dev. Neurosci.* **11,** 361–375.
196. Murphy, G. and Gavrilovic, J. (1999) Proteolysis and cell migration: creating a path? *Curr. Opin. Cell Biol.* **11,** 614–621.

197. Johnson, L. L., Dyer, R., and Hupe, D. J. (1998) Matrix metalloproteinases. *Curr. Opin. Chem. Biol.* **2,** 466–471.
198. Yong, V. W., et al. (1998) Matrix metalloproteinases and diseases of the CNS. *Trends Neurosci.* **21,** 75–80.
199. Vaillant, C., et al. (1999) Spatiotemporal expression patterns of metalloproteinases and their inhibitors in the postnatal developing rat cerebellum. *J. Neurosci.* **19,** 4994–5004.
200. Zuo, J., et al. (1998) Degradation of chondroitin sulfate proteoglycan enhances the neurite-promoting potential of spinal cord tissue. *Exp. Neurol.* **154,** 654–662.
201. Zuo, J., et al. (1998) Neuronal matrix metalloproteinase-2 degrades and inactivates a neurite-inhibiting chondroitin sulfate proteoglycan. *J. Neurosci.* **18,** 5203–5211.
202. Campbell, I. L. and Pagenstecher, A. (1999) Matrix metalloproteinases and their inhibitors in the nervous system: the good, the bad and the enigmatic. *Trends Neurosci.* **22,** 285–287.
203. Fukuda, M., Hiraoka, N., and Yeh, J. C. (1999) C-type lectins and sialyl Lewis X oligosaccharides. Versatile roles in cell-cell interaction. *J. Cell Biol.* **147,** 467–470.
204. Probstmeier, R. and Pesheva, P. (1999) I-type lectins in the nervous system. *Progr. Neurobiol.* **58,** 163–184.
205. Elgavish, S. and Shaanan, B. (1997) Lectin-carbohydrate interactions: different folds, common recognition principles. *Trends Biochem. Sci.* **22,** 462–467.
206. Brinkman-Van der Linden, E. C. and Varki, A. (2000) New aspects of siglec binding specificities, including the significance of fucosylation and of the sialyl-Tn epitope. Sialic acid-binding immunoglobulin superfamily lectins. *J. Biol. Chem.* **275,** 8625–8632.
207. Dallas, S. L. (2000) Measuring interactions between ECM and TGF beta-like proteins. *Methods Mol. Biol.* **139,** 231–243.
208. Murphy-Ullrich, J. E. and Poczatek, M. (2000) Activation of latent TGF-beta by thrombospondin-1: mechanisms and physiology. *Cytokine Growth Factor Rev.* **11,** 59–69.
209. Aktas, G. and Kayton, R. (2000) Ultrastructural immunolocalization of basic fibroblast growth factor in fibroblasts and extracellular matrix. *Histochem. Cell Biol.* **113,** 227–233.
210. Raines, E. W., Koyama, H., and Carragher, N. O. (2000) The extracellular matrix dynamically regulates smooth muscle cell responsiveness to PDGF. *Ann. NY Acad. Sci.* **902,** 39–51; discussion 51–52.
211. Ornitz, D. M. (2000) FGFs, heparan sulfate and FGFRs: complex interactions essential for development. *Bioessays* **22,** 108–112.
212. Selleck, S. B. (2000) Proteoglycans and pattern formation: sugar biochemistry meets developmental genetics. *Trends Genet.* **16,** 206–212.
213. Perrimon, N. and Bernfield, M. (2000) Specificities of heparan sulphate proteoglycans in developmental processes. *Nature* **404,** 725–728.
214. Lekmine, F., et al. (1999) Influence of laminin substratum on cell proliferation and CALC I gene expression in medullary thyroid carcinoma C cell lines. *Mol. Cell Endocrinol.* **157,** 181–189.
215. Seiffert, M., et al. (1998) Mitogenic and adhesive effects of tenascin-C on human hematopoietic cells are mediated by various functional domains. *Matrix Biol.* **17,** 47–63.
216. Taraboletti, G., et al. (1990) Platelet thrombospondin modulates endothelial cell adhesion, motility, and growth: a potential angiogenesis regulatory factor. *J. Cell Biol.* **111,** 765–772.
217. Engel, J. (1989) EGF-like domains in extracellular matrix proteins: localized signals for growth and differentiation? *FEBS Lett.* **251,** 1–7.
218. Grumet, M. (1991) Cell adhesion molecules and their subgroups in the nervous system. *Curr. Opin. Neurobiol.* **1,** 370–376.
219. Kasper, C., et al. (2000) Structural basis of cell-cell adhesion by NCAM. *Nat. Struct. Biol.* **7,** 389–393.

220. Chothia, C. and Jones, E. Y. (1997) The molecular structure of cell adhesion molecules. *Annu. Rev. Biochem.* **66,** 823–862.
221. Humphries, M. J. and Newham, P. (1998) The structure of cell-adhesion molecules. *Trends Cell Biol.* **8,** 78–83.
222. Brummendorf, T. and Rathjen, F. G. (1993) Axonal glycoproteins with immunoglobulin- and fibronectin type III-related domains in vertebrates: structural features, binding activities, and signal transduction. *J. Neurochem.* **61,** 1207–1219.
223. Grumet, M., et al. (1991) Structure of a new nervous system glycoprotein, Nr-CAM, and its relationship to subgroups of neural cell adhesion molecules. *J. Cell Biol.* **113,** 1399–1412.
224. Grumet, M. (1992) Structure, expression, and function of Ng-CAM, a member of the immunoglobulin superfamily involved in neuron-neuron and neuron-glia adhesion. *J. Neurosci. Res.* **31,** 1–13.
225. Murase, S. and Schuman, E. M. (1999) The role of cell adhesion molecules in synaptic plasticity and memory. *Curr. Opin. Cell Biol.* **11,** 549–553.
226. Kiss, J. Z. (1998) A role of adhesion molecules in neuroglial plasticity. *Mol. Cell Endocrinol.* **140,** 89–94.
227. Doherty, P., Fazeli, M. S., and Walsh, F. S. (1995) The neural cell adhesion molecule and synaptic plasticity. *J. Neurobiol.* **26,** 437–446.
228. Storms, S. D. and Rutishauser, U. (1998) A role for polysialic acid in neural cell adhesion molecule heterophilic binding to proteoglycans. *J. Biol. Chem.* **273,** 27,124–27,129.
229. Kiss, J. Z. and Rougon, G. (1997) Cell biology of polysialic acid. *Curr. Opin. Neurobiol.* **7,** 640–646.
230. Hynes, R. O. (1992) Integrins: versatility, modulation, and signaling in cell adhesion. *Cell* **69,** 11–25.
231. Ruoslahti, E. (1996) Integrin signaling and matrix assembly. *Tumour Biol.* **17,** 117–124.
232. Jones, L. S. (1996) Integrins: possible functions in the adult CNS. *Trends Neurosci.* **19,** 68–72.
233. Kansas, G. S. (1996) Selectins and their ligands: current concepts and controversies. *Blood* **88,** 3259–3287.
234. Whelan, J. (1996) Selectin synthesis and inflammation. *Trends Biochem. Sci.* **21,** 65–69.
235. Hynes, R. O. (1992) Specificity of cell adhesion in development: the cadherin superfamily. *Curr. Opin. Genet. Dev.* **2,** 621–624.
236. Yagi, T. and Takeichi, M. (2000) Cadherin superfamily genes: functions, genomic organization, and neurologic diversity. *Genes Dev.* **14,** 1169–1180.
237. Redies, C. (2000) Cadherins in the central nervous system. *Progr. Neurobiol.* **61,** 611–648.
238. Shapiro, L., et al. (1995) Structural basis of cell-cell adhesion by cadherins [see comments]. *Nature* **374,** 327–337.
239. Shapiro, L. and Colman, D. R. (1998) Structural biology of cadherins in the nervous system. *Curr. Opin. Neurobiol.* **8,** 593–599.
240. Troyanovsky, S. M. (1999) Mechanism of cell-cell adhesion complex assembly. *Curr. Opin. Cell Biol.* **11,** 561–566.
241. Pertz, O., et al. (1999) A new crystal structure, Ca2+ dependence and mutational analysis reveal molecular details of E-cadherin homoassociation. *EMBO J.* **18,** 1738–1747.
242. Shapiro, L. and Colman, D. R. (1999) The diversity of cadherins and implications for a synaptic adhesive code in the CNS. *Neuron* **23,** 427–430.
243. Juliano, R. L. and Haskill, S. (1993) Signal transduction from the extracellular matrix. *J. Cell Biol.* **120,** 577–585.
244. Menegoz, M., et al. (1997) Paranodin, a glycoprotein of neuronal paranodal membranes. *Neuron* **19,** 319–331.

245. Peles, E., et al. (1997) Close similarity between Drosophila neurexin IV and mammalian Caspr protein suggests a conserved mechanism for cellular interactions. *Cell* **88,** 745–746.
246. Giancotti, F. G. (1997) Integrin signaling: specificity and control of cell survival and cell cycle progression. *Curr. Opin. Cell Biol.* **9,** 691–700.
247. Humphries, M. J. (1996) Integrin activation: the link between ligand binding and signal transduction. *Curr. Opin. Cell Biol.* **8,** 632–640.
248. Dib, K. (2000) BETA 2 integrin signaling in leukocytes. *Front Biosci.* **5,** D438–451.
249. Critchley, D. R. (2000) Focal adhesions: the cytoskeletal connection. *Curr. Opin. Cell Biol.* **12,** 133–139.
250. Miyamoto, S., Akiyama, S. K., and Yamada, K. M. (1995) Synergistic roles for receptor occupancy and aggregation in integrin transmembrane function. *Science* **267,** 883–885.
251. Miyamoto, S., et al. (1995) Integrin function: molecular hierarchies of cytoskeletal and signaling molecules. *J. Cell Biol.* **131,** 791–805.
252. Yamada, K. M. and Miyamoto, S. (1995) Integrin transmembrane signaling and cytoskeletal control. *Curr. Opin. Cell Biol.* **7,** 681–689.
253. Dedhar, S. and Hannigan, G. E. (1996) Integrin cytoplasmic interactions and bidirectional transmembrane signalling. *Curr. Opin. Cell Biol.* **8,** 657–669.
254. Girault, J. A., et al. (1999) FAK and PYK2/CAKbeta in the nervous system: a link between neuronal activity, plasticity and survival? *Trends Neurosci.* **22,** 257–263.
255. Bader, B. L., et al. (1998) Extensive vasculogenesis, angiogenesis, and organogenesis precede lethality in mice lacking all alpha v integrins. *Cell* **95,** 507–519.
256. Hughes, P. E. and Pfaff, M. (1998) Integrin affinity modulation. *Trends Cell Biol.* **8,** 359–364.
257. Kolanus, W. and Seed, B. (1997) Integrins and inside-out signal transduction: converging signals from PKC and PIP3. *Curr. Opin. Cell Biol.* **9,** 725–731.
258. Hubbard, S. R. (1999) Structural analysis of receptor tyrosine kinases. *Prog. Biophys. Mol. Biol.* **71,** 343–358.
259. Schenk, P. W. and Snaar-Jagalska, B. E. (1999) Signal perception and transduction: the role of protein kinases. *Biochim. Biophys. Acta* **1449,** 1–24.
260. Robertson, S. C., Tynan, J. A., and Donoghue, D. J. (2000) RTK mutations and human syndromes when good receptors turn bad. *Trends Genet.* **16,** 265–271.
261. Schlaepfer, D. D., Jones, K. C., and Hunter, T. (1998) Multiple Grb2-mediated integrin-stimulated signaling pathways to ERK2/mitogen-activated protein kinase: summation of both c-Src- and focal adhesion kinase-initiated tyrosine phosphorylation events. *Mol. Cell Biol.* **18,** 2571–2585.
262. Stoker, A. and Dutta, R. (1998) Protein tyrosine phosphatases and neural development. *Bioessays* **20,** 463–472.
263. Zhang, Z. Y. (1998) Protein-tyrosine phosphatases: biological function, structural characteristics, and mechanism of catalysis. *Crit. Rev. Biochem. Mol. Biol.* **33,** 1–52.
264. Levy, J. B., et al. (1993) The cloning of a receptor-type protein tyrosine phosphatase expressed in the central nervous system. *J. Biol. Chem.* **268,** 10,573–10,581.
265. Margolis, R. U. and Margolis, R. K. (1997) Chondroitin sulfate proteoglycans as mediators of axon growth and pathfinding. *Cell Tissue Res.* **290,** 343–348.
266. Peles, E., Schlessinger, J., and Grumet, M. (1998) Multi-ligand interactions with receptor-like protein tyrosine phosphatase beta: implications for intercellular signaling. *Trends Biochem. Sci.* **23,** 121–124.
267. Barnea, G., et al. (1994) Receptor tyrosine phosphatase beta is expressed in the form of proteoglycan and binds to the extracellular matrix protein tenascin. *J. Biol. Chem.* **269,** 14,349–14,352.
268. Grumet, M., et al. (1994) Interactions with tenascin and differential effects on cell adhesion of neurocan and phosphacan, two major chondroitin sulfate proteoglycans of nervous tissue. *J. Biol. Chem.* **269,** 12,142–12,146.

269. Milev, P., et al. (1998) High affinity binding and overlapping localization of neurocan and phosphacan/protein-tyrosine phosphatase-zeta/beta with tenascin-R, amphoterin, and the heparin-binding growth-associated molecule. *J. Biol. Chem.* **273,** 6998–7005.
270. Milev, P., et al. (1998) The core protein of the chondroitin sulfate proteoglycan phosphacan is a high-affinity ligand of fibroblast growth factor-2 and potentiates its mitogenic activity. *J. Biol. Chem.* **273,** 21,439–21,442.
271. Canoll, P. D., et al. (1993) The expression of a novel receptor-type tyrosine phosphatase suggests a role in morphogenesis and plasticity of the nervous system. *Brain Res. Dev. Brain Res.* **75,** 293–298.
272. Snyder, S. E., et al. (1996) Comparison of RPTP zeta/beta, phosphacan, and trkB mRNA expression in the developing and adult rat nervous system and induction of RPTP zeta/beta and phosphacan mRNA following brain injury. *Brain Res. Mol. Brain Res.* **40,** 79–96.
273. Sakurai, T., Friedlander, D. R., and Grumet, M. (1996) Expression of polypeptide variants of receptor-type protein tyrosine phosphatase beta: the secreted form, phosphacan, increases dramatically during embryonic development and modulates glial cell behavior in vitro. *J. Neurosci. Res.* **43,** 694–706.
274. Canoll, P. D., et al. (1996) Three forms of RPTP-beta are differentially expressed during gliogenesis in the developing rat brain and during glial cell differentiation in culture. *J. Neurosci. Res.* **44,** 199–215.
275. Hu, S. and Reichardt, L. F. (1999) From membrane to cytoskeleton: enabling a connection. *Neuron* **22,** 419–422.
276. Sheetz, M. P., Felsenfeld, D. P., and Galbraith, C. G. (1998) Cell migration: regulation of force on extracellular-matrix-integrin complexes. *Trends Cell Biol.* **8,** 51–54.
277. Pfaff, M., et al. (1998) Integrin beta cytoplasmic domains differentially bind to cytoskeletal proteins. *J. Biol. Chem.* **273,** 6104–6109.
278. Sampath, R., Gallagher, P. J., and Pavalko, F. M. (1998) Cytoskeletal interactions with the leukocyte integrin beta2 cytoplasmic tail. Activation-dependent regulation of associations with talin and alpha-actinin. *J. Biol. Chem.* **273,** 33,588–33,594.
279. Retta, S. F., et al. (1998) beta1-integrin cytoplasmic subdomains involved in dominant negative function. *Mol. Biol. Cell* **9,** 715–731.
280. Schmidt, J. M., et al. (1999) Interaction of talin with actin: sensitive modulation of filament crosslinking activity. *Arch. Biochem. Biophys.* **366,** 139–150.
281. Bass, M. D., et al. (1999) Talin contains three similar vinculin-binding sites predicted to form an amphipathic helix. *Biochem. J.* **341(Pt 2),** 257–263.
282. Borowsky, M. L. and Hynes, R. O. (1998) Layilin, a novel talin-binding transmembrane protein homologous with C-type lectins, is localized in membrane ruffles. *J. Cell Biol.* **143,** 429–442.
283. Calderwood, D. A., et al. (1999) The Talin head domain binds to integrin beta subunit cytoplasmic tails and regulates integrin activation. *J. Biol. Chem.* **274,** 28,071–28,074.
284. Steimle, P. A., et al. (1999) Polyphosphoinositides inhibit the interaction of vinculin with actin filaments. *J. Biol. Chem.* **274,** 18,414–18,420.
285. Schoenwaelder, S. M. and Burridge, K. (1999) Bidirectional signaling between the cytoskeleton and integrins. *Curr. Opin. Cell Biol.* **11,** 274–286.
286. Weekes, J., Barry, S. T., and Critchley, D. R. (1996) Acidic phospholipids inhibit the intramolecular association between the N- and C-terminal regions of vinculin, exposing actin-binding and protein kinase C phosphorylation sites. *Biochem. J.* **314(Pt 3),** 827–832.
287. Gilmore, A. P. and Burridge, K. (1996) Regulation of vinculin binding to talin and actin by phosphatidyl-inositol-4-5-bisphosphate. *Nature* **381,** 531–535.
288. Loo, D. T., Kanner, S. B., and Aruffo, A. (1998) Filamin binds to the cytoplasmic domain of the beta1-integrin. Identification of amino acids responsible for this interaction. *J. Biol. Chem.* **273,** 23,304–23,312.

289. Ohta, Y., et al. (1999) The small GTPase RalA targets filamin to induce filopodia. *Proc. Natl. Acad. Sci. USA* **96,** 2122–2128.
290. Cunningham, C. C., et al. (1992) Actin-binding protein requirement for cortical stability and efficient locomotion. *Science* **255,** 325–327.
291. Barde, Y. A. (1989) Trophic factors and neuronal survival. *Neuron* **2,** 1525–1534.
292. Frost, E. E., et al. (1999) Integrins mediate a neuronal survival signal for oligodendrocytes. *Curr. Biol.* **9,** 1251–1254.
293. Henderson, C. E. (1996) Programmed cell death in the developing nervous system. *Neuron* **17,** 579–585.
294. Pettmann, B. and Henderson, C. E. (1998) Neuronal cell death. *Neuron* **20,** 633–647.
295. Kuan, C. Y., et al. (2000) Mechanisms of programmed cell death in the developing brain. *Trends Neurosci.* **23,** 291–297.
296. Lauffenburger, D. A. and Horwitz, A. F. (1996) Cell migration: a physically integrated molecular process. *Cell* **84,** 359–369.
297. Perris, R. (1997) The extracellular matrix in neural crest-cell migration. *Trends Neurosci.* **20,** 23–31.
298. Parnavelas, J. G. (2000) The origin and migration of cortical neurones: new vistas. *Trends Neurosci.* **23,** 126–131.
299. Anton, E. S., Kreidberg, J. A., and Rakic, P. (1999) Distinct functions of alpha3 and alpha(v) integrin receptors in neuronal migration and laminar organization of the cerebral cortex. *Neuron* **22,** 277–289.
300. Quaranta, V. (2000) Cell migration through extracellular matrix: membrane-type metalloproteinases make the way. *J. Cell Biol.* **149,** 1167–1170.
301. Debby-Brafman, A., et al. (1999) F-Spondin, expressed in somite regions avoided by neural crest cells, mediates inhibition of distinct somite domains to neural crest migration. *Neuron* **22,** 475–488.
302. Tucker, R. P., et al. (1999) Thrombospondin-1 and neural crest cell migration. *Dev. Dyn.* **214,** 312–322.
303. Landolt, R. M., et al. (1995) Versican is selectively expressed in embryonic tissues that act as barriers to neural crest cell migration and axon outgrowth. *Development* **121,** 2303–2312.
304. Perris, R. and Perissinotto, D. (2000) Role of the extracellular matrix during neural crest cell migration. *Mech. Dev.* **95,** 3–21.
305. Li, H., et al. (2000) Coordinate regulation of cadherin and integrin function by the chondroitin sulfate proteoglycan neurocan. *J. Cell Biol.* **149,** 1275–1288.
306. Miura, R., et al. (1999) The proteoglycan lectin domain binds sulfated cell surface glycolipids and promotes cell adhesion. *J. Biol. Chem.* **274,** 11,431–11,438.
307. Reichardt, L. F., et al. (1989) Integrins and cell adhesion molecules: neuronal receptors that regulate axon growth on extracellular matrices and cell surfaces. *Dev. Neurosci.* **11,** 332–347.
308. Tiveron, M. C., et al. (1992) Selective inhibition of neurite outgrowth on mature astrocytes by Thy-1 glycoprotein. *Nature* **355,** 745–748.
309. Doherty, P., et al. (1991) Neurite outgrowth in response to transfected N-CAM and N-cadherin reveals fundamental differences in neuronal responsiveness to CAMs. *Neuron* **6,** 247–258.
310. Doherty, P. and Walsh, F. S. (1991) The contrasting roles of N-CAM and N-cadherin as neurite outgrowth-promoting molecules. *J. Cell Sci.* Suppl, **15,** 13–21.
311. Frei, T., et al. (1992) Different extracellular domains of the neural cell adhesion molecule (N-CAM) are involved in different functions. *J. Cell Biol.* **118,** 177–194.
312. Bixby, J. L. and Zhang, R. (1990) Purified N-cadherin is a potent substrate for the rapid induction of neurite outgrowth. *J. Cell Biol.* **110,** 1253–1260.
313. Lagenaur, C. and Lemmon, V. (1987) An L1-like molecule, the 8D9 antigen, is a potent substrate for neurite extension. *Proc. Natl. Acad. Sci. USA* **84,** 7753–7757.

314. Ethell, I. M. and Yamaguchi, Y. (1999) Cell surface heparan sulfate proteoglycan syndecan-2 induces the maturation of dendritic spines in rat hippocampal neurons. *J. Cell Biol.* **144,** 575–586.
315. Pires Neto, M. A., Braga-de-Souza, S., and Lent, R. (1999) Extracellular matrix molecules play diverse roles in the growth and guidance of central nervous system axons. *Braz. J. Med. Biol. Res.* **32,** 633–638.
316. Condic, M. L., Snow, D. M., and Letourneau, P. C. (1999) Embryonic neurons adapt to the inhibitory proteoglycan aggrecan by increasing integrin expression. *J. Neurosci.* **19,** 10,036–10,043.
317. Dou, C. L. and Levine, J. M. (1995) Differential effects of glycosaminoglycans on neurite growth on laminin and L1 substrates. *J. Neurosci.* **15,** 8053–8066.
318. Morris, J. E. (1993) Proteoglycans and the modulation of cell adhesion by steric exclusion. *Dev. Dyn.* **196,** 246–251.
319. Garcia-Abreu, J., et al. (2000) Contribution of heparan sulfate to the non-permissive role of the midline glia to the growth of midbrain neurites. *Glia* **29,** 260–272.
320. Neugebauer, K. M., et al. (1991) Vitronectin and thrombospondin promote retinal neurite outgrowth: developmental regulation and role of integrins. *Neuron* **6,** 345–358.
321. Goodman, C. S. (1996) Mechanisms and molecules that control growth cone guidance. *Annu. Rev. Neurosci.* **19,** 341–377.
322. Emerling, D. E. and Lander, A. D. (1996) Inhibitors and promoters of thalamic neuron adhesion and outgrowth in embryonic neocortex: functional association with chondroitin sulfate. *Neuron* **17,** 1089–1100.
323. Powell, E. M., et al. (1997) Mechanisms of astrocyte-directed neurite guidance. *Cell Tissue Res.* **290,** 385–393.
324. Tisay, K. T. and Key, B. (1999) The extracellular matrix modulates olfactory neurite outgrowth on ensheathing cells. *J. Neurosci.* **19,** 9890–9899.
325. Brittis, P. A., Canning, D. R., and Silver, J. (1992) Chondroitin sulfate as a regulator of neuronal patterning in the retina. *Science* **255,** 733–736.
326. Son, Y. J., Patton, B. L., and Sanes, J. R. (1999) Induction of presynaptic differentiation in cultured neurons by extracellular matrix components. *Eur. J. Neurosci.* **11,** 3457–3467.
327. Faissner, A. and Steindler, D. (1995) Boundaries and inhibitory molecules in developing neural tissues. *Glia* **13,** 233–254.
328. Fitch, M. T. and Silver, J. (1997) Glial cell extracellular matrix: boundaries for axon growth in development and regeneration. *Cell Tissue Res.* **290,** 379–384.
329. Golding, J., Shewan, D., and Cohen, J. (1997) Maturation of the mammalian dorsal root entry zone—from entry to no entry. *Trends Neurosci.* **20,** 303–308.
330. Gonzalez, M. L. and Silver, J. (1994) Axon-glia interactions regulate ECM patterning in the postnatal rat olfactory bulb. *J. Neurosci.* **14,** 6121–6131.
331. Katoh-Semba, R., et al. (1995) Chondroitin sulphate proteoglycans in the rat brain: candidates for axon barriers of sensory neurons and the possible modification by laminin of their actions. *Eur. J. Neurosci.* **7,** 613–621.
332. Steindler, D. A., et al. (1990) Boundaries during normal and abnormal brain development: in vivo and in vitro studies of glia and glycoconjugates. *Exp. Neurol.* **109,** 35–56.
333. Geisert, E. E., Jr. and Bidanset, D. J. (1993) A central nervous system keratan sulfate proteoglycan: localization to boundaries in the neonatal rat brain. *Brain Res. Dev. Brain Res.* **75,** 163–173.
334. Robson, J. A. and Geisert, E. E., Jr. (1994) Expression of a keratin sulfate proteoglycan during development of the dorsal lateral geniculate nucleus in the ferret. *J. Comp. Neurol.* **340,** 349–360.
335. Seo, H. and Geisert, E. E., Jr. (1995) A keratan sulfate proteoglycan marks the boundaries in the cortical barrel fields of the adult rat. *Neurosci. Lett.* **197,** 13–16.

336. Watanabe, E., et al. (1995) Distribution of a brain-specific proteoglycan, neurocan, and the corresponding mRNA during the formation of barrels in the rat somatosensory cortex. *Eur. J. Neurosci.* **7,** 547–554.
337. Snow, D. M., et al. (1990) Sulfated proteoglycans in astroglial barriers inhibit neurite outgrowth in vitro. *Exp. Neurol.* **109,** 111–130.
338. Bovolenta, P., Wandosell, F., and Nieto-Sampedro, M. (1993) Characterization of a neurite outgrowth inhibitor expressed after CNS injury. *Eur. J. Neurosci.* **5,** 454–465.
339. Canning, D. R., et al. (1993) beta-Amyloid of Alzheimer's disease induces reactive gliosis that inhibits axonal outgrowth. *Exp. Neurol.* **124,** 289–298.
340. Mellitzer, G., Xu, Q., and Wilkinson, D. G. (1999) Eph receptors and ephrins restrict cell intermingling and communication. *Nature* **400,** 77–81.
341. Jhaveri, S., Erzurumlu, R. S., and Crossin, K. (1991) Barrel construction in rodent neocortex: role of thalamic afferents versus extracellular matrix molecules. *Proc. Natl. Acad. Sci. USA* **88,** 4489–4493.
342. Bonnet, F., et al. (1996) Structure and cellular distribution of mouse brain testican. Association with the postsynaptic area of hippocampus pyramidal cells. *J. Biol. Chem.* **271,** 4373–4380.
343. Carlson, S. S. and Wight, T. N. (1987) Nerve terminal anchorage protein 1 (TAP-1) is a chondroitin sulfate proteoglycan: biochemical and electron microscopic characterization. *J. Cell Biol.* **105(6 Pt 2),** 3075–3086.
344. Iwata, M., Wight, T. N., and Carlson, S. S. (1993) A brain extracellular matrix proteoglycan forms aggregates with hyaluronan. *J. Biol. Chem.* **268,** 15,061–15,069.
345. Cohen, N. A., et al. (1997) Expression of agrin in the developing and adult rat brain. *Neuroscience* **76,** 581–596.
346. Hilgenberg, L. G., Hoover, C. L., and Smith, M. A. (1999) Evidence of an agrin receptor in cortical neurons. *J. Neurosci.* **19,** 7384–7393.
347. Tanaka, H., et al. (2000) Molecular modification of N-cadherin in response to synaptic activity. *Neuron* **25,** 93–107.
348. Denzer, A. J., et al. (1997) Synaptic differentiation: the role of agrin in the formation and maintenance of the neuromuscular junction. *Cell Tissue Res.* **290,** 357–365.
349. Chamberlain, J. (1999) The dynamics of dystroglycan. *Nat. Genet.* **23,** 256–258.
350. Patton, B. L., Chiu, A. Y., and Sanes, J. R. (1998) Synaptic laminin prevents glial entry into the synaptic cleft. *Nature* **393,** 698–701.
351. Araque, A., et al. (1999) Tripartite synapses: glia, the unacknowledged partner. *Trends Neurosci.* **22,** 208–215.
352. Fields, R. D. and Itoh, K. (1996) Neural cell adhesion molecules in activity-dependent development and synaptic plasticity. *Trends Neurosci.* **19,** 473–480.
353. Landmesser, L. (1997) Synaptic plasticity: fastening synapses by adhesion. *Curr. Biol.* **7,** R28–30.
354. Bliss, T. V. and Collingridge, G. L. (1993) A synaptic model of memory: long-term potentiation in the hippocampus. *Nature* **361,** 31–39.
355. Soderling, T. R. and Derkach, V. A. (2000) Postsynaptic protein phosphorylation and LTP. *Trends Neurosci.* **23,** 75–80.
356. Schachner, M. (1997) Neural recognition molecules and synaptic plasticity. *Curr. Opin. Cell Biol.* **9,** 627–634.
357. Nakic, M., et al. (1998) Long-term potentiation in vivo increases rat hippocampal tenascin-C expression. *J. Neurobiol.* **37,** 393–404.
358. Kohmura, N., et al. (1998) Diversity revealed by a novel family of cadherins expressed in neurons at a synaptic complex. *Neuron* **20,** 1137–1151.
359. Nakagami, Y., et al. (2000) Laminin degradation by plasmin regulates long-term potentiation. *J. Neurosci.* **20,** 2003–2010.

360. Luthl, A., et al. (1994) Hippocampal long-term potentiation and neural cell adhesion molecules L1 and NCAM. *Nature* **372,** 777–779.
361. Muller, D., et al. (1996) PSA-NCAM is required for activity-induced synaptic plasticity. *Neuron* **17,** 413–422.
362. Martin, K. C. and Kandel, E. R. (1996) Cell adhesion molecules, CREB, and the formation of new synaptic connections. *Neuron* **17,** 567–570.
363. Goldman, S. A. and Luskin, M. B. (1998) Strategies utilized by migrating neurons of the postnatal vertebrate forebrain. *Trends Neurosci.* **21,** 107–114.
364. Tramontin, A. D. and Brenowitz, E. A. (2000) Seasonal plasticity in the adult brain. *Trends Neurosci.* **23,** 251–258.
365. Tramontin, A. D., Hartman, V. N., and Brenowitz, E. A. (2000) Breeding conditions induce rapid and sequential growth in adult avian song control circuits: a model of seasonal plasticity in the brain. *J. Neurosci.* **20,** 854–861.
366. Celio, M. R. and Blumcke, I. (1994) Perineuronal nets—a specialized form of extracellular matrix in the adult nervous system. *Brain Res. Brain Res. Rev.* **19,** 128–145.
367. Bruckner, G., et al. (1993) Perineuronal nets provide a polyanionic, glia-associated form of microenvironment around certain neurons in many parts of the rat brain. *Glia* **8,** 183–200.
368. Celio, M. R. (1993) Perineuronal nets of extracellular matrix around parvalbumin-containing neurons of the hippocampus. *Hippocampus* **3(Spec No),** 55–60.
369. Maleski, M. and Hockfield, S. (1997) Glial cells assemble hyaluronan-based pericellular matrices in vitro. *Glia* **20,** 193–202.
370. Celio, M. R., et al. (1998) Perineuronal nets: past and present. *Trends Neurosci.* **21,** 510–515.
371. Luth, H. J., Fischer, J., and Celio, M. R. (1992) Soybean lectin binding neurons in the visual cortex of the rat contain parvalbumin and are covered by glial nets. *J. Neurocytol.* **21,** 211–221.
372. Wintergerst, E. S., Faissner, A., and Celio, M. R. (1996) The proteoglycan DSD-1-PG occurs in perineuronal nets around parvalbumin-immunoreactive interneurons of the rat cerebral cortex. *Int. J. Dev. Neurosci.* **14,** 249–255.
373. Haunso, A., et al. (1999) Phosphacan immunoreactivity is associated with perineuronal nets around parvalbumin-expressing neurones. *Brain Res.* **834,** 219–222.
374. Koppe, G., et al. (1997) Characterization of proteoglycan-containing perineuronal nets by enzymatic treatments of rat brain sections. *Histochem. J.* **29,** 11–20.
375. Blumcke, I., Eggli, P., and Celio, M. R. (1995) Relationship between astrocytic processes and "perineuronal nets" in rat neocortex. *Glia* **15,** 131–140.
376. Wintergerst, E. S., et al. (1996) Temporal and spatial appearance of the membrane cytoskeleton and perineuronal nets in the rat neocortex. *Neurosci. Lett.* **209,** 173–176.
377. Eddleston, M. and Mucke, L. (1993) Molecular profile of reactive astrocytes: implications for their role in neurologic disease. *Neuroscience* **54,** 15–36.
378. Bahr, M. and Bonhoeffer, F. (1994) Perspectives on axonal regeneration in the mammalian CNS. *Trends Neurosci.* **17,** 473–479.
379. Aubert, I., Ridet, J. L., and Gage, F. H. (1995) Regeneration in the adult mammalian CNS: guided by development. *Curr. Opin. Neurobiol.* **5,** 625–635.
380. Fawcett, J. W. (1997) Astrocytic and neuronal factors affecting axon regeneration in the damaged central nervous system. *Cell Tissue Res.* **290,** 371–377.
381. Fawcett, J. W. and Asher, R. A. (1999) The glial scar and central nervous system repair. *Brain Res. Bull.* **49,** 377–391.
382. Norton, W. T. (1999) Cell reactions following acute brain injury: a review. *Neurochem. Res.* **24,** 213–218.
383. Fernaud-Espinosa, I., Nieto-Sampedro, M., and Bovolenta, P. (1998) A neurite outgrowth-inhibitory proteoglycan expressed during development is similar to that isolated from adult brain after isomorphic injury. *J. Neurobiol.* **36,** 16–29.

384. Fawcett, J. W. (1992) Intrinsic neuronal determinants of regeneration. *Trends Neurosci.* **15,** 5–8.
385. Zuo, J., Hernandez, Y. J., and Muir, D. (1998) Chondroitin sulfate proteoglycan with neurite-inhibiting activity is up-regulated following peripheral nerve injury. *J. Neurobiol.* **34,** 41–54.
386. Haas, C. A., et al. (1999) Entorhinal cortex lesion in adult rats induces the expression of the neuronal chondroitin sulfate proteoglycan neurocan in reactive astrocytes. *J. Neurosci.* **19,** 9953–9963.
387. McKeon, R. J., Jurynec, M. J., and Buck, C. R. (1999) The chondroitin sulfate proteoglycans neurocan and phosphacan are expressed by reactive astrocytes in the chronic CNS glial scar. *J. Neurosci.* **19,** 10,778–10,788.
388. Braunewell, K. H., et al. (1995) Up-regulation of a chondroitin sulphate epitope during regeneration of mouse sciatic nerve: evidence that the immunoreactive molecules are related to the chondroitin sulphate proteoglycans decorin and versican. *Eur. J. Neurosci.* **7,** 792–804.
389. Geisert, E. E., Jr., et al. (1996) Up-regulation of a keratan sulfate proteoglycan following cortical injury in neonatal rats. *Int. J. Dev. Neurosci.* **14,** 257–267.
390. McKeon, R. J., Hoke, A., and Silver, J. (1995) Injury-induced proteoglycans inhibit the potential for laminin-mediated axon growth on astrocytic scars. *Exp. Neurol.* **136,** 32–43.
391. Stichel, C. C., et al. (1995) Differential expression of the small chondroitin/dermatan sulfate proteoglycans decorin and biglycan after injury of the adult rat brain. *Brain Res.* **704,** 263–274.
392. Asher, R. A., et al. (2000) Neurocan is upregulated in injured brain and in cytokine-treated astrocytes. *J. Neurosci.* **20,** 2427–2438.
393. Bignami, A., Perides, G., and Rahemtulla, F. (1993) Versican, a hyaluronate-binding proteoglycan of embryonal precartilaginous mesenchyma, is mainly expressed postnatally in rat brain. *J. Neurosci. Res.* **34,** 97–106.
394. Yamada, H., et al. (1997) The brain chondroitin sulfate proteoglycan brevican associates with astrocytes ensheathing cerebellar glomeruli and inhibits neurite outgrowth from granule neurons. *J. Neurosci.* **17,** 7784–7795.
395. Selkoe, D. J. (1998) The cell biology of beta-amyloid precursor protein and presenilin in Alzheimer's disease. *Trends Cell Biol.* **8,** 447–453.
396. Vickers, J. C., et al. (2000) The cause of neuronal degeneration in Alzheimer's disease. *Prog. Neurobiol.* **60,** 139–165.
397. Podlisny, M. B., et al. (1998) Oligomerization of endogenous and synthetic amyloid beta-protein at nanomolar levels in cell culture and stabilization of monomer by Congo red. *Biochemistry* **37,** 3602–3611.
398. Neve, R. L. and Robakis, N. K. (1998) Alzheimer's disease: a re-examination of the amyloid hypothesis. *Trends Neurosci.* **21,** 15–19.
399. Tran, P. B. and Miller, R. J. (1999) Aggregates in neurodegenerative disease: crowds and power? *Trends Neurosci.* **22,** 194–197.
400. Wisniewski, T., et al. (1994) Acceleration of Alzheimer's fibril formation by apolipoprotein E in vitro. *Am. J. Pathol.* **145,** 1030–1035.
401. Wood, S. J., Chan, W., and Wetzel, R. (1996) Seeding of A beta fibril formation is inhibited by all three isotypes of apolipoprotein E. *Biochemistry* **35,** 12,623–12,628.
402. Bronfman, F. C., et al. (1998) Laminin blocks the assembly of wild-type A beta and the Dutch variant peptide into Alzheimer's fibrils. *Amyloid* **5,** 16–23.
403. Monji, A., et al. (1998) Laminin inhibits Abeta42 fibril formation in vitro. *Brain Res.* **788,** 187–190.
404. Alvarez, A., et al. (1995) Acetylcholinesterase, a senile plaque component, affects the fibrillogenesis of amyloid-beta-peptides. *Neurosci. Lett.* **201,** 49–52.

405. Castillo, G. M., et al. (1997) Perlecan binds to the beta-amyloid proteins (A beta) of Alzheimer's disease, accelerates A beta fibril formation, and maintains A beta fibril stability. *J. Neurochem.* **69,** 2452–2465.
406. Cotman, S. L., Halfter, W., and Cole, G. J. (2000) Agrin binds to beta-amyloid (Abeta), accelerates abeta fibril formation, and is localized to Abeta deposits in Alzheimer's disease brain. *Mol. Cell Neurosci.* **15,** 183–198.
407. Chintala, S. K., Tonn, J. C., and Rao, J. S. (1999) Matrix metalloproteinases and their biological function in human gliomas. *Int. J. Dev. Neurosci.* **17,** 495–502.
408. Koblinski, J. E., Ahram, M., and Sloane, B. F. (2000) Unraveling the role of proteases in cancer. *Clin. Chim. Acta.* **291,** 113–135.
409. Fidler, P. S., et al. (1999) Comparing astrocytic cell lines that are inhibitory or permissive for axon growth: the major axon-inhibitory proteoglycan is NG2. *J. Neurosci.* **19,** 8778–8788.
410. Zagzag, D., et al. (1996) Tenascin-C expression by angiogenic vessels in human astrocytomas and by human brain endothelial cells in vitro. *Cancer Res.* **56,** 182–189.
411. Faissner, A. and Kruse, J. (1990) J1/tenascin is a repulsive substrate for central nervous system neurons. *Neuron* **5,** 627–637.
412. Husmann, K., Faissner, A., and Schachner, M. (1992) Tenascin promotes cerebellar granule cell migration and neurite outgrowth by different domains in the fibronectin type III repeats. *J. Cell Biol.* **116,** 1475–1486.

6
Homeostatic Properties of Astrocytes

Wolfgang Walz and Bernhard H. J. Juurlink

1. INTRODUCTION

The most striking feature of the group of satellite cells collectively termed "astrocytes" is the fact that they seem to constitute a huge syncytium in the brain, since these cells are connected by gap junctions. They take up about 40% of the brain volume, but, since they are smaller than neurons, their cell number is about 10× larger. Other structural specializations are endfeet around the endothelial cells of capillaries, and cytoplasmic processes that surround synaptic clefts. These structural specializations enable the astrocytes to interact with all major elements in the central nervous system, and to integrate within the syncytium the available information from different areas. Indeed, astrocyte function seems to be twofold: First, they are support cells that respond to neuronal activity with more efficient delivery of supplies (such as lactate [lact]) and a more effective removal system of products (such as potassium [K] and transmitter substances) that would interfere with the normal operation of neurons; a second function is the detection of synaptic use, and the subsequent modification of the synaptic transmission. This is an involvement that has more to do with information processing than tissue homeostasis, although both are of course interrelated. This chapter deals with the homeostatic functions of astrocytes and the subsequent chapter focuses on the second functional role, the interaction of these cells with synaptic processing.

Early histological investigations in the brain clearly demonstrated the presence of two major types of astrocytes in the adult brain: fibrous and protoplasmic *(1)*. Fibrous astrocytes are located in the white matter, having long processes that run between myelinated fibers, and are the cells responsible for vascular endfeet production. Some of these processes also extend into the gray matter. Other more specialized astrocyte-related cells are also present in the adult central nervous system *(2)*: ependymal cells, that form a continuous lining around the ventricles and the central canal, are dealt with in a separate chapter. Tanycytes are cells found in patches in the circumventricular organs, *see* Chapter 12. Pituicytes are astrocyte-like cells located in the posterior pituitary (neural lobe or neurohypophysis); Bergmann glia appear to be radial glia that have survived into adulthood in the cerebellum only. Finally, Muller cells are specialized cells in the retina, where they occur along with astrocytes. Astrocyte classification is complicated by the fact that astrocytes in the gray matter seem to be divided into two

From: *The Neuronal Environment: Brain Homeostasis in Health and Disease*
Edited by: W. Walz © Humana Press Inc., Totowa, NJ

groups: those that express detectable glial fibrillary acidic protein (GFAP) and those that do not. There are indications that the two subgroups have different physiological properties, but this conclusion is controversial *(3)*: the GFAP distribution seems to coincide with the presence (GFAP-negative) and absence (GFAP-positive) of voltage-gated K and sodium (Na) channels in these cells. The function of voltage-gated ion channels in astrocytes is not clear, but there is speculation that they may play a role in proliferation *(4)*, K buffering *(5)*, or motility/size changes *(6)*.

Many ionotropic and metabotropic receptors for neuronal signaling substances are also present in astrocytes. Metabotropic receptors seem to be important in the integration of astrocytic function with neuronal activity level, but the role of the ionotropic receptors is less clear *(7)*. In cultured astrocytes, the exposure to serum factors induces the expression of virtually all known receptors, except the N-methyl-D-aspartate (NMDA)-type glutamate (Glu) receptor. Astrocytes *in situ* have fewer receptors and they are distributed in a more heterogeneous manner *(8)*. However, even in vivo, receptors for most major transmitter systems are present, e.g., the α-amino-3-hydroxy-5-methyl-4-isoazolepropionate (AMPA)-type Glu and the γ-aminobutyric acid A ($GABA_A$) receptor *(9,10)*. Activation of the ionotropic receptors will, in almost all cases, induce a membrane depolarization *(11)*. This transmitter-induced depolarization could play a role in enhancing K-mediated depolarization during neuronal activity, which is important for spatial buffering (*see* Subheading 2.).

2. K HOMEOSTASIS

One of the best-documented cases for functional astrocytic–neuronal interactions is the regulation of the K concentration in the extracellular space (ECS). This role prevents a build-up of K in the ECS during neuronal activity. The K ions leave the neuronal elements and enter the ECS, because of increased conductance and existing driving force during action and synaptic potentials. The increase is approx 1 m*M* above the resting level during the passage of a single action potential. During intense stimulation, or during seizures, the concentration in the ECS can reach a ceiling level of 12 m*M*. Only during injury (hypoxia/ischemia, trauma, hypoglycemia) is the ceiling level disrupted, and concentrations of up to 25 m*M* are reached *(3)*. An extreme case is the occurrence of spreading depression waves, which result in transient elevations of 30–80 m*M* (*see* Subheading 5.). EC K increases of up to 5 m*M* lead to hyperexcitability, as the result of effects on the membrane potential and direct modulation of ion channel gates *(12)*. K increases that exceed 5 m*M*, in addition, affect the efficacy of synaptic transmission *(13)*. Thus, in order to protect the neural circuits from distortions of information transfer, and to prevent hyperexcitability, mechanisms have to be in force to prevent all but a minor build-up of EC K during neuronal activity.

One would expect that immediate reuptake of K into neurons and diffusion in the ECS, would be sufficient to prevent a build-up of excess EC K. There must be K removal sites not resident in neurons, because iontophoretically applied EC K is removed as efficiently as K that was released from the neurons *(14)*. Only in the second case would there be simultaneous accumulation of neuronal Na that would stimulate the neuronal Na/K pump. If K is iontophoresed into the ECS, the diffusion curves appear normal, but the volume fraction routinely exceeds 100% *(15)*, which is physically impossible, and the only solution is that the exit of K from the ECS through adjacent cell membranes is facilitated by auxiliary mechanisms. The existence of two different K clearance principles residing in astrocytes is now an established fact: The removal of excess K by current loops through the astrocytic syncytium, which would not lead to K accumulation in astrocytes, since, for each ion entering, one would leave at a site distant from the active neurons (this mechanism is known as spatial buffer *[16]*); the removal of excess K by uptake and accumulation into adjacent astrocytes, including its transient storage and subsequent return *(17)*.

The spatial buffer concept was first advanced by Orkand et al. *(16)*. In a syncytium, the membrane of neighboring cells has a tendency to stay isopotential. Therefore, any region that is experiencing an increased EC K concentration will have a K equilibrium potential that is more positive than the membrane potential. This will lead to an inward driving force for K and, since the astrocytic membrane is highly permeable to K, it will enter the cell via a passive current. This current must be part of a closed loop, and the K current will be distributed to the other parts of the syncytium via the gap junctions. In regions further removed from the area exposed to high EC K concentrations, the K equilibrium potential is more negative than the membrane potential, because of the tendency of the syncytium to remain isopotential. Therefore, at these farther-removed areas, there is an outwardly directed driving force for K. Thus, the current into the cells (at regions with high EC K), inside the syncytium and out of the cells (at regions distant from high K locations), is almost completely carried by K. The loop is closed by a return current in the ECS, which is carried mainly by Na and chloride (Cl) ions (the bulk of the EC ions). Despite the closed current loop, K does not cycle, but is passively transported from an EC location with high K concentrations, and dispersed to EC areas with low K.

Nicholson and Philips *(18)* studied ion diffusion in rat cerebellum, using ion-selective microelectrodes and iontophoretic point sources. Assuming an ECS of 20%, they found, for a variety of anions and cations, a close correlation of their movement with the laws of macroscopic diffusion. However, there was one ion that did not fit into the scheme: The movement of K in an electrical gradient is more in keeping with an ECS that occupies more than 100% of the brain volume *(18–20)*. The anomalous nature of the K migration can be explained by assuming that the ion does not remain in the ECS, but is, in fact, the major current carrier across cell membranes. Hounsgaard and Nicholson *(21)* analyzed the possibility in more detail. They measured changes in the EC K concentration in the vicinity of Purkinje cells in guinea pig cerebellar slices. No EC K changes were seen when the cells were hyperpolarized. Only during spike activity was there a rise in EC K levels. In the vicinity of glial cells, however, a hyperpolarizing current, injected into glia, reduced the outside K concentration in a symmetrical manner; depolarizing current injection induced a rise in the outside K levels. These results demonstrate that the K ions are mainly using the glial cell membrane during their movement in an electrochemical gradient. Thanks to its easy accessibility and well-layered structure, the retina is part of the CNS that is best-characterized in terms of spatial buffering. The major glial element of the retina stretches through all layers, starting at the photoreceptors and ending with endfeet at the inner limiting membrane adjacent to the vitreous humor. The inner and outer plexiform layers are the areas of light-induced K increases: This is where K ions have to enter the Muller cells generating a current sink. The endfeet K channel density is 10-fold greater than in other regions *(22)*. Since the endfeet border at a large liquid reservoir, any K released will be quickly disperse, keeping the EC K concentration close to normal. If, during light stimulation, the plexiform layers accumulate excess EC K, and the Muller cell membranes in this region are depolarized, there is a driving force for K currents into the cell for these areas, and toward the hyperpolarized endfeet where K is released. Again, the return current will be carried by Na and Cl ions *(23)*. Because this mechanism is based on anatomically fixed loops resulting from the increased endfeet K channel density, it is also called "K siphoning" *(23)*. Whether such siphoning and endfeet specialization also applies to astrocytes in the cerebral cortex and other brain structures is still a matter of speculation. The most pressing argument against a buffer function is that the few measurements of the length constant of the glial processes proved the constant to be too short for K currents to reach neighboring cells via gap junctions *(24)*. The constant could be large enough, however, to disperse K within reach of the elaborate processes of a single astrocyte. There is direct evidence from hippocampal slices that glial ion channels are involved in K homeostasis. Treatment of these slices with cesium (Cs) and intense synaptic activation, together caused increases in EC K, which in turn caused abnormal discharged of neurons *(25)*. The effects were not caused by actions of Cs on neuronal ion channels, but resulted from action on glial inwardly rectifying K channels. Inhibition of this Cs-sensitive glial clearance mechanism caused seizure-like discharges because of K accumulation. However, such a Cs-sensitive glial inward-rectifier K channel could mediate Donnan-like accumulation of K and anions through channels for these ions (*see* below).

The operation of the spatial buffer current requires no storage of K ions: For every K ion entering the syncytium, one is leaving at the same time, although at a different location. Thus, no significant accumulation of K inside astrocytes will take place, and any observed accumulation can be seen as an indication of an active mechanism other than spatial buffering. In all preparations tested, astrocytes accumulate K ions when the EC concentration increases. They release K again as soon as the EC concentration is lowered. This transient accumulation or storage was observed when the EC K level was increased artificially in a uniform way (a situation in which special buffer currents should not arise), or when, in functional modules, the neighboring neurons were activated by physiological or by nonphysiological stimuli. This was shown for cultured astrocytes *(26)* and oligodendrocytes *(27)*, glial cells in brain slices *(28)*, glial cells of the drone retina *(29)*, glial cells of the leech CNS *(30)*, and reactive astrocytes in hippocampal scar tissue *(31)*. The increases are similar, if the neighboring neuronal elements are stimulated or the external K level is increased experimentally to levels observed during neuronal stimulation. This is an important point, since neuronal stimulation increases not only the external K concentration, but also significantly changes calcium (Ca) and hydrogen levels, as well as transmitter and metabolite levels *(32,33)*. The increase in glial intracellular (IC) K is therefore exclusively a response to the increased EC K levels, and the other changes may modify this response, but do not cause it.

Ransom et al. *(34)* compared the efficiency of reuptake into axons with uptake into astrocytes, after activity-dependent K accumulation in the rat optic nerve. They used a situation in which spatial buffering is not a factor, and found two distinct poststimulus K clearance processes. The results were compatible with a model of K removal that attributes the fast, initial phase of K removal to K uptake by glial Na/K pumps, and the slower, sustained decline to K uptake via axonal Na/K pumps. The astrocytes express an isoform of the Na/K pump that is sensitive to changes in the EC K concentration around the physiological concentration of 3 m*M* *(35)*. In contrast, axons express an isoform that is insensitive to changes in EC K in the physiological range, but is sensitive to IC changes to Na *(36)*. Because of the dimensions of the ECS periaxonal space, the EC K concentration exhibits larger changes than the axonal Na concentration during neuronal activity. Thus, the astrocytes are able to react first, and faster than the axons.

Another question is the mechanism by which astrocytes take up K. In contrast to the spatial buffering, this uptake mechanism must be electroneutral. The bulk of the K uptake has to be either in the form of a KCl (or $KHCO_3^-$) accumulation or of a Na–K exchange, or it is a combination of both.

It became clear that, although part of the K uptake is sensitive to ouabain (the Na/K pump blocker), at high uptake rates for K, there is simply not enough Na in the cells to account for a 1:1, or even 3:2, exchange of Na by K. In fact, the internal Na concentration of astrocytes, which is ~10–15 m*M* *(35)*, hardly changes during K uptake *(37,38)*, suggesting that a transmembrane Na cycle is in operation: K is pumped in by the Na/K pump, and the drop in the IC Na concentration is prevented by a simultaneous stimulation of the electroneutral K-Na-2Cl co-transporter. This transporter will passively follow the combined driving forces of all three ions, leading to an inward flow. Na, which is pumped out by the Na/K pump, is replenished by this transporter. In such a way, Na is practically cycling across the membrane with a stable IC concentration, providing the counterion for the Na/K pump and the main driving force for the carrier. The net effect is a KCl accumulation and swelling caused by the osmolyte increase *(39)*. The evidence for such a mechanism comes from radiotracer analysis in cultured astrocytes, and the fact that inhibition of either the pump or the transporter was not additive, and that there was an overlap. This was confirmed by Rose and Ransom *(35)*, who found the bumetanide-sensitive co-transporter to be a more efficient mechanism than the tetrodotoxin-sensitive Na channel, to replenish Na ions. The bumetanide-sensitive Na-K-Cl co-transporter exhibits general properties, in all cells, that would be well-suited for K uptake, namely, a strong inward driving force and a high affinity for EC K *(40)*. Recently, a neuron-specific K-Cl co-transport was characterized that could play a role in KCl uptake into depolarized neurons and axons *(41)*.

Another alternative is the passive uptake of KCl through ion channels, rather than carriers. The driving force would be provided by the Donnan forces, and would continue until both Cl and K equilibrium potential, as well as the membrane potential, are equal. This accumulation, however, is crucially dependent on a significant Cl permeability of the cell membrane. It therefore contradicts the paradigms of the spatial buffer mechanism. The question of the existence of a significant Cl permeability during resting conditions, or at depolarized potentials, is very controversial. Radiotracer analysis of ion permeabilities in cultured astrocytes established the lack of any significant resting permeabilities for Cl *(37)*. The use of patch-clamp techniques showed a more complicated picture: Kraig et al. *(42,43)* found a Cl conductance in cultured astrocytes that were swollen, rounded up, or stellate, but not in flat, polygonally shaped astrocytes. The conductance was significant at resting potentials. Cytoskeletal actin controlled the open-state probability, conductance, and degree of rectification. This role of actin seems to explain the effect of cell shape on the expression of the Cl conductance. Ballanyi et al. *(28)* recorded intracellularly from glial cells in slices of guinea pig olfactory cortex, and found clear evidence for a Cl conductance at and near resting potentials. They also found that the Cl equilibrium potential and membrane potential were almost identical, and that Cl therefore is passively distributed. This contradicts findings by MacVicar et al. *(44)* and Walz and Wuttke *(31)*, of an active Cl accumulation in reactive astrocytes of hippocampal slices. The first used the reversal potential of the $GABA_A$ response, and the latter found a resting Cl conductance, whose pharmacological block hyperpolarized the membrane. Indeed, Ballanyi et al.'s results are more compatible with the situation in cultured oligodendrocytes *(11)*: a distinct possibility, since their results were not accompanied by attempts to immunocytochemically identify their glial type, in contrast to MacVicar and Walz, who used GFAP-positive cells. On the other hand, those latter two authors used reactive astrocytes, which, because of the actin–Cl channel interaction, may have a Cl conductance that is not expressed in normal astrocytes. Walz and Wuttke *(31)* showed a K accumulation in these reactive astrocytes that were exposed to excess K levels (10 m*M*). Blocking either the Cl conductance or the Na/K pump alone had no effect on the accumulation. But, if both were blocked at the same time, no accumulation of K occurred. This indicates that, at least in reactive (but not necessarily normal) astrocytes, both the carrier- and the channel-operated mechanisms work in parallel, and one can replace the other. This would make sense, since the carrier-operated mechanism is more energy-dependent than the channel one. However, such a parallel system is not yet demonstrated in normal astrocytes. It is also not clear how this parallel operation would be compatible with the spatial buffer mechanism, for which there is clear evidence from *in situ* preparations. An alternative is that, at rest, the Cl channels are closed, and are only opened after depolarization (caused by excess external K accumulation). There is some in vitro evidence for this. In flat, cultured astrocytes, there is no evidence for a Cl conductance, as investigated with patch clamp and radiotracer technology. However, when a patch was excised from the cell-attached mode, there was a relatively large Cl outward conductance (mediating Cl influx), as though an internal inhibitor (maybe actin filaments) was removed *(24,45–48)*. The channel was permeable for other small anions. Most of those authors found that only massive depolarization could activate the channel. This would mean only pathological events (spreading depression, anoxic depolarization) could open this channel. And, indeed, it has been shown that, during the passage of a spreading depression wave *in situ*, a large anion conductance opens *(20)*.

Thus, taken together, there is evidence for both carrier- and channel-operated K accumulation, but their precise mode of operation, their interaction with the spatial buffer mechanism, and the exact situations in which each of these mechanisms is maximally activated, are still unknown.

3. OTHER IONS: HYDROGEN

All astrocytes react to depolarization with an alkalinization. This phenomenon is called "depolarization-induced alkalinization," and results from the action of a electrogenic Na-bicarbonate co-transporter *(49)*. Since the co-transporter is carrying an electric charge, and therefore changing the transmembrane potential, its operation is also dependent on the actual membrane potential. Thus, a depolarization leads to Na and bicarbonate influx, and therefore to an alkalinization. Membrane hyperpolarization would accomplish just the opposite, and therefore acidify the cytoplasm. Neuronal activity leads to EC K accumulation, and this in turn to astrocytic depolarization. In astrocytes in the frontal cortex, a neuronal stimulation produced a 20-mV depolarization in astrocytes, which in turn causes an alkalinization of 0.2 pH units *(50)*. This could constitute a crucial link in a neuronal–glial signal system *(32)*. The alkalinization is facilitating glucose (GLU) utilization and Glu uptake *(51)*, as is the formation of glutamine from Glu and ammonium. Therefore, the potential signal system could be used to take up Glu and GLU, and to return glutamine and lact to active neurons (*see* Subheading 7.1).

4. TRANSMITTER INACTIVATION

Current concepts suggest that astrocytes protect neurons from excess transmitter release, and also take up and metabolize transmitter substances, in order to return them later as inactive precursors *(52)*. There are two principally different ways in which astrocytes participate in transmitter inactivation: uptake and subsequent metabolism; and EC enzymatic degradation. Astrocytic processes closely surround synaptic terminals, and are therefore well-suited to inactivate any transmitter that would escape the neuronal uptake/inactivation systems *(53)*. The concept of glial transmitter uptake is complicated by evidence *(54)* that these carrier systems can, under certain circumstances, reverse and contribute to transmitter release, rather than transmitter inactivation. The focus of interest is obviously the amino acid uptake systems, because of the clear correlation with disease processes. The Glu system is dealt with in Chapter 7 in detail, because of its relevance for Ca signaling between neurons and astrocytes.

GABA is the most abundant transmitter in the brain, and usually has inhibitory actions. Astrocytes take up GABA with high- and low-affinity systems, although it appeared to be at a rate somewhat less than Glu *(52)*. The low-affinity system seems to represent nonspecific uptake by the high-affinity transport system for taurine *(55)*. There are four isolated cDNAs that encode for highly homologous high-affinity GABA transporters, termed GAT 1, GAT 2, GAT 3, and betaine glycine (Gly) transporter. The concentration of GABA-transaminase is high in astrocytes, but GAD is absent, which means that the GABA concentration in astrocytes is probably low, and not detectable, as several studies have shown *(56)*. GAT 2 is found in arachnoid and ependymal cells *(57)*. GAT 3 is only found in astrocytes *(58)*, and GAT 1 is found in axonal terminals, as well as in astrocytic processes *(59)*. All these transporters are high-affinity systems and Na/Cl-dependent. However, they have different pharmacological properties, ionic requirements, and tissue distributions. It is not clear if GAT 1 and 3 co-exist in astrocytic processes. In the cortex, as well as many other parts of the brain, there is a high correlation between axonal terminals that contain GABA and GAT 1 and neighboring astrocytic processes that contain either GAT 1 or 3, or maybe both *(58)*. This arrangement suggests that astrocytic processes limit diffusion of GABA out of the synaptic cleft. However, in cerebellar Purkinje cells and in neurons of the thalamus, there are no GABA transporters, although the cells themselves are GABAergic. Instead, astrocytic processes almost completely envelope the terminals and separate them from each other *(56,60)*. The processes contain a high density of GAT 1 and 3. Thus, in these cases, it seems that astrocytes are responsible for most of the transmitter removal. The reasons and the functional implications are unknown. It is known that all these transporters can reverse and release GABA, especially if depolarized to a certain extent *(54)*. This fact led to speculations concerning astrocytes being involved in GABA release into the synaptic cleft. However, one has to keep in mind that GABA taken up is acted upon by GABA transaminase, and, thus, the GABA concentration in astrocytes is so low that it is not detectable.

Gly plays a role as a transmitter at inhibitory glycinergic synapses, opening strychnine-sensitive Cl-permeable channels. In addition, it is a high-affinity co-agonist with Glu on NMDA receptors, found at glutamatergic synapses. There are two Gly transporters, the neuronal GlyT2a and the glial GlyT1b *(61)*, which are both Na/Cl-coupled, but with a different stoichiometry. This results in neurons accumulating Gly under all conditions; astrocytes can reverse the uptake when conditions change. The astrocytic release could be triggered by an increase in the IC Na concentration. Because the GlyT1b are localized on fine astrocytic processes throughout the CNS, whose volume is small, a small amount of Na accumulation would be enough to cause this concentration change. Mechanisms for such a Na increase might be astrocytic AMPA receptors, which are activated when neuronal pathways are heavily used *(62)*. Thus, astrocytes may fine-tune the amount of Gly, according to surrounding activity patterns, and that this in turn controls the function of NMDA receptors (for more detail, *see* Chapter 7).

Uptake of all three monoamines (norepinephrine [NE], dopamine [DA], and serotonin [5-HT]) has been demonstrated for cultured astrocytes. It seems that all three are rapidly degraded in astrocytes, to a variety of oxidized, deaminated, methylated, and sulphated products *(52)*; hence, accumulation and subsequent release of these monoamines are not to be expected, if the transporter's driving force changes. Dave and Kimelberg *(63)* suggested that astrocytes *in situ* share a common 5-HT transporter with the neurons. This transporter has high affinity for 5-HT, and is Na-dependent and fluoxetine-sensitive. The situation is not so clear for the other monoamines; furthermore, results are mostly available from cultured astrocytes. NE is taken up by two astrocytic systems: a "neuronal" Na-dependent transporter, and an "extraneuronal" Na-independent one *(64)*. Inazu et al. *(65)* demonstrated that DA is taken up by the same systems as is NE. In neurons, the neuronal uptake system is specific for NE, and does not transport significant amounts of DA. The existence of two systems in astrocytes seems to suggest that one (Na-dependent) mechanism could be perisynaptic, and that the other (Na-independent) mechanism may be responsible for clearing spillover into neighboring areas, and may also be operational in pathological situations, when the Na gradient is diminished. It also appears that many monoamine uptake inhibitors that are in use therapeutically are targeting neuronal and astrocytic sites.

Adenosine uptake and metabolism seem to exhibit some species specificity. Gu et al. studied cultured human fetal astrocytes, and found two subtypes of adenosine transporters. Both were nonconcentrative, low-affinity, high-capacity systems, one of which was inhibitor-sensitive, and the other nonsensitive to inhibitors. The second one seemed to remove adenosine at a faster rate then synaptoneurosomes. The adenosine is phosphorylated to nucleotides, rather than deaminated. Under conditions of heavy adenosine accumulation in mouse astrocytes, Bender et al. *(66)* found a third system, one that was a Na-dependent concentrative mechanism, and that could mediate adenosine release.

Histamine is taken up and metabolized in astrocytes. However, the uptake by a high-affinity, Na- and K-dependent mechanism can be reversed, if the cells are depolarized *(67)*. It was shown in vivo that astrocytes had a major contribution to histamine clearance *(68)*.

Astrocytes contain a nonspecific cholinesterase that degrades acetylcholine *(52)*. Choline is taken up by astrocytes *(69)*.

Astrocytes are involved to different degrees and in different ways in this transmitter inactivation, and, as it appears, in the occasional manipulation of external transmitter concentrations. There are not many studies that compare neuronal with astrocytic removal efficiencies, and those that exist seem to focus on Glu. Using the temperature-sensitivity of the astrocytic transporter, it was shown that, after stimulation of Schaffer's collateral-commissural synapses in the hippocampus, astrocytes were responsible for removing the majority of the released Glu *(62)*. One has to be careful, however, in generalizing this important observation to other transmitters, and one has also to keep in mind that extensive neuronal stimulation was used. With more physiological stimuli, the complex arrangements in and around the synapses may well be different.

5. INITIAL RESPONSE TO ACUTE INJURY

Astrocytes react with fast changes to metabolic disruptions in hypoxic/ischemic and traumatic injuries. Within minutes, they swell, and change their ion channel pattern and ionic composition. They also participate in spreading depression, a key ingredient of most focal injuries. It is now clear that astrocytes react differently to ischemic episodes, if they are in isolation (cell culture or acutely isolated) or *in situ*. The differences are caused by compounds (K, neurotransmitters) released from neurons *(70)*. Thus, results obtained exclusively with isolated astrocytes must be evaluated carefully.

If local cerebral blood flow falls below 20% of normal, energy depletion occurs in surrounding cells, and ion homeostasis is subsequently lost. This loss results in anoxic depolarization, a hallmark of infarct sites *(71)*. Within minutes of such an event, astrocytes swell, which involves mostly processes around capillaries and neurons, but cell bodies are also swollen *(72)*. This swelling is characterized by pale and watery cytoplasm and dissociation of filament bundles, which is different from the hypertrophy of reactive astrocytes that appears later (1–5 d), as part of the delayed response *(73)*. The mechanisms involved in this swelling are not well-understood. Astrocytes in isolation do not swell as a consequence of energy depletion: They do so only if neurons are involved *(74)*. Astrocytes swell within seconds, if exposed to high K, because of K and anion uptake *(75)*. They also swell within 30 min of Glu exposure, because of Na and Cl increase. Other factors that involve neurons and cause swelling in conjunction with ischemia are acid–base changes (NaCl accumulation), breakdown of selective membrane permeability caused by action of free radicals and fatty acids (all reviewed in ref. 72). Thus, the swelling in this initial phase probably consists of an ion-mediated fast phase and a slower, metabolically driven one *(74)*. Exact mechanisms are not yet elucidated. However, the consequences of the astrocytic swelling for neuronal survival are mostly negative: It will shrink the ECS and constrict exchange with the blood vessels. The ECS volume decrease will lead to a further increase in the concentration of excitatory substances (K, Glu) (*see* Chapter 3). The astrocytic swelling is eventually counteracted by regulatory volume decrease, which will result in release of amino acids *(76)*.

Astrocytes depolarize quickly by ~50 mV. This part of the anoxic depolarization is not caused by intrinsic ion channel changes, but by release of K and Glu from depolarized neurons *(70)*. The released Glu causes Ca release from IC glial stores, in addition to the action of voltage-gated Ca channels on the astrocytic membrane, which are activated by the depolarization. The ion changes lead to a reversal of the Glu transporter on the astrocytes *(77)*. The sustained increase in Ca reduces the gap junction permeability, but in itself is not sufficient to uncouple the astrocytes *(78)*. Indeed, the gap junctions stay open until loss of cell integrity *(79)*. This seems to be important, since the astrocytes seem to die in groups, with the death triggered by neighboring cells. Astrocytes become rapidly acidic, and the degree of the acidosis seems to depend on GLU availability, and is highest in ischemia accompanied by hyperglycemic conditions *(80)*. They represent the most acidic compartment in the infarct region. The mechanism is unknown, but it is clear that such a proton gradient can only exist if the cell membrane stays intact for a long time in these ischemic astrocytes *(81)*. This confirms the other abovementioned observations.

Spreading depression is a phenomenon that occurs in focal brain injury *(82)*. It involves the breakdown of ion gradients, and has the form of a wave that moves slowly through the central nervous system. After the wave leaves an area, the tissue appears normal again within a short time. It is created at the border zone between infarct and penumbra, and moves radially out in intervals, separated by time periods of approx 10 min duration. The waves leave the penumbra and invade healthy tissue. The waves do not damage neurons *per se*, and lead to upregulation of gliotic features in microglia and astrocytes in healthy tissue. However, in tissue with compromised blood flow, the repeated passage of these waves will damage neurons. Tissue bordering the infarct zone does not repolarize after the invasion of a wave, but remains depolarized and eventually is incorporated into the evolving infarct *(83)*. Therefore, the infarct grows stepwise after each episode of spreading depression *(84)*. Astrocytes seem to be involved in the propagation of such a wave, which is only observed in tissue consisting of neurons and astrocytes. Experiments have shown that astrocytic gap junctions are a prerequisite for such propagation *(85)*. The passage of the wave involves the opening of a large anion conductance *(86)*. Such a conductance has been described in astrocytes *(11)*. During the passage of such a wave, astrocytes transiently increase their volume and turn alkaline transiently *(82)*. However, there is as yet no complete concept of the spreading depression mechanism that would explain all features. Still, it is clear that it is based on an intensive collaborative effort between neurons and astrocytes, and that it has the characteristics of a signal system that is active in focal injury, with damaging (penumbra) and protective (gliosis in healthy areas) features.

6. DELAYED RESPONSE TO INJURY: REACTIVE ASTROCYTES

Reactive astrocytes appear within days of any disturbance to the brain. Such a disturbance may be caused by acute or chronic injury, or it may result from slow, intrinsic neuronal degeneration. Hallmarks are hypertrophy, increase in filaments and in processes, changes in the expression of enzymes, different EC matrix composition, and different secretion products. If neuronal loss is occurring, this reaction also involves proliferation *(87)*. Astrocytes seem to undergo a considerable change in their function, and therefore also in their interactions with their environment.

There seems to be a difference between the reactive response in diffuse and focal injury. In diffuse injury, such as neurodegenerative diseases, there is only evidence for the expression of the same reactive properties, albeit in different intensities. However, in focal injury, such as stroke, two types of reactive astrocytes are present: the local response type, and the remote response type. In both types, there is an increase of GFAP within days of the injury. This increase is permanent in the area local response type, and returns to control levels in the remote areas. However, in both types, there is a transition from $GFAP^-$ to $GFAP^+$ cells *(88)*, with the transition again being transient in the noninjured area. In the necrotic area, the astrocytes show persistent hypertrophy and an increase in processes. In the remote area, the hypertrophy is small and transient. The expression of vimentin represents the most dramatic difference between both types. In local-response-type astrocytes, it shows a clear correlation with necrosis in the core injury area; the astrocytes in the remote location do not express any detectable vimentin. Proliferation is also correlated with the presence of vimentin. The two regionally distinct populations differ in more aspects than just the vimentin content. The expression of EC matrix proteins *(89)* and intermediate filament-associated proteins *(90)*, as well as that of other molecules *(91)*, is different in both populations. This strongly suggests that the two populations serve different functions, especially regarding the guidance and support of regenerating axons.

The genesis of the two types of reactive astrocytes is probably triggered by different factors. The local response could be initiated by microglial secretion products, because microglia activate before there is any sign of reactive astrogliosis *(92)*. It could also be a combination of stress factors released from neurons (K, adenosine triphosphate [ATP], or Glu) with microglial cytokines *(93)*. There is evidence that the remote reactive gliosis is triggered by spreading depression waves, in conjunction with Ca signals and/or alkaline pH changes in the astrocytes *(94–96)*.

Are these reactive astrocytes neuroprotective, or are they detrimental for neuronal survival? There is evidence that the remote-type reactive astrocyte exerts a protective effect. If this kind of gliotic response is caused by spreading depression alone, any subsequent injury is followed by a better-than-normal neuronal survival rate *(97)*. The reason for this is not yet clear, but it could involve the secretion of neurotrophic factors, such as nerve growth factor, basic fibroblast growth factor, neurotrophin 3, and ciliary neurotrophic factor *(98)*. Such reactive astrocytes also possess two independent mechanisms for rapid clearance of excess EC K, which will aid in neuronal survival *(31)*. On the other hand, there is evidence for an inhibitory effect of reactive astrocytes on axonal regeneration. This inhibition seems to result from the expression of proteoglycans in the EC matrix of astrocytes, and not from a mechanical barrier function *(99)*. Not all reactive astrocytes show these increases of proteoglycans after injury. They seem to be restricted to astrocytes that demarcate the borders of cysts and the interface between activated macrophages–microglia within the necrotic core area and surrounding reactive astrocytes *(100)*. This is an indication that local response-type reactive astrocytes are responsible for the detrimental effect on neuronal regeneration. Their priority seems to be to mark the border of the damaged core area, and to prevent its spread into adjacent tissue. Reactive astrocytes at a distance from the core injury are more likely to have different priorities, and to focus their function on assisting neuronal survival.

7. NEURON–ASTROCYTE METABOLIC INTERACTIONS

7.1. Introduction

The rate of oxidative metabolism in neurons is approx 10× that of glial cells *(101)*. This observation, together with the appreciation that astrocytic endfeet surround the microvessels, suggests that there may be metabolic coupling between astrocytes and neurons, and, indeed, there is considerable evidence for this. Because ~3% of all oxygen is incompletely reduced to the superoxide anion *(102)*, the high rate of oxidative metabolism results in neurons producing more strong oxidants than would astrocytes. Neurons not only have a high rate of strong oxidant production, but they are also characterized by being vulnerable to oxidative damage, because of two factors: a large membrane surface area, which is comprised of a large proportion of polyunsaturated fatty acids *(103)* liable to peroxidative damage *(104)*, and large numbers of Glu receptors and ion channels that can allow Ca^{2+} to enter the cell. Strong oxidants, such as the hydroxyl radical *(105)* or peroxynitrous acid *(106)*, can cause polyunsaturated fatty acid peroxidation, which, in the presence of oxygen, initiates a self-perpetuating chain reaction of lipid peroxidation, causing alterations in membrane fluidity *(107)*, increased permeability of membranes *(108)*, and decreased membrane ATPase activity *(109)*. This results in increased Na^+ and Ca^{2+} influx, resulting in depletion of cellular ATP stores and all the problems associated with cellular Ca^{2+} overload *(110,111)*. Lipid radicals and peroxides also break down, forming proinflammatory isoprostanes *(112)* and isoleukotrienes *(113)*, or strong oxidants, including dicarbonyls, such as malondialdehyde *(114)* and α,β-unsaturated aldehydes *(115,116)*. The latter are strong oxidants that can interfere with critical cellular functions, such as Glu uptake *(115,117)*, alter membrane protein configuration *(108)*, and can interfere with maintenance of ion homeostasis *(118)*, as well as mitochondrial respiration *(119)*.

A high rate of oxidative metabolism suggests that the energy of a large proportion of the electrons temporarily stored in the form of reduced nicotinamide adenine dinucleotide phosphate (NAD[P]H) will be used in establishing the proton gradient across the inner mitochondrial membrane, rather than used in the form of redox buffers, such as glutathione (GSH): This results in a relatively oxidized cytosol in neurons, in contrast to the more reduced cytosol in astrocytes. So the curious situation exists that the mechanisms that enable neurons to have a high rate of oxidative metabolism also ensure that neurons have an intrinsic lessor capability to cope with the strong oxidants produced. It seems logical that there be some redox coupling between neurons and astrocytes.

7.2. Glycolysis

The brain is dependent on GLU for the formation of ATP *(101)*. GLU passes down its concentration gradient from the blood into the brain, to enter the cells using a variety of GLU transporters that are not dependent on the Na gradient. The endothelium, as well as astrocytes, use glucose transporter-1 (GLUT1); neurons use GLUT3 *(120,121)*. The endothelial GLUT1 is a highly glycosylated 55-kDa form; the astrocyte GLUT1 is a nonglycosylated 45 kDa form. A feature associated with the blood–brain barrier (BBB) is that astrocytic processes surround the microvasculature *(122)*.

Astrocytes have a more active glycolytic cycle than neurons *(123)*, with mouse astrocytes metabolizing ~1.6 μmol GLU/h/mg protein *(124)*, resulting in the release of

~2 μmol lact/h/mg protein for mouse astrocytes *(125)* and ~1 μmol lact/h/mg protein for rat astrocytes *(126)*. Since GLU moves down the concentration gradient, the consequence of having astrocytic processes surrounding the microvasculature, where GLU effluxes the capillaries, is that probably most GLU enters the astrocyte compartment *(127)*. The net influx K_m for GLU by GLUT1 is 1.6 m*M*, which is higher than the net efflux K_m *(128)*. This kinetic asymmetry is allosterically regulated by IC metabolites, ensuring net GLU influx where there is great metabolic demand and EC levels of GLU are low. GLUT3 has a lower K_m for GLU than GLUT1, but it has a lower net exchange capacity *(128)*. These features of GLUT1 and GLUT3 may allow movement of GLU, under conditions of low metabolic demand from blood to astrocyte to ECS to neuron, ensuring that neurons obtain sufficient GLU for the pentose phosphate cycle.

The relationship of astrocyte processes to the microvasculature and the asymmetric kinetic properties of GLUT1 and the lower K_m of GLUT3 for GLU suggests that astrocytes either provide GLU or some energy-rich intermediate to neurons. It has been known for some time that astrocytes undergo anaerobic glycolysis, resulting in release of lact *(125)*; this efflux is mediated by the monocarboxylate transporter 1 *(129)* and 2 *(130)*. Lact, a better substrate for oxidative metabolism in neurons than GLU *(131,132)*, is released by astrocytes, and appears to be taken up by adjacent neurons *(133–135)* via the monocarboxylate transporter-2 *(129)*. Microdialysis studies have demonstrated that neuronal activity increases lact in the ECS *(136)*; since, within a few days, the ECS around a microdialysis tubing is surrounded by reactive astrocytes, it provides evidence that neuronal activity causes astrocytes to increase production and release of lact. Does this metabolic interaction occur in vivo, or is this an artifact of the in vitro preparations being examined? A consideration of the regulation of glycolysis supports the idea of metabolic interaction.

Glycolysis is limited by the feedback inhibition of ATP and citrate on phosphofructokinase and pyruvate kinase *(137)*, and of NADH by feedback on glyceraldehyde-3-phosphate dehydrogenase *(138)*. NADH is formed by the reduction of NAD resulting from conversion of one molecule of glyceraldehyde-3-phosphate to one molecule of phosphoglyceroyl phosphate, which ultimately is converted to one molecule of pyruvate, causing four adenosine diphosphate molecules to be phosphorylated to four ATP molecules. Hence, pyruvate accumulation inhibits glycolysis. One means of overcoming the inhibition by NADH of glyceraldehyde-3-phosphate dehydrogenase, which is the main metabolic control of the rate of glycolysis, is to convert pyruvate to lact, with the concomitant oxidation of NADH to NAD. This reaction is catalyzed by the enzyme, lactate dehydrogenase (LDH). What governs the direction of this reaction is not only the concentration of the involved metabolites, but also the LDH isoenzyme composition.

In most cell types, there are two isoforms expressed: LDH-A (anaerobic isoform, also known as the muscle, or M, isoform) and LDH-B (aerobic isoform, also known as the heart, or H, isoform), which combine into a tetramer with five possible combinations: 4LDHA (LDH5), 3LDHA:1LDHB, 2LDHA:2LDHB, 1LDHA:3LDHB, 4LDHB (LDH1). The aerobic isoenzymes of LDH promote the formation of pyruvate from lact, with the concomitant reduction of NAD to NADH; the anaerobic isoenzymes promote the formation of lact from pyruvate with the concomitant oxidation of NADH to NAD, thereby allowing continued functioning of the glycolytic cycle *(138,139)*. The predominant LDH isoenzyme in astrocytes is the anaerobic LDH5 *(140)*; the predominant LDH isoenzyme in neurons is aerobic LDHI *(141)*. It is known that what governs the rate of glycolysis in astrocytes is not energy demand *(142)*; since astrocytes have predominantly the anaerobic LDH isoenzymes, it is likely that this feature governs glycolysis in astrocytes. With the high rate of oxidative metabolism in neurons, their high NAD:NADH ratios, as well as their aerobic LDH isoenzyme, promote the formation of pyruvate from lact, and concomitant reduction of NAD to NADH. Lact not only is the preferred substrate by neurons *(123,131,132)*, but it has the advantage, compared to GLU, because it bypasses the steps of glycolysis, and the conversion of one molecule of lact to pyruvate also generates one molecule of NADH. In exporting lact, astrocytes not only export a metabolite but also a reducing-equivalent. The result is that every two molecules of lact taken up by the neuron and converted to pyruvate yield two NADH molecules that can alter the neuronal redox status or be used to generate six ATP molecules; furthermore, the energy inherent in the two pyruvate molecules formed can give rise to 28 additional ATP molecules, for a total potential production of 34 ATP molecules from each two lact molecules formed from one GLU molecules. The theoretical potential number of ATP molecules that can be formed from the energy in one GLU molecule is 36; hence, it can be seen that it is much more efficient for neurons to rely on lact than GLU for its energy.

Astrocytes also store GLU in the form of glycogen *(122)*, which is formed, not only from GLU, but significant amounts are formed from Kreb cycle intermediates *(143)*, mediated by phosphenolpyruvate carboxykinase *(143)*. The key gluconeogenic enzyme, fructose-1,6-bisphosphatase, is found exclusively in astrocytes *(144)*. This gluconeogenic capacity of astrocytes indicates that glycogen stores play an important role in CNS function. Indeed, these astrocyte glycogen stores have been shown to promote neuronal survival when GLU levels drop in mixed neuronal–glial cultures *(145)*. The glycogen stores in astrocytes also form a source of releasable lact *(146)*, which has been shown to be involved in protecting axons during a period of GLU deprivation *(147)*. Astrocyte glycogen stores are also used during memory formation *(148)*. The link between the coupling of neuronal energy demand and astrocyte glycolytic rate appears to be neurotransmitters released from neurons acting on astrocyte receptors *(149)*. Thus, NE *(150)* and ATP *(151)* cause glycogenolysis. Glu stimulates glycolysis in astrocytes *(152,153)*; this stimulation is caused by the increased IC Na^+ resulting from the Na-dependent Glu uptake.

7.3. The Kreb Cycle

The Kreb cycle has two functions: It oxidizes acetyl moieties to water and carbon dioxide, yielding intermediate energy stores, such as NADH, which can be used to generate the proton gradient across the inner mitochondrial membrane; and it provide intermediates for synthetic purposes. The Kreb cycle starts with the 6-carbon citric acid, going through a series of intermediates, resulting, in two decarboxylation steps, to yield the three-carbon succinate, and ending with the four-carbon oxaloacetate; the latter four-carbon moiety can combine with an acetyl moiety to form citric acid. Hence, if an intermediate is removed for synthetic purposes, it stops the cycle. The only way for the cycle to continue, when intermediates are used for synthetic purposes, is to synthesize new intermediates: The major anaplerotic mechanism that facilitates this is the enzyme pyruvate carboxylase, which ligates a molecule of CO_2 to pyruvate, resulting in the formation of oxaloacetate. Pyruvate carboxylase is not present in neurons, but is present in astrocytes *(154,155)*. The consequence of this is that neurons are ultimately dependent on astrocytes for a source of Kreb cycle intermediates that are being used for synthetic purposes.

Kreb cycle intermediates can be transferred from astrocyte to neuron, using a variety of molecules. There is clear evidence that α-ketoglutarate, malate, and succinate are released from astrocytes *(156)*. These intermediates are taken up by synapses via high-affinity transport processes *(157–159)*. Another source of Kreb cycle intermediates is amino acids that can be released from astrocytes. A classical example is glutamine *(160–162)*. We have found that incorporating α-ketoglutarate into the culture medium greatly promotes neuronal survival in cell culture *(163)*, and now routinely incorporate α-ketoglutarate into the growth medium used for culturing neurons *(164)*.

7.4. *Ammonia Detoxification*

The major means by which ammonia is cleared from the CNS is by ligating it to Glu forming glutamine; this enzyme is localized to astrocytes *(165,166)*. This enzymatic activity subserves two critical functions: It clears ammonia, and it converts the excitotoxic amino acid into the neutral amino acid, glutamine, which can be returned to neurons as a source of Kreb cycle intermediate.

8. NEURON–ASTROCYTE REDOX INTERACTIONS

Neurons have a higher rate of oxidative metabolism than astrocytes, the consequence of which is that the neuronal cytosol is more oxidized than the astrocyte cytosol. This suggests that there be redox coupling between neurons and astrocytes. Because of the high rate of anaerobic glycolysis, and the high activity of GLU-6-phosphate dehydrogenase *(167)* and the NADP-dependent isoforms of isocitrate dehydrogenase *(168)*, astrocytes have a great capacity to form NAD(P)H. NADPH is used by reactions that reduce oxidized-glutathione *(169)*, oxidized-thioredoxin *(170)*, and other redox-active compounds. Because of the presence of transhydrogenases *(171)*, NAD(H) is interconvertible with NADP(H) *(171)*. Astrocytes can transfer reducing equivalents to neurons in several ways. As noted above, one way is release lact, which is taken up by neurons: Every lact molecule transferred is the equivalent of the transfer of one reducing equivalent (i.e., NAD[P]H).

GSH is the electron donor in the scavenging of a variety of strong oxidants, as well as in the reduction of oxidized-ascorbate (ASC) *(172)*, and therefore plays a central role in many of the mechanisms used to reduce oxidative stress. Neurons have a somewhat lower content of GSH than do astrocytes *(173–176)*. Small increases in GSH can result in marked increases in the ability of cells to cope with oxidative stress *(177)*. The rate-limiting amino acid in GSH synthesis is cysteine *(178)*. Cysteine is readily autooxidized to cystine *(179)*, and 90% of cysteine equivalents is in the form of cystine, which is taken up by the X_c-antiporter in exchange with Glu *(179)*. Neurons do not take up cystine, but do take up cysteine *(180)*. Astrocytes promote neuronal GSH synthesis by release of cysteine *(180)*, as well as by release of cysteinyl-Gly *(181)*.

Neurons have a 10-fold higher level of ASC (10 m*M*) than do astrocytes *(174,182)*. Why is this? One possible functional reason is that ASC plays an important role in reducing the vitamin E radical formed by the action of vitamin E scavenging a lipid peroxyl radical. The very large membrane surface area, containing an abundance of polyunsaturated fatty acids, results in neurons being susceptible to lipid peroxidation. For reviews of the role of vitamin E and C in inhibiting chain propagation of lipid peroxidation, *see* refs. *172* and *183*. Because GSH plays an important role in the reduction of oxidized ASC (dehydroascorbate [DHA]) *(184,185)*, the question arises as to where the reducing power for reduction of oxidized ASC (DHA) resides: the astrocyte or neuron? Both ASC and DHA are taken by astrocytes in vitro *(186)*. DHA passes down its concentration gradient, from the ECS through the GLUT1, into the cytosol, where it is quickly reduced to ASC. Both GLUT1 and GLUT3 have been shown to transport DHA *(187)*. In contrast, ASC is taken up into astrocytes in vitro via a Na-dependent process *(186)*; however, this Na-dependent uptake mechanism is expressed mostly, if not exclusively, in vivo in neurons *(188)*. In vitro, Glu stimulates ASC release from astrocytes *(189)*. Low EC Na^+ and K^+-induced depolarization also caused ASC release from astrocytes *(190)*, which suggests the possibility that DHA is released from neurons and taken up by astrocytes, with the driving force being the DHA concentration gradient; the DHA then being reduced to ASC in astrocytes, and released to be taken up principally by neurons, because of the higher concentration of the Na-dependent ASC transporters in neurons than astrocytes. This is still speculation, and requires further experimental investigation.

In addition to the GSH-dependent ASC reductase activity, the selenoenzyme, thioredoxin reductase (TrxRed), can also reduce DHA *(191)*, as well as the ascorbyl radical *(192)*, using NADPH as the electron donor. We have recently demonstrated that, although, under normal culture conditions, astrocytes have only a slightly greater (~25%) TrxRed activity than neurons, it is inducible in astrocytes, but not neurons, by a phase-2 enzyme inducer *(176)*. TrxRed activity can also indirectly enable reducing equivalents to shuttle from astrocytes to neurons. TrxRed reduces lipoic acid to dihydrolipoate *(193)*. Lipoic acid is efficiently taken up by a variety of cell types, such as human diploid fibroblasts and C6 glioma cells, reduced to dihydrolipoate, and released back into the culture medium, where it can reduce cystine to cysteine *(194,195)*. There is evidence that uptake of dihydrolipoate added to the culture medium minimizes GSH loss and cell death in neuronal cultures *(196)*, and protects against Glu excitoxicity *(197)*. Thus, one possible consequence of increased TrxRed activity in astrocytes is to promote reduction lipoic acid and release of dihydrolipic acid, which can be taken up by adjacent neurons. This is speculative, and requires experimental verification.

9. CONCLUSIONS

In his book, *The Neuron and The Glial Cell*, Santiago Ramón y Cajal discusses the role of neuroglia, which is now referred to specifically as astrocytes, in nervous system function *(198)*. He discusses Golgi's nutritive function for astrocytes, but discounts it on several grounds, including that dendrites do not contact neuroglia. With the introduction of electron microscopic approaches many decades ago, astrocytic processes are now known to closely approximate dendrites. A second function, first put forth by his brother, P. Ramón y Cajal, is that astrocytes function to isolate the conducting parts of the neuron from neighboring neurons, thereby enhancing information transmission.

Ramón y Cajal was correct, in part, in that an important role of astrocytes is to facilitate communication among neurons. The K-buffering and neurotransmitter uptake mechanisms of astrocytes serve to enhance both temporal and spatial resolution in information transmission, and to decrease the chances of crosstalk-related noise in information transmission. Golgi was also correct, in part, in that astrocytes play an important role in the nutritional support of neurons. Undoubtedly, Santiago Ramón y Cajal would be delighted with what is now known with respect to the role of astrocytes in ion homeostasis, neurotransmitter inactivation, and responses to injury. And, undoubtedly, Camillo Golgi would be equally delighted with what is now known with respect to astrocyte–neuron metabolic and redox interactions.

ACKNOWLEDGMENTS

W. W. and B. H. J. J. are supported by research grants from the Canadian Institutes of Health Research.

REFERENCES

1. Privat, A. and Rataboul, P. (1986) Fibrous and protoplasmic astrocytes, in *Astrocytes* (Fedoroff, S. and Vernadakis, A., eds.), Academic, San Diego, CA, pp. 105–130.
2. Vernadakis, A. and Roots, B. I., eds. (1995) *Neuron-Glia Interactions During Phylogeny*. Humana, Totowa, NJ.
3. Walz, W. (2000) Controversy surrounding the existence of discrete functional classes of astrocytes in adult gray matter. *Glia* **31,** 95–103.
4. Pappas, C. A., Ullrich, N., and Sontheimer, H. (1994) Reduction of glial proliferation by K^+ channel blockers is mediated by changes in pHi. *Neuroreport* **6,** 193–196.
5. Ransom, C. B. and Sontheimer, H. (1995) Biophysical and pharmacological characterization of inwardly rectifying K^+ currents in rat spinal cord astrocytes. *J. Neurophysiol.* **73,** 333–346.
6. Soroceanu, L., Manning, T. J., Jr., and Sontheimer, H. (1999) Modulation of glioma cell migration and invasion using $Cl(^-)$ and $K(^+)$ ion channel blockers. *J. Neurosci.* **19,** 5942–5954.
7. Kettenmann, H. and Ransom, B. R., eds. (1995) *Neuroglia.* Oxford University Press, New York.
8. Kimelberg, H. K., Cai, Z., Rastogi, P., Charniga, C. J., Goderie, S., Dave, V., and Jalonen, T. O. (1997) Transmitter-induced calcium responses differ in astrocytes acutely isolated from rat brain and in culture. *J. Neurochem.* **68,** 1088–1098.
9. Bekar, L. K., Jabs, R., and Walz, W. (1999) $GABA_A$ receptor agonists modulate K^+ currents in adult hippocampal glial cells in situ. *Glia* **26,** 129–138.
10. Condorelli, D. F., Conti, F., Gallo, V., Kirchhoff, F., Seifert, G., Steinhauser, C., Verkhratsky, A., and Yuan, X. (1999) Expression and functional analysis of glutamate receptors in glial cells. *Adv. Exp. Med. Biol.* **468,** 49–67.
11. Walz, W. (1995) Distribution and transport of chloride and bicarbonate ions across glial cell membranes, in *Neuroglia* (Kettenmann, H. and Ransom, B. R., eds.), Oxford University Press, New York, pp. 221–229.
12. Leech, C. A. and Stanfield, P. R. (1981) Inward rectification in frog skeletal muscle fibres and its dependence on membrane potential and external potassium. *J. Physiol.* **319,** 295–309.
13. Balestrino, M., Aitken, P. G., and Somjen, G. G. (1986) The effects of moderate changes of extracellular K^+ and Ca^{2+} on synaptic and neural function in the CA1 region of the hippocampal slice. *Brain Res.* **377,** 229–239.

14. Heinemann, U., Neuhaus, S., and Dietzel, I. (1883) Aspects of potassium regulation in normal and gliotic brain tissue, in *Current Problems in Epilepsy* (Baldy-Moulinier, M., Ingvar, D., and Beldrum, B., eds.), John Libby, London, pp. 271–277.
15. Nicholson, C. and Rice, M. E. (1991) Diffusion of ions and transmitters in the brain cell microenvironment, in *Volume Transmission in the Brain* (Fuxe, K. and Agnati, L. F., eds.), Raven, New York, pp. 279–294.
16. Orkand, R. K., Nicholls, J. G., and Kuffler, W. (1966) Effect of nerve impulses on the membrane potential of glial cells in the central nervous system of amphibia. *J. Neurophysiol.* **29,** 788–806.
17. Hertz, L. (1965) Possible role of neuroglia: a potassium-mediated neuronal-neuroglial-neuronal impulse transmission system. *Nature* **206,** 1091–1094.
18. Nicholson, C. and Phillips, J. M. (1981) Ion diffusion modified by tortuosity and volume fraction in the extracellular microenvironment of the rat cerebellum. *J. Physiol. (Lond.)* **321,** 225–257.
19. Gardner-Medwin, A. R. (1983) Analysis of potassium dynamics in mammalian brain tissue. *J. Physiol. (Lond.)* **335,** 393–426.
20. Nicholson, C., Phillips, J. M., and Gardner-Medwin, A. R. (1979) Diffusion from an iontophoretic point source in the brain: role of tortuosity and volume fraction. *Brain Res.* **169,** 580–584.
21. Hounsgaard, J. and Nicholson, C. (1983) Potassium accumulation around individual Purkinje cells in cerebellar slices from the guinea-pig. *J. Physiol. (Lond.)* **340,** 359–388.
22. Newman, E. A. (1986) High potassium conductance in astrocyte endfeet. *Science* **233,** 453–454.
23. Newman, E. A. (1995) Glial cell regulation of extracellular potassium, in *Neuroglia* (Kettenman, H. and Ransom, B. R., eds.), Oxford University Press, New York, pp. 717–731.
24. Barres, B. A., Koroshetz, W. J., Chun, L. L., and Corey, D. P. (1990) Ion channel expression by white matter glia: the type-1 astrocyte. *Neuron* **5,** 527–544.
25. Janigro, D., Gasparini, S., D'Ambrosio, R., McKhann, G., 2nd, and DiFrancesco, D. (1997) Reduction of K+ uptake in glia prevents long-term depression maintenance and causes epileptiform activity. *J. Neurosci.* **17,** 2813–2824.
26. Walz, W., Wuttke, W., and Hertz, L. (1984) Astrocytes in primary cultures: membrane potential characteristics reveal exclusive potassium conductance and potassium accumulator properties. *Brain Res.* **292,** 367–374.
27. Kettenmann, H. (1987) K^+ and Cl^- uptake by cultured oligodendrocytes. *Can. J. Physiol. Pharmacol.* **65,** 1033–1037.
28. Ballanyi, K., Grafe, P., and ten Bruggencate, G. (1987) Ion activities and potassium uptake mechanisms of glial cells in guinea- pig olfactory cortex slices. *J. Physiol. (Lond.)* **382,** 159–174.
29. Coles, J. A. and Orkand, R. K. (1983) Modification of potassium movement through the retina of the drone (Apis mellifera male) by glial uptake. *J. Physiol. (Lond.)* **340,** 157–174.
30. Wuttke, W. A. (1990) Mechanism of potassium uptake in neuropile glial cells in the central nervous system of the leech. *J. Neurophysiol.* **63,** 1089–1097.
31. Walz, W. and Wuttke, W. A. (1999) Independent mechanisms of potassium clearance by astrocytes in gliotic tissue. *J. Neurosci. Res.* **56,** 595–603.
32. Ransom, B. R. (1992) Glial modulation of neural excitability mediated by extracellular pH: a hypothesis. *Progr. Brain Res.* **94,** 37–46.
33. Ballanyi, K. (1995) Modulation of glial potassium, sodium and chloride activities by the extracellular milieu, in *Neuroglia* (Kettenmann, H. and Ransom, B. R., eds.), Oxford University Press, New York, pp. 289–298.
34. Ransom, C. B., Ransom, B. R., and Sontheimer, H. (2000) Activity-dependent extracellular K^+ accumulation in rat optic nerve: the role of glial and axonal Na^+ pumps. *J. Physiol. (Lond.)* **522(Pt 3),** 427–442.

35. Rose, C. R. and Ransom, B. R. (1996) Intracellular sodium homeostasis in rat hippocampal astrocytes. *J. Physiol. (Lond.)* **491,** 291–305.
36. Rose, C. R. and Ransom, B. R. (1997) Regulation of intracellular sodium in cultured rat hippocampal neurones. *J. Physiol. (Lond.)* **499,** 573–587.
37. Walz, W. and Hertz, L. (1983) Intracellular ion changes of astrocytes in response to extracellular potassium. *J. Neurosci. Res.* **10,** 411–423.
38. Walz, W. and Hinks, E. C. (1986) A transmembrane sodium cycle in astrocytes. *Brain Res.* **368,** 226–232.
39. Walz, W. (1987) Swelling and potassium uptake in cultured astrocytes. *Can. J. Physiol. Pharmacol.* **65,** 1051–1057.
40. Payne, J. A., Xu, J. C., Haas, M., Lytle, C. Y., Ward, D., and Forbush, B., 3rd (1995) Primary structure, functional expression, and chromosomal localization of the bumetanide-sensitive Na-K-Cl cotransporter in human colon. *J. Biol. Chem.* **270,** 17,977–17,985.
41. Payne, J. A. (1997) Functional characterization of the neuronal-specific K-Cl cotransporter: implications for $[K^+]_o$ regulation. *Am. J. Physiol.* **273,** C1516–1525.
42. Lascola, C. D. and Kraig, R. P. (1996) Whole-cell chloride currents in rat astrocytes accompany changes in cell morphology. *J. Neurosci.* **16,** 2532–2545.
43. Lascola, C. D., Nelson, D. J., and Kraig, R. P. (1998) Cytoskeletal actin gates a Cl- channel in neocortical astrocytes. *J. Neurosci.* **18,** 1679–1692.
44. MacVicar, B. A., Tse, F. W., Crichton, S. A., and Kettenmann, H. (1989) GABA-activated Cl- channels in astrocytes of hippocampal slices. *J. Neurosci.* **9,** 3577–3583.
45. Gray, P. T. and Ritchie, J. M. (1986) A voltage-gated chloride conductance in rat cultured astrocytes. *Proc. R Soc. Lond. B Biol. Sci.* **228,** 267–288.
46. Nowak, L., Ascher, P., and Berwald-Netter, Y. (1987) Ionic channels in mouse astrocytes in culture. *J. Neurosci.* **7,** 101–109.
47. Sonnhof, U. (1987) Single voltage-dependent K^+ and Cl^- channels in cultured rat astrocytes. *Can. J. Physiol. Pharmacol.* **65,** 1043–1050.
48. Jalonen, T. (1993) Single-channel characteristics of the large-conductance anion channel in rat cortical astrocytes in primary culture. *Glia* **9,** 227–237.
49. Deitmer, J. W. and Schneider, H. P. (1995) Voltage-dependent clamp of intracellular pH of identified leech glial cells. *J. Physiol. (Lond.)* **485,** 157–166.
50. Chesler, M. and Kraig, R. P. (1989) Intracellular pH transients of mammalian astrocytes. *J. Neurosci.* **9,** 2011–2019.
51. Brookes, N. (1997) Intracellular pH as a regulatory signal in astrocyte metabolism. *Glia* **21,** 64–73.
52. Martin, D. L. (1995) The role of glia in the inactivation of neurotransmitters, in *Neuroglia* (Kettenmann, H. and Ransom, B. R., eds.), Oxford University Press, New York, pp. 732–745.
53. Kosaka, T. and Hama, K. (1986) Three-dimensional structure of astrocytes in the rat dentate gyrus. *J. Comp. Neurol.* **249,** 242–260.
54. Levi, G. and Raiteri, M. (1993) Carrier-mediated release of neurotransmitters. *Trends Neurosci.* **16,** 415–419.
55. Martin, D. L. and Shain, W. (1979) High affinity transport of taurine and beta-alanine and low affinity transport of gamma-aminobutyric acid by a single transport system in cultured glioma cells. *J. Biol. Chem.* **254,** 7076–7084.
56. De Biasi, S., Vitellaro-Zuccarello, L., and Brecha, N. C. (1998) Immunoreactivity for the GABA transporter-1 and GABA transporter-3 is restricted to astrocytes in the rat thalamus. A light and electron-microscopic immunolocalization. *Neuroscience* **83,** 815–828.
57. Durkin, M. M., Smith, K. E., Borden, L. A., Weinshank, R. L., Branchek, T. A., and Gustafson, E. L. (1995) Localization of messenger RNAs encoding three GABA transporters in rat brain: an in situ hybridization study. *Brain Res. Mol. Brain Res.* **33,** 7–21.

58. Minelli, A., DeBiasi, S., Brecha, N. C., Zuccarello, L. V., and Conti, F. (1996) GAT-3, a high-affinity GABA plasma membrane transporter, is localized to astrocytic processes, and it is not confined to the vicinity of GABAergic synapses in the cerebral cortex. *J. Neurosci.* **16,** 6255–6264.
59. Minelli, A., Brecha, N. C., Karschin, C., DeBiasi, S., and Conti, F. (1995) GAT-1, a high-affinity GABA plasma membrane transporter, is localized to neurons and astroglia in the cerebral cortex. *J. Neurosci.* **15,** 7734–7746.
60. Ribak, C. E., Tong, W. M., and Brecha, N. C. (1996) Astrocytic processes compensate for the apparent lack of GABA transporters in the axon terminals of cerebellar Purkinje cells. *Anat. Embryol. (Berl.)* **194,** 379–390.
61. Roux, M. J. and Supplisson, S. (2000) Neuronal and glial glycine transporters have different stoichiometries. *Neuron* **25,** 373–383.
62. Bergles, D. E., Dzubay, J. A., and Jahr, C. E. (1997) Glutamate transporter currents in bergmann glial cells follow the time course of extrasynaptic glutamate. *Proc. Natl. Acad. Sci. USA* **94,** 14821–14825.
63. Dave, V. and Kimelberg, H. K. (1994) Na(+)-dependent, fluoxetine-sensitive serotonin uptake by astrocytes tissue-printed from rat cerebral cortex. *J. Neurosci.* **14,** 4972–4986.
64. Kimelberg, H. K. and Goderie, S. K. (1993) Effect of ascorbate on Na(+)-independent and Na(+)-dependent uptake of [^{3}H]norepinephrine by rat primary astrocyte cultures from neonatal rat cerebral cortex. *Brain Res.* **602,** 41–44.
65. Inazu, M., Kubota, N., Takeda, H., Zhang, J., Kiuchi, Y., Oguchi, K., and Matsumiya, T. (1999) Pharmacological characterization of dopamine transport in cultured rat astrocytes. *Life Sci.* **64,** 2239–2245.
66. Bender, A. S., Woodbury, D. M., and White, H. S. (1994) Ionic dependence of adenosine uptake into cultured astrocytes. *Brain Res.* **661,** 1–8.
67. Huszti, Z. (1998) Carrier-mediated high affinity uptake system for histamine in astroglial and cerebral endothelial cells. *J. Neurosci. Res.* **51,** 551–558.
68. Huszti, Z., Prast, H., Tran, M. H., Fischer, H., and Philippu, A. (1998) Glial cells participate in histamine inactivation in vivo. *Naunyn Schmiedebergs Arch. Pharmacol.* **357,** 49–53.
69. Walz, W. (1995) Acetylcholine and serotonin receptor activation, in *Neuroglia* (Kettenmann, H. and Ransom, B. R., eds.), Oxford University Press, New York, pp. 346–353.
70. Duffy, S. and MacVicar, B. A. (1996) In vitro ischemia promotes calcium influx and intracellular calcium release in hippocampal astrocytes. *J. Neurosci.* **16,** 71–81.
71. Iadecola, C. (1999) Mechanisms of cerebral ischemic damage, in *Cerebral Ischemia* (Walz, W., ed.), Humana, Totowa, NJ, pp. 3–32.
72. Kimelberg, H. K. (1999) Cell swelling in cerebral ischemia, in *Cerebral Ischemia* (Walz, W., ed.), Humana, Totowa, NJ, pp. 45–67.
73. Rumpel, H., Nedelcu, J., Aguzzi, A., and Martin, E. (1997) Late glial swelling after acute cerebral hypoxia-ischemia in the neonatal rat: a combined magnetic resonance and histochemical study. *Pediatr. Res.* **42,** 54–59.
74. Walz, W., Klimaszewski, A., and Paterson, I. A. (1993) Glial swelling in ischemia: a hypothesis. *Dev. Neurosci.* **15,** 216–225.
75. Walz, W. (1992) Mechanism of rapid K(+)-induced swelling of mouse astrocytes. *Neurosci. Lett.* **135,** 243–246.
76. Rutledge, E. M. and Kimelberg, H. K. (1996) Release of [^{3}H]-D-aspartate from primary astrocyte cultures in response to raised external potassium. *J. Neurosci.* **16,** 7803–7811.
77. Szatkowski, M. and Attwell, D. (1994) Triggering and execution of neuronal death in brain ischaemia: two phases of glutamate release by different mechanisms. *Trends Neurosci.* **17,** 359–365.
78. Wang, Z., Tymianski, M., Jones, O. T., and Nedergaard, M. (1997) Impact of cytoplasmic calcium buffering on the spatial and temporal characteristics of intercellular calcium signals in astrocytes. *J. Neurosci.* **17,** 7359–7371.

79. Cotrina, M. L., Lin, J. H., Alves-Rodrigues, A., Liu, S., Li, J., Azmi-Ghadimi, H., et al. (1998) Connexins regulate calcium signaling by controlling ATP release. *Proc. Natl. Acad. Sci. USA* **95,** 15,735–15,740.
80. Kraig, R. P. and Chesler, M. (1990) Astrocytic acidosis in hyperglycemic and complete ischemia. *J. Cereb. Blood Flow Metab.* **10,** 104–114.
81. Kraig, R. P., Lascola, C. D., and Caggiano, A. (1995) Glial responses to brain ischemia, in *Neuroglia* (Kettenmann, H. and Ransom, B. R., eds.), Oxford University Press, New York, pp. 964–976.
82. Irwin, A. and Walz, W. (1999) Spreading depression waves as mediators of secondary injury and of protective mechanisms, in *Cerebral Ischemia* (Walz, W., ed.), Humana, Totowa, NJ, pp. 35–44.
83. Hossmann, K.-A. (1994) Viability thresholds and the penumbra of focal ischemia. *Ann. Neurol.* **36,** 557–565.
84. Mies, G., Iijima, T., and Hossmann, K. A. (1993) Correlation between peri-infarct DC shifts and ischaemic neuronal damage in rat. *Neuroreport* **4,** 709–711.
85. Nedergaard, M., Cooper, A. J. L., and Goldman, S. A. (1995) Gap junctions are required for the propagation of spreading depression. *J. Neurobiol.* **28,** 433–444.
86. Phillips, J. M. and Nicholson, C. (1979) Anion permeability in spreading depression investigated with ion-sensitive microelectrodes. *Brain Res.* **173,** 567–571.
87. Jabs, R. K., Bekar, L., and Walz, W. (1999) Reactive astrogliosis in the injured and postischemic brain, in *Cerebral Ischemia* (Walz, W., ed.), Humana, Totowa, NJ, pp. 233–249.
88. Petito, C. K. (1986) Transformation of postischemic perineuronal glial cells. I. Electron microscopic studies. *J. Cereb. Blood Flow Metab.* **6,** 616–624.
89. David, S. and Ness, R. (1993) Heterogeneity of reactive astrocytes, in *Biology and Pathology of Astrocyte-Neuron Interactions* (Fedoroff, S., Juurlink, B. H. J., and Doucette, R., eds.), Plenum, New York, pp. 303–312.
90. Yang, H. Y., Lieska, N., Kriho, V., Wu, C. M., and Pappas, G. D. (1997) A subpopulation of reactive astrocytes at the immediate site of cerebral cortical injury. *Exp. Neurol.* **146,** 199–205.
91. Malhotra, S. K., Shnitka, T. K., and Elbrink, J. (1990) Reactive astrocytes: a review. *Cytobios* **61,** 133–160.
92. Kato, H. and Walz, W. (2000) The initiation of the microglial response. *Brain Pathol.* **10,** 137–143.
93. del Zoppo, G., Ginis, I., Hallenbeck, J. M., Iadecola, C., Wang, X., and Feuerstein, G. Z. (2000) Inflammation and stroke: putative role for cytokines, adhesion molecules and iNOS in brain response to ischemia. *Brain Pathol.* **10,** 95–112.
94. Kraig, R. P., Dong, L. M., Thisted, R., and Jaeger, C. B. (1991) Spreading depression increases immunohistochemical staining of glial fibrillary acidic protein. *J. Neurosci.* **11,** 2187–2198.
95. Gehrmann, J., Mies, G., Bonnekoh, P., Banati, R., Iijima, T., Kreutzberg, G. W., and Hossmann, K. A. (1993) Microglial reaction in the rat cerebral cortex induced by cortical spreading depression. *Brain Pathol.* **3,** 11–17.
96. Cotrina, M. L., Kang, J., Lin, J. H., Bueno, E., Hansen, T. W., He, L., Liu, Y., and Nedergaard, M. (1998) Astrocytic gap junctions remain open during ischemic conditions. *J. Neurosci.* **18,** 2520–2537.
97. Matsushima, K., Hogan, M. J., and Hakim, A. M. (1996) Cortical spreading depression protects against subsequent focal cerebral ischemia in rats. *J. Cereb. Blood Flow Metab.* **16,** 221–226.
98. Rudge, J. S. (1993) Astrocyte-derived neurotrophic factors, in *Astrocytes* (Murphy, S., ed.), Academic, San Diego, pp. 267–307.
99. Canning, D. R., Hoke, A., Malemud, C. J., and Silver, J. (1996) A potent inhibitor of neurite outgrowth that predominates in the extracellular matrix of reactive astrocytes. *Int. J. Dev. Neurosci.* **14,** 153–175.

100. Fitch, M. T. and Silver, J. (1997) Glial cell extracellular matrix: boundaries for axon growth in development and regeneration. *Cell Tissue Res.* **290,** 379–384.
101. Siesjö, B. K. (1978) *Brain Energy Metabolism.* John Wiley, Chichester, UK, p. 607.
102. Fridovich, I. (1986) Biological effects of the superoxide radical. *Arch Biochem. Biophys.* **247,** 1–11.
103. Agranoff, B. W., Benjamins, J. A., and Hajra, A. K. (1999) Lipids, in *Basic Neurochemistry. Molecular, Cellular and Medical Aspects* (Sigel, J., Agranoff, B. W., Albers, R. W., Fisher, S. K., and Uhler, M. D., eds.), Raven, New York, pp. 47–67.
104. White, B. C., Daya, A., DeGracia, D. J., O'Neil, B. J., Skjaerlund, J. M., Trumble, S., Krause, G. S., and Rafols, J. A. (1993) Fluorescent histochemical localization of lipid peroxidation during brain reperfusion following cardiac arrest. *Acta Neuropathol.* **86,** 1–9.
105. Braughler, J. M. and Hall, E. D. (1989) Central nervous system trauma and stroke. I. Biochemical considerations for oxygen radical formation and lipid peroxidation. *Free Rad. Biol. Med.* **6,** 289–301.
106. Rubbo, H., Radi, R., Trujillo, M., Telleri, R., Kalyanaraman, B., Barnes, S., Kirk, M., and Freeman, B. A. (1994) Nitric oxide regulation of superoxide and peroxynitrite-dependent lipid peroxidation. Formation of novel nitrogen-containing oxidized lipid derivatives. *J. Biol. Chem.* **269,** 26,066–26,075.
107. McGrath, L. T., Douglas, A. F., McClean, E., Brown, J. H., Doherty, C. C., Johnston, G. D., and Archbold, G. P. (1995) Oxidative stress and erythrocyte membrane fluidity in patients undergoing regular dialysis. *Clin. Chim. Acta* **235,** 179–188.
108. Subramaniam, R., Roediger, F., Jordan, B., Mattson, M. P., Keller, J. N., Waeg, G., and Butterfield, D. A. (1997) The lipid peroxidation product, 4-hydroxy-2-trans-nonenal, alters the conformation of cortical synaptosomal membrane proteins. *J. Neurochem.* **69,** 1161–1169.
109. Rauchova, H., Ledvinkova, J., Kalous, M., and Drahota, Z. (1995) The effect of lipid peroxidation on the activity of various membrane-bound ATPases in rat kidney. *Int. J. Biochem. Cell Biol.* **27,** 251–255.
110. Hossmann, K. A. (1994) Glutamate-mediated injury in focal cerebral ischemia: the excitotoxin hypothesis revised. *Brain Pathol.* **4,** 23–36.
111. Siesjö, B. K., Zhao, Q., Pahlmark, K., Siesjo, P., Katsura, K., and Folbergrova, J. (1995) Glutamate, calcium, and free radicals as mediators of ischemic brain damage. *Ann. Thorac. Surg.* **59,** 1316–1320.
112. Liu, T.-Z., Stern, A., and Morrow, J. D. (1998) The isoprostanes: unique bioactive products of lipid peroxidation. *J. Biomed. Sci.* **5,** 415–420.
113. Harrison, K. A. and Murphy, R. C. (1995) Isoleukotrienes are biologically active free radical products of lipid peroxidation. *J. Biol. Chem.* **270,** 17,273–17,278.
114. Esterbauer, H., Zollner, H., and Schauer, R. J. (1990) Aldehydes formed by lipid peroxidation: mechanisms of formation, occurrence, and determination, in *Membrane Lipid Peroxidation* (Vigo-Pelfrey, C., ed.), CRC, Boca Raton, FL, pp. 239–268.
115. Springer, J. E., Azbill, R. D., Mark, R. J., Begley, J. G., Waeg, G., and Mattson, M. P. (1997) 4-hydroxynonenal, a lipid peroxidation product, rapidly accumulates following traumatic spinal cord injury and inhibits glutamate uptake. *J. Neurochem.* **68,** 2469–2476.
116. Comporti, M. (1998) Lipid peroxidation and biogenic aldehydes: from the identification of 4- hydroxynonenal to further achievements in biopathology. *Free Rad. Res.* **28,** 623–635.
117. Blanc, E. M., Keller, J. N., Fernandez, S., and Mattson, M. P. (1998) 4-hydroxynonenal, a lipid peroxidation product, impairs glutamate transport in cortical astrocytes. *Glia* **22,** 149–160.
118. Mark, R. J., Lovell, M. A., Markesbery, W. R., Uchida, K., and Mattson, M. P. (1997) A role for 4-hydroxynonenal, an aldehydic product of lipid peroxidation, in disruption of ion homeostasis and neuronal death induced by amyloid β-peptide. *J. Neurochem.* **68,** 255–264.

119. Picklo, M. J., Amarnath, V., McIntyre, J. O., Graham, D. G., and Montine, T. J. (1999) 4-Hydroxy-2(E)-nonenal inhibits CNS mitochondrial respiration at multiple sites. *J. Neurochem.* **72,** 1617–1624.
120. Morgello, S., Uson, R. R., Schwartz, E. J., and Haber, R. S. (1995) The human blood–brain barrier glucose transporter (GLUT1) is a glucose transporter of gray matter astrocytes. *Glia* **14,** 43–54.
121. Vannucci, S. J., Maher, F., and Simpson, I. A. (1997) Glucose transporter proteins in brain: delivery of glucose to neurons and glia. *Glia* **21,** 2–21.
122. Peters, A., Palay, S. L., and Webster, H. D. (1991) *The Fine Structure of the Nervous System. Neurons and Their Supporting Cells*, 3rd ed. Oxford University Press, New York, p. 494.
123. Peng, L., Zhang, X., and Hertz, L. (1994) High extracellular potassium concentrations stimulate oxidative metabolism in a glutamatergic neuronal culture and glycolysis in cultured astrocytes but have no stimulatory effect in a GABAergic neuronal culture. *Brain Res.* **663,** 168–172.
124. Devon, R. M. and Juurlink, B. H. J. (1989) Dynamic morphological responses of mouse astrocytes in primary cultures following medium changes. *Glia* **2,** 266–272.
125. Walz, W. and Mukerji, S. (1988) Lactate production and release in cultured astrocytes. *Neurosci. Lett.* **86,** 296–300.
126. Juurlink, B. H. J. (1994) Type-2 astrocytes have much greater susceptibility to heat stress than type-1 astrocytes. *J. Neurosci. Res.* **38,** 196–201.
127. Forsyth, R. J. (1996) Astrocytes and the delivery of glucose from plasma to neurons. *Neurochem. Int.* **28,** 231–241.
128. Gould, G. W. and Holman, G. D. (1993) The glucose transporter family: structure, function and tissue-specific expression. *Biochem. J.* **295,** 329–341.
129. Broer, S., Rahman, B., Pellegri, G., Pellerin, L., Martin, J. L., Verleysdonk, S., Hamprecht, B., and Magistretti, P. J. (1997) Comparison of lactate transport in astroglial cells and monocarboxylate transporter 1 (MCT 1) expressing *Xenopus laevis* oocytes. Expression of two different monocarboxylate transporters in astroglial cells and neurons. *J. Biol. Chem.* **272,** 30,096–30,102.
130. Gerhart, D. Z., Enerson, B. E., Zhdankina, O. Y., Leino, R. L., and Drewes, L. R. (1998) Expression of the monocarboxylate transporter MCT2 by rat brain glia. *Glia* **22,** 272–281.
131. Larrabee, M. G. (1996) Partitioning of CO2 production between glucose and lactate in excised sympathetic ganglia, with implications for brain. *J. Neurochem.* **67,** 1726–1734.
132. Larrabee, M. G. (1995) Lactate metabolism and its effects on glucose metabolism in an excised neural tissue. *J. Neurochem.* **64,** 1734–1741.
133. Poitry-Yamate, C. L., Poitry, S., and Tsacopoulos, M. (1995) Lactate released by muller glial cells is metabolized by photoreceptors from mammalian retina. *J. Neurosci.* **15,** 5179–5191.
134. Tsacopoulos, M. and Magistretti, P. J. (1996) Metabolic coupling between glia and neurons. *J. Neurosci.* **16,** 877–885.
135. Schurr, A., Miller, J. J., Payne, R. S., and Rigor, B. M. (1999) An increase in lactate output by brain tissue serves to meet the energy needs of glutamate-activated neurons. *J. Neurosci.* **19,** 34–39.
136. Korf, J. (1996) Intracerebral trafficking of lactate in vivo during stress, exercise, electroconvulsive shock and ischemia as studied with microdialysis. *Dev. Neurosci.* **18,** 405–414.
137. Scrutten, M. C. and Utter, M. F. (1968) The regulation of glycolysis and gluconeogenesis in animal tissues. *Ann. Rev. Biochem.* **37,** 249–302.
138. Tilton, W. M., Seaman, C., Carriero, D., and Piomelli, S. (1991) Regulation of glycolysis in the erythrocyte: role of the lactate/pyruvate and NAD/NADH ratios. *J. Lab. Clin. Med.* **118,** 146–152.

139. Dawson, D. M., Goodfriend, T. L., and Kaplan, N. O. (1964) Lactate dehydrogenase: functions of the two types. *Science* **143,** 929–933.
140. Marrif, H. and Juurlink, B. H. J. (1999) Astrocytes respond to hypoxia by increasing glycolytic capacity. *J. Neurosci. Res.* **57,** 255–260.
141. Bittar, P. G., Charnay, Y., Pellerin, L., Bouras, C., and Magistretti, P. J. (1996) Selective distribution of lactate dehydrogenase isoenzymes in neurons and astrocytes of human brain. *J. Cereb. Blood Flow Metab.* **16,** 1079–1089.
142. Swanson, R. A. and Benington, J. H. (1996) Astrocyte glucose metabolism under normal and pathological conditions in vitro. *Dev. Neurosci.* **18,** 515–521.
143. Schmoll, D., Fuhrmann, E., Gebhardt, R., and Hamprecht, B. (1995) Significant amounts of glycogen are synthesized from 3-carbon compounds in astroglial primary cultures from mice with participation of the mitochondrial phosphoenolpyruvate carboxykinase isoenzyme. *Eur. J. Biochem.* **227,** 308–315.
144. Schmoll, D., Cesar, M., Fuhrmann, E., and Hamprecht, B. (1995) Colocalization of fructose-1,6-bisphosphatase and glial fibrillary acidic protein in rat brain. *Brain Res.* **677,** 341–344.
145. Swanson, R. A. and Choi, D. W. (1993) Glial glycogen stores affect neuronal survival during glucose deprivation in vitro. *J. Cereb. Blood Flow Metab.* **13,** 162–169.
146. Dringen, R., Gebhardt, R., and Hamprecht, B. (1993) Glycogen in astrocytes: possible function as lactate supply for neighboring cells. *Brain Res.* **623,** 208–214.
147. Ransom, B. R. and Fern, R. (1997) Does astrocytic glycogen benefit axon function and survival in CNS white matter during glucose deprivation? *Glia* **21,** 134–141.
148. Odowd, B. S., Gibbs, M. E., Ng, K. T., Hertz, E., and Hertz, L. (1994) Astrocytic glycogenolysis energizes memory processes in neonate chicks. *Brain Res. Dev. Brain Res.* **78,** 137–141.
149. Magistretti, P. J., Sorg, O., Yu, N. C., Martin, J. L., and Pellerin, L. (1993) Neurotransmitters regulate energy metabolism in astrocytes: implications for the metabolic trafficking between neural cells. *Dev. Neurosci.* **15,** 306–312.
150. Odowd, B. S., Barrington, J., Ng, K. T., Hertz, E., and Hertz, L. (1995) Glycogenolytic response of primary chick and mouse cultures of astrocytes to noradrenaline across development. *Brain Res. Dev. Brain Res.* **88,** 220–223.
151. Sorg, O., Pellerin, L., Stolz, M., Beggah, S., and Magistretti, P. J. (1995) Adenosine triphosphate and arachidonic acid stimulate glycogenolysis in primary cultures of mouse cerebral cortical astrocytes. *Neurosci. Lett.* **188,** 109–112.
152. Pellerin, L. and Magistretti, P. J. (1994) Glutamate uptake into astrocytes stimulates aerobic glycolysis: a mechanism coupling neuronal activity to glucose utilization. *Proc. Natl. Acad. Sci. USA* **91,** 10,625–10,629.
153. Magistretti, P. J. and Pellerin, L. (1996) Cellular mechanisms of brain energy metabolism. Relevance to functional brain imaging and to neurodegenerative disorders. *Ann. NY Acad. Sci.* **777,** 380–387.
154. Yu, A. C. H., Hertz, L., and Schousboe, A. (1983) Pyruvate carboxylase activity in primary cultures of astrocytes and neurons. *J. Neurochem.* **41,** 1484–1487.
155. Cesar, M. and Hamprecht, B. (1995) Immunocytochemical examination of neural rat and mouse primary cultures using monoclonal antibodies raised against pyruvate carboxylase. *J. Neurochem.* **64,** 2312–2318.
156. Westergaard, N., Sonnewald, U., and Schousboe, A. (1994) Release of alpha-ketoglutarate, malate and succinate from cultured astrocytes: possible role in amino acid neurotransmitter homeostasis. *Neurosci. Lett.* **176,** 105–109.
157. Shank, R. P. and Bennett, D. J. (1993) 2-Oxoglutarate transport: a potential mechanism for regulating glutamate and tricarboxylic acid cycle intermediates in neurons. *Neurochem. Res.* **18,** 401–410.

158. Shank, R. P., Baldy, W. J., and Ash, C. W. (1989) Glutamine and 2-oxoglutarate as metabolic precursors of the transmitter pools of glutamate and GABA: correlation of regional uptake by rat brain synaptosomes. *Neurochem. Res.* **14,** 371–376.
159. Shank, R. P. and Campbell, G. L. (1984) Alpha-ketoglutarate and malate uptake and metabolism by synaptosomes: further evidence for an astrocyte-to-neuron metabolic shuttle. *J. Neurochem.* **42,** 1153–1161.
160. Peng, L., Hertz, L., Huang, R., Sonnewald, U., Petersen, S. B., Westergaard, N., Larsson, O., and Schousboe, A. (1993) Utilization of glutamine and of TCA cycle constituents as precursors for transmitter glutamate and GABA. *Dev. Neurosci.* **15,** 367–377.
161. Westergaard, N., Sonnewald, U., and Schousboe, A. (1995) Metabolic trafficking between neurons and astrocytes: the glutamate glutamine cycle revisited. *Dev. Neurosci.* **17,** 203–211.
162. Mckenna, M. C., Tildon, J. T., Stevenson, J. H., and Huang, X. L. (1996) New insights into the compartmentation of glutamate and glutamine in cultured rat brain astrocytes. *Dev. Neurosci.* **18,** 380–390.
163. Juurlink, B. H. J., Munoz, D. G., and Ang, L. C. (1991) Motoneuron survival in vitro: effects of pyruvate, α-ketoglutarate, gangliosides and potassium. *Neurosci. Lett.* **133,** 25–28.
164. Juurlink, B. H. J. and Walz, W. (1998) Neural cell culture techniques, in *Neuromethods: Cell Neurobiology Techniques* (Boulton, A. A., Baker, G. B., and Bateson, A. N., eds.), Humana, Totowa, NJ, pp. 53–104.
165. Martinez-Hernandez, A., Bell, K. P., and Norenberg, M. D. (1977) Glutamine synthetase: glial localization in brain. *Science* **195,** 1356–1358.
166. Norenberg, M. D. and Martinez-Hernandez, A. (1979) Fine structural localization of glutamine synthetase in astrocytes of rat brain. *Brain Res.* **161,** 303–310.
167. Rust, R. S., Jr., Carter, J. G., Martin, D., Nerbonne, J. M., Lampe, P. A., Pusateri, M. E., and Lowry, O. H. (1991) Enzyme levels in cultured astrocytes, oligodendrocytes and Schwann cells, and neurons from the cerebral cortex and superior cervical ganglia of the rat. *Neurochem. Res.* **16,** 991–999.
168. Juurlink, B. H. J. (1993) NADP-linked isozymes are the major forms of isocitrate dehydrogenase in mouse type-1-like astrocytes. *Life Sci.* **52,** 1087–1090.
169. Schirmer, R. H., Krauth-Siegel, R. L., and Schilz, G. E. (1989) Glutathione reductase, in *Glutathione. Chemical, Biochemical, and Medical Aspects* (Dolphin, D., Poulson, R., and Avramovic, O., eds.), Krieger, Malabar, FL, pp. 553–596.
170. Björnstedt, M., Kumar, S., Bjorkhem, L., Spyrou, G., and Holmgren, A. (1997) Selenium and the thioredoxin and glutaredoxin systems. *Biomed. Environ. Sci.* **10,** 271–279.
171. Andres, A., Satrustegui, J., and Machado, A. (1983) Development of NADPH-consuming pathways in heart, brain and liver of the rat. *Biol. Neonate* **43,** 198–204.
172. Juurlink, B. H. J. (1999) Management of oxidative stress in the CNS: the many roles of glutathione. *Neurotox. Res.* **1,** 119–140.
173. Kranich, O., Hamprecht, B., and Dringen, R. (1996) Different preferences in the utilization of amino acids for glutathione synthesis in cultured neurons and astroglial cells derived from rat brain. *Neurosci. Lett.* **219,** 211–214.
174. Rice, M. E. and Russo-Menna, I. (1998) Differential compartmentalization of brain ascorbate and glutathione between neurons and glia. *Neuroscience* **82,** 1213–1233.
175. Iwata-Ichikawa, E., Kondo, Y., Miyazaki, I., Asanuma, M., and Ogawa, N. (1999) Glial cells protect neurons against oxidative stress via transcriptional up-regulation of the glutathione synthesis. *J. Neurochem.* **72,** 2334–2344.
176. Eftekharpour, E., Holmgren, A., and Juurlink, B. H. J. (2000) Thioredoxin reductase and glutathione synthesis is upregulated by *t*-butylhydroquinone in cortical astrocytes but not in cortical neurons. *Glia* **31,** 241–248.

177. Thorburne, S. K. and Juurlink, B. H. J. (1996) Low glutathione and high iron govern the susceptibility of oligodendroglial precursors to oxidative stress. *J. Neurochem.* **67,** 1014–1022.
178. Meister, A. (1983) Glutathione. *Ann. Rev. Biochem.* **79,** 711–760.
179. Bannai, S. (1984) Transport of cystine and cysteine in mammalian cells. *Biochim. Biophys. Acta* **779,** 289–306.
180. Sagara, J.-I., Miura, K., and Bannai, S. (1993) Maintenance of neuronal glutathione by glial cells. *J. Neurochem.* **61,** 1672–1676.
181. Dringen, R., Pfeiffer, B., and Hamprecht, B. (1999) Synthesis of the antioxidant glutathione in neurons: supply by astrocytes of CysGly as precursor for neuronal glutathione. *J. Neurosci.* **19,** 562–569.
182. Rice, M. E. (1999) Ascorbate compartmentalization in the CNS. *Neurotoxicity Res.* **1,** 81–90.
183. Juurlink, B. H. J. (1996) Central role of glutathione in governing the response of astroglial and oligodendroglial cells to ischemia-related insults. *Recent Res. Develop Neurochem.* **1,** 179–192.
184. Winkler, B. S., Orselli, S. M., and Rex, T. S. (1994) The redox couple between glutathione and ascorbic acid: a chemical and physiological perspective. *Free Rad. Biol. Med.* **17,** 333–349.
185. Wells, W. W., Xu, D. P., and Washburn, M. P. (1995) Glutathione: dehydroascorbate oxidoreductases. *Methods Enzymol.* **252,** 30–38.
186. Siushansian, R., Tao, L., Dixon, S. J., and Wilson, J. X. (1997) Cerebral astrocytes transport ascorbic acid and dehydroascorbic acid through distinct mechanisms regulated by cyclic AMP. *J. Neurochem.* **68,** 2378–2385.
187. Rumsey, S. C., Kwon, O., Xu, G. W., Burant, C. F., Simpson, I., and Levine, M. (1997) Glucose transporter isoforms GLUT1 and GLUT3 transport dehydroascorbic acid. *J. Biol. Chem.* **272,** 18,982–18,989.
188. Berger, U. V. and Hediger, M. A. (2000) The vitamin C transporter SVCT2 is expressed by astrocytes in culture but not *in situ*. *Neuroreport* **11,** 1395–1399.
189. Wilson, J. X., Peters, C. E., Sitar, S. M., Daoust, P., and Gelb, A. W. (2000) Glutamate stimulates ascorbate transport by astrocytes. *Brain Res.* **858,** 61–66.
190. Korcok, J., Yan, R., Siushansian, R., Dixon, S. J., and Wilson, J. X. (2000) Sodium-ascorbate cotransport controls intracellular ascorbate concentration in primary astrocyte cultures expressing the SVCT2 transporter. *Brain Res.* **881,** 144–151.
191. May, J. M., Mendiratta, S., Hill, K. E., and Burk, R. F. (1997) Reduction of dehydroascorbate to ascorbate by the selenoenzyme thioredoxin reductase. *J. Biol. Chem.* **272,** 22,607–22,610.
192. May, J. M., Cobb, C. E., Mendiratta, S., Hill, K. E., and Burk, R. F. (1998) Reduction of the ascorbyl free radical to ascorbate by thioredoxin reductase. *J. Biol. Chem.* **273,** 23,039–23,045.
193. Arner, E. S., Nordberg, J., and Holmgren, A. (1996) Efficient reduction of lipoamide and lipoic acid by mammalian thioredoxin reductase. *Biochem. Biophys. Res. Commun.* **225,** 268–274.
194. Handelman, G. J., Han, D., Tritschler, H., and Packer, L. (1994) Alpha-lipoic acid reduction by mammalian cells to the dithiol form, and release into the culture medium. *Biochem. Pharmacol.* **47,** 1725–1730.
195. Han, D., Handelman, G., Marcocci, L., Sen, C. K., Roy, S., Kobuchi, H., et al. (1997) Lipoic acid increases de novo synthesis of cellular glutathione by improving cystine utilization. *Biofactors* **6,** 321–338.
196. Tirosh, O., Sen, C. K., Roy, S., Kobayashi, M. S., and Packer, L. (1999) Neuroprotective effects of alpha-lipoic acid and its positively charged amide analogue. *Free Radic. Biol. Med.* **26,** 1418–1426.

197. Muller, U. and Krieglstein, J. (1995) Prolonged pretreatment with alpha-lipoic acid protects cultured neurons against hypoxic, glutamate-, or iron-induced injury. *J. Cereb. Blood Flow Metab.* **15,** 624–630.
198. Ramón y Cajal, S. (1909) *The Neuron and the Glial Cell.* Charles C Thomas, Springfield, IL.

7
Glutamate-Mediated Astrocyte–Neuron Communication in Brain Physiology and Pathology

Micaela Zonta and Giorgio Carmignoto

1. INTRODUCTION

The presence of astrocytes in nervous tissue was initially thought to reflect the necessity of a mechanical supporting system for neurons, a sort of aspecific glue keeping neuronal cells together. If this were the case, the human brain would exhibit the best set of neuronal packing ever seen throughout the phylogenetic tree. Evolution, however, does not deal with glues.

Results obtained during the past two decades have contributed to overcoming, at least in part, this initial misconception. Indeed, increasing attention has been paid to the understanding of the real astrocytic function, and active roles have been ascribed to these cells in many important aspects of brain function. The ability of astrocytes to influence neuronal development was thus recognized, as witnessed by their ability to guide axonal growth *(1)*, and enhance the formation *(2)* and strengthening *(3)* of the synapse. Furthermore, astrocytes provide the energetic substrates necessary for neuronal activity *(4)*, and participate in the clearance of excessive concentration of potassium ions *(5)* and, most importantly, of neurotransmitters, such as the excitatory amino acid, glutamate (Glu), in the extracellular space (ECS) *(6,7)*. Although these functions assign an evident importance to astrocytes, the depicted role for glial cells is still subservient to neurons. In other words, astrocyte roles should be limited, to enable neurons to perform their own, exclusive task, i.e., transferring and processing of information.

In contrast to this view, recent findings showed that astrocytes are capable of finely sensing neuronal activity, and, in response to neuronal signals, of talking back to neurons. This chapter focuses on the intriguing possibility that astrocytes are qualified partners of neurons in events which we always refer to as "neuronal" communication. As to the properties of this partnership, a common "language," i.e., the neurotransmitter Glu, is used for this bidirectional communication, although the signal may differ in the two cells, in terms of temporal and spatial features. The last part of this chapter discusses the possible participation of the glutamatergic crosstalk between astrocytes and neurons in several neurodegenerative diseases of the central nervous system. Noteworthy is that the neuronal loss occurring in the brain under pathological

From: *The Neuronal Environment: Brain Homeostasis in Health and Disease*
Edited by: W. Walz © Humana Press Inc., Totowa, NJ

conditions is, at least to some extent, owing to an abnormally elevated concentration of Glu in the ECS *(8)*. The possibility is discussed that an impairment of specific Glu transporters in astrocytes, as well as unregulated processes of Glu release from these cells, contribute to the excitotoxic action of Glu.

2. ASTROCYTE–NEURON CROSSTALK

2.1. *Someone Is Listening to Neuronal Tete-A-Tete: Eavesdropper Astrocytes*

The analysis of the relative location of neurons and astrocytes in the architecture of nervous tissue reveals a frequent intimate association between astroglial processes and synapses *(9)*. In the hippocampus, for instance, astrocytes are associated with 57% of neuronal synaptic connections *(10)*. Because of this physical contiguity, a neuron-to-astrocyte signaling was always considered possible. A coherent directional communication, however, demands as a prerequisite the release from neurons of signaling molecules that are specifically recognized by astrocytes, thereby allowing these cells to "listen" to neuronal activity.

Clues for the existence of this specific neuron-to-astrocyte communication are represented by observations that astrocytes express on their plasma membrane a variety of receptors for neurotransmitters, including Glu *(11)*. Probably because Glu is the main excitatory neurotransmitter in the brain, the action of Glu in astrocytes, and thus its role in neuron–glia interactions, has been intensely investigated. Other features, however, emphasize the relevance of Glu in neuron–astrocyte interactions: Astrocytes are equipped with an efficient Glu-uptake system that gives an important contribution to the control of Glu concentration in the synaptic cleft *(6,7)*; under certain conditions, the reverse operation of the transporter may result in nonvesicular release of Glu from astrocytes *(12)*; astrocytes provide the substrate that allows neurons to sustain glutamatergic signaling (astrocytes convert Glu released at the synapse to glutamine, which is then hydrolized back to Glu by a neuronal glutaminase, and reused by neurons as neurotransmitter) *(13)*. This chapter, therefore, centers attention on glutamatergic communication, keeping in consideration that other neurotransmitters may be also involved in astrocyte–neuron signaling.

Astrocytes possess both ionotropic (α-amino-3-hydroxy-5-methyl-4-isoxazole-propionic acid [AMPA]/kainate) *(14,15)* and metabotropic Glu receptors (mGluRs) *(16,17)*, whose activation by Glu results in cationic inward current and Ca^{2+} release from intracellular (IC) stores, respectively. The first evidence for Glu-sensitivity in astrocytes was obtained in in vitro preparations: Cultured hippocampal astrocytes responded to Glu application with elevations in cytosolic Ca^{2+} concentration ($[Ca^{2+}]_i$), which could take the form of Ca^{2+} oscillations, and show different patterns, depending on the agonist dose *(16,17)*. Increasing concentration of Glu resulted in increasing frequency of $[Ca^{2+}]_i$ oscillations; very high doses were found to be associated with a sustained Ca increase *(18)*.

Intercellular waves of $[Ca^{2+}]_i$ elevations, with a propagation velocity of about 20 μm/s, are also frequently observed in cultured astrocytes, following various types of stimuli *(19,20)*. The possibility exists that they are present also in astrocytes *in situ*. In such a case, $[Ca^{2+}]_i$ waves may represent a pathway for extraneuronal long-distance signaling within the brain *(16,21)*.

The first indication for the existence of a neuron-to-astrocyte signaling was provided in 1992, by two studies that demonstrated that neuronal activity, both in cortical cultures *(22)* and in organotypically cultured hippocampal slices *(23)*, could trigger, in glial cells $[Ca^{2+}]_i$ oscillations and $[Ca^{2+}]_i$ waves comparable, although less regular, to those seen after Glu stimulation of cultured astrocytes. In organotypically cultured brain slices, tetrodotoxin blocked these Ca^{2+} responses *(23)*, pointing to the involvement of neuronal activity in glial cell activation. These findings added a more physiological perspective to the presence of GluRs in astrocytes, hinting at the presence of glutamatergic signaling conveying information from neurons to astrocytes. Indeed, the authors' interpretation of their results is that Glu released from axonal terminals is responsible for $[Ca^{2+}]_i$ elevations in nearby astrocytes *(23)*.

This hypothesis was validated in experiments carried out in a model more physiological than cells in culture, such as acute hippocampal slices *(24,25)*. In this preparation, we found that stimulation of Schaffer collaterals resulted in $[Ca^{2+}]_i$ elevations, not only in CA1 pyramidal neurons, but also in other small cells of the stratum radiatum, subsequently identified as astrocytes by glial fibrillary acidic protein immunostaining. On continuous neuronal stimulation, astrocytes could display relatively regular $[Ca^{2+}]_i$ oscillations. Neuronal and astrocytic responses to electrical stimulation could be selectively blocked by using ionotropic and metabotropic GluR antagonists, respectively. The mGluR antagonist, α-methyl-4-carboxyphenylglycine, abolished the $[Ca^{2+}]_i$ increases in astrocytes, without altering neuronal response, thus demonstrating that synaptically released Glu *in situ* can activate glial mGluRs and initiate IC responses *(25)*.

High, but not low, frequency stimulation was able to trigger astrocyte responses, suggesting that the more Glu is released, the higher is the probability that the neurotransmitter reaches astrocyte receptors *(25)*. Although astrocytic processes are interposed between axon terminals and dendrites, no synaptic-like membrane apposition has been documented between glutamatergic neurons and astrocytes, except for immature oligodendrocytes in the rat hippocampus *(26)*. It is therefore plausible that only a considerable release of Glu from neurons can result in neurotransmitter escaping from the synaptic cleft and activating glial receptors. Astrocytes can thus be recruited as additional, postsynaptic targets of synaptically released Glu, only under conditions of high neuronal activity.

2.1.1. Can Astrocytes Change According to What They Hear from Neurons?

Plasticity is one of the most notable neuronal features, and underlines important processes, such as learning and memory. This plasticity relies on various processes, such as changes in synaptic strength, changes in membrane properties and protein expression levels, reorganization of synaptic connections, and dendritic spine extension and retraction *(27)*. All these events are activity-dependent, i.e., they occur in strict dependence on the level and pattern of experienced stimulation. Because astrocytes can also experience different degrees of neuronal activity *(25)*, the question arises about how carefully they are listening to interneuronal communication, and whether they can discern different levels of neuronal activity, and accordingly organize their response.

A series of observations suggest that activation of astrocytes by synaptically released Glu is not simply an on/off event. We found that an increase in the frequency, or in the intensity, of Schaffer collateral stimulation in acute hippocampal slices resulted in an increased frequency of $[Ca^{2+}]_i$ oscillations in astrocytes *(25)*. This final effect may be mediated by two diverse events: a higher firing rate of neuronal afferents, which determines an increased concentration of Glu in the ECS; and the recruitment of neuronal fibers (which were not initially stimulated), which results in a greater Glu release in the vicinity of astrocytic processes. These observations demonstrate that astrocytes can discriminate the level of neuronal activity, and integrate neuronal signals in a dynamically regulated Ca^{2+} response.

Furthermore, astrocytes have been shown to possess a form of memory by virtue of which the frequency of their $[Ca^{2+}]_i$ oscillations increases upon successive episodes of stimulation with Glu *(18)*. This phenomenon is relatively long-lasting, because the frequency of oscillations in the second, with respect to the first, challenge with Glu was observed to be significantly higher, even when performed after 3 h. This ability to "remember" the history of previous stimulation endows astrocytes with a property that was supposed to be peculiar to neurons. The increase in oscillation frequency represents a potentiation of the response of astrocytes to Glu, which may, indeed, resemble another Glu-mediated event, such as the well-documented process of long-term potentiation (LTP) in neurons *(28,29)*. Noteworthy is that one of the IC messengers that is thought to play a role in LTP, i.e., nitric oxide (NO) *(30)*, is also involved in the potentiation of the astrocyte response. The induction of this astrocyte property was indeed abolished by the NO synthase inhibitor, Nω-nitro-L-arginine methyl ester, and subsequently restored by NO donors *(18)*. The plasticity of the astrocyte response was initially reported in primary cultures *(18)*, then confirmed in hippocampal astrocytes from acute brain slices *(25)*, which displayed an increased frequency of oscillations upon multiple episodes of neuronal stimulation. The neuron-induced change in astrocyte responsiveness may, therefore, represent a mode of transferring information from neurons to astrocytes.

In addition to this form of functional plasticity, astrocytes can also alter their physical association with neurons in concurrence with diverse patterns of neuronal stimulation. This morphological plasticity was first evidenced in rat hypothalamus, in which the reorganization of synaptic connection, occurring after appropriate stimuli, appears to be linked to changes in the level to which magnocellular neuroendocrine cells are ensheathed by nearby astrocytes. Stimulation of the hypothalamic nuclei by water deprivation, or by induction of maternal behavior, determines a retraction of astrocytic processes, resulting in a transient de-ensheathment of neuroendocrine cells *(31)*. This change has been suggested to lead to the uncovering of additional postsynaptic sites, and eventually to mediate an increase in the synaptic efficacy. As to the possible mechanisms accounting for these structural modifications, several lines of evidence suggest that they depend on the activation of astrocytic β-adrenergic receptors by neurotransmitter released by adrenergic innervation *(32)*.

The ability of glial processes to undergo similar configurational changes is not restricted to hypothalamus. Cultured hippocampal astrocytes have been shown to extend filopodia, upon exogenous application of Glu *(33)*; an increased association of astroglial processes with synapses is coupled to long-term potentiation in hippocampal dentate gyrus *(34)*, and to the increase in synaptic connections in rats raised under enriched conditions, compared to individually caged rats *(35)*.

Whether these structural modifications represent a feature of astrocytes with a distinct functional significance, and are necessary for the modulation of synaptic activity, remains to be fully investigated. Undoubtedly, the refined sensitivity to neuronal signals endows astrocytes to be something more than an inert audience to the neuronal performance.

2.2. *Astrocytes Join Neurons in Their Glu-Mediated Conversation*

The ability of astrocytes to sense neuronal communication, and to dynamically change their functional and structural features, raises important questions about the physiological role of this phenomenon. Given the importance of Ca^{2+} ions as second messengers, the neuronal activity-dependence of $[Ca^{2+}]_i$ elevations in astrocytes implies that neurons can affect $[Ca^{2+}]_i$-dependent events in these cells. In such a case, do these include a Ca^{2+}-regulated output from astrocytes? And, if it does exist, can it eventually have some bearings on astrocyte-to-neuron communication?

The pioneering experiments in such an exciting area were performed independently in 1994 by the groups led by Haydon and Nedergaard. Both groups investigated neuronal response to astrocyte stimulation in primary neuron–astrocyte mixed cultures, monitoring IC Ca^{2+} concentration by means of fluorescent indicators and Ca^{2+}-imaging techniques. Nedergaard et al. *(36)* triggered a $[Ca^{2+}]_i$ wave through confluent astrocytes, by providing a mechanical stimulation to a single astrocyte; Haydon et al. *(37)* induced a similar response in astrocytes, with bradykinin, or with selective photostimulation. In all cases, the response of glial cells was followed by a $[Ca^{2+}]_i$ increase in adjacent, co-cultured neurons. The authors' explanation of these corresponding results, however, was different: Nedergaard found that neuronal responses could be blocked by gap junctions inhibitors, but not by Glu receptor antagonists; Haydon, beside measuring a release of Glu from neuron-free cultures of astrocytes, showed that $[Ca^{2+}]_i$ increases in neurons resulted from activation of neuronal ionotropic Glu receptors. The conclusions drawn by these two studies are thus substantially different, and hard to reconcile. The most likely explanation for this contradiction may reside in different experimental conditions, which can modify the properties of cultured astrocytes. It is also possible that both phenomena are part of a complex system of interactions between neurons and astrocytes. Recently, evidence for gap junction communication between these two cell types has indeed been provided *(38)*, in the noradrenergic nucleus locus ceruleus.

The results by Haydon's group, on the other hand, initiated a sequence of related studies that ultimately confirmed a Ca^{2+}-dependent Glu release from astrocytes. Hippocampal neurons, plated on confluent astrocyte cultures, were tested for their response to the passage of $[Ca^{2+}]_i$ waves in underlying astrocytes *(39)*. Neurons showed delayed responses, consisting of Glu-dependent $[Ca^{2+}]_i$ elevations and slow, depolarizing, inward currents. These effects in neurons were mediated by activation of N-methyl-D-aspartic acid (NMDA) and AMPA Glu receptors.

The field of investigation then moved to brain slice preparations, which better preserve the connections physiologically existing between neurons and astrocytes. In this preparation, we performed selective stimulation of mGluRs *(25)*. This stimulation induced the well-known $[Ca^{2+}]_i$ response in astrocytes, and resulted in secondary $[Ca^{2+}]_i$ increases in neurons. Although a direct activation of neuronal mGluRs could account for part of these responses, most of them were completely blocked by ionotropic receptors inhibitors, demonstrating that they resulted from an induced release of Glu. Whether this Glu was released by neurons or by astrocytes was assessed by applying the stimulus after 1 h incubation in tetanus neurotoxin, which selectively blocks neuronal exocytosis. Under these conditions, neuronal response to astrocyte activation was still present, thus supporting the view that Glu is indeed released from astrocytes upon mGluR-mediated $[Ca^{2+}]_i$ elevations *(25)*.

In addition to these Ca^{2+}-imaging studies, parallel application of an enzymatic fluorimetric assay, originally developed by Nicholls et al. *(40)*, allowed measurement of the total amount of Glu released from both cultured astrocytes and astrocytes in brain slices, after stimulation with glutamatergic agonists or prostaglandin E_2 (PGE_2) *(41)*. The same enzymatic method was recently applied to single cell measurements of Glu release: Waves of extracellular (EC) Glu released by cultured astrocytes could thus be observed to occur in strict correlation with mechanically induced $[Ca^{2+}]_i$ waves in astrocyte monolayer *(42)*.

Taken together, the described results demonstrate that astrocytes, besides understanding neuronal signaling, can feed back to neurons using the same signal, i.e., the transmitter, Glu, and consequently alter neuronal Ca^{2+} homeostasis. Glu-induced astrocytic signaling, and hence the initiation of astrocytic Glu release, is much slower than the neuronal one, because it depends on Ca^{2+} release from internal stores and propagation of $[Ca^{2+}]_i$ waves, rather than on membrane Ca^{2+} channels opening and action potential propagation.

Nonetheless, this capability can potentially result in an unexpected role of astrocytes in joining neurons in glutamatergic transmission, thereby influencing information transfer and processing in the brain.

2.2.1. Ca^{2+}-Dependency and Mechanism of Glu Release from Astrocytes

Astrocytes can release Glu via the reversal of Glu transporters *(12)*, and through a swelling-induced process *(43)*, but none of these mechanisms can account for ligand-induced, Ca^{2+}-dependent release of Glu.

Increases in IC Ca^{2+} concentration have been shown to be both sufficient and necessary to obtain Glu release from cultured astrocytes. Indeed, diverse protocols of stimulation, all resulting in $[Ca^{2+}]_i$ elevations in glial cells, have been applied to trigger the release process: application of chemical agonists *(25,37,41,44–46)*, electrical/mechanical stimuli *(37,42,47–52)*, or photoactivation *(37,50,53)*. A recent study developed by Parpura and Haydon *(53)* used photolysis of caged Ca^{2+} to investigate the concentration of IC Ca^{2+}, which must be reached to induce the release of Glu from cultured astrocytes. The experimental conditions allowed for monitoring both the $[Ca^{2+}]_i$ changes occurring in UV-stimulated astrocytes and the Glu-induced inward currents in single neurons grown on astrocyte microislands. This approach revealed that substantial glutamatergic currents can be evoked in neurons, upon modest changes in astrocytic $[Ca^{2+}]_i$, i.e., upon an increase of less than 100 n*M*. These data need to be verified in more intact preparations, but, since physiological stimuli, such as Glu or norephinephrine, raise astrocytic Ca^{2+} levels to much higher values (over 1 μ*M*), this finding is consistent with the idea that Glu release from astrocytes can be a physiological event within the brain *(53)*.

The possibility of a $[Ca^{2+}]_i$ threshold in the pathway leading to Glu release in astrocytes has been investigated in another recent study *(46)*. Human embryonic kidney (HEK) cells, transiently expressing the NMDA receptor were plated on cultured astrocytes, and were successfully used to detect the episodes of transmitter release upon stimulation of astrocytic mGluRs. Under these conditions, we found a significant degree of correlation between the amplitude of astrocytic $[Ca^{2+}]_i$ response, measured in the cell soma, and the probability of Glu release, detectable as a NMDA receptor (NMDAR)-mediated $[Ca^{2+}]_i$ increase in transfected HEK cells adjacent to stimulated astrocytes. Glutamate release was always observed to occur following $[Ca^{2+}]_i$ elevations of an amplitude above 550 n*M*; below this value, it could be detected only in a few occasions *(46)*.

Another exciting point in the investigation of the mechanism of Glu release concerns the role of mGluR-mediated $[Ca^{2+}]_i$ oscillations, which, as reported above, display peculiar features of plasticity, in terms of changes in their frequency. If $[Ca^{2+}]_i$ is critical for the occurrence of Glu release, is there any related event that depends on the frequency of $[Ca^{2+}]_i$ oscillations? This point was further analyzed in the same study *(46)*. Since the increase of IC Ca^{2+} concentration controls the release of Glu from astrocytes, an intriguing possibility was that every transient in $[Ca^{2+}]_i$ oscillation could result in a distinct episode of Glu release. NMDAR-transfected HEK cells were thus used to test the hypothesis of a pulsatile release of Glu retaining the oscillatory behavior of $[Ca^{2+}]_i$ signaling. In correlation with 1-aminocyclopentane-1,3-dicarboxylic acid-induced $[Ca^{2+}]_i$ oscillations in astrocytes, we could indeed observe repetitive NMDAR-mediated $[Ca^{2+}]_i$ elevations in nearby, transfected HEK cells. These results imply that $[Ca^{2+}]_i$ oscillations are the coding system that regulates a pulsatile release of Glu from astrocytes, and that an increase in oscillation frequency necessarily corresponds to an increased frequency of transmitter release from glial cells. The relevance of this finding is underlined by the accepted evidence that the frequency of $[Ca^{2+}]_i$ oscillation in astrocytes is dynamically regulated by the level of neuronal activity *(25)*. Therefore, the process of Glu release may represent, in astrocytes, an output finely regulated in its frequency by the level of neuronal activity.

The mechanism underlying Ca^{2+}-dependent release of transmitter has been investigated in astrocytes, by means of various pharmacological tools. Glu release was determined, in acute hippocampal slices, by means of both a fluorimetric assay and Ca^{2+}-imaging techniques *(41)*. The results thus obtained showed that PG formation is an essential step for the release process, because it can be abolished by inhibiting the enzymes involved in PG synthesis, i.e., phospholipase A_2 and cyclo-oxygenase. Further support for the involvement of PGs in the IC pathway, leading to Glu release, was provided by the finding that application of PGE_2 could mimic the releasing action mediated by GluR stimulation. An autocrine- or paracrine-positive loop, mediated by Glu itself, may have a role in enhancing the process of release, since inhibition of mGluRs determines a substantial decrease of Glu release, revealed as the absence of Glu-mediated responses in CA1 neurons, upon astrocyte stimulation with PGE_2 *(41)*.

The Ca dependency of Glu release from astrocytes led the researchers to explore the similarities of this event with the Ca^{2+}-dependent exocytosis of neurotransmitter occurring in neurons, and, eventually, to consider the possibility that a vesicular mechanism accounts for transmitter release in astrocytes. To explore this possibility, astrocytes were tested for the expression of the proteins involved in the exocytotic pathway. It is now widely accepted that a major role in the regulated secretion of exocytotic vesicles in neurons is played by the proteins of the SNARE complex, which are essential for the events of vesicle docking and fusion *(54,55)*. Astrocytes have been shown to express the SNARE proteins, syntaxin, synaptobrevin II *(56)*, and SNAP-23, a homolog of neuronal SNAP-25 *(57)*, and the small GTPase rab3, also supposed to be involved in vesicular exocytosis *(58)*. Moreover, synaptobrevin and synaptophysin were identified by immunoelectron microscopy in association with the membrane of vesicular organelles *(58)*.

To further unravel the mechanism of release, various authors took advantage of the reported ability of different neurotoxins to either stimulate or block the vesicular fusion event in neurons, and investigated their effects on Ca^{2+}-dependent release of Glu from astrocytes. The results of these experiments showed a substantial similarity between neuronal and astrocytic sensitivity to these toxins: Glu release from astrocytes was stimulated by the potent segretagogue, α-latrotoxin *(45,59)*, and was abolished by tetanic and botulinic toxins, highly selective proteases that exert their toxicity on neurons by cleaving SNARE proteins, and thus preventing vesicle exocytosis *(46,51,60)*.

We recently obtained further information on the mechanism of Glu release from astrocytes *(46)*. By using the experimental model already described, i.e., co-cultures of astrocytes and NMDAR-transfected HEK cells, and patch-clamp techniques, we could record, from these latter cells, repetitive, inward currents, upon induction of $[Ca^{2+}]_i$ oscillations in astrocytes. These responses were mediated by Glu released from astrocytes, because they could be blocked by the specific NMDAR antagonist, D-2-amino-5-phosphonopentanoic acid. The kinetics of these currents resemble NMDA-excitatory postsynaptic currents, indicating that astrocytes can release Glu through a mechanism similar to the exocytosis of neurotransmitters in neurons. Indeed, the fast kinetics of the NMDA response reflect a rapid activation of the NMDAR in HEK cells, which derives from a steep increase in the concentration of Glu in the ECS. The only process known to result in such a rapid increase in concentration of a neurotransmitter is the fusion of vesicles with the plasma membrane.

Further evidence that supports the existence in astrocytes of a Glu release process, dependent on vesicle exocytosis, derives from the results obtained with the vacuolar adenosine triphosphatase (v-ATPase) inhibitor, bafilomycin A1. The filling of synaptic vesicles with Glu requires the presence of an electrochemical gradient through the vesicular membrane, which is provided by the activity of a v-ATPase. Dissipation of this gradient, achievable by incubation with the v-ATPase inhibitor, bafilomycin A1, prevents Glu uptake into vesicles, thereby blocking neuronal transmission *(51,61–64)*. Following incubation of astrocytes with bafilomycin A1, the release of Glu from stimulated astrocytes was strongly reduced, thus providing additional support for the presence in astrocytes of Glu-filled vesicles *(46,51)*.

Taken together, evidence presented in this chapter indicates that, although ultrastructural evidence of Glu-containing vesicles must be provided, in order to unambiguously demonstrate the existence of an exocytotic pathway, Ca^{2+}-dependent Glu release from astrocytes can occur via a mechanism that shares several common features with neurotransmitter exocytosis in neurons.

2.3. Modulation of Neuronal Transmission

The existence in astrocytes of a Ca^{2+}-dependent release of Glu raised a great interest in the possibility that astrocytes could be involved in the modulation of glutamatergic neuronal transmission. Over the past years, evidence for such a modulation was provided by a number of studies performed both in neuron–astrocyte co-cultures and in brain slices.

In the first of these studies *(49)*, Glu release from cultured astrocytes (triggered by a mechanical or electrical stimulation) was found to have multiple effects on neuronal transmission. Indeed, it resulted, first, in a reduction in the amplitude of evoked excitatory, as well as inhibitory, postsynaptic currents recorded from neurons *(49)*. As a plausible explanation, the authors suggest that Glu released from astrocytes may act on presynaptic mGluRs, thus resulting in the reduction of neurotransmitter release *(65–67)*. Second, it resulted in a slow inward current, mediated by both NMDA RECs and non-NMDA receptors *(49,50,53)*, and, third, in an increase in the frequency of both spontaneous excitatory and inhibitory postsynaptic events *(50)*. Because the amplitude of the spontaneous events was not significantly changed, the presynaptic terminal is once again the most likely target for Glu action. The authors *(50)* suggested that NMDARs located in the presynaptic terminals (but extrasynaptically) are good candidates. As to the slow inward current, it appeared not to be linked to astrocyte-mediated modulation of evoked synaptic transmission, since it was observed in most, but not all, of the cases. Its possible effects on synaptic transmission remain to be investigated fully.

The capacity of astrocytes to modulate neuronal transmission was confirmed in the retina *(68)*. In an intact preparation, such as the retina eye cup, ganglion cells were observed to change their firing rate (in response to flashes of light) in temporal correlation with the arrival of the $[Ca^{2+}]_i$ wave in astrocytes and Müller cells. Depending on the neuron being recorded, the $[Ca^{2+}]_i$ wave triggered either a decrease or an increase in the firing rate. Support for the possibility that Glu, released by glial cells, mediates the observed change in the neuronal response, derives from the observation that, in the presence of AMPA and NMDAR antagonists, the modulatory effect on neuronal firing associated with the $[Ca^{2+}]_i$ wave was almost abolished. The γ-aminobutyric acid (GABA) receptor antagonists, strychnine and bicuculline, also blocked the effect on neuronal firing associated with the $[Ca^{2+}]_i$ wave. This latter observation led the authors to suggest that the observed inhibition of ganglion cell firing may result from excitation of inhibitory interneurons by Glu released from astrocytes.

A role for astrocytes in the modulation of inhibitory synaptic transmission, mediated by gabaergic interneurons, was confirmed in acute hippocampal slice preparations *(52)*. By performing dual recordings from pairs of cells, i.e., either a pyramidal neuron and an interneuron, or a pyramidal neuron and an astrocyte, Kang et al. *(52)* investigated the possible involvement of astrocytes in the potentiation of inhibitory transmission mediated by repetitive firing of interneurons. This experimental approach allowed those authors to stimulate astrocytes without directly interfering with neurons. By applying depolarizing current pulses through the patch pipet, they could trigger $[Ca^{2+}]_i$ increases in the patched astrocyte, as well as in other electrically coupled astrocytes. Such a selective astrocyte stimulation resulted in an increased frequency and amplitude of miniature inhibitory postsynaptic currents in the pyramidal neuron, thus suggesting that astrocytes can affect inhibitory transmission.

Additional observations, provided further support for the possibility that astrocytes mediated the activity-dependent decrease in the failure rate of inhibitory transmission onto pyramidal neurons, which the authors observed following stimulation of interneurons: interneuron firing elicited in surrounding astrocytes $GABA_B$-mediated $[Ca^{2+}]_i$ elevations; the increase in frequency and amplitude of miniature inhibitory postsynaptic currents, triggered in pyramidal neurons by the $GABA_B$ receptor agonist, baclofen, was accompanied by $[Ca^{2+}]_i$ elevations in astrocytes; antagonists of ionotropic Glu receptors were found to block both the astrocyte- and the baclofen-induced increases of miniature inhibitory postsynaptic currents. This latter result strongly supports the possibility that the release of Glu from astrocytes is involved in the astrocyte-mediated potentiation of inhibitory synaptic transmission. That study provides compelling, though indirect, evidence that a well-known, physiologically relevant phenomenon, such as the activity-dependent potentiation of inhibitory synaptic transmission between interneurons and pyramidal neurons in the CA1 hippocampal region, is critically dependent on astrocytes.

The release of Glu from astrocytes can therefore exert complex effects on neuronal transmission: It can trigger either a decrease or an increase of neurotransmission, presumably by activating different presynaptic Glu receptors; it can bind to GluRs on the postsynaptic neuronal membrane, and trigger slow inward currents, as well as $[Ca^{2+}]_i$ elevations; and it can potentiate the action of inhibitory interneurons on CA1 pyramidal neurons in the hippocampus. These multiple actions, even of opposite signs, can hardly be reconciled into a coherent view for the role of astrocyte Glu release. To define its specific participation in glutamatergic transmission, therefore, more information is needed about the connectivity between astrocytes and neurons.

Astrocyte-derived Glu may ultimately result in either a potentiation or depression of the excitatory neuronal output, depending on what type of neurons astrocyte processes impinge upon. In addition, different astrocytes may establish different anatomical connections, which characterize "that" specific type of astrocyte–neuron interaction. Support for this hypothesis derives from the finding that significant heterogeneity exists in the astrocyte population. A representative example of such heterogeneity comes from a recent study *(69)* that described the distinctive complexity of processes from individual cerebellar Bergmann glia. The three-dimensional reconstruction revealed a relatively organized, complex, subcellular structure of Bergamnn glia processes, which enwrap synapses between parallel fibers and Purkinje neurons. Despite the rapid advance in understanding of the sensitivity of astrocytes to neurotransmitter release from active synaptic terminals, more must also be known about neuron-to-astrocyte communication and, particularly, through what factors and signaling mechanism(s) neurons exert a control on astrocyte release of Glu. In this respect, we have recently provided evidence that the release of Glu from astrocytes is pulsatile and regulated by $[Ca^{2+}]_i$ oscillations: The higher the frequency of $[Ca^{2+}]_i$ oscillations, the higher is the number of episodes of Glu release. Given that the frequency of $[Ca^{2+}]_i$ oscillations, as mentioned above, is under the direct, dynamic control of neuronal activity, our results give significant support to the possibility that astrocyte Glu release is also under the control of neurons. A question then arises: Under what conditions can Glu released from axonal terminals trigger $[Ca^{2+}]_i$ oscillations in astrocytes?

In acute brain slices, it was found that a single pulse applied to Schaffer collaterals could not evoke a response from astrocytes *(25,70)*. In contrast, 100–200-ms trains of stimuli at high frequency (20–50 Hz) evoked Glu-mediated $[Ca^{2+}]_i$ oscillations, the frequency of which was observed to progressively increase at increasing firing rate of axonal afferents *(25,52)*. Neurons may, therefore, activate the Ca^{2+}-dependent release of Glu from astrocytes only under selected conditions, as a feedback mechanism, with a timing coded by the pattern and periodicity of Glu-mediated $[Ca^{2+}]_i$ oscillations.

Beside Glu, astrocytes may affect glutamatergic transmission by releasing other agents, such as glycine (Gly) *(71)*, which acts as a co-agonist on the NMDA receptor *(72,73*; Table 1). Occupation of the Gly site is required for Glu to gate the channel *(74)*. Glu released from synaptic terminals may thus activate non-NMDA receptors on astrocyte processes, to trigger the release of Gly, which then acts on the Gly site of NMDA receptor. Radioactively labeled D-serine (Ser) is released from cultured astrocytes upon stimulation with non-NMDA receptor agonists, but not NMDA agonists or high EC K^+-induced depolarization *(75)*. In the brain, D-Ser was found to be localized predominantly in astrocyte processes in proximity of NMDA receptor 2A/B subtypes *(76)*, and the racemase enzyme, catalyzing the formation of D-Ser from L-Ser, was immunocytochemically identified in astrocytes *(77)*. Given that D-Ser can substitute Gly in activating the NMDAR *(78)*, it seems plausible that astrocytes may enhance NMDA-mediated postsynaptic responses by releasing D-Ser in response to Glu released from synaptic terminals.

A huge amount of experimental and theoretical work needs to be figured out for the characterization of the modulatory role of astrocytes in neuronal transmission. As already mentioned, $[Ca^{2+}]_i$ oscillations, mediated by mGluRs, trigger in astrocytes the release of Glu, by a mechanism similar in its pharmacological sensitivity, and, at least in part, in its kinetic characteristics to the vesicle-mediate release of the neurotransmitter in neurons. If Glu release from astrocytes indeed occurs through an exocytotic mechanism, its physiological consequences on neuronal excitability and glutamatergic transmission would be far more complex than those suggested on the basis of available data.

Given its physiological features, Glu release from astrocytes could not be expected to represent a mode of signaling as rapid as neuronal synaptic transmission. It is certainly possible, however, that it does carry some information. After the initial activation of an astrocyte in contact with active synaptic sites, excitation may travel as a $[Ca^{2+}]_i$ wave to other astrocytes, triggering Glu release from these astrocytes, and ultimately resulting in the spreading of excitation (Fig. 1). The frequency of neuronal activity-dependent $[Ca^{2+}]_i$ oscillations in astrocytes, as well as the selective spreading of astrocyte excitation through gap junctions, may, therefore, represent key features of a novel mode of transferring excitation in a spatially and temporally determined neuron–astrocyte domain. To what extent this process indeed carries information, and what type of event in brain function may depend on it, represent exciting challenges for glial cell research in years to come.

Table 1
Summary of Evidence for Release of Glutamatergic Transmission-Related Compounds from Astrocytes

Experimental model	Stimulus to astrocytes	Released compound	General description	Refs.
Neuron–astrocyte co-cultures	Mechanical, bradykinin, photostimulation	Glu	$[Ca^{2+}]_i$ elevations in neurons	*(37)*
	Electrical, mechanical	Glu	$[Ca^{2+}]_i$ elevations in neurons	*(47)*
	Electrical, mechanical	Glu	$[Ca^{2+}]_i$ elevations and depolarization in neurons	*(39)*
	Electrical, mechanical	Glu	Slow inward currents in neurons and inhibition of evoked EPSCs	*(49)*
	Electrical, mechanical, UV photolysis	Glu	Increased frequency of spontaneous EPSCs	*(50)*
Neuron–astrocyte Microislands	Mechanical	Glu	Slow inward currents in neurons and inhibition of excitatory and inhibitory evoked EPSCs	*(51)*
	UV photolysis of caged Ca^{2+}	Glu	$[Ca^{2+}]_i$ increase of less than 100 n*M* can trigger the release	*(53)*
Astrocyte cultures	Bradykinin, α-latrotoxin	Glu	—	*(59)*
	Bradykinin	Glu Aspartate	—	*(45)*
	AMPA/mGluR activation or PGE_2	Glu	—	*(41)*
	Electrical, mechanical	Glu	—	*(42)*
	Glu	D-Ser	—	*(75,77)*
Astrocyte cultures and NMDAR-transfected HEK cells	AMPA/*t*ACPD, quisqualate	Glu	Pulsatile release and fast inward currents in HEK cells	*(46)*
HC slices	mGluR activation	Glu	$[Ca^{2+}]_i$ elevations in pyramidal neurons	*(25)*
	AMPA/*t*ACPD activation or PGE_2	Glu	$[Ca^{2+}]_i$ elevations in pyramidal neurons	*(41)*
	Intracellular depolarizing pulses	Glu	Increased frequency and amplitude of spontaneous IPSCs in pyramidal neurons	*(52)*

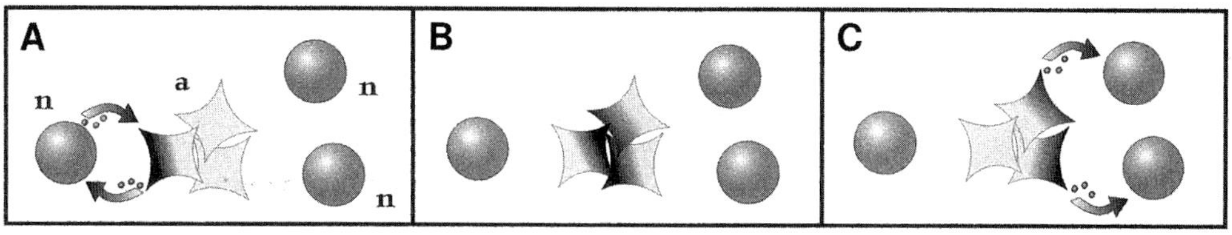

Fig. 1. Astrocytes may mediate intercellular spreading of excitation. **(A)** Glu released by an active synapse of a neuron (n) can reach nearby astrocyte (a), and induce mGluR-mediated $[Ca^{2+}]_i$ elevations. This $[Ca^{2+}]_i$ increase can lead to Glu release from the astrocyte, and potentially result in a modulation of the activity of that specific synapse. **(B)** In addition, astrocyte excitation may travel as a $[Ca^{2+}]_i$ wave throughout connected astrocytes, and **(C)** result in Glu release from astrocytes that were not initially activated, thus signaling to neurons located in other regions.

3. CAN ASTROCYTIC GLU CONTRIBUTE TO NEUROPATHOLOGIES?

Astrocytes can be directly involved in pathological alterations of neuronal activity, not only by virtue of their ability to modify the EC environment, in terms of clearance of K^+ ions and neurotransmitters, but also through the release of neurotransmitters, such as Glu. The neurodegenerative events that indeed occur in diverse pathologies such as spongiform encephalopathies, Alzheimer's disease (AD), and ischemic insult, display Glu excitotoxicity as a common hallmark *(79)*.

Excitotoxic action of Glu on neurons *(8,80)* is linked mostly, although not exclusively, to NMDA receptors, whose excessive stimulation leads to cell death or damage, by dramatically increasing IC Ca^{2+} concentration *(81)*, and thus initiating a cascade of Ca^{2+}-dependent events that include accumulation of IC reactive oxygen species *(82)*, and consequent oxidative damage to membranes, proteins, and nucleic acids.

An excess of Glu in the ECS may derive from both increased release and/or impaired uptake. So far, most attention has been focused on alterations of the high-affinity Glu transporter of astrocytes, which may give a crucial contribution to the massive increase in EC Glu concentration. Because astrocytes play an important role in the clearance of Glu in the synaptic vicinity *(6,7,83,84)*, impairment of glial function can lead to excitotoxicity, as a consequence of abolished Glu uptake or of reversal of the transporters, resulting in neurotransmitter release from astrocytes themselves *(12,85)*.

Less attention has been paid to the possible role of Ca^{2+}-dependent Glu release from astrocytes, in the development or exacerbation of neurodegenerative disorders. The possibility exists that pathological agents involved in these disorders affect Ca^{2+} signaling in astrocytes, and lead to nonphysiological events of Glu release, thus altering the ability of glial cells to sustain a coherent, Glu-mediated communication with neurons. The next subheading gathers available information on astrocytic contribution to Glu excitotoxicity in prion diseases, AD, and ischemia.

3.1. Prion Diseases

Transmissible spongiform encephalopathies, also known as prion diseases, are neurodegenerative disorders in which the cellular isoform of prion protein (PrP^c) is converted to the scrapie isoform (PrP^{Sc}), by an as-yet unknown mechanism. Apparently, PrP^{Sc} is not the translation product of a specific gene, but it is believed to derive from the normal cellular isoform via a posttranslational conformational transition, which would be induced by PrP^{Sc} itself *(86)*.

PrP^c is a cell surface glycoprotein that is expressed by many cell types, including neurons and astrocytes. Unlike PrP^c, PrP^{Sc} is proteinase K-resistant, and aggregates in EC deposits within the brain. In the progression of prion disease, characteristic accumulation of PrP^{Sc} and activation of microglia are followed by astrocytosis and neurodegeneration *(87,88)*.

The development of transgenic mice lacking PrP^c ($Prnp^{0/0}$), together with the use of the neurotoxic prion peptide, PrP106-126, have allowed investigation of the events that lead to neurodegeneration in prion disease.

The identification of the exact cell type responsible for the pathogenesis and propagation of the disease is still a matter of debate: Susceptibility to prion disease has been shown in transgenic mice that express PrP^c, either only in astrocytes *(89)*, or only in neurons *(90)*. Neurotoxicity of PrP106-126 may therefore result from interaction of the peptide with different cell types, and may lead to additive or converging toxic effects.

PrP106-126 was reported to inhibit Glu uptake in $Prnp^{+/+}$, but not in $Prnp^{0/0}$ astrocytes *(91)*: It is plausible that PrP106-126 directly interacts with PrP^c, determining its functional loss, thus converting $Prnp^{+/+}$ astrocytes in functionally $Prnp^{0/0}$ astrocytes, which indeed have a lower value of Glu uptake. The higher-uptake K_m observed in $Prnp^{0/0}$ astrocytes suggests a decrease in the substrate–transporter affinity. Since chelation of EC copper abolishes the K_m increase produced by PrP106-126, the hypothesis is supported that PrP^c is involved in Cu binding or uptake. PrP^c loss could thus impair Cu binding, and cause alterations of the Glu transporter molecule and/or imbalancing of IC redox reactions *(91)*.

By using an elegant co-culture system, the same author *(92)* demonstrated that toxicity of PrP106-126 to $Prnp^{0/0}$ neurons requires an excess number of $Prnp^{+/+}$ astrocytes, and is Glu-mediated *(92)*. Cerebellar neurons were plated in normal multiwell trays; astrocytes and/or microglia were plated in tissue culture inserts, which could be easily added or removed from neuron-containing wells. $Prnp^{0/0}$ neuron survival was shown to be significantly affected by PrP106-126, only in the presence of a large number of $Prnp^{+/+}$ astrocytes, or in the presence of both $Prnp^{+/+}$ microglia and astrocytes. Indeed, PrP106-126 can stimulate astrocyte proliferation through factors released by activated microglia *(93)*, which would explain the temporal correlation existing between astrocytosis and the rapid increase in neurodegeneration that is characteristic of prion disease.

PrP106-126 neurotoxicity was inhibited by NMDAR antagonists *(92)*. Since PrP106-126 inhibits Glu uptake by $PrP^{+/+}$ astrocytes, application of the peptide to neuron–astrocyte co-cultures withdraws the protective role of glial uptake, and enhances Glu toxicity.

These data relate to $Prnp^{0/0}$ neurons, and thus do not rule out a direct action of PrP106-126 on $Prnp^{+/+}$ neurons, as other authors suggest: For instance, PrP106-126 has been shown to elevate $[Ca^{2+}]_i$ levels in hypothalamic GnRH neuronal cell line *(94)*, and to cause a decrease in neuronal IC glutathione level *(95)*.

In cultured astrocytes, prion fragment PrP106–126 has been shown to induce $[Ca^{2+}]$; increased through voltage-dependent Ca^{2+} channels (VOCCs) *(96)*. Although astrocytes *in situ*, under normal conditions, appear not to express VOCCs *(97)*, reactive astrocytes express significant levels of VOCCs *(98)*, thus supporting the possibility that, in pathological conditions, their activation contributes to the increase in EC Glu concentration, by inducing the Ca^{2+}-dependent release of transmitter from astrocytes.

3.2. Alzheimer's Disease

The most evident feature of AD is the presence of EC protein deposits forming senile plaques and fibrils in the brain *(99)*. The main constituents of these deposits are amyloid β-proteins (Aβ), β-sheeted peptides that originate from proteolytic degradation of the IC amyloid precursor protein (APP). Under physiological conditions, APP holoprotein is constitutively cleaved within the Aβ domain, to release the soluble portion, APPS, in the ECS, but pathological conditions, such as AD and Down's syndrome, are characterized by an amyloidogenic processing of APP *(100)*.

Another major feature of AD is the abundant presence of reactive microglia and astrocytes surrounding amyloid plaques, which may secrete inflammatory molecules and plaque constituents. Indeed, Aβ-activated astrocytes overexpress various proteins, such as interleukin-1, inducible NO synthase *(101)*, and APP itself *(102)*. Stimuli that mediate an increase in cyclic adenosine monophosphate, such as PGE_2, have been shown to increase APP levels in astrocytes *(103)*, and cyclic adenosine monophosphate-regulated transcription has been reported also for apolipoprotein E, another component of amyloid deposits *(104)*.

A crucial role in the neuropathology of AD is ascribed to Aβ peptide, although its pathogenic mechanism is still not clear. Application of this peptide has toxic effects on primary cultures of rat hippocampal neurons *(105)*, and exacerbate Glu excitotoxicity in human cultured cortical neurons *(106)*. The mechanism of Aβ peptide-mediated neuronal loss has thus been suggested to involve Glu excitotoxicity. Spatial distribution of AD indeed matches the regions of the brain that use Glu as a major neurotransmitter *(107,108)*, and this pathology is characterized by the loss of glutamatergic neurons, which presumably accounts for the mental dysfunctions reported in AD. Oxidative stress has also been shown to occur in the brain of AD patients, and could significantly contribute to the development of this pathology *(109)*.

The involvement of astrocytes in AD-related Glu excitotoxicity derives from the observation that Aβ peptide can inhibit Na^+-dependent Glu uptake in cultured astrocytes. The ability of antioxidant agents, such as Trilax, to protect against this inhibition, indicates a role for oxidative stress in Aβ-induced impairment of glial Glu transporter *(110)*.

In vivo infusion of Aβ fragments was recently performed in rat nucleus basalis *(111)*. Under those conditions, the authors observed an elevation in EC Glu concentration, an increased Ca^{2+} influx in the infusion area, comparable to that obtained with NMDA infusion, and NMDAR-mediated loss of neuronal fibers *(111)*. These results are consistent with the existence of an excitotoxic mechanism underlying AD pathology and, together with in vitro evidences of Aβ-induced inhibition of astrocytic Glu transporters, suggest that astrocytes could be involved in the AD-associated excitotoxicity.

As to the possibility of a Ca^{2+}-dependent Glu release from astrocytes, upon Aβ application, Harris et al. *(110)*, in their study describing the inhibition of Glu uptake, did not observe any $[Ca^{2+}]_i$ increase in treated astrocytes. In contrast, a more recent paper *(112)* showed that astrocytes responded to Aβ application with a transient $[Ca^{2+}]_i$ rise, and that the probability of occurrence of this response was dramatically increased in the absence of EC Ca^{2+}. These results support the hypothesis that Aβ peptide binds to the receptor that senses EC [Ca^{2+}], as previously reported for hippocampal neurons *(113)*. Activation of this receptor by Aβ peptide in astrocytes, particularly under pathological conditions that lead to reduced EC Ca^{2+} concentration, may result in exacerbation of $[Ca^{2+}]_i$ increases, and hence of Ca^{2+}-dependent events, such as the release of Glu.

3.3. *Ischemia*

Interruptions of cerebral blood flow, such as those occurring after stroke or embolism formation, cause localized ischemia in the brain tissue, leading to neuronal damage. The degree of this damage is higher at the center of the ischemic region, where neuronal death is irreversibly induced within several minutes; in the surrounding area, called the "penumbra," neurons can survive for up to 24 h, and undergo a delayed death. During these 24 h, penumbral areas experience conditions of hypoperfusion *(114)*, which continue to imperil neuronal viability. The presence of this temporal window, before definitive loss of neurons, allows possible therapies to be applied, in order to prevent at least part of the ischemia-related damage.

The cascade of events leading to neuronal death has not been clearly established. Several studies indicate a key role for excessive elevation of EC Glu, and consequent Ca^{2+} overload in neurons *(8,81)*. NO formation *(115)*, hyperexcitability *(116)*, and apoptosis *(117,118)* have been suggested as important downstream events involved in delayed neuronal death.

The identification of the cell type(s) that release Glu in ischemic conditions, and the mechanism accounting for this process, has obvious relevance for the research of target-specific therapies.

Since energy failure results in the impairment of ATP-dependent processes, and the filling of vesicles with neurotransmitter and their fusion with plasma membrane require ATP hydrolysis, Glu release during severe ischemia is probably not the result of vesicular release *(12,85,116)*. On the other hand, a fall in GTP and ATP levels can also compromise the function of neuronal and/or glial Glu transporters, which require Na^+/K^+-ATPase activity to sustain Na^+-dependent uptake of neurotransmitter *(119)*. Impairment or reversal of Glu transporters may thus account for the increase in EC Glu concentration.

A recent study *(120)* provided evidence for a major role for the reversal of Glu transporters in Glu release in severe brain ischemia. Those authors examined this release, upon perfusion of EC slices with an EC solution mimicking severe ischemia, by monitoring the occurrence of anoxic depolarization and Glu-induced inward currents in CA1 neurons. Both these events were abolished by the Glu transporter inhibitor, L-*trans*-pyrrolidine-2,4-dicarboxylate, but not by other agents inhibiting Ca^{2+} or swelling-dependent release. Since blockade of the main glial transporter, GLT-1, by dihydrokainate had no effect on the described neuronal depolarizations, these results indicate a more relevant role for neuronal transporters in this process *(120)*.

The participation of compromised glial transporters in the development of neurodegeneration cannot, however, be excluded. The effects of the loss of glial or neuronal Glu transporter were examined in organotypic spinal cord cultures, and in intact brains, upon administration of chronic antisense oligonucleotides *(121)*. The loss of glial Glu transporter, GLT-1 or astrocyte-specific Glu transporter (GLAST), but not of the neuronal EAAC1, was sufficient to determine a marked elevation in EC Glu concentration. The correlated degree of neurotoxicity, evaluated both on the basis of histological preparations and through observation of animal behavior, was significantly higher in the absence of glial transporters. These data, obviously, do not demonstrate that the impairment of Glu uptake systems in astrocytes is involved in ischemia-related neurotoxicity. However, they show the dramatic implications for neuronal survival of any pathological condition that leads to dysfunction of glial uptake systems. Evidence suggests that ischemia may be one of these conditions: A reduction of Glu uptake was indeed observed *(122)* in cultured astrocytes subjected to hypoxia combined with acidosis, conditions that mimic an incomplete ischemia. Another study on cultured astrocytes *(123)* reported a transient increase in both Na^+/K^+-ATPase and Glu uptake in the first hours of ischemic treatment, followed by a loss of both activities and significant cell death after 24 h.

Taken together, these results suggest that a dysfunction in glial Glu transporters may occur in brain ischemia, and contribute to the massive increase in EC Glu concentration, thus contributing to the Glu-dependent events that ultimately lead to neuronal loss.

ACKNOWLEDGMENTS

The original work presented in this chapter was supported by grants from Armenise-Harvard University Foundation and Telethon-Italy (Grant No. 1095). We would like to thank Prof. Tullio Pozzan for the critical reading of the manuscript.

REFERENCES

1. Hatten, M. E. (1990) Riding the glial monorail: a common mechanism for glial-guided neuronal migration in different regions of the developing mammalian brain. *Trends Neurosci.* **13,** 179–184.
2. Nakanishi, K., Okouchi, Y., Takatoshi, U., Kiyofumi, A., Ichiro, I., Eksioglu, Y. Z., et al. (1994) Astrocytic contributions to functioning synapse formation estimated by spontaneous neuronal intracellular Ca^{2+} oscillations. *Brain Res.* **659,** 169–178.
3. Pfrieger, F. W. and Barres, B. A. (1997) Synaptic efficacy enhanced by glial cells in vitro. *Science* **277,** 1684–1687.
4. Magistretti, P. J. and Pellerin, L. (1996) Cellular bases of brain energy metabolism and their relevance to functional brain imaging: evidence for a prominent role of astrocytes. *Cereb. Cortex* **6,** 50–61.
5. Orkand, R. K., Coles, J. A., and Tsacopoulos, M. (1986) The role of glial cells in ion homeostasis in the retina of the honeybee drone. *Exp. Brain Res.* **14,** 404–413.
6. Rothstein, J. D., Martin, L., Levey, A. I., Dykes-Hoberg, M., Jin, L., Wu, D., Nash, N., and Kuncl, R. W. (1994) Localization of neuronal and glial glutamate transporters. *Neuron* **13,** 713–725.
7. Barbour, B., Brew, H., and Attwell, D. (1988) Electrogenic glutamate uptake in glial cells is activated by intracellular potassium. *Nature* **335,** 433–435.
8. Choi, D. W. and Rothman, S. M. (1990) The role of glutamate neurotoxicity in hypoxy-ischemic neuronal death. *Ann. Rev. Neurosci.* **13,** 171–182.
9. Peters, A., Palay, S. L., and Webster, H. de F. (1991) *The Fine Structure of the Central Nervous System: Neuron and Their Supportive Cells*, Oxford University Press, New York.
10. Harris, K. M. and Ventura, R. (1999) Three-dimensional relationships between hippocampal synapses and astrocytes. *J. Neurosci.* **19,** 6897–6906.

11. Gallo, V. and Ghiani, C. A. (2000) Glutamate receptors in glia: new cells, new inputs and new functions. *Trends Pharmacol. Sci.* **21,** 252–258.
12. Szatkowski, M., Barbour, B., and Attwell, D. (1990) Non-vesicular release of glutamate from glial cells by reverse electrogenic glutamate uptake. *Nature* **348,** 443–446.
13. Pfrieger, F. W. and Barres, B. A. (1996) New views on synapse-glia interactions. *Curr. Opin. Neurobiol.* **6,** 615–621.
14. Bowman, C. L. and Kimelberg, H. K. (1984) Excitatory aminoacids directly depolarize rat brain astrocytes in primary culture. *Nature* **311,** 655–659.
15. Kettenmann, H., Backus, K. H., and Schachner, M. (1985) Aspartate, glutamate and gamma-aminobutyric acid depolarize cultured astrocytes. *Neurosci. Lett.* **52,** 25–29.
16. Cornell-Bell, A. H., Finkbeiner, S. M., Cooper, M. S., and Smith, S. J. (1990) Glutamate induces calcium waves in cultured astrocytes: long-range glial signaling. *Science* **247,** 470–473.
17. Glaum, S. R., Holzwarth, J. A., and Miller, R. J. (1990) Glutamate receptors activate Ca^{2+} mobilization and Ca^{2+} influx into astrocytes. *Proc. Natl. Acad. Sci. USA* **87,** 3454–3458.
18. Pasti, L., Pozzan, T., and Carmignoto, G. (1995) Long-lasting changes of calcium oscillations in astrocytes. A new form of glutamate-mediated plasticity. *J. Biol. Chem.* **270,** 15,203–15,210.
19. Cornell-Bell, A. H. and Finkbeiner, S. M. (1991) Ca^{2+} waves in astrocytes. *Cell Calcium* **12,** 185–204.
20. Finkbeiner, S. (1992) Calcium waves in astrocytes-filling in the gaps. *Neuron.* **8,** 1101–1108.
21. Smith, S. J. (1994) Neuromodulatory astrocytes. *Curr. Biol.* **4,** 807–810.
22. Murphy, T. H., Blatter, L. A., Wier, W. G., and Baraban, J. M. (1992) Rapid communication between neurones and astrocytes in primary cortical cultures. *J. Neurosci.* **13,** 2672–2679.
23. Dani, J. W., Chernjavsky, A., and Smith, S. J. (1992) Neuronal activity triggers calcium waves in hippocampal astrocyte network. *Neuron* **8,** 429–440.
24. Porter, J. T. and McCarthy, K. D. (1995) GFAP-positive hippocampal astrocytes *in situ* respond to glutamatergic neuroligand with increased in $[Ca^{2+}]_i$. *Glia* **13,**101–112.
25. Pasti, L., Pozzan, T., and Carmignoto, G. (1997) Intracellular calcium oscillations in astrocytes: a highly plastic, bidirectional form of communication between neurones and astrocytes in situ. *J. Neurosci.* **17,** 7817–7830.
26. Bergles, D. E., Roberts, J. D., Somogyi, P., and Jahr, C. E. (2000) Glutamatergic synapses on oligodendrocyte precursor cells in the hippocampus. *Nature* **405,** 187–191.
27. Kandel, E. R. (2000) Cellular mechanisms of learning and the biological basis of individuality, in *Principles of Neural Science* (Kandel, E. R., Schartz, J. H., and Jessel, T. M., eds.), McGraw-Hill, New York, pp. 1247–1279.
28. Bliss, T. V. P. and Lomo, T. (1973) Long term potentiation of synaptic transmission in the dentate area of anaesthetized rabbit following stimulation of the perforant path. *J. Physiol. (Lond.)* **232,** 331–356.
29. Bliss, T. V. P. and Collingridge, G. L. (1993) A synaptic model of memory: long term potentiation in the hippocampus. *Nature* **361,** 31–39.
30. Schuman, E. M. and Madison, D. V. (1993) Nitric oxide as an intercellular signal in long-term potentiation. *Semin. Neurosci.* **5,** 207–215.
31. Tweedle, C. D. and Hatton, G. I. (1977) Ultrastructural changes in rat hypothalamic neurosecretory cells and their associative glia during minimal dehydration and rehydration. *Cell Tissue Res.* **181,** 59–72.
32. Laming, P. R., Kimelberg, H., Robinson, S., Salm, A., Hawrylak, N., Müller, C., Roots, B., and Ng, K. (2000) Neuronal-glial interactions and behaviour. *Neurosci. Biobehav. Rev.* **24,** 295–340.
33. Cornell-Bell, A. H., Thomas, P. G., and Smith, S. J. (1990) The excitatory neurotransmitter glutamate causes filopodia formation in cultured hippocampal astrocytes. *Glia* **3,** 322–334.

34. Wenzel, J., Lammert, G., Meyer, U., and Krug, M. (1991) The influence of long-term potentiation on the spatial relationship between astrocytic processes and potentiated synapses in the dentate gyrus neuropil of the rat brain. *Brain Res.* **560,** 122–131.
35. Jones, T. A. and Greenough, W. T. (1996) Ultrastructural evidence for increased contact between astrocytes and synapses in rats reared in a complex environment. *Neurobiol. Mem. Learn.* **65,** 45–56.
36. Nedergaard, M. (1994) Direct signaling from astrocytes to neurones in cultures of mammalian brain cells. *Science* **263,** 1768–1771.
37. Parpura, V., Basarky, T. A., Liu, F., Jeftinija, K., Jeftinija, S., and Haydon, P. G. (1994) Glutamate-mediated astrocyte-neurone signalling. *Nature* **369,** 744–747.
38. Alvarez-Maubecin, V., García-Hernández, F., Williams, J. T., and Van Bockstaele, E. J. (2000) Functional coupling between neurons and glia. *J. Neurosci.* **20,** 4091–4098.
39. Hassingher, T. D., Atkinson, P. B., Strecker, G. J., Whalen, R. L., Dudek, F. E., Kossel, A. H., and Kater, S. B. (1995) Evidence for glutamate-mediated activation of hippocampal neurons by glial calcium wave. *J. Neurobiol.* **28,** 159–170.
40. Nicholls, D. G. and Sihra, T. S. (1986) Synaptosomes possess an exocytotic pool of glutamate. *Nature* **321,** 772–773.
41. Bezzi, P., Carmignoto, G. , Pasti, L., Vesce, S., Rossi, D., Lodi Rizzini, B., Pozzan, T., and Volterra, A. (1998) Prostaglandins stimulate calcium-dependent glutamate release in astrocytes. *Nature* **391,** 281–285.
42. Innocenti, B., Parpura, V., and Haydon P. G. (2000) Imaging extracellular waves of glutamate during calcium signaling in cultured astrocytes. *J. Neurosci.* **20,** 1800–1808.
43. Kimelberg, H. K., Goderie, S. K., Higman, S., and Waniewski, R. A. (1990) Swelling-induced release of glutamate, aspartate, and taurine from astrocyte cultures. *J. Neurosci.* **10,** 1583–1591.
44. Sanzgiri, R. P., Araque, A., and Haydon, P. G. (1999) Prostaglandin $E_{(2)}$ stimulates glutamate receptor-dependent astrocyte neuromodulation in cultured hippocampal cells. *J. Neurobiol.* **41,** 221–229.
45. Jeftinija, S. D., Jeftinija, K. V., Stefanovic, G., and Liu, F. (1996) Neuroligand-evoked calcium-dependent release of excitatory amino acids from cultured astrocytes. *J. Neurochem.* **66,** 676–684.
46. Pasti, L., Zonta, M., Pozzan, T., Vicini, S., and Carmignoto, G. (2001) Cytosolic calcium oscillations in astrocytes may regulate exocytotic release of glutamate. *J. Neurosci.* **21,** 477–484.
47. Charles, A. C. (1994) Glia-neuron intercellular calcium signaling. *Dev. Neurosci.* **16,** 196–206.
48. Hassingher, T. D., Guthrie, P. B., Atkinson, P. B., Bennett, M. V. L., and Kater, S. (1996) An extracellular signaling component in propagation of astrocytic calcium waves. *Proc. Natl. Acad. Sci USA* **93,** 13,268–13,273.
49. Araque, A., Parpura, V., Sanzgiri, R. P., and Haydon, P. G. (1998) Glutamate-dependent astrocyte modulation of synaptic transmission between cultured hippocampal neurones. *Eur. J. Neurosci.* **10,** 2129–2142.
50. Araque, A., Sanzgiri, R. P., Parpura, V., and Haydon, P. G. (1998) Calcium elevation in astrocytes causes an NMDA receptor-dependent increase in the frequency of miniature synaptic currents in cultured hippocampal neurones. *J. Neurosci.* **18,** 6822–6829.
51. Araque, A., Li, N., Doyle, R. T., and Haydon, P. G. (2000) Snare protein-dependent glutamate release from astrocytes. *J. Neurosci.* **20,** 666–673.
52. Kang, J., Jiang, L., Glodman S. A., and Nedergaard, M. (1998) Astrocyte-mediated potentiation of inhibitory synaptic transmission. *Nature Neurosci.* **1,** 683–692.
53. Parpura, V. and Haydon, P. G. (2000) Physiological astrocytic calcium levels stimulate glutamate release to modulate adjacent neurons. *Proc. Natl. Acad. Sci. USA* **97,** 8629–8634.

54. Söllner, T., Bennet, M. K., Whiteheart, S. W., Scheller, R. H., and Rothman, J. E. (1993) A protein assembly-disassembly pathway in vitro that may correspond to sequential steps of synaptic vesicle docking, activation and fusion. *Cell* **75,** 409–418.
55. Söllner, T., Whiteheart, S. W., Brunner, M., Erdjument-Bromage, H., Geromanos, S., Tempst, P., and Rothman, J. E. (1993) SNAP receptors implicated in vesicle targeting and fusion. *Nature* **362,** 318–324.
56. Parpura, V., Fang, Y., Basarky, T. A., Jahn, R., and Haydon, P. G. (1995) Expression of synaptobrevin II, cellubrevin and syntaxin but not SNAP-25 in cultured astrocytes. *FEBS Lett.* **377,** 489–492.
57. Hepp, R., Perraut, M., Chasserot-Golaz, S., Galli, T., Aunis, D., Langley, K., and Grant, J. L. N. (1999) Cultured glial cells express the SNAP-25 analogue SNAP-23. *Glia* **27,** 181–187.
58. Maienschein, V., Marxen, N., Volknandt, W., and Zimmermann, H. (1999) A plethora of presynaptic proteins associated with ATP-storing organelles in cultured astrocytes. *Glia* **26,** 233–244.
59. Parpura, V., Liu, F., Brethorst, S., Jeftinija, K., Jeftinija, S., and Haydon, P. G. (1995) α-latrotoxin stimulates glutamate release from cortical astrocytes in cell culture. *FEBS Lett.* **360,** 266–270.
60. Schiavo, G., Benfenati, F., Poulain, B., Rossetto, O., Polverino de Laureto, P., DasGupta, B. R., and Montecucco, C. (1992) Tetanus and botulinum-B neurotoxins block neurotransmitter release by proteolitic cleavage of synaptobrevin. *Nature* **359,** 832–834.
61. Bowman, E. J., Siebers, A., and Altendorf, K. (1988) Bafilomycins: a class of inhibitors of membrane ATPases from microorganisms, animal cells and plant cells. *Proc. Natl. Acad. Sci. USA* **85,** 7972–7976.
62. Maycox, P. R., Hell, J. W., and Jahn, R. (1990) Amino acid neurotransmission: spotlight on synaptic vesicles. *Trends Neurosci.* **13,** 83–87.
63. Stevens, T. H. and Forgac, M. (1997) Structure, function and regulation of the vacuolar (H^+)-ATPase. *Annu. Rev. Cell. Dev. Biol.* **13,** 779–808.
64. Zhou, Q., Petersen, C. C. H., and Nicoll, R. A. (2000) Effects of reduced vesicular filling on synaptic transmission in rat hippocampal neurons. *J. Physiol.* **525,** 195–206.
65. Forsythe, I. D. and Clements, J. D. (1990) Presynaptic glutamate receptors depress excitatory monosynaptic transmission between mouse hippocampal neurons. *J. Physiol.* **429,** 1–16.
66. Baskys, A. and Malenka, R. C. (1991) Agonists at metabotropic glutamate receptors presynaptically inhibit EPSCs in neonatal rat hippocampus. *J. Physiol.* **444,** 687–701.
67. Forsythe, I. D. and Barnes-Davies, M. (1997) Synaptic transmission: well-placed modulators. *Curr. Biol.* **7,** 362–365.
68. Newman, E. A. and Zahs, K. R. (1998) Modulation of neuronal activity by glial cells in the retina. *J. Neurosci.* **18,** 4022–4028.
69. Grosche, J., Matyash, V., Möller, T., Verkhratsky, A., Teichenbach, A., and Kettenmann, H. (1999) Microdomains for neuron-glia interaction: parallel fiber signaling to Bergmann glial cells. *Nature Neurosci.* **2,** 139–143.
70. Porter, J. T. and McCarthy, K. D. (1996) Hippocampal astrocytes *in situ* respond to glutamate released from synaptic terminals. *J. Neurosci.* **16,** 5073–5081.
71. Levi, G. and Patrizio, M. (1992) Astrocyte heterogeneity: endogenous aminoacid levels and release evoked by non-N-methyl-D-aspartate receptor agonists and by potassium-induced swelling in type-1 and type-2 astrocytes. *J. Neurochem.* **58,** 1943–1952.
72. McBain, C. J. and Mayer, M. L. (1994) N-methyl-D-aspartic receptor structure and function. *Physiol. Rev.* **74,** 723–760.
73. Cull-Candy, S. (1995) NMDA receptors: do glia hold the key? *Curr. Biol.* **5,** 841–843.
74. Kleckner, N. W. and Dingledine, R. (1993) Requirement for glycine in activation of NMDA receptors expressed in *Xenopus oocytes*. *Science* **241,** 835–837.

75. Schell, M. J., Molliver, M. E., and Snyder, S. H. (1995) D-serine, an endogenous synaptic modulator: localization to astrocytes and glutamate-stimulated release. *Proc. Natl. Acad. Sci. USA* **92,** 3948–3952.
76. Schell, M. J., Brady, R. O., Jr., Molliver, M. E., and Snyder, S. H. (1997) D-serine as a neuromodulator: regional and developmental localizations in rat brain glia resemble NMDA receptors. *J. Neurosci.* **17,** 1604–1615.
77. Wolosker, H., Blackshaw, S., and Snyder, S. H. (1999) Serine racemase: a glial enzyme synthesizing D-serine to regulate glutamate N-methyl-D-aspartate neurotransmission. *Proc. Natl. Acad. Sci. USA* **96,** 13,409–13,414.
78. Kemp, J. A. and Leeson, P. D. (1993) The glycine site of the NMDA receptor: five years on. *Trends Pharmacol. Sci.* **14,** 20–25.
79. Michaelis, E. K. (1998) Molecular biology of glutamate receptors in the central nervous system and their role in excitotoxicity, oxidative stress and aging. *Progr. Neurobiol.* **54,** 369–415.
80. Nicotera, P., Leist, M., and Manzo, L. (1999) Neuronal cell death: a demise with different shapes. *Trends Pharmacol. Sci.* 20**,** 46–51.
81. Rothman, S. M. and Olney, J. W. (1995) Excitotoxicity and the NMDA receptor: still lethal after eight years. *Trends Neurosci.* 18**,** 57–58.
82. Lafon-Cazal, M., Pietri, S., Culcasi, M., and Bockaert, J. (1993) NMDA-dependent superoxide production and neurotoxicity. *Nature* **364,** 535–537.
83. Bergles, D. E. and Jahr, C. E. (1998) Glial contribution to glutamate uptake at Schaffer collateral-commissural synapses in the hippocampus. *J. Neurosci.* **18,** 7709–7716.
84. Bergles, D. E. and Jahr, C. E. (1997) Synaptic activation of glutamate transporters in hippocampal astrocytes. *Neuron* **19,** 1297–1308.
85. Attwell, D., Barbour, B., and Szatkowski, M. (1993) Nonvesicular release of neurotransmitter. *Neuron.* **11,** 401–407.
86. Prusiner, S. B. (1982) Novel proteinaceous infectious particles cause scrapie. *Science* **216,** 136–144.
87. Prusiner, S. B. and DeArmond, S. J. (1994) Prion diseases and neurodegeneration. *Annu. Rev. Neurosci.* **17,** 311–339.
88. Williams, A., Lucassen, P. J., Ritchie, D., and Bruce, M. (1997) PrP deposition, microglial activation and neuronal apoptosis in murine scrapie. *Exp. Neurol.* **144,** 433–438.
89. Raeber, A. J., Race, R. E., Brandner, S., Priola, S. A., Sailer, A., Bessen, R. A., et al. (1997) Astrocyte-specific expression of hamster prion protein (PrP) renders PrP knockout mice susceptible to hamster scrapie. *EMBO J.* **16,** 6057–6065.
90. Race, R. E., Priola, S. A., Bessen, R. A., Ernst, D., Dockter, J., Rall, G. F., et al. (1995) Neuron-specific expression of a hamster prion protein minigene in transgenic mice induces susceptibility to hamster scrapie agent. *Neuron* **15,** 1183–1191.
91. Brown, D. R. and Mohn C. M. (1999) Astrocytic glutamate uptake and prion protein expression. *Glia* **25,** 282–292.
92. Brown, D. R. (1999) Prion protein peptide neurotoxicity can be mediated by astrocytes. *J. Neurochem.* **73,** 1105–1113.
93. Brown, D. R., Schmidt, B., and Kretzschmar, H. A. (1996) A neurotoxic prion protein fragment enhances proliferation of microglia but not astrocytes in culture. *Glia* **18,** 59–67.
94. Kawahara, M., Kuroda, Y., Arispe, N., and Rojas, E. (2000) Alzheimer's beta-amyloid, human islet amylin and prion protein fragment evoke intracellular free Ca^{2+} elevations by a common mechanism in a hypothalamic GnRH neuronal cell line. *J. Biol. Chem.* **275,** 14,077–14,083.
95. Schroder, H. C., Perovic, S., Kavsan, V., Ushijima, H., and Muller, W. E. (1998) Mechanisms of prionSc and HIV-1 gp120 induced neuronal cell death. *Neurotoxicology* **19,** 683–688.

96. Florio, T., Thellung, S., and Schettini, G. (1996) Intracellular calcium rise through L-type calcium channels, as molecular mechanism for prion protein fragment 106-126-induced astroglial proliferation. *Biochem. Biophys. Res. Comm.* **228,** 397–405.
97. Carmignoto, G., Pasti, L., and Pozzan, T. (1998) On the role of voltage-dependent calcium channels in calcium signaling of astrocytes *in situ*. *J. Neurosci.* **18,** 4637–4645.
98. Westenbroek, R. E., Baush, S. B., Lin, R. C. S., Frank, J. E., Noebels, J. L., and Catterall, W. A. (1998) Upregulation of L-type Ca^{2+} channels in reactive astrocytes after brain injury, hypomyelination, and ischemia. *J. Neurosci.* **18,** 2321–2334.
99. Alzheimer, A. (1907) Über eine eigenartige Erkrankung der Hirnrinde. *Allg. Z. Psychiatr. Psychol-Gericht. Med.* **64,** 146–148.
100. Selkoe, D. J. (1991) The molecular pathology of Alzheimer's disease. *Neuron* **6,** 487–498.
101. Hu, J., Akama, K. T., Krafft, G. A., Chromy, B. A., and Van Eldik, L. J. (1998) Amyloid-beta peptide activates cultured astrocytes: morphological alterations, cytokine induction and nitric oxide release. *Brain Res.* **785,** 195–206.
102. Fernaud-Espinosa, I., Nieto-Sampedro, M., and Bovolenta, P. (1993) Differential activation of microglia and astrocytes in aniso- and isomorphic gliotic tissue. *Glia* **8,** 277–291.
103. Lee, R. K., Knapp, S., and Wurtman, R. J. (1999) Prostaglandin E_2 stimulates amyloid precursor protein gene expression: inhibition by immunosuppressants. *J. Neurosci.* **19,** 940–947.
104. Garcia, M. A., Vazquez, J., Gimenez, C., Valdivieso, F., and Zafra, F. (1996) Transcription factor AP-2 regulates human apolipoprotein E gene expression in astrocytoma cells. *J. Neurosci.* **16,** 7550–7556.
105. Yankner, B. A., Duffy, L. K., and Kirschner, D. A. (1990) Neurotrophic and neurotoxic effects of amyloid beta protein: reversal by tachykinin neuropeptides. *Science* **250,** 279–282.
106. Mattson, M. P., Cheng, B., Davis, D., Bryant, K., Lieberburg, I., and Rydel, R. E. (1992) beta-Amyloid peptides destabilize calcium homeostasis and render human cortical neurons vulnerable to excitotoxicity. *J. Neurosci.* **12,** 376–389.
107. Pearson, R. C. A., Esiri, M. M., Hiorns, R. W., Wilcock, G. K., and Powell, T. P. S. (1985) Anatomical correlates of the pathological changes in the neocortex in Alzheimer's disease. *Proc. Natl. Acad. Sci. USA* **82,** 4531–4534.
108. Rogers, J. and Morrison, J. H. (1985) Quantitative morphology and regional laminar distributions of senile plaques in Alzheimer's disease. *J. Neurosci.* **5,** 2801–2808.
109. Miranda, S., Opazo, C., Larrondo, L. F., Muñoz, F. J., Ruiz, F., Leighton, F., and Inestrosa, N. C. (2000) The role of oxidative stress in the toxicity induced by amyloid β-peptide in Alzheimer's disease. *Progr. Neurobiol.* **62,** 633–648.
110. Harris, M. E., Wang, Y., Pedigo, N. W., Jr., Hensley, K., Butterfield, D. A., and Carney, J. M. (1996) Amyloid β peptide (25-35) inhibits Na^+ dependent glutamate uptake in rat hippocampal astrocyte cultures. *J. Neurochem.* **67,** 277–286.
111. Harkany, T., Abrahàm, I., Timmerman, W., Laskay, G., Tòth, B., Sasvàri, M., et al. (2000) β-Amyloid neurotoxicity is mediated by a glutamate-triggered excitotoxic cascade in rat nucleus basalis. *Eur. J. Neurosci.* **12,** 2735–2745.
112. Stix, B. and Reiser, G. (1998) β-Amyloid peptide 25-35 regulates basal and hormone-stimulated Ca^{2+} levels in cultured rat astrocytes. *Neurosci. Lett.* **243,** 121–124.
113. Ye, C., Ho-Pao, C. L., Kanazirska, M., Quinn, S., Rogers, K., Seidman, C. E., et al. (1997) Amyloid-β proteins activate Ca^{2+}-permeable channels through calcium-sensing receptors. *J. Neurosci. Res.* **47,** 547–554.
114. Hossmann, K. A. and Kleihues, P. (1973) Reversibility of ischemic brain damage. *Arch. Neurol.* **29,** 375–384.
115. Huang, Z., Huang, P. L., Panahian, N., Dalkara, T., Fishman, M. C., and Moskowitz, M. A. (1994) Effects of cerebral ischemia in mice deficient in neuronal nitric oxide synthase. *Science* **265,** 1883–1885.

116. Szatkowski, M. and Attwell, D. (1994) Triggering and execution of neuronal death in brain ischaemia: two phases of glutamate release by different mechanisms. *TINS* **17,** 359–365.
117. Martinou, J. C., Dubois-Dauphin, M., Staple, J. K., Rodriguez, I., Frankowski, H., Missotten, M., et al. (1994) Overexpression of BCL-2 in transgenic mice protects neurons from naturally occurring cell death and experimental ischemia. *Neuron* **13,** 1017–1030.
118. Choi, D. W. (1996) Ischemia-induced neuronal apoptosis. *Curr. Opin. Neurobiol.* **6,** 667–672.
119. Nicholls, D. and Attwell, D. (1990) The release and uptake of excitatory amino acids. *TiPS* **11,** 462–468.
120. Rossi, D. J., Oshima, T., and Attwell, D. (2000) Glutamate release in severe brain ischaemia is mainly by reversed uptake. *Nature* **403,** 316–321.
121. Rothstein, J. D., Dykes-Hoberg, M., Pardo, C. A., Bristol, L. A., Jin, L., Kuncl, R. W., et al. (1996) Knockout of glutamate transporters reveals a major role for astroglial transport in excitotoxicity and clearance of glutamate. *Neuron* **16,** 675–686.
122. Swanson, R. A., Farrell, K., and Stein, B. A. (1997) Astrocyte energetics, function, and death under conditions of incomplete ischemia: a mechanism of glial death in the penumbra. *Glia* **21,** 142–153.
123. Stanimirovic, D. B., Ball, R., and Durkin, J. P. (1997) Stimulation of glutamate uptake and Na,K-ATPase activity in rat astrocytes exposed to ischemia-like insults. *Glia* **19,** 123–134.

8
Axonal Conduction and Myelin

Jeffery D. Kocsis

1. INTRODUCTION

The association of myelin with the axon is perhaps the most intimate and well-studied functional relationship between a glial element and a neuron. Myelin is formed as membranous wraps around an axon in the peripheral nervous system by Schwann cells, which make a single segment of myelin, and in the central nervous system by oligodendrocytes that make multiple myelin segments. The membranous wraps formed by myelin confers a high resistance and low capacitance to a considerable length of a myelinated axon (internodal region) that greatly increases the conduction velocity and security of impulse conduction. In addition to myelinated axons there are nonmyelinated axons, which do not have myelin, are typically smaller in diameter and conduct at much reduced velocities.

The action potential of myelinated axons is generated at the relatively narrow node of Ranvier and "skips" from node to node providing for saltatory conduction *(1)*. Photomicrographs of teased peripheral nerve fibers are shown in Fig. 1 in order to give a sense of scale between the relative size of the node of Ranvier (arrows) and the internodal region (between arrows). An internode of a large myelinated axon can be as long as a millimeter, but the node of Ranvier is typically only several microns in length. The inward current generated at a given node can depolarize the axon membrane at the next node without wasting charge or time by depolarizing the internodal membrane that is covered by myelin (Fig. 2A). The majority of the current generated by a given node passes across the axon membrane at the next node by local current flow, thereby allowing for a high current density to be generated at a site distal to the previous node. Activation of sodium (Na) channels at this previously quiescent node will now generate an action potential. However, if the myelin is disrupted as illustrated in Fig. 2B, current generated by a given node will now be distributed longitudinally and "leak" across the internodal membrane. The current density at the next node will be reduced and the time to charge the membrane capacitance increased. The consequence of this longitudinal redistribution of current, subsequent to myelin disruption, will be either impulse failure because insufficient potential develops across the node to activation Na channels, or a delay in action potential activation from the increased membrane charging time. These two changes, impulse failure or conduction slowing are two hallmark signs of conduction impairment in demyelinating disease.

From: *The Neuronal Environment: Brain Homeostasis in Health and Disease*
Edited by: W. Walz © Humana Press Inc., Totowa, NJ

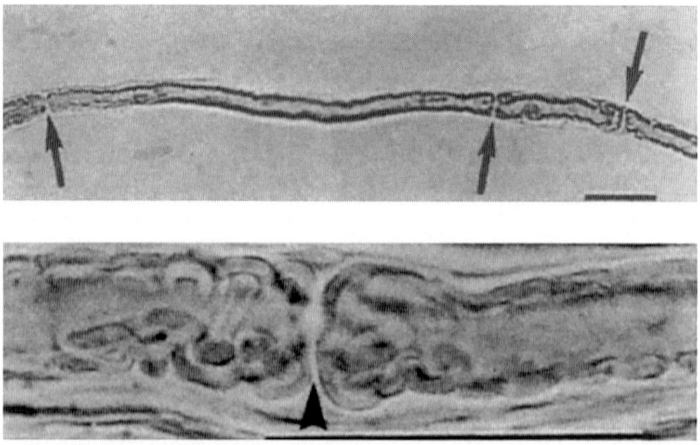

Fig. 1. Teased regenerated sciatic nerve fibers showing relative size of nodal and internodal regions. Arrows indicate nodes of Ranvier, and the regions between areas show myelinated internodes. The internode on the left is conventional, but the one on the right is relatively short which is typical in regenerated axons. Scale bars: top = 40 μm. (Modified with permission from ref. *86*.)

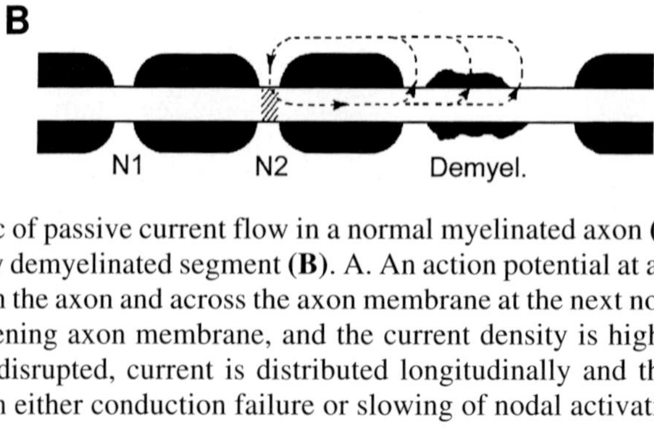

Fig. 2. Schematic of passive current flow in a normal myelinated axon **(A)** and a myelinated axon with a partially demyelinated segment **(B)**. A. An action potential at a node (N1) results in current flow through the axon and across the axon membrane at the next node (N2). The myelin insulates the intervening axon membrane, and the current density is high at N2. (B) When a myelin segment is disrupted, current is distributed longitudinally and the current density is reduced, resulting in either conduction failure or slowing of nodal activation.

Early experimental evidence for saltatory conduction in nerve, by Huxley and Stampfli *(2)*, is illustrated in Fig. 3. Longitudinal currents were recorded at small increments along the course of a single nerve fiber. The recordings obtained at three points along an internode are very similar. However, when the next internode is reached there is a discontinuity in the current and an incremental jump in latency. This is very evident in Fig. 3C where the latency of the responses from the shock artifact "jump," i.e., increase when a node is reached. Saltatory conduction can greatly increase the speed of

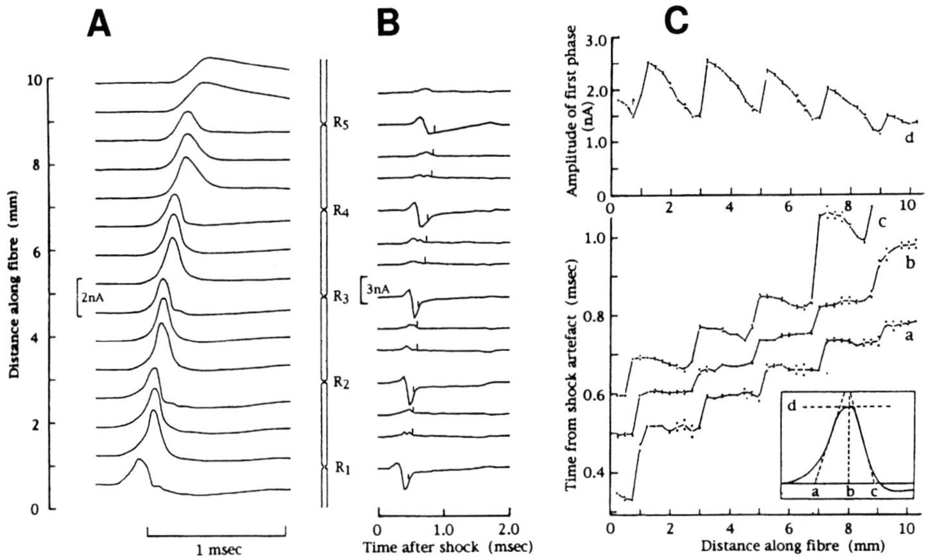

Fig. 3. (A) Longitudinal current along a myelinated frog axon during a propagated action potential. **(B)** Transmembrane current at different points along a single myelinated fiber; inward current is downward. **(C)** The latency from stimulus onset to upstroke (a), peak (b), downstroke (c), and peak amplitude of the currents in A at different sites along the fiber. Inset shows how points in C were determined. (Adapted with permission from ref. *86*.)

an action potential and preserve space by allowing for smaller diameter axons with rapid conduction. For example, the conduction velocity of a 500 μm giant squid axon is about the same as a 10 μm myelinated axon. It has been suggested that oligodendrocytes in the CNS developed from selection pressure to further economize space and maintain conduction velocity. A given oligodendrocyte can form several myelinated segments, compared to only one for a peripheral Schwann cell. Thus, axons myelinated by oligodendrocytes in brain and spinal cord have the advantage of rapid conduction where space constraints are critical to keep the size of the brain and head manageable. However, disruption of a single oligodendrocyte will have a greater adverse effect on conduction than will disruption of a single Schwann cell because more myelin segments are disrupted. A number of important anatomical and biophysical specializations occur at and near the node of Ranvier and its myelin-forming partner. In this chapter key features of these specializations are discussed as are experimental strategies to repair myelin, and to improve conduction following demyelination.

2. IONIC BASIS OF ACTION POTENTIAL IN MYELINATED NERVE

Early studies extensively studied the ionic basis of the action potential using the nodal voltage-clamp at the amphibian node of Ranvier *(3–5)*. Similar to the squid axon the action potential in myelinated nerve fibers is generated by both Na and potassium (K) currents. A full quantitative description of the action potential for the squid axon has been reviewed many times and is not repeated here (*see* refs. *6* and *7* for a complete formalization). However, the membrane potential of an axon, when it is voltage

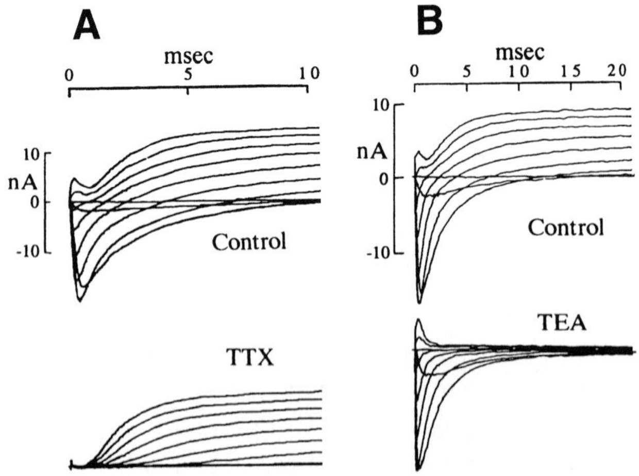

Fig. 4. Ionic currents obtained from frog nodes after leakage and capacity current subtraction in normal Ringer's solution, and in the presence of 300 n*M* TTX (**A**) and 6 m*M* TEA (**B**). (Adapted with permission from ref. *86*.)

clamped to achieve uniform potential changes, can be described by a simple first-order differential equation:

$$C_m dE_m/dt = -I_{\text{ionic}} + I_m$$

where $I_m = (I_{\text{ionic}} + I_c)$, $I_{\text{ionic}} = (I_{\text{Na}} + I_{\text{K}} + I_{\text{L}})$, C_m is the membrane capacitance and I_m, I_{ionic}, and I_{C} are the total membrane current, and ionic and capacitative currents, respectively. I_{Na}, I_{K}, and I_{L} are the Na, K, and leak currents, respectively.

Hodgkin and Huxley *(6)* integrated this equation, and an associated set of equations, with a set of voltage- and time-dependent dimensionless parameters (*m*, *h*, and *n*), and found that the computed action potential was almost identical to the recorded action potential. Thus, the initial inward Na current is responsible for the depolarizing phase of the action potential and a delayed outward K current for repolarization (Fig. 4A). These currents on amphibian nodes can be separated pharmacologically by application of tetrodotoxin (TTX) to block Na currents and tetraethylammonium (TEA) to block K currents (Fig. 4). Although there are a number of small conductance differences between squid and amphibian currents, they are generally very similar *(4)*.

3. IONIC CHANNELS ORGANIZATION OF THE MAMMALIAN MYELINATED AXON

3.1. Na Channels

Much work indicates that Na channels are present at the node in relatively high concentration. Experiments using radiolabeled saxitoxin (STX) have estimated the relative density of Na channel at the mammalian node and internode *(8)*. In that initial binding experiment, estimates of about 10,000 Na channels/μm of nodal membrane were obtained. However, a number of corrections were made with subsequent experiments such as accounting for Schwann cell Na channels, and current data suggests that nodal Na channel density is about 1000–2000 Na channels/μm^2 of nodal membrane, and that internodal membrane has a density of about 25 Na channels/μm^2 of *(9)*. The density of Na channels is very high at the node, but the absolute number of channels at the internodal axonal membrane is of the same order of magnitude because of its extensive area. A number of other studies using immunohistochemical techniques *(10)*, freeze-fracture *(11,12)* and electrophysiological techniques *(13–15)* also indicate a high density of nodal Na current. Shrager *(16)* used a loose patch clamp technique and estimated an internodal Na channel density of 20 to 25/μm^2. The high density of voltage-gated Na channels at the node is important because synchronous activation of these channels will provide substantial current to assure the efficacy of activation of the next set of nodes and high-fidelity saltatory conduction.

Recent work indicates that some myelinated axons have kinetically and pharmacologically distinct Na channels. For example, in addition to a fast TTX-sensitive Na current, a kinetically slow voltage-dependent Na^+ channel is present on large-diameter myelinated cutaneous afferent axons; myelinated muscle afferent and motor axons do not have the slow current *(17–19)*, which indicates that a diversity of Na currents can occur on different myelinated axons. Figure 5 shows the change in waveform of intra-axonally recorded action potentials of a cutaneous and muscle afferent fiber, following application of 4-aminopyridine (4-AP). The cutaneous afferent develops a large depolarizing potential (Fig. 5A,B), referred to as the "delayed depolarization" *(20)*, but the muscle afferent shows only a slight broadening in the action potential (Fig. 5C). These recordings were obtained from young rats, in which some of the 4-AP can reach the paranodal region. Much evidence now indicates that the large cutaneous afferent neurons and their axons have a kinetically distinct and slow Na current, and it has been suggested that the delayed depolarization is generated by activation of this current *(17)*. The fact that the current is on the cell bodies and axons of cutaneous, but not muscle or motor fibers of the same general size indicates that myelinated axons can show distinct ion channel organization, depending on functional class.

Peripheral nerve injury modifies expression of regulatory genes, chemical and mechanical receptors in axons proximal to the injury site, and their neuronal cell bodies of origin, which changes functional phenotypes in cutaneous afferent neurons *(21–24)*. The kinetically slow Na^+ currents in 40–50 μm diameter DRG neurons and their peripheral processes (large diameter myelinated axons) are reduced following the peripheral nerve injury *(19,25,26)*. In addition, the refractory period of the action potential in the dorsal roots was shortened after sciatic nerve injury, even though other standard measures of axonal membrane excitability, such as resting potential and the action potential amplitude were not significantly changed. Biophysical studies of small C-type nociceptive dorsal root ganglion neurons *(27)* and large cutaneous afferent neurons *(28)* indicate that the fast TTX-sensitive Na^+ currents "reprime" faster after sciatic nerve injury. Our results showing a reduction in the refractory period of the action potential following injury, could be the result of this injury-induced change in fast Na^+ channel kinetics. Thus, changes in Na^+ currents of cutaneous afferent axons after nerve injury might influence the patterning of action potential activity. These changes in Na^+ channel properties following peripheral injury could allow inappropriate sensory signaling to the central nervous system (CNS).

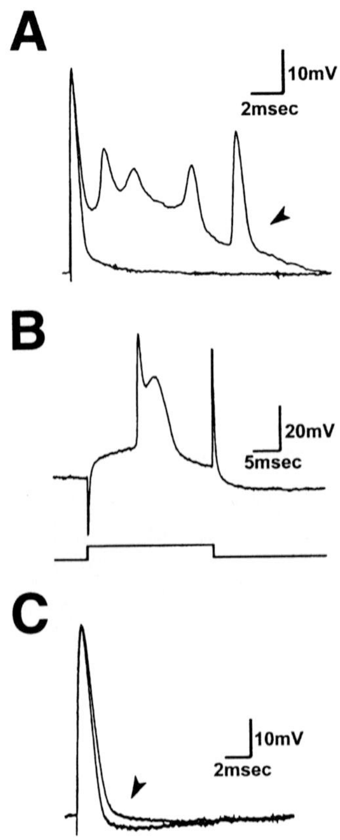

Fig. 5. Intra-axonal recordings from a myelinated cutaneous afferent axons (**A** and **B**) and a myelinated muscle afferent axon of similar conduction velocity before and after (arrows) 4-AP application. Note the distinct delayed depolarization following the action potential in the cutaneous afferent (A and B) in the presence of 4-AP. A and C were activated by nerve stimulation and B by a direct depolarizing current pulse. The muscle afferent was only minimally affected by 4-AP and did not have a delayed depolarization. (Modified with permission from ref. *69*.)

The changes in the delayed depolarization of dorsal root axons after two types of sciatic nerve injury provide clues to a possible peripheral signal that regulates these changes. Following axotomy by nerve ligation, which prevents target reinnervation and isolates the cut nerve from the distal nerve segment, the injury-induced decrement in the delayed depolarization was maximal and did not recover with time. However, the changes are considerably ameliorated if the nerve is damaged by crush, which transects the axons, but allows re-establishment of peripheral target (i.e., skin) connections, subsequent to injury. These results suggest that, in response to peripheral nerve injury of afferent neurons, a component of the peripheral target tissue assumes a significant role in regulating slow Na channel expression. Nerve growth factor (NGF) regulates the expression of TTX-resistant action potentials in cutaneous afferent neurons during development *(29,30)* and in adults *(31)*. Exogenous NGF has also been observed to influence Na current expression in vitro *(32–34)*. Molecular biological studies also support the idea that Na channel gene expression is regulated by NGF *(35,36)*. In addition, nerve injury modulates the phenotypic expression of the Na^+ channel. The expression of type III Na^+ channels increases, and sensory-neuron-specific Na^+ channel transcripts decrease following sciatic nerve transection in some DRG neurons *(37,38)*. Na^+ channel mRNA expression reverts to an embryonic pattern after axotomy, with the re-emergence of type III Na^+ channel mRNA that is expressed at high levels embryonically, but is undetectable in normal adult neurons *(35)*. Moreover, NGF has recently been shown to regulate slow Na^+ currents on axotomized large cutaneous afferent neurons *(25)*, and α-SNS Na^+ channel expression is rescued in DRG neurons after axotomy by NGF in vivo *(38,39)*. These arguments strongly suggest that there is a switch in the mode of Na^+ channel expression in at least some cutaneous afferent DRG neurons, following axotomy, and it seems likely that NGF may modulate axonal Na^+ channels.

Thus, trophic support from the peripheral target is important, to maintain the slow Na^+ channel expression in adult DRG neurons, even though most DRG neurons no longer require target-derived trophic support for survival in adults *(40)*. Distinct trophic factor support may have an important role in maintaining the slow Na^+ channel expression on the afferent neuron and axon in adult mammalians. This interaction between Schwann cells and regenerating axons emphasizes the importance of glial–axonal interactions through the microenvironment of nerve, for the appropriate maintenance of axonal ion channel organization.

3.2. K Channels

A convergence of biochemical, morphological and biophysical studies indicate that the relative distribution of Na and K channels is not homogenous along mammalian myelinated axons. As mentioned above, Na channels are clustered at the node of Ranvier. Horackova et al. *(41)* first suggested that the mammalian node of Ranvier indeed has voltage-gated Na currents, but a paucity of outward K current. This difference was confirmed and extended by Chiu et al. *(42)* in rabbit nodes and Brismar *(43)* in rat nodes. The paucity of K conductance in mammalian nodes represents a fundamental distinguishing feature of the mammalian node of Ranvier. Figure 6 illustrates a comparison of ionic currents between rabbit and frog node of Ranvier. The essential absence of outward K current is evident. Moreover, pharmacological studies on dorsal column axons in the rat indicate a paucity of K currents at the nodes of these central nervous system axons *(44)*. However, when the node is acutely disrupted while being voltage-clamped, a pronounced outward K current emerged *(14,45)*, which suggested that K channels were present at paranodal or internodal regions. When the K channel blocking agent, 4-AP is applied to a normal mature rat ventral root axon there is no change in the action potential waveform, suggesting that the node does not have K channels that contribute to repolarization (Fig. 7A). However, when 4-AP is applied to regenerating axons (Fig. 7B), before myelination is complete, and to demyelinated axons (Fig. 7C) there is a clear delay in repolarization, with broadening of the action potential *(46,47)*. From these and other experiments a model has been presented that shows the node has primarily Na channels and that K channels are present at the internode *(48)*. In this model, repolarization of the action potential has been attributed to rapid Na channel inactivation and a large leakage current *(13,41–43)*.

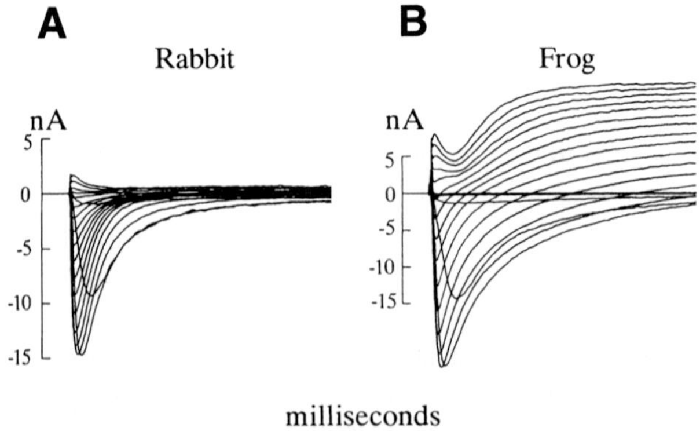

Fig. 6. Comparison of ionic currents in rabbit **(A)** and frog **(B)** nodes after leakage subtraction. A series of depolarizing pulses from a holding potential of –80 to –55 mv was applied. Note the relative absence of outward K current in the rabbit node. (Modified with permission from ref. *49*.)

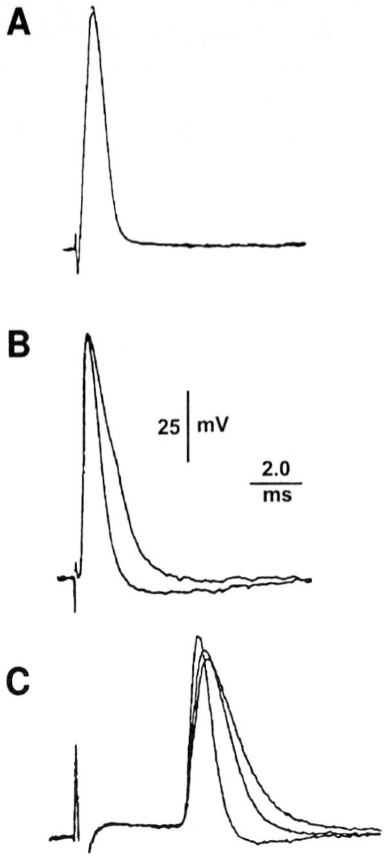

Fig. 7. Action potentials showing the effect of 4-aminopyridine on normal mature **(A)** premyelinated **(B)**, and demyelinated ventral root axons. 4-AP had virtually no effect on a normal axon, but broadened the action potential in both the immature myelinated axon and the demyelinated axon, supporting data indicating that fast 4-AP-sensitive K channels are located under the myelin.

An exciting advance over the past two decades has been the discovery that K channels on mammalian axons display pharmacologic, kinetic, and regional diversity. Two prominent classes of K channels exist on mammalian myelinated axons: a fast 4-AP-sensitive and a slower TEA-sensitive current *(49,50–52)*. While it is clear as discussed above that adult mammalian nodes have a paucity of voltage-gated K current, additional studies *(50)* indicate that the TEA-sensitive slow current has some representation at the node, and demonstrate that a considerable amount of the conductance is inactivated at resting potential. Moreover, there is a consensus that the fast K currents are virtually absent at the node, but are represented at the paranodal and internodal regions. The presence of the 4-AP-sensitive current at regions under the myelin let to studies showing that 4-AP application to demyelinated fibers could widen the action potential, thereby generating a larger current from a given action potential and overcome conduction *(53–56)*. Clinical studies have been carried out with 4-AP in patients with multiple sclerosis to examine its effectiveness as a symptomatic therapy *(57–59)*. These studies have shown an improvement in motor functions, reduction in scotoma and improved critical flicker fusion in MS patients.

3.3. *Functional Implications of Heterogeneous Channel Distribution*

Myelinated axons in both the CNS and peripheral nervous system (PNS) have a complementary distribution of Na and K channels at the node and internode, but the functional significance of this organization is not clear. An important functional consideration is the very high density of Na channels at the node, as described previously (~1000–2000/μm^2). This high density assures a high safety factor for impulse conduction. Conversely, it has been determined by electrophysiological experiments *(16,60)*, saxitoxin-binding studies *(8)* and immunohistochemical studies *(10)* that only a few Na channels are present at the paranode and internode. This paucity of Na channels suggests that it is not only the high resistance of the internodal membrane, which is conferred by myelin, that prevents action potential generation at the internodal membrane, but the low density of Na channels may also contribute. Unlike the squid axon, in which K channels are important in the repolarization of the action potential, the mammalian node, which has a paucity of K channels at the node, repolarizes from rapid Na channel inactivation and a large leak current *(42,61)*. This is in marked contrast to amphibian myelinated axons in which the delayed rectifying K current is important in repolarization.

The fast K channels, which are sensitive to 4-AP, are present at the paranodal and internodal regions of the axon. In the paranodal region, there are axo–glial junctions formed by the paranodal loops of myelin-forming cells, and these junctions may form a barrier to isolate the fast K channels at this region of the axon *(62)*. It has been suggested that the function of these paranodal fast 4-AP-sensitive currents is to prevent re-excitation of the node following the action potential *(14,46)*. However, when the paranodal region is disrupted during demyelination the fast channels are exposed *(14,53,54)* and may redistribute over broader regions of the axon (possible including the node) *(63)*. Experiments taking advantage of this reorganization of K channels after demyelination showed that external application of 4-AP can result in the overcoming of conduction block in demyelinated axons. The rationale is that the fast K currents may contribute to action potential repolarization in the demyelinated axons, and that

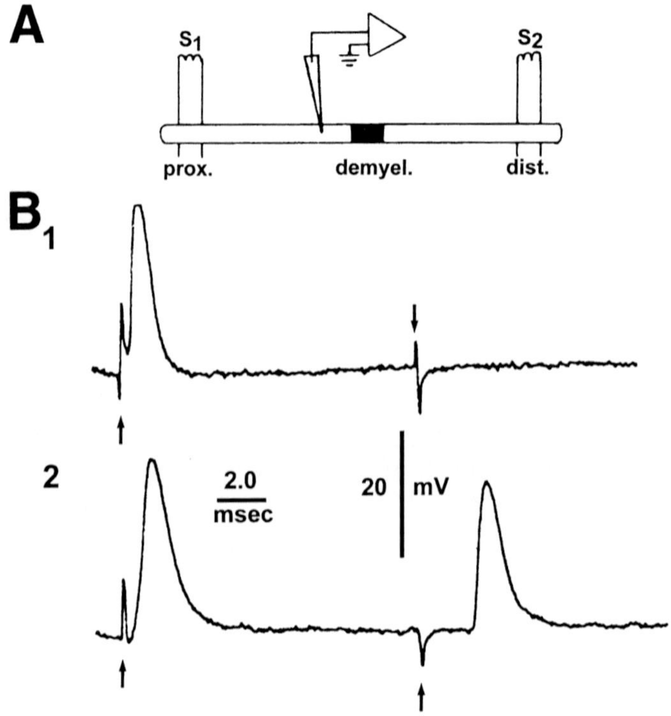

Fig. 8. A. Schematic showing a focal site of demyelination in the sciatic nerve and single axon recording and stimulation sites (S1 and S2). The single axon recording in B-1 indicates that an action potential was propagated over the nerve region proximal to the lesion site, but failed to propagate through the demyelination zone. After application of 4-AP (B-2) the action potential broadened and was able to propagate through the site of demyelination indicating the overcoming of conduction block by pharmacologically blocking fast K currents. (Adapted with permission from ref. *53*.)

blockade of these channels broadens the spike, which then generates greater inward current and increases the probability of activating the next nodal region. An example of this approach is shown in Fig. 8.

Although most investigators agree on the essential absence of fast K channels at the mammalian node, there are some data *(64)* to suggest that the node may have slow K channels (100/μm^2). The density is reduced to about 1/30 at paranodal and internodal axon membrane *(52)*. Much of this current is activated at normal resting potential, therefore it would function similar to leakage current, and could contribute to repolarization of the action potential. Chui and Ritchie *(14)* have suggested that internodal slow K channels could contribute to maintenance of resting potential. Another potentially important function of the TEA-sensitive slow K channels could be their contribution to spike frequency adaptation *(51)*. During long lasting depolarizations, myelinated axons typically generate only a single action potential or only a few. Moreover, multiple stimulus presentation results in activation of a prolonged hyperpolarization, which is blocked by TEA. Blockade of slow K channels with TEA blocks the post-tetanic hyperpolarization, and leads to repetitive firing *(49,51,65)*.

4. CONDUCTION IN DEMYELINATED AND REMYELINATED AXONS

The loss of myelin in both the CNS and PNS can result in conduction block and slowing. However, a number of observations suggest that chronically demyelinated axons can conduct impulses, albeit at a low velocity. There are a number of changes that occur on the axon membrane that may account for stable, but slowed conduction. Moreover, recent work indicates that transplantation of a variety of myelin-forming cells into the demyelinated CNS can both anatomically and electrophysiologically repair the demyelinated axon.

4.1. Demyelinated Axons

Acute demyelination is associated with conduction block and conduction slowing. This is probably the result of longitudinal current shunting, as shown in Fig. 2, and the fact that the internodal membrane has a paucity of Na channels. However, in chronically demyelinated axons, the demyelinated regions display a slow, but continuous mode of conduction *(66,67)*. Those investigators suggest that Na channels can cluster in the internodal region after demyelination. In most experimental models of demyelination, conduction can occur, but is slowed. For example, conduction is slowed, but secure in the myelin-deficient rat *(68)* and following chemically induced demyelination *(69,70)*. Figure 9 illustrates conduction, from intra-axonal recordings in the experimentally demyelinated rat spinal cord. Lesions were made by X-irradiation and ethidium bromide injections *(69)*. The demyelinated zone covered most of the central region of dorsal columns for 6–8 mm along the longitudinal axis of the spinal cord. Figure 10B shows a field of demyelinated axons in the model system and can be compared to normal dorsal column axons in Fig. 10A. The action potentials in Fig. 9B-1 are from a normal myelinated segment of the cord (S3-S4) stimulation sites. Note the short latencies of the responses indicative of rapid impulse conduction. When stimuli were presented over the demyelinated segment (S1-S2), and conduction distance was the same, a considerable increase in latency is observed (Fig. 9B-2). The reduction in conduction velocity for this fiber from its myelinated to nonmyelinated segment was from ~8 m/s to ~0.8 m/s.

Another feature of the demyelinated axons is that their conduction velocity does not significantly increase with increasing axonal size. Conduction velocity was plotted for the demyelinated segment versus the corresponding velocity for the myelinated portion of the same axon (Fig. 9D). Note, in the figure that, as velocity (and therefore axon diameter) increases for the normal segment (abscissa: open circles), from ~3.4 to 7 m/s, that the conduction velocity through the demyelinated segment (ordinate) remains stable. This suggests that the axon diameter of the demyelinated axons does not scale with conduction velocity and that all of the demyelinated axons, even with considerable variation in diameter conduction, at a similar velocity of less than 1 m/s. In addition to general slowing, the uniformity of conduction of demyelinated axons with sizes and functional properties would reduce the fidelity and integrative properties of the information conveyed by these fibers.

4.2. Remyelinated Axons

Following experimentally induced demyelination in the CNS, both oligodendrocytes and Schwann cells can spontaneously remyelinate at least a portion of the demyeli-

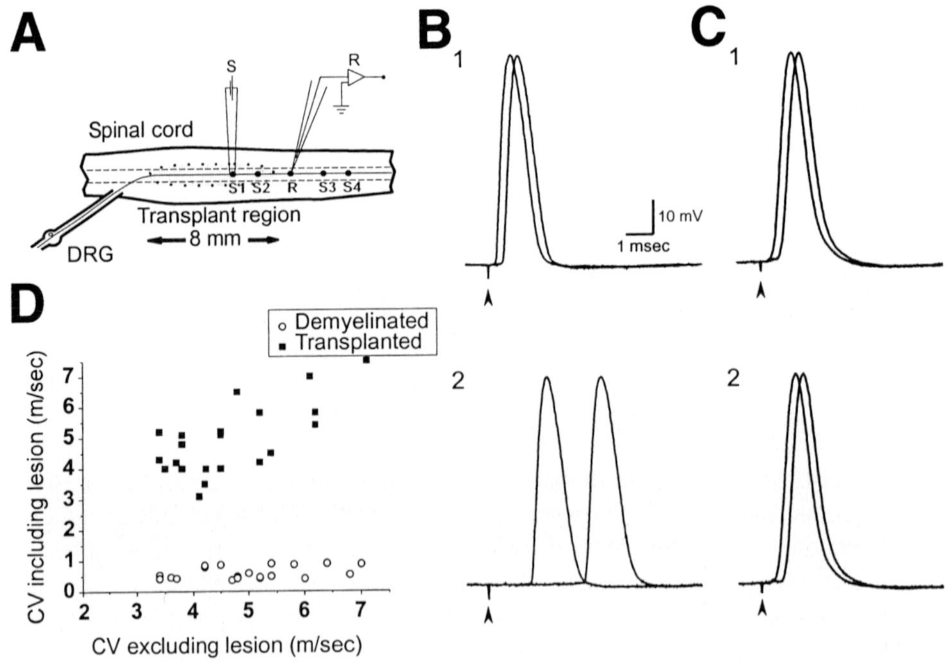

Fig. 9. (A) Schematic showing the region of demyelination in the spinal cord (dorsal columns) and the recording and stimulating sites. Intra-axonal recordings were obtained just outside of the lesion zone. Two sets of stimulating electrodes were positioned equidistant within (S1-S2) and outside (S3-S4) the lesion zone. **(B)** Intra-axonally recorded action potentials from S3-S4 **(B-1)** and S1-S2 **(B-2)** stimulation in the demyelinated condition. Note the long latency responses for conduction through the demyelination zone (B-2). **(C)** Similar recordings as in B, but following Schwann cell transplantation. Note the reduction in latency for conduction through the remyelinated zone (B-2). Pooled data in **(D)** show the improvement in conduction following cell transplantation, compared to the demyelinated condition. (Adapted with permission from ref. *69*.)

nated axons. In addition, the introduction of myelin-forming cells into the lesion can facilitate remyelination and improve conduction. Figure 10C is an electron micrograph of the demyelinated dorsal columns of the spinal cord that were transplanted with cultured Schwann cells *(69)*. The myelinated axons are typical of peripheral Schwann cell myelin. They have a large nuclear and cytoplasmic domain, a basement membrane and the spacing between the axons is large as occurs in peripheral nerve. The transplanted cells were transfected with LacZ and the Gal-X products are shown in Fig. 9D, indicating that indeed the donor cells were responsible for the remyelination.

Intra-axonal recordings (Fig. 9 C,D) indicate that conduction velocity was restored after anatomical repair of the axons by cell transplantation. Notice the short latencies of the responses recorded through the lesion zone in Fig. 9D. The pooled data in Fig. 9C indicate that conduction was increased in the remyelinated axons, and that it increased proportionately to increasing axon diameter. Thus, unlike the demyelinated fibers, which did not increase conduction velocity with increasing axon diameter, the remyelinated fibers re-established the important relationship between size and veloc-

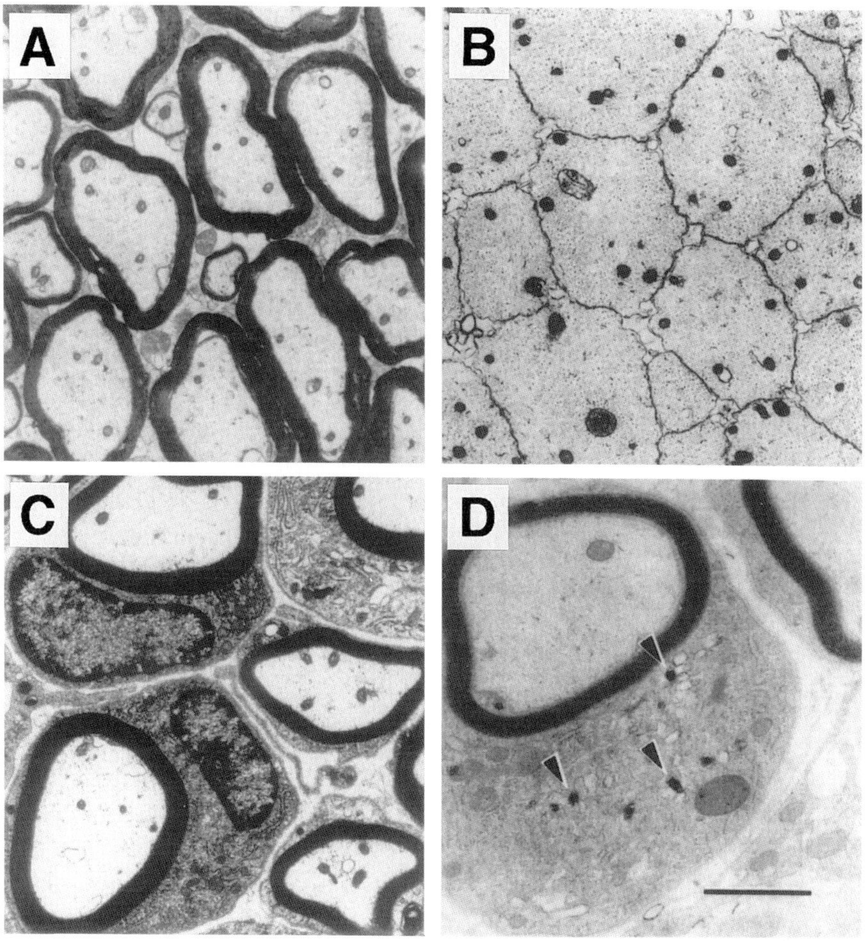

Fig. 10. Electron micrographs of normal **(A)**, demyelinated **(B)**, and Schwann cell transplant-induced remyelinated **(C)** dorsal columns. **(D)** The transplanted cells transfected with LacZ and Gal-X products. Scale bar (A–C) = 4 μm, D, 2 μm.

ity. Additionally, the axons were able to conduct impulses at frequencies similar to normal *(69)*.

A number of myelin-forming cell types can re-establish functional myelin when transplanted into the demyelinated CNS, including cryopreserved adult Schwann cells *(71)*. Advances in stem cell research and the development of human clonal neural precursor cells, derived from either embryonic or adult CNS *(72)* may allow for an abundant source of myelin-forming cells. Learish et al. *(73)* have demonstrated that fetal neural stem cells can be treated to establish self-renewing pre-O2-A progenitors. These cells form extensive oligodendrocyte myelination, when transplanted into the myelin-deficient neonatal rat. Recently, Brustle et al. *(74)* further demonstrated that human embryonic stem cell-derived glial precursors can be used as a source of myelinating transplants. Advances in the cell biology of progenitor cells, derived from embryonic, fetal, or adult CNS, opens the prospect of developing cell lines as a potential source of cell therapy for demyelinating diseases. The demonstration that cryopreserved adult human Schwann cells can remyelinate and restore conduction in the demyelinated rat spinal cord indicates the feasibility of an autologous cell therapy approach in human.

5. TRANSPLANTATION OF PERIPHERAL MYELIN-FORMING CELLS INTO TRANSECTED SPINAL CORD ENHANCES AXONAL REGENERATION

Electrophysiological and histological examination of transected dorsal support the conclusion that transplantation of either olfactory ensheathing cells (OECs) or Schwann cells is capable of facilitating long distance regeneration of functionally remyelinated axons within the spinal cord *(75,76)*. Compound action potentials (CAPs) can be detected at least 15 mm (recording chamber length) rostral to histologically complete dorsal column transections that had received transplantations of neonatal OEC or adult rat Schwann cells (Fig. 11). Conduction velocity and frequency-response characteristics of the axons extending across the lesion are indicative of myelinated axons, and histological sections reveal numerous Schwann cell-like myelinated axon profiles within the dorsal column area rostral to the lesion.

Although the electrophysiological data were consistent with functionally remyelinated axons, it is noteworthy that both OEC and Schwann cell-induced regenerated axons had conduction velocities greater than control (18.9 ± 7.2 m/s and 19 ± 9.0 m/s for OEC and Schwann cells, respectively, compared to 13.8 ± 1.0 m/s for controls). An analysis of axon size indicated that the regenerated axons were larger, suggesting that either larger axons preferentially regenerated, or that the transplanted cells induced a larger axonal size of the regenerated axons. The regenerated axons were also able to follow repetitive stimuli at frequencies up to 200 Hz, similar to control axons. Taken together, these results indicate that the regenerated axons induced by OEC or Schwann cell transplantation, allow a limited number of rapidly and securely conducting axons to extend a considerable distance across the transection site.

Although the conduction properties of regenerated axons, following either OEC or Schwann cell transplantation were similar, they displayed morphological differences. Both showed a peripheral-type myelination pattern, with large cytoplasmic and nuclear components outside of the myelin, but the myelinated axons following OEC transplantation tended to cluster in bundles, whereas myelinated axons were more dispersed following Schwann cell transplantation. When OECs are transplanted into demyelinated dorsal columns, the remyelination pattern was not clustered *(70)*. The bundling of regenerated myelinated axons which was typical in the OEC transplants, suggests that there may be some differences in the way the axons grow in the presence of OECs and Schwann cells. In spite of these differences, we found no difference in the electrophysiological properties following OEC or Schwann cell transplantation.

Li et al. *(77)* report a similar bundling of myelinated axons following transplantation of OECs in a corticospinal lesion site, but, at greater distances from the lesion zone an oligodendrocyte pattern of myelination was established. We did not morphologically characterize the myelination pattern at distant sites within the dorsal columns. However, one proposal is that the OECs and Schwann cells may serve as guide cells for the axons to extend across the lesion by providing a hospitable regenerative environment. These cells do not express inhibitory proteins that facilitate growth cone collapse, as do

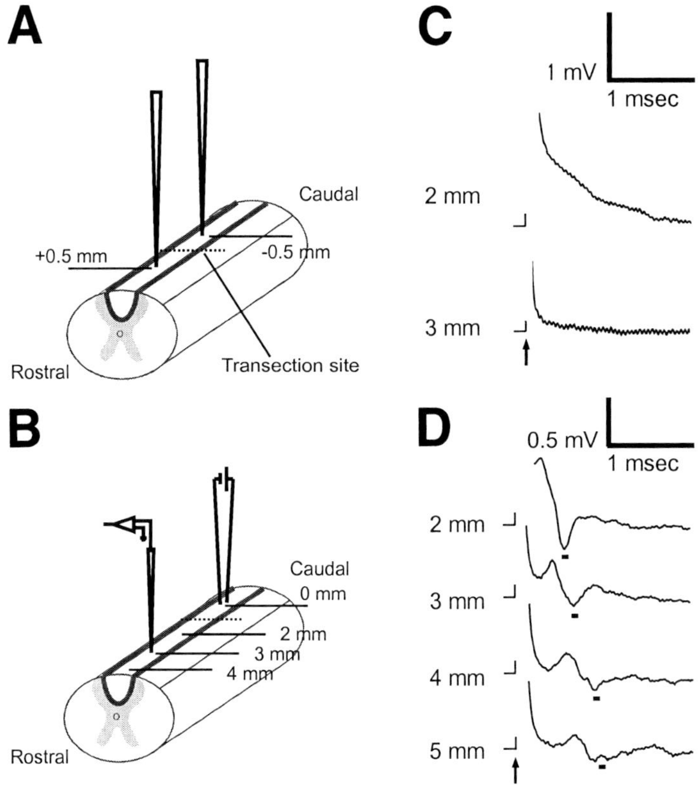

Fig. 11. (A) Schematic showing transection site in the dorsal columns of the spinal cord and position of cell injections. **(B)** Positions of stimulating and recording electrodes. Stimulating electrodes were positioned 1 mm caudal to the transection site (0 mm), and recordings were obtained at 1-mm intervals beyond the lesion site; the schematic shows recordings up to 4 mm, but recordings were obtained up to 15 mm. **(C)** CAP recordings for control lesion. At 2 mm rostral to the stimulating site (i.e., 1 mm beyond the transection site), in a transected spinal cord without cell transplantation, no clear response could be obtained. At 3 mm (2 mm beyond the transection site, no CAP could be detected; note the flat baseline following the stimulus artifact. **(D)** CAP recording for OEC transplanted lesion. In a transected spinal cord that had been transplanted with OECs, propagating CAPs could be detected several mm beyond the transection site; note the distinct negatives recorded at various distances beyond the lesion site. The peak negativities are indicated by solid squares. (Adapted with permission from ref. *75*.)

oligodendrocytes. Moreover, a number of studies have shown that CNS axons can regenerate through peripheral nerve grafts *(78)* or in an environment where glial inhibitory proteins have been neutralized by antibodies *(79)*. Transplanted OECs and Schwann cells may provide a regeneration environment free of inhibitory proteins or trophic support to encourage axonal regeneration. It is well established that both OECs *(80)* and Schwann cells *(81)* can express NGF P^{75} receptors, and Schwann cells are known to express NGF *(82)* after axotomy. Whether NGF and NGF receptor interactions or other neurotrophins contribute to the enhanced regenerative capacity of CNS axons after OEC and Schwann cell transplantation is yet to be determined.

An important concern regarding evaluating the potential effectiveness of procedures such as cell transplantation to induce axonal regeneration is to determine if the axons regenerate through the appropriate denervated host tract. Li and Raisman *(83)* point out that Schwann cell transplantation into small electrolytic lesions of the corticospinal tract, or into the ascending fibers in the fasciculus gracilis, encouraged axonal sprouting and regeneration, but did not lead to extensive longitudinal regeneration within the denervated tract. However, Li et al. *(77)* report that OEC transplantation into an electrolytically ablated corticospinal tract does result in elongative regeneration into the appropriate pathway. We found that longitudinal conduction spanned the same distance for OEC or Schwann cell transplantation. Moreover, we injected the neurons of origin whose axons extend into the dorsal columns (L4-L5) with a fluorescent marker; axons from these neurons were observed crossing the lesion and ascending in the dorsal column indicating regeneration through the appropriate tract.

The apparent discrepancy between the long distance growth of sensory axons in Schwann cell transplanted rats in this study, compared to the more localized ramification of regenerating axons reported by others *(83,84)*, may be attributable to differences in Schwann cell isolation approaches. Previous experiments in this laboratory investigating remyelination of demyelinated dorsal column axons by transplanted Schwann cells indicated that highly purified Schwann cell cultures (differential cell attachment procedures, mitotic inhibitors, and repeated replating) resulted in Schwann cell preparations that did not migrate, proliferate, or remyelinate dorsal column axons as readily as freshly isolated cells. Acutely dissociated Schwann cells used in this study may therefore have been more capable of facilitating axonal regeneration than the populations of Schwann cells used by other investigators. Since degenerated axonal membranes are believed to serve as a stimulus for inducing Schwann cell proliferation and migration *(85)*, impurities in the freshly isolated Schwann cell preparation may stimulate the transplanted Schwann cells to proliferate and interact with the transected axons.

A limiting factor, with current experimental glial cell transplantation approaches to induce axonal regeneration, is the relative paucity of regenerated axons that traverse the transection site; new methods to increase the number of regenerating axons will certainly be important. However, despite this limitation, some degree of functional improvement in forepaw reaching behavior in the rat has been reported following OEC transplantation *(77)*. Our results indicate that, although limited in number, the regenerated axons provide a secure and rapidly conducting novel information line, which can extend a considerable distance across the transected spinal cord.

5. CONCLUDING REMARKS

The association of an axon and myelin establishes an important functional unit for economy of space and rapid impulse conduction. Myelinated axons have a complex heterogeneous distribution of Na and K channels that contributes to the fidelity of impulse transmission. The very high density of Na channels at the node of Ranvier, and the paucity of these channels at the internodal axolemma, establishes a molecular basis for secure conduction; inward depolarizing current at a given node spreads passively to the next node, where an abundance of Na channels are available for activation. The insulating effect of the myelin prevents current leak across the lengthy internodal membrane, and high transmission can result, through saltatory conduction, i.e., the skipping of the action potential from node to node. It has been estimated that internodal activation time for a large myelinated fiber, with an internodal length of 1 mm can be as fast as 20 μs, which corresponds to a conduction velocity of 50 m/s. The diameter of the myelinated fiber would be ~10 μm or less, but a corresponding nomyelinated fiber would need to be as >500 μm.

The relative paucity of K channels at the node is more enigmatic, with respect to function. One suggestion is that there simply may not be enough membrane space to accommodate the high Na channel. The rapid inactivation of mammalian nodal Na channels along with leak current and some K current near resting potential is thought to be responsible for action potential repolarization. The near-absence of fast K currents at node, with their disposition at the paranode, may dampen residual depolarization and stabilize the node. This may be functionally important because the optimal stimulus for a myelinated axon at the node appears to be a rapid depolarizing voltage stimulus, such as the action potential. The relative inability of the node under normal circumstances, to fire to slow depolarizations preserves the integrative property of the myelinated axon.

Myelin can serve a number of other functions within the CNS and PNS. CNS myelin is imbued with inhibitory proteins, such as NOGO, which elicit growth cone collapse and prevent CNS axonal regeneration. It has been speculated that this inhibitory role of myelin may be developmentally linked to decussation of pathways in the CNS. Such inhibitory proteins are not present on PNS myelin, and Schwann cells may facilitate axonal regeneration both by the lack of such proteins, and by positive trophic support, such as from NGF. Indeed, transplantation of peripheral myelin-forming cells into the transected spinal cord can facilitate CNS axonal regeneration. The role of myelin on impulse conduction, and the trophic and inhibitory effects of various components of myelin-forming cells, emphasizes the dynamic interaction of axon and myelin in normal physiological function, and in developmental and plastic changes of nerve and CNS white matter.

REFERENCES

1. Stampfli, R. (1954) Saltatory conduction in nerve. *Physiol. Rev.* **34,** 10.
2. Huxley, A. F. and Stampfli, R. (1949) Evidence for saltatory conduction in peripheral myelinated nerve fibres. *J. Physiol. (Lond.)* **08,** 315–339.
3. Dodge, F. A. and Frankenhaeuser, B. (1958) Membrane currents in isolated frog nerve fibre under voltage-clamp conditions. *J. Physiol. (Lond.)* **143,** 76–90.
4. Frankenhaeuser, B. and Huxley, A. F. (1964) The action potential in the myelinated nerve fibres of *Xenopus laevis* as computed on the basis of voltage-clamp data. *J. Physiol. (Lond.)* **171,** 302–315.
5. Stampfli, R. and Hille, B. (1976) Electrophysiology of the peripheral myelinated nerve, in *Frog Neurobiology* (Llinas, R. and Precht, W., eds.), Springer-Verlag, Berlin, pp. 3–32.
6. Hodgkin, A. L and Huxley, A. F. (1952) A quantitative description of membrane current and its application to conduction and excitation in nerve. *J. Physiol. (Lond.)* **117,** 500.
7. Hille, B. (1992) Ionic Channels of Excitable Membranes, 2nd ed., Sinauer, Sunderland, MA.
8. Ritchie, J. M. and Rogart, R. B. (1977) The density of sodium channels in mammalian myelinated nerve fibers and the nature of the axonal membrane under the myelin sheath. *Proc. Natl. Acad. Sci. USA* **74,** 211–215.
9. Howe, J. R. and Ritchie, J. M. (1990) Sodium currents in Schwann cells from myelinated and nonmyelinated nerves of neonatal and adult rabbits. *J. Physiol. (Lond.)* **425,** 169–210.

10. Black, J. A., Friedman, B., Waxman, S. G., Elmer, L. W., and Angelides, K. J. (1989) Immuno-ultrastructural localization of sodium channels at nodes of Ranvier and perinodal astrocytes in rat optic nerve. *Proc. R. Soc. Lond. B* **238,** 38–37.
11. Black, J. A., Foster, R. E., and Waxman, S. G. (1982) Rat optic nerve: Freeze-fracture studies during development of myelinated axons. *Brain Res.* **250,** 1–10.
12. Rosenbluth, J. (1985) Intramembranous particle patches in myelin-deficient rat mutant. *Neurosci. Lett.* **62,** 19–24.
13. Chiu, S. Y. and Ritchie, J. M. (1980) Potassium channels in nods and internodal axonal membrane in mammalian myelinated fibers. *Nature* **284,** 170–171.
14. Chiu, S. Y. and Ritchie, J. M. (1981) Evidence for the presence of potassium channels in the paranodal region of acutely demyelinated mammalian nerve fibres. *J. Physiol. (Lond.)* **313,** 415–437.
15. Neumcke, B. and Stampfli, R. (1982) Sodium currents and sodium-current fluctuations in rat myelinated nerve fibres. *J. Physiol. (Lond.)* **329,** 163–184.
16. Shrager, P. (1989) Sodium channels in single demyelinated mammalian axons. *Brain Res.* **483,** 149–154.
17. Honmou, O., Utzschneider, D., Rizzo, M., Bowe, C. M., Waxman, S. G., and Kocsis, J. D. (1994) Delayed depolarization and slow sodium currents in cutaneous afferents. *J. Neurophysiol.* **71,** 1627–638.
18. Kocsis, J. D., Gordon, T. R., and Waxman, S. G. (1986) Mammalian optic nerve fibers display two pharmacologically distinct potassium channels. *Brain Res.* **393,** 357–361.
19. SakaI, J., Honmou, O. K, Kocsis, J. D., and Hashi, K. (1998) The delayed depolarization in rat cutaneous afferent axons is reduced following nerve transection and ligation, but not crush: implications for injury-induced axonal Na^+ channel reorganization. *Muscle Nerve* **21,** 1040–1047.
20. Kocsis, J. D., Ruiz, J. A., and Waxman, S. G. (1983) Maturation of mammalian myelinated fibers: changes in action-potential characteristics following 4-aminopyridine application. *J. Neurophysiol.* **50,** 449–463.
21. Kingery, W., Fields, D., and Kocsis, J. D. (1988) Diminished dorsal root GABA sensitivity following chronic peripheral nerve injury. *Exp. Neurol.* **100,** 478–490.
22. Himes, B. T. and Tessler, A. (1989) Death of some dorsal root ganglion neurons and plasticity of others following sciatic nerve section in adult and neonatal rats. *J. Comp. Neurol.* **284,** 215–230.
23. Kenny, A. M. and Kocsis, J. D. (1998) Peripheral axotomy induces long-term JNK activation and AP-1 binding activity by c-jun and junD in adult rat dorsal root ganglia in vivo. *J. Neurosci.* **18,** 1318–1328.
24. Titmus, M. J. and Faber, D. S. (1990) Axotomy-induced alterations in the electrophysiological characteristics of neurons. *Progr. Neurobiol.* **35,** 1–51.
25. Oyelese, A. A., Rizzo, M. A., Waxman, S. G., and Kocsis, J. D. (1997) Differential effects of NGF and BDNF on axotomy-induced changes in GABA receptor-mediated conductance and sodium currents in cutaneous afferent neurons. *J. Neurophysiol.* **78,** 31–42.
26. Rizzo, M. A., Kocsis, J. D., and Waxman, S. G. (1995) Selective loss of slow and enhancement of fast Na+ currents in cutaneous afferent dorsal root ganglion neurons following axotomy. *Neurobiol. Dis.* **2,** 87–96.
27. Cummins, T. R. and Waxman, S. G. (1997) Downregulation of tetrodotoxin-resistant sodium currents and upregulation of a rapid repriming tetrodotoxin-sensitive sodium current in small spinal sensory neurons after nerve injury. *J. Neurosci.* **17,** 3503–3514.
28. Everill, B. K, Cummins, T. R., Waxman, S. G., and Kocsis, J. D. (1999) Effects of axotomy on sodium conductance in large adult cutaneous afferent dorsal root ganglion neurons. *Soc. Neurosci. Abstr.* **25,** 730.
29. Lewin, G. R., Ritter, A. M., and Mendell, L. M. (1992) On the role of nerve growth factor in the development of myelinated nociceptors. *J. Neurosci.* **12,** 1896–1905.

30. Ritter, A. M. and Mendell, L. M. (1992) Soma membrane properties of physiologically identified sensory neurons in the rat: effects of nerve growth factor. *J. Neurophysiol.* **68,** 2033–2041.
31. Aguayo, L. G., Weight, F. F., and White, G. (1991) TTX-sensitive action potentials and excitability of adult rat sensory neurons in serum- and exogenous nerve growth factor-free medium. *Neurosci. Lett.* **121,** 88–92.
32. D'Arcangelo, G., Paradiso, K., Shepard, D., Brehm, P., Halegoua, S., and Mandel, G. (1993) Neuronal growth factor regulation of two different sodium channel types through distinct signal transduction. *J. Cell Biol.* **122,** 915–921.
33. Fanger, G. R., Jones, J. R., and Maue, R. A. (1995) A differential regulation of neuronal sodium channel expression by endogenous and exogenous tyrosine kinase receptors expressed in rat pheochromocytoma cells. *J. Neurosci.* **15,** 202–213.
34. Kalman, D., Wong, B., Horvai, E. A., Cline, M. J., and O'Lague, P. H. (1990) Nerve growth factor acts through cAMP-dependent kinase to increase the number of sodium channels in PC12 cells. *Neuron* **2,** 355–366.
35. Waxman, S. G., Kocsis, J. D., and Black, J. A. (1994) Type III sodium channel mRNA is expressed in embryonic but not adult spinal sensory neurons, and is reexpressed following axotomy. *J. Neurophysiol.* **72,** 466–470.
36. Zur, K. B., Oh, Y., Waxman, S. G., and Black, J. A. (1995) Differential up-regulation of sodium channel alpha- and beta 1-subunit mRNAs in cultured embryonic DRG neurons following exposure to NGF. *Mol. Brain Res.* **30,** 97–105.
37. Black, J. A., Langworthy, K., Hinson, A. W., Dib-hajj, S. D., and Waxman, S. G. (1997) NGF has opposing effects on Na+ channel III and SNS gene expression in spinal sensory neurons. *Neuroreport* **8,** 2331–2335.
38. Dib-hajj, S., Black, J. A., Felts, P., and Waxman, S. G. (1996) Down-regulation of transcripts for Na channel alpha-SNS in spinal sensory neurons following axotomy. *Proc. Natl. Acad. Sci. USA* **93,** 14950–14954.
39. Dib-Hajj, S. D., Black, J. A., Cummins, T. R., Kenney, A. M., Kocsis, J. D., and Waxman, S. G. (1998) Rescue of alpha-SNS sodium channel expression in small dorsal root ganglion neurons after axotomy by nerve growth factor in vivo. *J. Neurophysiol.* **79,** 2668–2676.
40. Lindsay, R. M. (1988) Nerve growth factors (NGF, BDNF) enhance axonal regeneration but are not required for survival of adult sensory neurons. *J. Neurosci.* **8,** 2394–2405.
41. Horackova, M., Nonner, W., and Stampfli, R. (1968) Action potentials and voltage-clam currents of single Ranvier nodes. *Proc. Int. Union Physiol.* **7,** 198.
42. Chiu, S. Y., Ritchie, J. M., Rogart, R. B., and Stagg, D. (1979) A quantitative description of membrane currents in rabbit myelinated nerve. *J. Physiol. (Lond.)* **292,** 149–166.
43. Brismar, T. (1979) Potential clamp analysis on myelinated nerve fibres from alloxan diabetic rats. *Acta Physiol. Scand.* **105,** 384–386.
44. Kocsis, J. D. and Waxman, S. G. (1980) Absence of potassium conductance in central myelinated axons. *Nature* **287,** 348–349.
45. Ritchie, J. M. and Chiu, S. Y. (1981) Distribution of sodium and potassium channels in mammalian myelinated nerve. *Adv. Neurol.* **31,** 329–342.
46. Kocsis, J. D., Waxman, S. G., Hildebrand, C., and Ruiz, J. A. (1982) Regenerating mammalian nerve fibers: Changes in action potential waveform and firing characteristics following blockage of potassium conductance. *Proc. Roy. Soc. B* **217,** 277–288.
47. Bowe, C. M., Kocsis, J. D., Targ, E. F., and Waxman, S. G. (1987) Physiological effects of 4-aminopyridine on demyelinated mammalian motor and sensory fibers. *Ann. Neurol.* **22,** 264–268.
48. Waxman, S. G. and Ritchie, J. M. (1985) Organization of ion channels in the myelinated nerve fiber. *Science* **228,** 1502–1507.

49. Baker, M., Bostock, H., Grafe, P., and Martius, P. (1987) Function and distribution of three types of rectifying channel in rat spinal root myelinated axons. *J. Physiol. (Lond.)* **383,** 45–67.
50. Brismar, T. and Schwarz, J. R. (1985) Potassium permeability of rat myelinated nerve fibres. *Acta Physiol. Scand.* **124,** 141–148.
51. Kocsis, J. D., eng, D. L., Gordon, T. R., and Waxman, S. G. (1987) Functional differences between 4-AP and TEA-sensitive potassium channels in mammalian axons. *Neurosci. Lett.* **75,** 193–198.
52. Roper, J. and Schwarz, J. R. (1989) Heterogeneous distribution of fast and slow potassium channels in myelinated rat nerve fibers. *J. Physiol. (Lond.)* **416,** 93–110.
53. Targ, E. F. and Kocsis, J. D. (1985) 4-Aminopyridine leads to restoration of conduction in demyelinated rat sciatic nerve. *Brain Res.* **328,** 358–361.
54. Targ, E. F. and Kocsis, J. D. (1986) Action potential characteristics of demyelinated rat sciatic nerve following application of 4-aminopyridine. *Brain Res.* **363,** 1–9.
55. Bostock, H., Sears, T. A., and Sherratt, R. M. (1981) The effects of 4-aminopyridine and tetraethylammonium ions on normal and demyelinated mammalian nerve fibers. *J. Physiol. (Lond.)* **313,** 301–315.
56. Blight, A. R. and Young, W. (1989) Central axons in injured cat spinal cord recover electrophysiological function following remyelination by Schwann cells. *J. Neurol. Sci.* **91,** 15–34.
57. Davis, F. A., Stefoski, D., and Rush, J. (1990) Orally administered 4-aminopyridine improves clinical signs in multiple sclerosis. *Ann. Neurol.* **27,** 186–192.
58. Stefoski, D., Davis, F. A., Faut, M., and Schauf, C. L. (1987) 4-Aminopyridine improves clinical signs in multiple sclerosis. *Ann. Neurol.* **21,** 71–77.
59. Jones, R. E., Heron, J. R., Foster, D. H., Snelgar, R. S., and Mason, R. J. (1983) Effects of 4-aminopyridine in patients with multiple sclerosis. *J. Neurol. Sci.* **60,** 353–362.
60. Chiu, S. Y. and Schwartz, W. (1987) Sodium and potassium currents in acutely demyelinated internodes of rabbit sciatic nerves. *J. Physiol. (Lond.)* **391,** 631–649.
61. Schwarz, J. R. and Eikhof, G. (1987) Na currents and action potentials in rat myelinated nerve fibres at 20 and 37°C. *Pflugers Arch.* **409,** 567–577.
62. Berthold, C. H., and Rydmark, M. (1995) Morphology of normal peripheral axons, in The Axon Structure, Function and Pathophysiology (Waxman, S. G., Kocsis, J. D., and Stys, P., eds.), Oxford University, New York.
63. Schwarz, J. R., Corrette, B. J., Mann, K., and Wietholter, H. (1991) Changes of ionic channel distribution in myelinated nerve fibres from rats with experimental allergic neuritis. *Neurosci. Lett.* **122,** 205–209.
64. Safronov, B. V., Kampe, K., and Vogel, W. (1993) Single voltage dependent potassium channels in rat peripheral nerve membrane. *J. Physiol. (Lond.)* **460,** 675–691.
65. Eng, D. L., Gordon, T. R., Kocsis, J. D., and Waxman, S. G. (1988) Development of 4-AP and TEA sensitivities in mammalian myelinated nerve fibers. *J. Neurophysiol.* **60,** 2168–2179.
66. Bostock, H. and Sears, T. A. (1976) Continuous conduction in demyelinated mammalian nerve fibres. *Nature* **263,** 786–787.
67. Bostock, H. and Sears, T. A. (1978) The internodal axon membrane: Electrical excitability and continuous conduction in segmental demyelination. *J. Physiol. (Lond.)* **280,** 273–301.
68. Utzschneider, D. A., Archer, D. R., Duncan, I. R., et al. (1994) Cells enhances impulse conduction in amyelinated spinal cord axons in the myelin-deficient rat. *Proc. Natl. Acad. Sci. USA* **91,** 53–57.
69. Honmou, O., Felts, P. A., Waxman, S. G., and Kocsis, J. D. (1996) Restoration of normal conduction properties in demyelinated spinal cord axons in the adult rat by transplantation of exogenous Schwann cells. *J. Neurosci.* **16,** 3199–3208.
70. Imaizumi, T., Lankford, K. L., Waxman, S. G., Greer, C. A., and Kocsis, J. D. (1998) Transplanted olfactory ensheathing cells remyelinate and enhance axonal conduction in demyelinated dorsal column of the rat spinal cord. *J. Neurosci.* **18,** 6176–6185.

71. Kohama, I., Lankford, K. L., Preininderova, J., White, F. A., Vollmer, T. L., and Kocsis, J. D. (2001) Transplantation of cryopreserved adult human Schwann cells enhances axonal conduction in demyelinated spinal cord. *J. Neurosci.* **21,** 944–950.
72. Akiyama, Y., Honmou, O., Kato, T., Uede, T., Hashi, K., and Kocsis, J. D. (2001) Transplantation of clonal neural precursor cells derived from adult human brain establishes functional peripheral myelin in the rat spinal cord. *Exp. Neurol.* **167,** 27–39.
73. Learish, R. D., Brustle, O., Zhang, S. C., and Duncan, I. D. (1999) Intraventricular transplantation of oligodendrocyte progenitors into a fetal myelin mutant results in widespread formation of myelin. *Ann Neurol.* **45,** 716–722.
74. Brustle, O., Maskos, U., and McKay, D. (1995) Host-guided migration allows targeted introduction of neurons into the embryonic brain. *Neuron* **15,** 1275–1285.
75. Imaizumi, T., Lankford, K. L., and Kocsis, J. D. (2000) Transplantation of olfactory ensheathing cells or Schwann cells restores rapid and secure conductions across the transected spinal cord. *Brain Res.* **854,** 70–78.
76. Imaizumi, T., Lankford, K. L., Burton, W. V., Foder, W. L., and Kocsis J. D. (2000) Xenotransplantation of transgenic pig olfactory ensheathing cells promotes axonal regeneration in rat spinal cord. *Nature Biotechnol.* **18,** 949–953.
77. Li, Y., Field, P. M., and Raisman, G. (1997) Repair of adult rat corticospinal tract by transplants of olfactory ensheathing cells. *Science* **277,** 200–2002.
78. David, S. and Aguayo, A. (1981) Axonal elongation into peripheral nervous system "bridges" after central nervous system injury rat. *Science* **214,** 913–933.
79. Schwab, M. E., Kapfhammer, J. P., and Bandtlow, C. E. (1993) Inhibitors of neurite growth. *Annu. Rev. Neurosci.* **16,** 565–595.
80. Ramon-Cueto, A. and Valverde, F. (1995) Olfactory bulb ensheathing glia: a unique cell type with axonal growth-promoting properties. *Glia* **14,** 163–173.
81. Taniuchi, M., Clark, H. B., Schweitzer, J. B., and Johnson, E. M. (1988) Expression of nerve growth factor receptors by Schwann cells of axotomized peripheral nerves: ultrastructural location, suppression by contact, and binding properties. *J. Neurosci.* **8,** 664–681.
82. Windebank, J. A. and Poduslo, J. F. (1986) Neuronal growth factors produced by adult peripheral nerve after injury. *Brain Res.* **385,** 197–200.
83. Li, Y. Raisman, G. (1994) Schwann cells induce sprouting in motor and sensory axons in the adult at spinal cord. *J. Neurosci.* **14,** 4050–4063.
84. Kuhlengal, K. R., Bunge, M. B., Bunge, R. P., and Burton, H. (1990) Implantation of cultured sensory neurons and Schwann cells into lesioned neonatal rat spinal cord: II. Implant characteristics and examination of corticospinal tract growth. *J. Comp. Neurol.* **293,** 73–91.
85. Salzer, J. L., Williams, A. K., Glaser, I., and Bunge, R. P. (1980) Studies of Schwann cell proliferation: II. Characterization of the stimulation and specificity of the response to a neurite membrane fraction. *J. Cell Biol.* **84,** 739–766.
86. Hildebrand, C., Kocsis, J. D., Berglund, S., and Waxman, S. G. (1985) Remodelling of internodes in regenerated rat sciatic nerve. *Brain Res.* **358,** 163–170.

9
Coupling of Blood Flow to Neuronal Excitability

Albert Gjedde

1. INTRODUCTION

At least four mechanisms underlie the link between brain function and brain energy metabolism. These mechanisms tie the function of the brain to the work carried out in the brain (the function–work couple), the work of the brain to the cells that carry out the work (energetic compartmentation), the cells that carry out the work to the relative and absolute magnitudes of oxidative and nonoxidative energy metabolism of brain (energy–metabolism couple), and the energy metabolism of the brain to its blood supply (metabolism–flow couple).

The sodium (Na) theory defined the work of the excited brain as mostly electrochemical, i.e., subserving the transport of ions against their concentration gradients to maintain these gradients. According to the Na theory, depolarized neurons accumulate extra Na and lose potassium (K), and thus stimulate the hydrolysis of adenosine triphosphate (ATP) and generation of adenosine diphosphate (ADP) by the membrane-bound adenosine triphosphatase (ATPase). The Na theory assigned this work to neurons, and explained how the generation of ADP in turn would stimulate the oxidative phosphorylation of ADP to ATP. It also suggested that the change of ions, ATP, ADP, or brain metabolites would alter the resistance of the brain vasculature, and hence increase blood flow in proportion to the need for oxidative phosphorylation.

A recent alternative theory *(1)* places the work of the excited brain under the control of neurotransmitter cycling, which includes the import and amination of the excitatory amino acid, glutamate, by astrocytes. According to the theory, the import stimulates the rate of aerobic glycolysis in astrocytes, and leads to generation of lactate, which cannot be oxidized *in situ*, but must be exported to neurons, where it undergoes oxidation to carbon dioxide (CO_2). This theory revives an ancient, now abandoned, claim that astrocytic foot processes play a nutritive role in brain: The theory claimed that the circulation supplies glucose only to astrocytes, which in turn supply lactate to neurons. It is a condition of this claim that the glycolytic and oxidative components of brain metabolism are strictly compartmentalized, aerobic glycolysis taking place only in astrocytes, in proportion to the magnitude of glutamatergic neurotransmission, and the combustion of lactate taking place only in neurons, in proportion to the rate of glycolysis in astrocytes *(2)*. The theory is based on a selective interpretation of

From: *The Neuronal Environment: Brain Homeostasis in Health and Disease*
Edited by: W. Walz © Humana Press Inc., Totowa, NJ

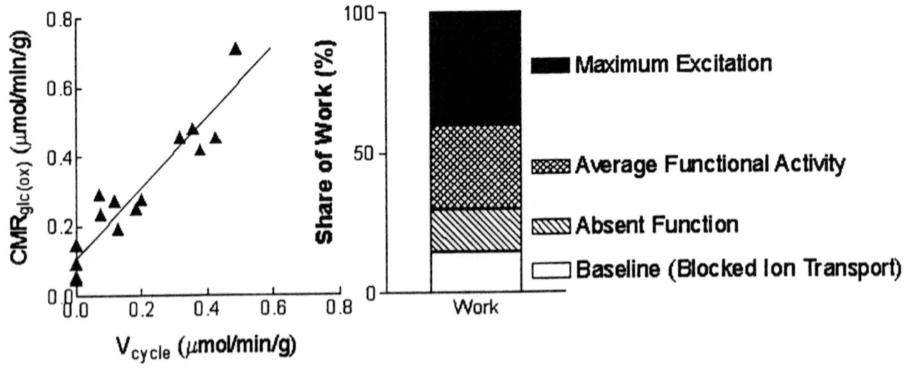

Fig. 1. Stages of brain metabolism, compared with rate of glutamate cycling and corresponding rate of brain oxidative metabolism measured by Shulman and Rothman *(1)*.

the evidence discussed below, and it is a question to what extent it is open to rejection by alternative interpretation.

2. THE BRAIN'S WORK

2.1. Work of Cognition

Energy is the measure of an ability to do work. The ability to carry out the work of information processing in the brain is linked to the concept of entropy, according to which information can be used to create order in a system. The order can be increased only by work that requires a supply of energy and leads to a loss of entropy. Information thus stimulates a loss of entropy by leading to a reduction of the number of ways in which the state of the system can be reached, i.e., by reducing the probability that the state occurred randomly. The lower this probability, the higher the information content and the lower the entropy.

Not all of the brain's work subserves the processing of information: As much as 50% of the daily average supports general cellular functions. The proportions of cellular and excitatory work in the brain were estimated in numerous studies *(3)*, ranging from electrical stimulation of brain slices to measurements of anesthesia in animals, and coma and stupor in humans. These findings were recently supplemented with magnetic resonance studies of rats at various stages of activity *(1)*. A summary of the stages of brain activation gleaned from these studies is presented in Fig. 1, which shows that the basic cellular work of brain tissue may represent as little as 10% of the maximum-achievable metabolic rate.

2.2. Energy Cost of Depolarization

There is general agreement that cognition itself is not energy-requiring, but the maintenance of the steady-state underlying cognition requires work, because cogitation depends on the alternating de- and repolarization of neurons. Depolarization is the loss of membrane potential by increased permeability of the neuronal membranes to Na ions, and repolarization is the subsequent reestablishment of the membrane potential by increased permeability of the membranes to K ions. The permeabilities allow Na ions to leak into neurons and K ions to leak out of neurons.

Table 1
Ion Concentrations in Nerve Cells

		Ion				
		Na		K		Cl
Variable	Unit	E&S	M	E&S	M	M
Equilibrium potential	mV	+41	+40	–84	–100	–75
Intracellular concentration	m*M*	27	30	80	140	8
Extracellular concentration	m*M*	133	150	3	3	130

Data from refs. *5* ("E&S"), and *6* ("M").

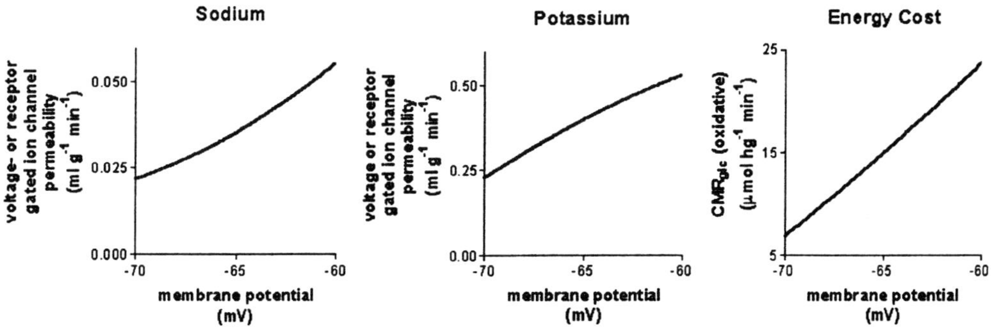

Fig. 2. Relationship between Na and K ion permeabilities, membrane potential, and corresponding requirement for oxidative glucose metabolism, calculated by Gjedde *(7)*, on condition of a constant Cl ion permeability of 0.549 mL g^{-1} min^{-1} (*see* Table 2).

2.2.1. Ion Concentrations

The extracellular (EC) and intracellular (IC) concentrations of Na and K ions, established by the leakage, stimulate the phosphorylation-type (P-type) Na^+-K^+-ATPase to exchange three IC Na ions for two EC K ions *(4)*, a discovery for which Aarhus University professor Jens C. Skou received the 1997 Nobel Prize in chemistry. The average steady-state concentrations in vivo are known with some uncertainty, as shown in Table 1.

2.2.2. Ion Permeabilities

The ion concentrations stimulate the ATPase to convert ATP to ADP, which in turn must be rephosphorylated. The membrane permeabilities of Na and K are associated with a specific ion leakage, and hence with a specific turnover of ATP, as shown in Fig. 2.

Figure 2 predicts that the ATP requirements change with the degree of membrane polarization, when the chloride (Cl) permeability is kept constant. Although there is no doubt that the ion fluxes and resulting field potentials change during the functioning of the brain, as evidenced by evoked potential recordings and electro- or magneto-encephalography, the question is whether the steady-state ATP requirements also change as a function of the potential changes. The relationship between the membrane potential and the ATP turnover was calculated by Gjedde *(7)*, with the Goldman-

Table 2
Ion Movements Across Nerve Cell Membranes

		Ion					
Variable	Unit	Na		K		Cl	
Transmembrane leakage	μmol g^{-1} min^{-1}	15	36	10	24	5	12
PS product at –65 mV	mL g^{-1} min^{-1}	0.038		0.404		0.549	
PS product at –55 mV	mL g^{-1} min^{-1}	0.044	0.082	0.285	0.617	0.246	0.549

From ref. 7, assuming 50% of ATP turnover dedicated to ion transport, calculated from the concentrations measured by McCormick *(6)*. PS is the permeability-surface-area product of the cell membranes. To estimate the Cl permeability, it was necessary to use a simplified form of the Goldman-Hodgkin-Katz constant field equation.

Hodgkin-Katz constant field equation, the steady-state 3:2 ratio between the fluxes of Na and K, and a constant Cl permeability, as listed in Table 2.

The analysis underlying Fig. 2 shows that it is possible to depolarize the membranes, without a change of the ion fluxes, if the permeabilities of all three ions are allowed to change in a coordinated manner, that of Na increasing and those of K and Cl decreasing. Table 2 lists two membrane potentials with the same fluxes of the Na, K, and Cl ions. In other words, with certain changes of the membrane permeability, neurons may undergo de- and repolarization, without actually having to change their ATP turnover, and hence without a need to change the rate of oxidative phosphorylation of ADP.

The claim that it is possible to maintain different levels of membrane polarization with no change of ATP requirement is of interest to the discussion of where and how the changes of membrane permeability are effected, and when neurotransmitter action leads to increased metabolism. The magnitude of the ion fluxes, and hence energy costs associated with the action potential changes of axonal membranes (AC potentials), is similar to the fluxes and energy costs associated with the graded potentials of pre-and postsynaptic, dendritic, and somatic membranes (DC potentials), each accounting for about one-half of the energy budget of the brain *(8)*.

2.3. *Energy Cost of Neurotransmitter Cycling*

AC and DC potential changes are effected in different ways, the AC potentials by means of voltage-gated ion channels, the DC potentials by means of receptor-gated ion channels linked to neurotransmitter receptors. The DC potentials are excitatory or inhibitory. The predominant excitatory neurotransmitter in cerebral cortex is glutamate. This excitatory amino acid is stored presynaptically in vesicles, and released to the synaptic cleft, upon adequate stimulation. From the synaptic cleft, it is reimported into neurons and astrocytes in symport with Na ions, three Na ions for each glutamate molecule.

The import leads to IC accumulation of Na ions, which must be extruded at the expense of ATP, one ATP molecule per glutamate molecule imported. If imported into astrocytes, glutamate must undergo amination to glutamine. Otherwise, the neurotransmitter could not be returned to neurons, because cell membranes are mostly impermeable to glutamate, but permeable to glutamine. The amination, by the glutamine synthase reaction with ammonium ions, occurs at the expense of yet another ATP molecule. In

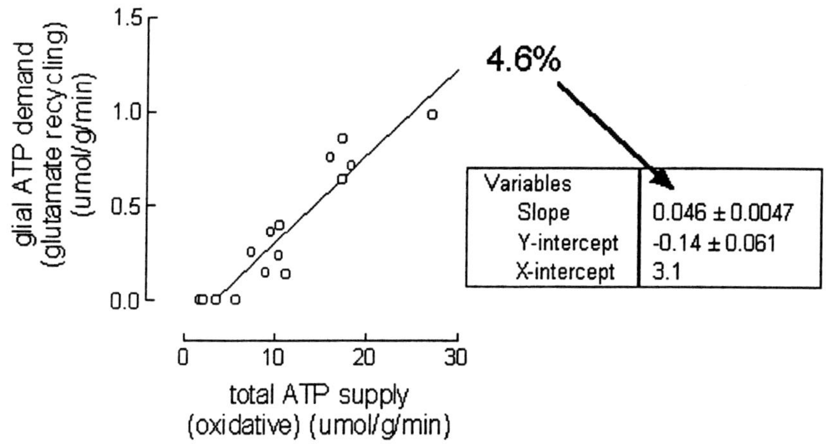

Fig. 3. Fraction of total oxidative metabolism of brain issue in vivo subserving neurotransmitter cycling of glutamate. Data replotted from ref. *1*.

addition, there are the costs of vesicular storage and release, perhaps of a magnitude similar to the reuptake and amination of the transmitter.

The study of Shulman and Rothman *(1)* indicates that the oxidative cost of the glutamate reuptake and amination is 5% of the total oxidative metabolism of the rat brain (Fig. 3). Because the need for amination only applies to uptake of glutamate into nonneuronal cells, it is a reasonable estimate that 5–10% of the total energy cost associated with excitation subserves the cost of excitatory neurotransmitter cycling; the remaining 90–95% subserves the cost of depolarization.

3. METABOLIC COMPARTMENTATION

3.1. Large and Small Compartments

The two major cell populations of the brain tissue are the neurons and the astrocytes, the latter extending their processes from the perivascular space to the space immediately adjacent to the synaptic cleft. The exact role of the astrocytes in the metabolism and function of the brain has remained a mystery, inviting much speculation about the significance of the peculiar connection between the brain's microvessels and the synapses, its functional elements. One such speculation *(9)* postulated that astrocytes feed neuron terminals by extracting a steady stream of nutrients directly from the microvessels through the foot processes and delivering these nutrients to the immediate vicinity of the terminals. After lingering unconvincingly for decades, this postulate was finally and definitively rejected when Brightman and Reese *(10)* proved that the foot processes form no part of the blood–brain barrier (BBB), even to the largest macromolecules, and thus could not possibly extract nutrients directly from the capillaries *(11)*. Newman and Paulson *(12)* speculated that astrocytes could carry out the reverse function of siphoning K ions from the extrasynaptic to the perivascular space. They reasoned that the inside route through the foot processes would permit the K to arrive at the perivascular space with a much lower time constant (66 ms) than if it were left to diffuse through the much more voluminous extracellular space (2.5 s).

Traditionally, the concept of compartmentation refers to the exchange between glutamate and glutmine in large and small compartments of brain tissue *(13,14)*, now generally believed to represent neurons and glial cells, respectively. The compartmentation ensures that glutamate is converted to glutamine in astrocytes. The intriguing question is whether the concept of separation of reactions can be extended to a host of other processes, including some involved in the linking of brain function and metabolism.

In addition to the specific enzymes and transporters involved in the glutamate–glutamine compartmentation, a number of other enzyme and transporter subtypes are differentially expressed in neurons and astrocytes. The most important to the present discussion include the glucose transporters (GLUT3 vs 45 kDa GLUT1) *(15)*, monocarboxylic acid transporters (MCT-1 vs MCT-2) *(16,17)*, excitatory amino acid transporters (EAAT-2 vs EAAT-3 in humans) *(18)*, EAAC-1 vs GLAST and GLT-1 in rats *(19)*, and lactate dehydrogenases (LDHs) and their associated mRNA (LD_1 vs LD_5) *(20–22)*.

3.2. Compartmentation of Metabolism

Recent measurements indicate that about 30% of the total glucose consumption occurs in glial cells, of which one-sixth (5% of the tissue total) is converted to lactate; neurons are the sites of 70% of the glucose consumption, of which only one-fourteenth (also 5% of the total) is converted to lactate *(23,24)*. The capacity of oxidative metabolism is now believed to reflect the habitual level of energy turnover of the cells, averaged over longer periods of time, and is thought to be unchangeable, except by sustained stimulation *(25)*.

Oxidative phosphorylation proceeds from pyruvate in near-equilibrium with lactate. If neurons are responsible for 72% (65/90) of this process, the issue of metabolic compartmentation is therefore a question of the origin of the pyruvate that is oxidized in neuronal mitochondria: Which are the relative contributions of neuronal and glial glycolysis to neuronal oxidative phosphorylation? Given the magnitudes of the oxidative and nonoxidative metabolic rates in the two populations of cells, the estimate of the glial origin of the pyruvate oxidized in neurons is 3% (2/65). This percentage may rise during excitation of nervous tissue, of course, depending on the degree of differential activation of the two populations of cells.

Oxidative metabolism is 19 times more efficient than nonoxidative metabolism. For this reason, it is an intriguing coincidence that the fraction of oxidative brain metabolism in the service of neurotransmitter cycling in glia (4.6%) is not unlike the fraction of efficiency ascribed to nonoxidative metabolism (5.3%). Thus, if the fueling of neurotransmitter cycling in glia were entirely nonoxidative, and neuronal metabolism entirely oxidative, glycolysis and subsequent oxidative phosphorylation could occur only in response to neurotransmitter cycling.

This coincidence has led to the claim that all glucose supplied to the brain undergoes glycolysis to lactate in astrocytes, which in turn supply all their lactate to neurons, which themselves take up no glucose *(2)*. The metabolism of neurons would be entirely oxidative, because all glycolysis would take place in astrocytes, which would have to feed the neurons with all the pyruvate they need.

This is an extension of the early claim that astrocytic endfeet serve to siphon glucose from the circulation. The claim was ultimately abandoned when the BBB was shown not to include the endfeet. However, there is little direct evidence in favor of the basic oxidativeness of metabolism being substantially different in neurons and glia, nor of a substantial net transfer of lactate to neurons in vivo. A difference could be imposed by excitation, during which the firing frequency may require commensurate time constants of removal of neurotransmitter, which may outstrip the habitually established capacity of the oxidative phosphorylation in glial cells.

3.3. *Lactate Dehydrogenase*

The equilibrium lactate–pyruvate ratio depends on the affinities of the prevailing isozymes of lactate dehydrogenase (LDH) (LD_1–LD_5) toward its two substrates, as influenced by the NAD^+: NADH ratio and the pH, because the K_m of LDH for pyruvate varies with the subtype, the aerobic heart form LD_1 (H_4) having the lowest K_m^{lact}/K_m^{pyr} ratio and the anaerobic muscle and liver form LD_5 (M_4) having the highest K_m^{lact}/K_m^{pyr} ratio *(26)*.

The brain has both the H_4 (LD_1), which causes pyruvate to rise quickly for a given increase of glycolysis, and the M_4 (LD_5), which causes pyruvate to rise more slowly *(20–22)*. Thus, the higher the ratio between LDH's affinities (i.e., the lower the ratio between the Michaelis constants) for lactate and pyruvate, the more rapid the approach to a new steady state, with a time constant of $[1 + (K_m^{lact}/K_m^{pyr})] / [\Sigma T_{max}/K_t]$ (*see* Eq. 3). LDH is also present in mitochondria, apparently with the same distribution of subtypes as in the cytosol, although additional subtypes also have been identified *(27)*.

3.4. *Transporters*

3.4.1. *Excitatory Amino Acid Transporters*

Astrocytes remove glutamate from excitatory synaptic clefts, where the astrocytic processes engulf the synapses, in humans, by means of the glutamate transporters, EAAT-1-5 *(18)*. glutamate transporters reside also on neurons, but recent gene knockout studies suggest that the glial EAAT-3 is indispensable for normal brain function, unlike the neuronal EAAT-2 *(28)*. Knockout of the transporter gene, or blockade of the transport, eliminates the increase of glucose phosphorylation in the tissue caused by simple somatosensory stimulation of the rat whisker barrels in vivo, although it is not known in which tissue compartment or population of cells the inhibition occurs *(29)*.

3.4.2. *Monocarboxylic Acid Transporters*

The near-equilibrium between pyruvate and lactate and the abundance of monocarboxylate transporters (MCTs) ensures rapid and reversible exchange between the cellular and EC pools of pyruvate and lactate. Pyruvate and lactate cross the membranes of brain tissue by means of facilitative proton-dependent transport catalyzed by the MCT family of membrane- spanning proteins *(30–33)*. In brain tissue, the important transporters appear to be MCT-1 and MCT-2. The MCT-1 protein spans the membranes of the capillary endothelium. Gerhart et al. *(16,17)* claim that MCT-1 also resides on neurons; the MCT-2 protein belongs to astrocytes, apparently particularly their foot processes, in disagreement with Broer et al. *(34)*, who originally assigned MCT-1 to astrocytes and MCT-2 to neurons. As proton symport, the transport is influenced by the pH of the cells. The MCT-2 is of higher affinity than the MCT-1, indicating that it is saturated by pyruvate and lactate at much lower concentrations than the MCT-1. For this reason, the MCT-2 may be more efficient at transporting pyruvate than MCT-1.

3.4.3. *Glucose Transporters*

Glucose is the source of pyruvate, and enters brain tissue, neurons, and astrocytes by means of facilitative insulin and Na insensitive transport by several members of the GLUT family of membrane-spanning proteins *(35)*. In brain, the important members

Table 3
Average Physiological Variables of Human Cerebral Cortex

Somatosensory cortex	Average	Ref.
J_{glc} (μmol g^{-1} min^{-1})	0.30	*(41)*
J_{O2} (μmol g^{-1} min^{-1})	1.60	*(41)*
CBF (mL g^{-1} min^{-1})	0.45	*(41)*
ATP turnover (μmol g^{-1} min^{-1})	10	Eq. 1
Lactate flux (μmol g^{-1} min^{-1})	0.05	Eq. 2

are GLUT1 and GLUT3 *(15)*. The 55-kDa GLUT1 resides in the membranes of the capillary endothelium; the slightly modified 45-kDa GLUT1 resides in the membranes of astrocytes and choroid plexus. The GLUT3 protein resides in the membranes of neurons. The exchange of glucose among the issue compartments of the brain is essentially equilibrative *(36)*, ensuring that the glucose concentration everywhere in brain tissue is the same substantial fraction of the plasma glucose *(37,38)*. The significance of the difference between the subtypes of glucose transporters in neurons and glia is not known.

4. OXIDATIVE AND NONOXIDATIVE METABOLISM

4.1. Average Cerebral Metabolic Rate in Awake Humans

Kety *(39)* and Lassen *(40)* first measured the magnitudes of resting brain energy metabolism and blood flow in human brain. More recently determined resting or average steady-state values of energy metabolism and blood flow of the human brain are listed in Table 3, together with the steady-state turnover rates of ATP and lactate, calculated from the stoichiometric relationships,

$$J_{ATP} = 2\,J_{glc} + 6\,J_{O2} \tag{1}$$

and

$$J_{lact} = 2\,J_{glc} - \frac{1}{3}\,J_{O2} \tag{2}$$

where J_{ATP} is the ATP production, J_{glc} the glucose consumption, and J_{lact} the lact production. Blood flow and metabolic rates were all measured by positron emission tomography; blood flow by iv bolus injection of [^{15}O]water, according to the method of Ohta et al. *(42)*; oxygen consumption by single-breath inhalation of [^{15}O]O_2, according to another method of Ohta et al. *(43)*; and glucose consumption by iv bolus injection of [^{18}F]fluorodeoxyglucose, according to the method of Kuwabara et al. *(44)*.

In the baseline, total glucose consumption of cerebral cortex is ~30 μmol hg^{-1} min^{-1}. The 10% nonoxidative metabolism leads to a lactate production of 5 μmol hg^{-1} min^{-1}, of which 50% is generated in neurons and 50% in glia, according to the oxidative efficiency of the two populations of cells. The lactate flux is about 25% of the T_{max} of the BBB MCT-1, consistent with a tissue lactate concentration of 1 m*M*, slightly higher than measured with magnetic resonance spectroscopy.

Table 4
Neuronal Activation of Brain Metabolism

Stimulus	Duration (min)	Supply: ΔCBF	Supply: ΔJ_{glc} (%)	Supply: ΔJ_{O2}	Products: ΔJ_{ATP} (μmol g^{-1} min^{-1})	Products: ΔJ_{lact} (μmol g^{-1} min^{-1})
Primary						
Somatosensory	1[a]	28	[17]	9*	0.96*	0.05
	1[b]	31	[18]	13*	1.35*	0.03
	1[c]	18	[8]	0	0.04	0.04
	20[d]	18	8	[0]	0.04	0.04
	20[c]	18	[8]	0	0.04	0.04
	45[e]	27	17	[0]	0.10	0.10
Visual (photic)	30[f]	43	27	0	0.15	0.15
	45[g]	49	51	5*	0.76*	0.26
Secondary						
Visual (checkerboard)	5[h]	25	[28]	28	2.83	0.008
	10[h]	26	[29]	29	2.93	0.009
	4[i] (1 Hz)	32	[10]	10	1.07	0.004
	4[i] (4 Hz)	38	[16]	16	1.71	0.006
	4[i] (8 Hz)	42	[6]	6	0.64	0.002
Thalamic stimulation	8[j]	88	[47]	47	5.02	0.019
Internal visualization	1[k]	31	[37]	37	3.95	0.015
Tactile learning	1[l]	23	—	—	—	—
Motor						
Hand grip	8[m]	30	[40]	40	4.27	0.016
Sequential finger touching	4.5[n] (1.5 Hz, M1)	22	[8]	8	1.14	0.002
	7[o] (3 Hz, M1)	15	[8]	8	0.91	0.002
	7[o] (3 Hz, ant.cing.)	10	[13]	13	1.48	0.004
	7[o] (3 Hz, putamen)	0	[16]	16	1.82	0.005

From [a]Fox & Raichle *(45)*, [b]Seitz & Roland *(46)*, [c]Fujita et al. *(47)*, [d]Kuwabara et al. *(41)*, [e]Ginsberg et al. *(48)*, [f]Ribeiro et al. *(49)*, [g]Fox et al. *(50)*, [h]Marrett & Gjedde *(51)*, [i]Vafaee & Gjedde *(52)*, [j]Katayama *(53)*, [k]Roland et al. *(54)*, [l]Roland et al. *(55)*, [m]Raichle et al. *(56)*, [n]Iida et al. *(57)*, [o]Vafaee & Gjedde *(58)*; values in brackets are estimates; *J_{O2} increase not significant.

4.2. Cerebral Metabolic Rate During Brain Activation

4.2.1. Glycolysis

Recent measurements of oxygen consumption during simple primary somatosensory stimulation of human cerebral cortex, summarized in Table 4, generally show little or no change of oxygen consumption of the human brain *(45–47,50,59)*. For example, with the single-inhalation method of measuring oxygen consumption, changes of blood flow and oxygen consumption were compared during 30 min of vibrotactile stimulation of one hand's fingers. In primary sensory cortex, the blood flow change was 18% both at the onset of stimulation, and still 11% after 20 min of stimulation, but the oxygen consumption failed to increase for as long as 30 min *(47)*. Yet, increases of as much as 50% of the rate of glucose phosphorylation were measured during the primary somatosensory stimulations listed in Table 2. Because many of these studies were complicated by the long circulation of tracer fluorodeoxyglucose required to determine glucose consumption accurately *(60)*, Kuwabara et al. *(41)* and Ribeiro et al. *(49)* shortened the method from 45 to 20 min, with the same result *(61)*.

When not coupled to oxidative phosphorylation, even tiny increases of energy demand must of course be accompanied by substantial increases of the glucose supply *(48)*. When maximally stimulated, the rate of pyruvate generation can rise to 3–4 μmol g^{-1} min^{-1} *(62,63)*, which is several fold the calculated maximum velocity (V_{max}) of pyruvate oxidation, and close to the calculated T_{max} of the mMCT symporter in mitochondria.

In the absence of an increase of oxidative phosphorylation, a 50% increase of the glucose phosphorylation rate causes the lactate generation to rise to as much as 35 μmol hg^{-1} min^{-1}, and the fraction of nonoxidative metabolism to rise from the 10% baseline to as much as 50%, although the total ATP flux rises by a mere 5%. When the rate of generation of lactate exceeds the T_{max} of the BBB MCT-1, lactate concentration continues to rise, and the concentration of pyruvate rises with it, until the transport into mitochondria by the mMCT and the rate of the reaction catalyzed by the pyruvate dehydrogenase complex match the rate of generation.

The reason for the failure of the oxygen consumption to increase is not known, but it has long been surmised that oxidative phosphorylation cannot match the sevenfold increase of pyruvate production seen under the most extreme circumstances of glycolytic stimulation of the mammalian brain *(64)*. The time constant for the LDH reaction is on the order of milliseconds, depending on the LDH subtype; the time constant of the pyruvate symporter is on the order of seconds. From the presence of the LDH subtypes, LD_1 and LD_5, it is predicted that the increases of pyruvate and lactate would occur with a time constant that is lower for LD_1 than for LD_5.

Given the experimentally observed increases of lactate, it is possible to determine the steady-state lactate–pyruvate concentration ratio by regression of the following equation to the data,

$$\Delta C_{\text{lact}}(t) = 2\left[\frac{K_{\text{m}}^{\text{lact}}}{K_{\text{m}}^{\text{pyr}}}\right]\frac{\Delta J_{\text{glc}}}{\left[\sum \frac{T_{\max}}{K_{\text{t}}}\right]}\left(1 - e^{-t\left[\sum \frac{T_{\max}}{K_{\text{t}}}\right]\Big/\left[1+\left[\frac{K_{\text{m}}^{\text{lact}}}{K_{\text{m}}^{\text{pyr}}}\right]\right]}\right) \tag{3}$$

where ΔJ_{glc} represents the increase of glycolysis, $\Sigma T_{max}/K_t$ the sum of the clearances of pyruvate into the mitochondrial matrix and across all membranes, including the endothelium (assuming $\overline{C}_{pyr}$ always to be small relative to K_t), and K_m^{lact}/K_m^{pyr} the equilibrium lactate-pyruvate ratio. The equilibrium $\Sigma T_{max}/K_t$ ratio can be calculated from the ratio between the average net rate of glucose consumption and the steady-state pyruvate concentration, so that $\Sigma T_{max}/K_t = 2\ J_{glc}/\overline{C}_{pyr}$. With a steady-state pyruvate concentration of 0.1 μmol g^{-1}, the average $\Sigma T_{max}/K_t$ ratio is 6 min^{-1}. Shram et al. *(65)* used lact-sensitive electrodes to measure the increase of lactate shown in Fig. 4, as it occurred after application of the glutamate receptor agonist N-methyl-D-aspartate to brain tissue in vivo. The change is consistent with a steady-state lactate–pyruvate ratio of 55, and thus with the LDH subtype LD_5.

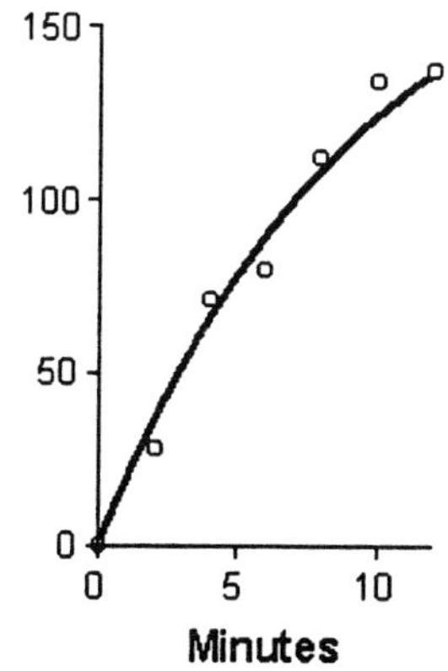

Fig. 4. Estimation of rate constant of lactate increase (% of baseline) calculated from measurements reported by Shram et al. *(65)*. The rate constant of 0.11 min^{-1} is consistent with a lactate–pyruvate concentration ratio of 55 (6/0.11), and hence with the glycolytic (or "white") subtype of LDH.

4.2.2. Oxidative Phosphorylation

Contrary to the results of no change of O_2 consumption upon simple primary somatosensory stimulation, both motor stimulation and more complex stimulation of visual cortex with a reversing checkerboard pattern for 5 or 10 min, caused significant increases of oxygen consumption *(51,52,56,66,67)*. The reported values of J_{O2} and the calculated ATP flux are listed in Table 4.

The reason for the slow initial rise of oxygen consumption may be the slow rise of the pyruvate concentration in the presence of a substantial lactate sink, when the bulk of the acceleration of glycolysis occurs in astrocytes. If pyruvate transport were indeed rate-limiting in state-3 activation *(68)*, the consumption of oxygen would depend on the cytosolic pyruvate concentration, and hence on the rate of glycolysis as described by the equation,

$$\Delta J_{O2}\,(t) = 5.6\;\Delta J_{glc}\,(1 - e^{-k\,t}) \tag{4}$$

where k is the rate constant of the pyruvate and lactate accumulations shown in Eq. 3 above. The predicted time-course of the increase of oxygen consumption is shown in Fig. 5, left panel, according to which the half-life of change is 2 min for a lactate–pyruvate ratio of 15 (LD_1, $k = 0.375$), but as much as 15 min for a ratio of 100 (LD_5, $k = 0.06$).

Table 4 identifies two kinds of oxidative responses, both relative to baseline: the primary somatosensory response, in which the rise of oxygen consumption averages only 10% of the rise of blood flow, and the motor and secondary somatosensory response, in which the rise of oxygen consumption averages 75% of the rise of blood flow. The average primary somatosensory response is a 3% increase of cerebral metabolism rate of O_2 for a 29% increase of cerebral blood flow; the average motor and secondary somatosensory response is a 27% increase of cerebral metabolism rate of O_2 for a 39% increase of cerebral blood flow (*see* Fig. 6).

From the oxidativeness of the complex secondary somatosensory and motor responses, it is to be expected that the rise of oxygen consumption after an acute stimu-

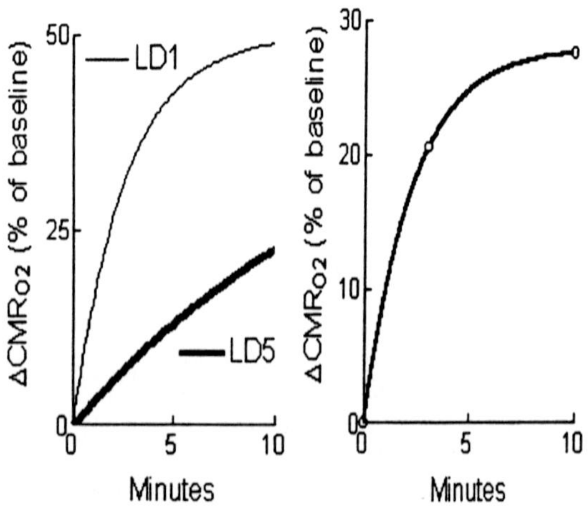

Fig. 5. (Left panel) Estimation of rate of increase of oxidative metabolism for two different rate constants dictated by different subtypes of LDH, according to Eq. 4. LD_1 subtype was modeled with rate constant $k = 0.375\ min^{-1}$, LD_5 subtype with rate constant $k = 0.050/min$. **(Right panel)** Estimation of rate constant of increase of oxidative metabolism calculated from measurements reported by Marrett and Gjedde *(51)*. The rate constant of $0.44\ min^{-1}$ is consistent with a lactate–pyruvate ratio of 13 (6/0.44), and hence with the "aerobic" (or "red") subtype of LDH (subtype LD_1).

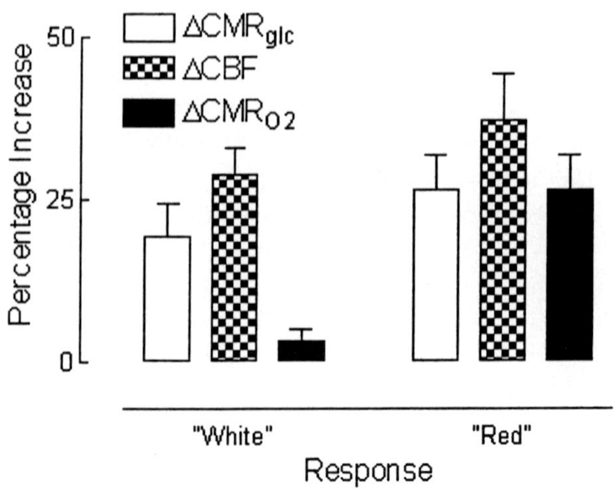

Fig. 6. Relative increases of glycolysis, blood flow, and oxidative metabolism, estimated for the two categories of stimulation (simple primary somatosensory, "white"; and complex secondary somatosensory or motor, "red") listed in Table 4.

lation of glycolysis would reflect a rise of pyruvate concentration indicative of LD_1, the subtype of LDH claimed to be characteristic of tissues with a comparatively high ratio of oxidative to glycolytic metabolism. Marrett and Gjedde *(51)* determined the time-course of increase of oxygen consumption shown in Fig. 5, right panel: Upon complex checkerboard stimulation of the visual system, the oxygen consumption rose at a rate of 0.44 min^{-1}, which is consistent with a steady-state lactate–pyruvate ratio of 13, and hence with the presence of LD_1 in the cells so stimulated.

5. BLOOD FLOW REGULATION

5.1. Putative Blood Flow Regulators

It is puzzling that oxygen consumption sometimes does not rise at all upon stimulation of brain tissue (primary somatosensory stimulation, e.g., vibrotactile stimulation), despite an increase of blood flow, suggesting either that additional factors prevent neurons from using the available O_2, or that the blood flow rises to satisfy other needs than a demand for O_2. Of course, blood flow supplies substances other than O_2, including glucose and amino acids, and removes substances such as water, lactate, pyruvate, hydrogen ions, and CO_2. The mechanism responsible for the change of blood flow may be understood only in relation to the role of the blood flow increase during excitation.

The regulation of blood flow in response to neuronal excitation is poorly understood, despite a century of investigation. The main issue, raised above, is the unknown nature of the regulator, which ties neuronal events to the magnitude of blood flow. Classically, the main contenders are the hydrogen and K ions, adenosine, CO_2 and nitric oxide (NO), and O_2 itself, of which K ions, adenosine, and NO are on the current short list of the most intensely studied molecules.

NO is involved in blood flow increases elicited by parallel and climbing fiber stimulation of cerebellar Purkinje cells *(69,70)*, but additional factors also play a role: adenosine during the climbing fiber stimulation, and K during the parallel fiber stimulation. The increase of blood flow in the cerebellar cortex was correlated closely to the summed EC field potentials elicited in the vicinity of the Purkinje cells, rather than to the efferent simple spike frequency of the Purkinje cells. The field potentials represent all current flows in the vicinity of synapses, generated by excitatory and inhibitory stimuli, and thus persist even when the efferent simple spike activity is fully inhibited *(71)*. However, the summed field potentials and associated flow changes can be eliminated completely by blockade of glutamatergic α-amino-3-hydroxy-5-methyl-4-isoazolepropionate receptors, suggesting that the primary stimulus for the BF increase is increased glutamatergic transmission, which subsequently may activate nitric oxide synthase (NOS) and Na–K ATPase in astrocytes, generating NO and accumulating K in these cells. The accumulated K subsequently is redistributed to other parts of the extracellular space, perhaps in the vicinity of microvessels *(12)*, where it may or may not cause vasodilatation.

5.2. Blood Flow in Service of O_2 Delivery and Water Removal

Recent findings *(3,72)* have revealed that the regulation of blood flow has important consequences for the delivery of O_2 for oxidative metabolism. Novel hypotheses claim that blood flow must rise to supply more O_2 during excitation of brain tissue, to maintain a constant mitochondrial O_2 tension in brain tissue. At a constant mitochondrial O_2 tension, oxygen consumption depends solely on the mean capillary O_2 tension for a given capillary density. Sudden changes of brain function must then be subserved by changes of blood flow that adjust the mean capillary O_2 tension upward. O_2 extraction fraction declines with higher blood flow rates, thus establishing a higher average capillary partial pressure of O_2 and a higher oxygen saturation of hemoglobin.

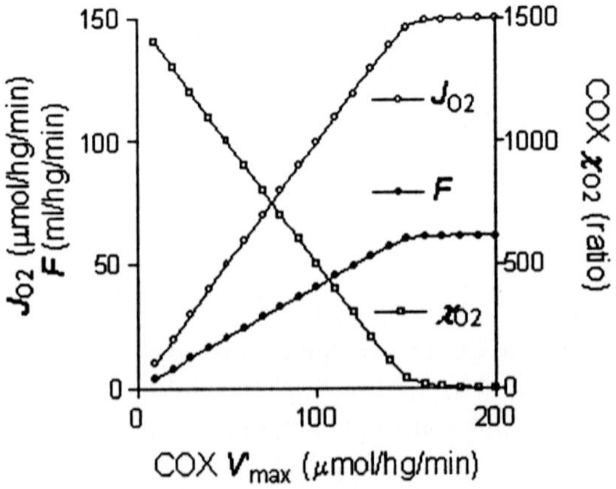

Fig. 7. O_2 diffusion capacity ceiling of oxidative metabolism in brain, calculated from Eqs. 5, 6, and 12. Abscissa is cytochrome oxidase activity ($V_{max} = \sigma_e V_{max}$), ordinates are (**left**) oxygen consumption and blood flow a constant ratio, and (**right**) mitochondrial O_2 tension relative to cytochrome oxidase half-saturation tension (χ_{O2}). The following values were used to solve Eq. 12: $J_{O_2}^{max} = 150$ µmol hg^{-1} min^{-1}, and $J_{50}^{cytox} = 150.1$ µmol hg^{-1} min^{-1}.

5.2.1. Michaelis–Menten Nature of Cytochrome Oxidase Reaction

The rate of oxygen consumption can be expressed as a simple Michaelis–Menten relationship between the mitochondrial O_2 tension and the kinetic properties of the cytochrome oxidase,

$$J_{O_2} = V_{max}^{cytox} \sigma_e \sigma_{O_2} \tag{5}$$

where J_{O2} is the net oxygen consumption, V_{max}^{cytox} is the maximum cytochrome *c* oxide reaction rate, and σ_e and σ_{O2} are the cytochrome *c* oxide saturation fractions by electrons and O_2, respectively. The O_2 tension in mitochondria (P_{O2}^{mit}) can then be derived as the tension consistent with the prevailing oxygen consumption rate and the half-saturation tension (P_{50}^{cytox}),

$$P_{O2}^{mit} = \frac{P_{50}^{cytox} J_{O2}}{\left(\sigma_e V_{max}^{cytox}\right) - J_{O2}} \tag{6}$$

which shows that the mitochondrial O_2 tension drops when the effective enzyme activity increases (assuming constant oxygen consumption). Figure 7 shows that the rate of oxygen consumption fails to rise above a certain threshold, despite further increases of the cytochrome oxide activity. The threshold is dictated by the mitochondrial O_2 tension, and is reached when the tension declines below the level associated with sufficient O_2 saturation of the cytochrome oxide. Only increases of the O_2 diffusion capacity (recruitment) or the mean capillary O_2 tension (increased blood flow) allow the rate of oxygen consumption to rise above this threshold.

5.2.2. Diffusion-Limited O_2 Delivery

Cytochrome *c* oxide cannot remain saturated when the mitochondrial P_{O2} declines, relative to the average capillary P_{O2}. This decline must occur when an imbalance exists

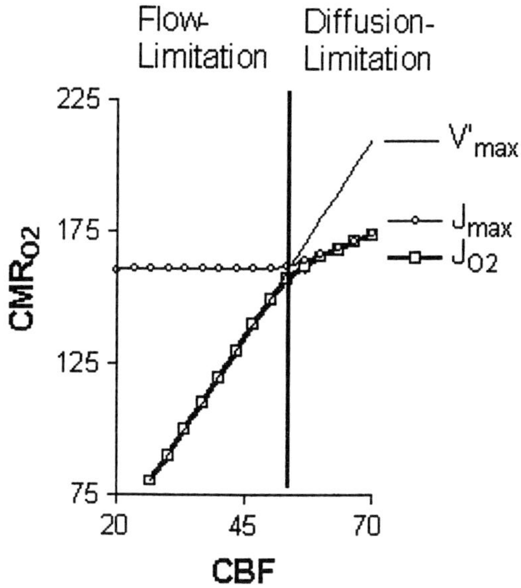

Fig. 8. Model of maximum O_2 delivery capacity ($J_{O_2}^{max}$) in nonlinear relation to blood flow, as described by Eq. 12 *(3,52)*. The maximum O_2 delivery capacities were calculated by nonlinear optimization of Eq. 12, on condition of a constant relation between $\sigma_e V_{max}^{cytox}$ and blood flow.

between the delivery of O_2 and the cytochrome oxide activity. From the description of the oxygen consumption as the product of the apparent O_2 diffusion capacity and the tension gradient between the capillaries and the mitochondria, it follows that the average capillary O_2 tension ($P_{O_2}^{cap}$) driving the delivery is given by

$$P_{O_2}^{cap} = J_{O_2}\left[\frac{1}{L} + \frac{P_{50}^{cytox}}{\left(\sigma_e V_{max}^{cytox}\right) - J_{O_2}}\right] \tag{7}$$

where L is the oxygen diffusion capacity. The equation is illustrated in Fig. 8, which shows the resulting mitochondrial O_2 tension and rate of oxygen consumption for a range of cytochrome oxidase activities ($\sigma_e V_{max}^{cytox}$), given a constant capillary O_2 tension, established by a constant O_2 extraction fraction, and hence a constant ratio between oxygen consumption and blood flow (linear flow–metabolism coupling).

5.2.3. *Flow-Limited O_2 Delivery*

The kinetic analysis of cytochrome oxidase activity shown in Fig. 8 revealed that increases of blood flow above the increase of oxygen consumption may deliver additional O_2 during excitation, when a decline of the mitochondrial O_2 tension threatens to reduce the O_2 saturation of cytochrome oxidase. At this threshold, oxygen consumption depends solely on the mean capillary O_2 tension for a given capillary density, and sudden changes of brain function must be subserved by changes of blood flow which adjust the mean capillary O_2 tension in the required direction.

We previously presented a simple one-dimensional model of O_2 diffusion to brain tissue *(3, 52)* to answer the question of whether blood flow must rise to deliver more O_2 during functional activation. The model is based on the claim that the average capillary O_2 saturation of hemoglobin is a function of the net extraction of O_2 assuming a reasonably even distribution of the O_2 delivery along the length of all capillaries,

$$\bar{S}_{O_2} = 1 - \frac{E_{O_2}}{2} \tag{8}$$

where $\bar{S}_{O_2}$ is the average capillary oxygen saturation of hemoglobin and E_{O_2} is the unidirectional O_2 extraction fraction, equal to the net O_2 extraction fraction when the tissue O_2 tension is negligible. The average capillary O_2 tension and average capillary hemoglobin saturation with O_2 are also related by the equation for the O_2 dissociation curve,

$$\bar{S}_{O_2} = \frac{1}{1 + \left[\frac{P_{50}^{hb}}{P_{O_2}^{cap}}\right]^h} \tag{9}$$

where P_{50}^{hb} is the hemoglobin half-saturation O_2 tension. The resulting equation establishes the inverse correlation between the net extraction fraction and the average capillary O_2 tension,

$$\bar{P}_{O_2}^{cap} = P_{50}^{hb} \sqrt[h]{\frac{2}{E_{O_2}} - 1} \tag{10}$$

where h is the Hill coefficient and E_{O_2} is the ratio $J_{O_2}/(F\ C_{O_2})$, where F is the blood flow and C_{O_2} is the arterial O_2 concentration. When the maximum delivery capacity is reached, the delivery is a function of the average capillary O_2 tension, according to the relationship,

$$J_{O_2}^{max} = L\bar{P}_{O_2}^{cap} \tag{11}$$

where $J_{O_2}^{max}$ is the maximum O_2 delivery capacity and L is the average tissue O_2 diffusion capacity between the capillary lumen and the mitochondria. Entering Eqs. 10 and 11 into, and solving, Eq. 7, the following relationship is obtained,

$$J_{O_2}^2 - J_{O_2}^2\left(J_{O_2}^{max} + \sigma_e V_{max}^{cytox} + J_{50}\right) + J_{O_2}^{max}\ \sigma_e\ V_{max}^{cytox} = 0 \tag{12}$$

where $J_{O_2}^{max}$ is the maximum O_2 delivery capacity $L\bar{P}_{O_2}^{cap}$ derived as $LP_{50}^{hb}\sqrt[h]{2/E_{O_2} - 1}$ above, and J_{50} is the product LP_{50}^{cytox}. This equation was graphed as Fig. 8 for a constant ratio between $\sigma_e V_{max}^{cytox}$ and blood flow, as expected for postsynaptic excitation when depolarization causes calcium to activate the mitochondrial enzymes.

Equation 12 can also be solved under the condition of a constant magnitude of $\sigma_e V_{max}^{cytox}$, as expected for the absence of neuronal excitation. In this case, the equation was fitted to pairs of measurements of J_{O_2} and E_{O_2}, to yield estimates of the parameters, L, $\sigma_e V_{max}^{cytox}$, and P_{50}^{cytox}, assuming these to be the same constants during all the

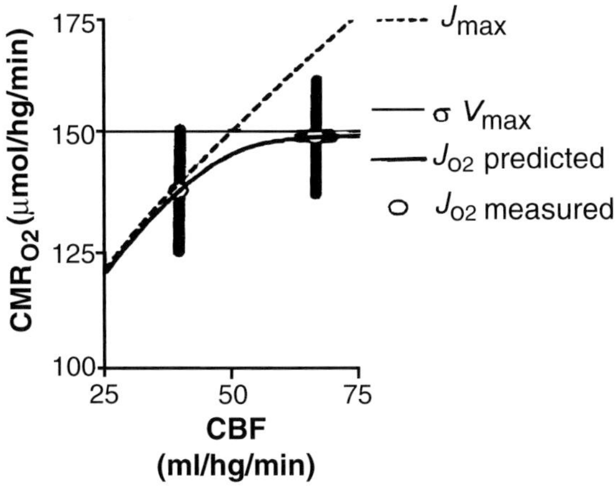

Fig. 9. Model of oxygen consumption (J_{O_2}) and maximum oxygen delivery ($J^{max}_{O_2}$) capacities in nonlinear relation to blood flow, as described by Eq. 12, for constant cytochrome oxidase activity. The oxygen consumption and maximum oxygen delivery capacities were calculated by nonlinear optimization of Eq. 12 on condition of a constant value of $\sigma_e V^{cytox}_{max}$ of 151.10 µmol hg^{-1} min^{-1} and measurements published in *(73)*. The value of *L* was determined to be 3.27 µmol hg^{-1} min^{-1} mmHg^{-1} for a P^{cytox}_{50} value of 0.05 mmHg, consistent with values obtained in humans in vivo by measurements of capillary density *(74)*.

measurements. In one experiment, Gjedde *(73)* determined blood flow and oxygen consumption during administration of indomethacin, which lowers blood flow in the brain. The results are shown in Fig. 9.

The decline of the oxygen saturation of hemoglobin resulting from the extraction of O_2 to the brain is measurable by magnetic resonance imaging as the blood-oxygen-ation-level-dependent contrast caused by the presence of deoxyhemoglobin. The measurements confirm that brain activity generally leads to increases of venous hemoglobin oxygenation, as evidenced by the decline of signal losses caused by the paramagnetic deoxyhemoglobin *(75,76)*.

5.2.4. Nitric Oxide

Blood flow regulation is important to the maintenance of a constant O_2 tension in mitochondria, and neuronal activation can mediate blood flow changes by means of the endothelium-derived relaxation factor NO. Several of the classic blood flow stimulators act by means of NO, including CO_2 and hydrogen ions, which are products of oxidative metabolism *(77–80)*. NO causes vasodilatation of brain resistance vessels, in addition to other effects. It is synthesized in astrocytes, neurons, and endothelial cells, in proportion to the cytosolic concentration of unbound Ca, and is generated in reactions catalyzed by cell-specific NOS, either endothelial (eNOS) or neuronal-astrocytic (nNOS), of which nNOS is by far the more abundant.

It is not clear to what extent only nNOS activation is involved in functionally induced increases of cerebral blood flow. It is known that blocking of the vascular receptors, suspected of being involved in the synthesis of NO by eNOS abolishes functionally induced blood flow increases, and focal changes of cortical blood flow induced by sensory stimulation can be eliminated by blocking endothelial acetylcholine receptors *(81)*, including those involved in mediating synthesis of NO. Yet the underlying cellular activation (whether glial or neuronal) is unimpeded, as indicated by increased glycolysis *(81)*, suggesting either that nNOS activation is not involved in the increase of glycolysis, or that eNOS activation is not required for neuronal excitation to proceed unimpeded. Other evidence suggests that the cerebral vasodilatation associated with simple somatosensory stimulation in rodents is mediated by nNOS activity *(82–84)*.

5.3. Blood Flow in the Service of Glc Delivery and Lact Removal

The changes of blood flow in response to stimulation of brain tissue show that the flow change, on average, is independent of the oxidativeness of the metabolic response. They also show that the blood flow response generally exceeds the response of oxidative metabolism. Thus, there seems to be little direct relation between the rise of blood flow and the demand for O_2. In contrast, the relationship between the rise of blood flow and the rise of glucose consumption appears to be almost linear, as shown in Fig. 6.

The findings suggest that a signal from a factor that depends on the rate of glycolysis elicits the blood flow increase, also when no additional O_2 is used. The absent use of additional O_2 may be physiological (absent postsynaptic depolarization in cases of inhibition, absent Ca ion accumulation in mitochondria), as in the cases of primary somatosensory stimulation, or pathological, as in cases of ischemia or hypoxemia. This raises the questions of the agent that mediates the increase of blood flow in the absence of an elevated need for O_2, and of the situations in which the demand for glucose is increased, but the demand for O_2 is not. There is evidence that this agent may be K ions, in situations such as parallel fiber stimulation in the cerebellum, in which the EC depolarization is induced by a glutamate release *(70)*, or may be adenosine, in situations such as climbing fiber stimulation, in which the simple efferent spike activity of the Purkinje cells is inhibited *(69)*. In these situations, the energy needed for glutamate transport into glial cells may depend on glycolysis.

5.3.1. Potassium

Neuronal excitation raises EC K and glutamate ion concentrations. Both are imported by glia, but in different ways. K enters astrocytes by several nonenergy-requiring routes, as well as in response to glutamate import. The glutamate symport with Na activates the P-type ATPase, which exchanges the IC Na ions with EC K ions. Only the process linked to glutamate transport represents an energy-requiring net transfer of K ions from neurons to astrocytes via the extracellular space *(85,86)*.

Glutamate is released when excitatory neurons fire, but the degree of resulting postsynaptic depolarization and activation depends on the sum of excitatory and inhibitory impulses converging on the postsynaptic neurons. In the conditions of postsynaptic inhibition, postsynaptic loss of K can be prevented when opposite changes of the permeabilities of Na, and K and Cl, ions clamp the ion fluxes to some resting average value, as exemplified in Table 2. The primary energy-demanding process in this situation would be the removal of the glutamate released from the excitatory presynaptic terminals, which leads to accumulation of K in astrocytes. Because of the high membrane permeability to K ions, this K escapes passively to the extracellular space, including the perivascular space surrounding capillaries.

Newman and Paulson *(12)* speculated that K released passively to the perivascular space may cause relaxation of smooth muscle and dilatation of resistance vessels, and hence increased blood flow. They reasoned that the time constant of delivery of K to the perivascular space is much smaller when the K is siphoned to the vessels inside the foot processes (66 ms) than when the K is left to diffuse through the extracellular space (2.5 s). The lower time constant may assure an immediate increase of blood flow to match the increase of glycolysis induced by the removal of glutamate.

The role of K ions in mediating functionally induced increases of blood flow in the brain was tested by Caesar et al. *(70)*, who found that the relative contributions of extracellularly applied K ions and adenosine to the regulation of blood flow in cerebellum varied among the cell populations, K (and NO) having the greatest effect in parallel fiber connections, and adenosine (and NO) having the greatest effect in climbing fiber connections.

5.3.2. *Lactate*

Minimal postsynaptic depolarization would have two important consequences: Postsynaptic mitochondrial dehydrogenases would not be activated by Ca, but astrocytes nevertheless would be stimulated to remove glutamate rapidly, hence possibly exceeding their oxidative capacity and generating lactate to be metabolized neither in neurons nor in astrocytes. Lactate generation could reflect the situation in which glutamate is released, but excitation is subliminal and postsynaptic depolarization is prevented by parallel inhibitory input. Mathiesen et al. *(71)* showed that stimulation of cerebellar neurons in some cases led to increased blood flow despite partly GABAergic inhibition of postsynaptic spiking activity, presumably by prevention of postsynaptic depolarization.

Thus, in inhibited or otherwise uncommonly excited states, it is possible that rapid removal of glutamate *(87)*, achievable only by glycolysis, may generate pyruvate in excess of the average oxidative capacity of glial cells. Thus, although recent measurements of the relative contributions of oxidative phosphorylation to the energy turnover in neurons and astrocytes suggest that only 30% of the total resting average energy conversion of brain tissue takes place in astrocytes, the astrocytes may contribute significantly more to the increase of nonoxidative metabolism when excitation of postsynaptic neurons is prevented by parallel inhibition. Skeletal muscle studies *(88)* confirm that a correlation exists between the increase of blood flow and the increase of lactate and Laptook et al. *(89)* showed that lactate may contribute to the regulation of cerebral blood flow. The increased blood flow would contribute to the removal of the lactate from brain tissue, but the generated lactate would not be oxidized in neurons, unless their oxidative metabolism were proportionately stimulated.

5.3.3. *Adenosine*

Adenosine is the product of dephosphorylation of adenosine nucleotides, and its concentration is thus believed to reflect the balance between energy demand and supply in nervous tissue *(90)*. As in the cases of lactate generation, it is possible that the demand imposed by certain types of experimental stimulation and excitation of nervous tissue exceeds the oxidative capacity of that tissue, and hence establishes an imbalance, characterized not only by increased lact, but also by increased adenosine *(69)*.

6. CONCLUSIONS

There are few facts in evidence of a rigid association in vivo between changes of oxygen consumption, glucose combustion, and blood flow in the human brain. The claim that blood flow must somehow equally satisfy the demands for O_2 and glucose during excitation is therefore without foundation. Roy and Sherrington *(91)* measured neither glucose nor oxygen consumption. They observed pial vasodilatation after administration of postmortem brain extracts to living animals. They surmised that a metabolite had caused the dilatation, and that that metabolite might be lactate.

The theoretical analysis suggests that the cerebral energy demand reflects both the steady-state level of postsynaptic membrane depolarization and axonal spike frequency. Substantial increases of net energy turnover do not appear to occur in response to neuronal inhibition caused by increased Cl conductance. Increased energy turnover is not required to sustain hyperpolarization caused by decreased conductance of Na or increased conductance of K. Increased energy supply is required to maintain graded dendritic and/or somatic membrane depolarization in the state of increased Na and K conductances.

Substantial lactate generation may occur when glutamate release is not followed by postsynaptic depolarization. In these cases, there may be a demand for glutamate removal of a magnitude exceeding the sum of the average glial and neuronal oxidative capacities. This lactate can be removed only by the circulation. The suggestion is consistent with the observation that simple primary somatosensory stimulation elicits a rapid "white" metabolic response, which may be of glial origin; the bulk of the motor or complex secondary somatosensory excitation leads to a "red" metabolic response, which is primarily of postsynaptic neuronal origin.

Combining the evidence from recent studies of the putative agents of blood flow regulation, it may be hypothesized that rises of both NO and K reflect afferent excitatory neurotransmission, leading to glutamate release and reuptake; rises of lactate and adenosine reflect the absence of an increased efferent spike activity accompanied by increased oxidative metabolism. By this account, both lactate and adenosine may indicate the degree to which the stimulus may be judged to be unphysiological, in comparison with the established oxidative capacity of the tissue.

REFERENCES

1. Shulman, R. G. and Rothman, D. L. (1998) Interpreting functional imaging studies in terms of neurotransmitter cycling. *Proc. Natl. Acad. Sci. USA* **95,** 11,993–11,998.
2. Magistretti, P. J., Pellerin, L., Rothman, D. L., and Shulman, R. G. (1999) Energy on demand. *Science* **283,** 496–497.
3. Gjedde, A. (1996) The relation between brain function and cerebral blood flow and metabolism, in *Cerebrovascular Disease* (Batjer, H. H., ed.), Lippincott-Raven, Philadelphia, pp. 23–40.
4. Skou, J. C. (1960) Further investigations on a Mg^{++}–Na^{+}-activated adenosine-triphosphatase, possibly related to the active, linked transport of Na^{+} and K^{+} across the nerve membrane. *Biochim. Biophys. Acta* **42,** 6–23.
5. Erecinska, M. and Silver, I. (1989) ATP and brain function. *J. Cereb. Blood Flow Metab.* **9,** 2–19.
6. McCormick, D. A. (1990) Membrane properties and neurotransmitter actions, in *The Synaptic Organization of the Brain*, 3rd ed. (Shepherd, G., ed.), Oxford University Press, New York, pp. 32–66.
7. Gjedde, A. (1993) The energy cost of neuronal depolarization, in *Functional Organization of the Human Visual Cortex* (Gulyas, B., Ottoson, D., and Roland, P. E., eds.), Pergamon, Oxford, pp. 291–306.

8. Laughlin, S. B., de Ruyter van Steveninck, R. R., and Anderson, J. C. (1998) The metabolic cost of neural information. *Nature Neurosci.* **1,** 36–41.
9. Andriezen, W. L. (1893) The neuroglia elements in the human brain. *Br. Med. J.* **ii,** 227–230.
10. Brightman, M. W. and Reese, T. S. (1969) Junctions between intimately apposed cell membranes in the vertebrate brain. *J. Cell. Biol.* **40,** 648–677.
11. Brightman, M. W., Klatzo, I., Olsson, Y., and Reese, T. S. (1970) The blood-brain barrier to proteins under normal and pathological conditions. *J. Neurol. Sci.* **110,** 215–239.
12. Paulson, O. B. and Newman, E. A. (1987) Does the release of potassium from astrocyte endfeet regulate cerebral blood flow? *Science* **237,** 896–898.
13. Cremer, J. E. (1976) The influence of liver-bypass on transport and compartmentation in vivo. *Adv. Exp. Med. Biol.* **69,** 95–102.
14. Cremer, J. E., Cunningham, V. J., Pardridge, W. M., Braun, L. D., and Oldendorf, W. H. (1979) Kinetics of blood–brain barrier transport of pyruvate, lactate and glucose in suckling, weanling and adult rats. *J. Neurochem.* **33,** 439–446.
15. Drewes, L. (1999) Transport of brain fuels, glucose and lactate, in *Brain Barrier Systems* (Paulson, O. B., Knudsen, G. M., and Moos, T., eds.), Alfred Benzon Symposium 45, Munksgaard, Copenhagen, pp. 285–295.
16. Gerhart, D. Z., Enerson, B. E., Zhdankina, O. Y., Leino, R. L., and Drewes, L. R. (1997) Expression of monocarboxylate transporter MCT1 by brain endothelium and glia in adult and suckling rats. *Am. J. Physiol.* **273,** E207–213.
17. Gerhart, D. Z., Enerson, B. E., Zhdankina, O. Y., Leino, R. L., and Drewes, L. R. (1998) Expression of the monocarboxylate transporter MCT2 by rat brain glia. *Glia* **22,** 272–281.
18. Vandenberg, R. J. (1998) Molecular pharmacology and physiology of glutamate transporters in the central nervous system. *Clin. Exp. Pharmacol. Physiol.* **25,** 393–400.
19. Rothstein, J. D., Dykes-Hoberg, M., Pardo, C. A., Bristol, L. A., Jin, L., Kuncl, R. W., et al. (1996) Knockout of glutamate transporters reveals a major role for astroglial transport in excitotoxicity and clearance of glutamate. *Neuron* **16,** 675–686.
20. Tholey, G., Roth-Schechter, B. F., and Mandel, P. (1981) Activity and isoenzyme pattern of lactate dehydrogenase in neurons and astroblasts cultured from brains of chick embryos. *J. Neurochem.* **36,** 77–81.
21. Bittar, P. G., Charnay, Y., Pellerin, L., Bouras, C., and Magistretti, P. J. (1996) Selective distribution of lactate dehydrogenase isoenzymes in neurons and astrocytes of human brain. *J. Cereb. Blood Flow Metab.* **16,** 1079–1089.
22. Laughton, J. D., Charnay, Y., Belloir, B., Pellerin, L., Magistretti, P. J., and Bouras, C. (2000) Differential messenger RNA distribution of lactate dehydrogenase LDH-1 and LDH-5 isoforms in the rat brain. *Neuroscience* **96,** 619–625.
23. Silver, I. A. and Erecinska, M. (1997) Energetic demands of the Na^+/K^+ ATPase in mammalian astrocytes. *Glia* **21,** 35–45.
24. Sokoloff, L. (1999) Energetics of functional activation in neural tissues. *Neurochem. Res.* **24,** 321–329.
25. Hevner, R. F., Liu, S., and Wong-Riley, M. T. (1995) A metabolic map of cytochrome oxidase in the rat brain: histochemical, densitometric and biochemical studies. *Neuroscience* **65,** 313–342.
26. Kaplan, N. O. and Everse, J. (1972) Regulatory characteristics of lactate dehydrogenases. *Adv. Enzyme Regul.* **10,** 323–336.
27. Tanaka, K., Watase, K., Manabe, T., Yamada, K., Watanabe, M., Takahashi, K., et al. (1997) Epilepsy and exacerbation of brain injury in mice lacking the glutamate transporter GLT-1. *Science* **276,** 1699–1702.
28. Brandt, R. B., Laux, J. E., Spainhour, S. E., and Kline, E. S. (1987) Lactate dehydrogenase in rat mitochondria. *Arch. Biochem. Biophys.* **259,** 412–422.

29. Hertz, L., Swanson, R. A., Newman, G. C., Marrif, H., Juurlink, B. H., and Peng, L. (1998) Can experimental conditions explain the discrepancy over glutamate stimulation of aerobic glycolysis? *Dev. Neurosci.* **20,** 339–347.
30. Oldendorf, W. H. (1973) Carrier-mediated blood-brain barrier transport of short-chain monocarboxylic organic acids. *Am. J. Physiol.* **224,** 1450–1453.
31. Halestrap, A. P. (1975) The mitochondrial pyruvate carrier. *Biochem. J.* **148,** 85–96.
32. Poole, R. C. and Halestrap, A. P. (1993) Transport of lactate and other monocarboxylates across mammalian plasma membranes. *Am. J. Physiol.* **264,** C761–82.
33. Halestrap, A. P. and Price, N. T. (1999) The proton-linked monocarboxylate transporter (MCT) family: structure, function and regulation. *Biochem. J.* **343,** 281–299.
34. Broer, S., Rahman, B., Pellegri, G., Pellerin, L., Martin, J. L., Verleysdonk, S., Hamprecht, B., and Magistretti, P. J. (1997) Comparison of lactate transport in astroglial cells and monocarboxylate transporter 1 (MCT 1) expressing Xenopus laevis oocytes. Expression of two different monocarboxylate transporters in astroglial cells and neurons. *J. Biol. Chem.* **272,** 30,096–30,102.
35. Gjedde, A. (1992) Blood-brain glucose transfer, in *Physiology and Pharmacology of the Blood-Brain Barrier, Handbook of Experimental Pharmacology* (Bradbury, M. W. B., ed.), Springer-Verlag, Berlin, pp. 65–115.
36. Diemer, N. H., Benveniste, H., and Gjedde, A. (1985) In vivo cell membrane permeability to deoxyglucose in rat brain. *Acta Neurol. Scand.* **72,** 87.
37. Gjedde, A. and Diemer, N. H. (1983) Autoradiographic determination of regional brain glucose content. *J. Cereb. Blood. Flow Metab.* **3,** 303–310.
38. Silver, I. A. and Erecinska, M. (1994) Extracellular glucose concentration in mammalian brain: continuous monitoring of changes during increased neuronal activity and upon limitation in oxygen supply in normo-, hypo-, and hyperglycemic animals. *J. Neurosci.* **14,** 5068–5076.
39. Kety, S. S. (1949) The physiology of the human cerebral circulation. *Anesthesiology* **10,** 610–614.
40. Lassen, N. A. (1959) Cerebral blood flow and oxygen consumption in man. *Physiol. Rev.* **39,** 183–238.
41. Kuwabara, H., Ohta, S., Brust, P., Meyer, E., and Gjedde, A. (1992) Density of perfused capillaries in living human brain during functional activation. *Progr. Brain Res.* **91,** 209–215.
42. Ohta, S., Meyer, E., Fujita, H., Reutens, D. C., Evans, A., and Gjedde, A. (1996) Cerebral [O-15] water clearance in humans determined by PET. I. Theory and normal values. *J. Cereb. Blood Flow Metab.* **16,** 765–780.
43. Ohta, S., Meyer, E., Thompson, C. J., and Gjedde, A. (1992) Oxygen consumption of the living human brain measured after a single inhalation of positron emitting oxygen. *J. Cereb. Blood Flow Metab.* **12,** 179–192.
44. Kuwabara, H., Evans, A. C., and Gjedde, A. (1990) Michaelis-Menten constraints improved cerebral glucose metabolism and regional lumped constant measurements with [^{18}F]fluoro-deoxyglucose. *J. Cereb. Blood Flow Metab.* **10,** 180–189.
45. Fox, P. T. and Raichle, M. E. (1986) Focal physiological uncoupling of cerebral blood flow and oxidative metabolism during somatosensory stimulation in human subjects. *Proc. Natl. Acad. Sci. USA* **83,** 1140–1144.
46. Seitz, R. J. and Roland, P. E. (1992) Vibratory stimulation increases and decreases the regional cerebral blood flow and oxidative metabolism: a positron emission tomography (PET) study. *Acta Neurol. Scand.* **86,** 60–67.
47. Fujita, H., Kuwabara, H., Reutens, D. C., and Gjedde, A. (1999) Oxygen consumption of cerebral cortex fails to increase during continued vibrotactile stimulation. *J. Cereb. Blood Flow Metab.* **19,** 266–271.

48. Ginsberg, M. D., Chang, J. Y., Kelley, R. E., Yoshii, F., Barker, W. W., Ingento, G., and Boothe, T. E. (1988) Increases in both cerebral glucose utilization and blood flow during execution of a somatosensory task. *Ann. Neurol.* **23,** 152–160.
49. Ribeiro, L., Kuwabara, H., Meyer, E., Fujita, H., Marrett, S., Evans, A., and Gjedde, A. (1993) Cerebral blood flow and metabolism during nonspecific bilateral visual stimulation in normal subjects, in *Quantification of Brain Function: Tracer Kinetics and Image Analysis in Brain PET* (Uemura, K., Lassen, N. A., Jones, T., and Kanno, I., eds.), Elsevier, Amsterdam, pp. 217–224.
50. Fox, P. T., Raichle, M. E., Mintun, M. A., and Dence, C. E. (1988) Nonoxidative glucose consumption during focal physiological activity. *Science* **241,** 462–464.
51. Marrett, S. and Gjedde, A. (1997) Changes of blood flow and oxygen consumption in visual cortex of living humans. *Adv. Exp. Med. Biol.* **413,** 205–208.
52. Vafaee, M. S. and Gjedde, A. (2000) Model of blood-brain transfer of oxygen explains non-linear flow–metabolism coupling during stimulation of visual cortex. *J. Cereb. Blood Flow Metab.* **20,** 747–754.
53. Katayama, Y., Tsubokawa, T., Hirayama, T., Kido, G., Tsukiyama, T., and Iio, M. (1986) Response of regional cerebral blood flow and oxygen metabolism to thalamic stimulation in humans as revealed by positron emission tomography. *J. Cereb. Blood Flow Metab.* **6,** 637–641.
54. Roland, P. E., Eriksson, L., Widen, L., and Stone-Elander, S. (1989) Changes in regional cerebral oxidative metabolism induced by tactile learning and recognition in man. *Eur. J. Neurosci.* **7,** 2373–2389.
55. Roland, P. E., Eriksson, L., Stone-Elander, S., and Widen, L. (1987) Does mental activity change the oxidative metabolism of the brain? *J. Neurosci.* **8,** 2373–2389.
56. Raichle, M. E., Grubb, R. L., Jr., Gado, M. H., Eichling, J. O., and Ter-Pogossian, M. M. (1976) Correlation between regional cerebral blood flow and oxidative metabolism. In vivo studies in man. *Arch. Neurol.* **33,** 523–526.
57. Iida, H., Jones, T., and Miura, S. (1993) Modeling approach to eliminate the need to separate arterial plasma in oxygen-15 inhalation positron emission tomography. *J. Nucl. Med.* **34,** 1333–1340.
58. Vafaee, M. S. and Gjedde, A. (2001) Oxygen consumption and blood flow changes in motor cortex, anterior cingulate, and putamen during sequential finger touching: evidence for inverse flow–metabolism coupling. *Personal Communication.*
59. Ohta, S., Reutens, D. C., and Gjedde, A. (1999) Brief vibrotactile stimulation does not increase cortical oxygen consumption when measured by single inhalation of positron emitting oxygen. *J. Cereb. Blood Flow Metab.* **19,** 260–265.
60. Sokoloff, L., Reivich, M., Kennedy, C., DesRosiers, M. H., Patlak, C. S., Pettigrew, K. D., Sakurada, O., and Shinohara, M. (1977) The [^{14}C]deoxyglucose method for the measurement of local cerebral glucose utilization: theory, procedure, and normal values in the conscious and anesthetized albine rat. *J. Neurochem.* **28,** 897–916.
61. Murase, K., Kuwabara, H., Yasuhara, Y., Evans, A. C., and Gjedde, A. (1996) Mapping of change in cerebral glucose utilization using fluorine-18 fluorodeoxyglucose double injection and the constrained weighted-integration method. *IEEE Transact. Med. Imag.* **15,** 824–835.
62. Robin, E. D., Murphy, B. J., and Theodore, J. (1984) Coordinate regulation of glycolysis by hypoxia in mammalian cells. *J. Cell Physiol.* **118,** 287–290.
63. Gjedde, A. (1984) On the measurement of glucose in brain. *Neurochem. Res.* **9,** 1667–1671.
64. van den Berg, C. J. and Bruntink, R. (1983) Glucose oxidation in the brain during seizures: experiments with labeled glucose and deoxyglucose, in *Glutamine, Glutamate and GABA in the Central Nervous System* (Hertz, L., Kvamme, E., McGeer, E. G., and Schousboe, A., eds.), Alan R. Liss, New York, pp. 619–624.

65. Shram, N. F., Netchiporouk, L. I., Martelet, C., Jaffrezic-Renault, N., Bonnet, C., and Cespuglio, R. (1998) In vivo voltammetric detection of rat brain lactate with carbon fiber microelectrodes coated with lactate oxidase. *Anal. Chem.* **70,** 2618–2622.
66. Vafaee, M., Meyer, E., Marrett, S., Evans, A. C., and Gjedde, A. (1998) Increased oxygen consumption in human visual cortex: respond to visual stimulation. *Acta Neurol. Scand.* **98,** 85–89.
67. Vafaee, M. S., Meyer, E., Marrett, S., Paus, T., Evans, A. C., and Gjedde, A. (1999) Frequency-dependent changes in cerebral metabolic rate of oxygen during activation of human visual cortex. *J. Cereb. Blood Flow Metab.* **19,** 272–277.
68. Shearman, M. S. and Halestrap, A. P. (1894) The concentration of the mitochondrial pyruvate carrier in rat liver and heart mitochondria determined with alpha-cyano-beta-(1-phenylindol-3-yl) acrylate. *Biochem. J.* **223,** 673–676.
69. Akgören, N, Mathiesen, C., Rubin, I., and Lauritzen, M. (1997) Laminar analysis of activity-dependent increases of CBF in rat cerebellar cortex: dependence on synaptic strength. *Am. J. Physiol.* **273,** H1166–H1176.
70. Caesar, K., Akgoren, N., Mathiesen, C., and Lauritzen, M. (1999) Modification of activity-dependent increases in cerebellar blood flow by extracellular potassium in anaesthetized rats. *J. Physiol. (Lond.)* **520,** 281–292.
71. Mathiesen, C., Caesar, K., Akgoren, N., and Lauritzen, M. (1998) Modification of activity-dependent increases of cerebral blood flow by excitatory synaptic activity and spikes in rat cerebellar cortex. *J. Physiol. (Lond.)* **512,** 555–566.
72. Gjedde, A. (1996) PET criteria of cerebral tissue viability in ischemia. *Acta Neurol. Scand* **93,** 3–5.
73. Gjedde, A., Vafaee, M. S., Østergaard, L., et al. (2001) Decline of cerebral oxygen consumption during indomethacin-induced flow reduction is consistent with diffusion-limited oxygen delivery. *J. Cereb. Blood Flow Metab.* **21 Suppl 1,** S110.
74. Vafaee, M. S. and Gjedde, A. (2001) CMR02-CBF changes by finger motion: No evidence of flow-metabolism coupling in putamen. *NeuroImage* **13,** S1012.
75. Ogawa, S., Lee, T. M., Nayak, A. S., et al. (1990) Oxygenation-sensitive contrast in magnetic resonance imaging of rodent brain at high magnetic fields. *Magn. Reson. Med.* **14,** 68–78.
76. Kwong, K. K., Belliveau, J. W., Chesler, D. A., Goldberg, I. E., Weisskoff, R. M., Poncelet, B. P., et al. (1992) Dynamic magnetic resonance imaging of human brain activity during primary sensory stimulation. *Proc. Natl. Acad. Sci. USA* **89,** 5675–5679.
77. Iadecola, C. (1992) Does nitric oxide mediate the increases in cerebral blood flow elicited by hypercapnia? *Proc. Natl. Acad. Sci. USA* **89,** 3913–3916.
78. Iadecola, C., Pelligrino, D. A., Moskowitz, M. A., and Lassen, N. A. (1994) Nitric oxide synthase inhibition and cerebrovascular regulation. *J. Cereb. Blood Flow Metab.* **14,** 175–192.
79. Fabricius, M. and Lauritzen, M. (1994) Examination of the role of nitric oxide for the hypercapnic rise of cerebral blood flow in rats. *Am. J. Physiol.* **266,** H1457–1464.
80. Villringer, A. and Dirnagl, U. (1995) Coupling of brain activity and cerebral blood flow: basis of functional neuroimaging. *Cerebrovasc. Brain Metab. Rev.* **7,** 240–276.
81. Ogawa, M., Magata, Y., Ouchi, Y., Fukuyama, H., Yamauchi, H., Kimura, J., Yonekura, Y., and Konishi, J. (1994) Scopolamine abolishes cerebral blood flow response to somatosensory stimulation in anesthetized cats: PET study. *Brain Res.* **650,** 249–252.
82. Ayata, C., Ma, J., Meng, W., Huang, P., Moskowitz, M. A. (1996) L-NA-sensitive rCBF augmentation during vibrissal stimulation in type III nitric oxide synthase mutant mice. *J. Cereb. Blood Flow Metab.* **16,** 539–541.
83. Ma, J., Ayata, C., Huang, P. L., Fishman, M. C., and Moskowitz, M. A. (1996) Regional cerebral blood flow response to vibrissal stimulation in mice lacking type I NOS gene expression. *Am. J. Physiol.* **270,** H1085–1090.

84. Cholet, N., Seylaz, J., Lacombe, P., and Bonvento, G. (1997) Local uncoupling of the cerebrovascular and metabolic responses to somatosensory stimulation after neuronal nitric oxide synthase inhibition. *J. Cereb. Blood Flow Metab.* **17,** 1191–1201.
85. Takahashi, S., Shibata, M., and Fukuuchi, Y. (1997) Effects of increased extracellular potassium on influx of sodium ions in cultured rat astroglia and neurons. *Brain Res. (Dev. Brain Res.)* **104,** 111–117.
86. Longuemare, M. C., Rose, C. R., Farrell, K., Ransom, B. R., Waxman, S. G., and Swanson, R. A. (1999) K(+)-induced reversal of astrocyte glutamate uptake is limited by compensatory changes in intracellular Na+. *Neuroscience* **93,** 285–292.
87. Kojima, S., Nakamura, T., Nidaira, T., Nakamura, K., Ooashi, N., Ito, E., et al. (1999) Optical detection of synaptically induced glutamate transport in hippocampal slices. *J. Neurosci.* **19,** 2580–2588.
88. Connett, R. J., Gayeski, T. E., and Honig, C. R. (1985) Energy sources in fully aerobic rest-work transitions: a new role for glycolysis. *Am. J. Physiol.* **248,** H922–H929.
89. Laptook, A. R., Peterson, J., and Porter, A. M. (1988) Effects of lactic acid infusions and pH on cerebral blood flow and metabolism. *J. Cereb. Blood Flow Metab.* **8,** 193–200.
90. Berne, R. M., Knabb, R. M., Ely, S. W., and Rubio, R. (1983) Adenosine in the local regulation of blood flow: a brief overview. *Fed. Proc.* **42,** 3136–3142.
91. Roy, C. S. and Sherrington, C. S. (1890) On the regulation of the blood supply of the brain. *J. Physiol. (Lond.)* **11,** 85–108.

III

Brain Macroenvironment

10

Choroid Plexus and the Cerebrospinal–Interstitial Fluid Systems

Roy O. Weller

1. INTRODUCTION

In most organs of the body, other than the central nervous system (CNS), extracellular fluid is derived from the blood, and drains along lymphatics to regional lymph nodes. Such a system of lymphatic drainage has a well-characterized role in B- and T-lymphocyte-mediated immune reactions *(1)*. By contrast, extracellular fluid associated with the CNS falls into two main categories: cerebrospinal fluid (CSF) and interstitial fluid (ISF), which are, to varying degrees, separate. The relationship between the CNS and the immune system is not as well-characterized as in other organs. CSF is produced by the choroid plexus and fills the ventricular system of the brain and the craniospinal subarachnoid space. In human, CSF drains mostly into the blood, by bulk flow through arachnoid villi and granulations associated with the major venous sinuses *(2)*; ISF drains along perivascular channels surrounding cerebral and leptomeningeal arteries, and is thus separated from CSF. Much of the CSF and the ISF in the rat drains via nasal lymphatics to regional lymph nodes, although these fluids also enter directly into the blood through primitive arachnoid villi *(3,4)*. Although ISF drains from the human CNS along perivascular pathways that are separated from the CSF in the subarachnoid space, it is not clear how much drains to regional lymph nodes. The complex relationship between the CNS and the immune system is reflected by the "immunological privilege" of the CNS, whereby tissue allografts survive for substantially longer periods in the CNS than in other organs of the body *(3)*.

This chapter is a review of the anatomical and functional aspects of the cerebrospinal fluid, its production by the choroid plexus and the anatomical pathways by which CSF drains from the intracranial and intraspinal cavities. In addition, the importance of drainage pathways for interstitial fluid from the CNS is emphasized. Historically, there has been considerable emphasis on the study of CSF physiology and drainage; ISF and its drainage from the CNS has been less extensively investigated *(5)*. This chapter seeks to redress the balance, particularly because ISF drainage pathways appear to play key roles in immune reactions (IRs) in the brain and in the development of dementias in the elderly *(5,6)*.

From: *The Neuronal Environment: Brain Homeostasis in Health and Disease*
Edited by: W. Walz © Humana Press Inc., Totowa, NJ

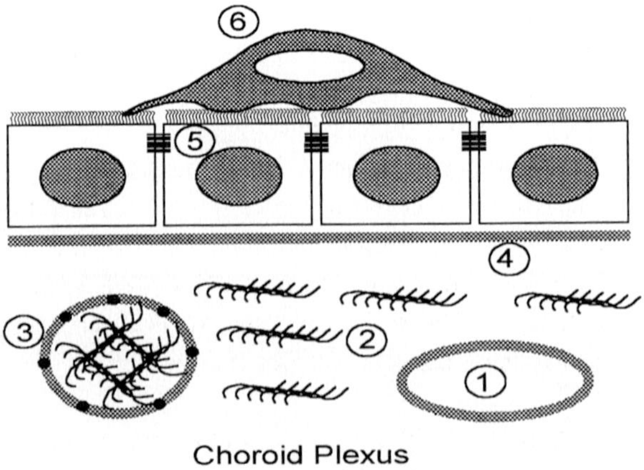

Fig. 1. Choroid plexus: Fluid and protein pass from blood vessels (1) into the stroma populated by leptomeningeal cells and bundles of collagen (2). Aggregations of collagen (3) calcify to form calcospherites in the aging human choroid plexus. The epithelium is the site of the blood–CSF barrier, and is separated from the stroma by a basement membrane (4). Individual epithelial cells are joined by tight junctions (5), and bear microvilli to which resident choroid plexus macrophages (6) cling.

2. CHOROID PLEXUS AND THE PRODUCTION OF CSF

Embryologically, the choroid plexus in both rat and human is derived from invaginations of blood vessels and leptomeninges covered by a layer of modified ependyma *(2)*. In the rat, the choroid plexus is a relatively uncomplicated, folded structure that virtually fills the normal lateral ventricles (LVs), in human, the choroid plexus is an extensive structure extending for some 7–8 cm in the LVs, from the foramen of Monro into the temporal horns of the LVs. Choroid plexus is also present in the third ventricles, the fourth ventricle, and extrudes into the subarachnoid space through the foramina of Luschka in human *(7)*.

Although differing in size and shape, the choroid plexus has a similar histological structure in the rat and in human. Blood vessels in the stroma of the choroid plexus (Fig. 1) are derived in human from the anterior and posterior choroidal arteries, which supply the LV and third ventricular choroid plexuses, and from the inferior cerebellar arteries for the fourth ventricular choroid plexus *(7)*. The stroma contains collagen bundles and leptomeningeal cells. A common feature of the aging human choroid plexus is the presence of calcospherites, which are, in effect, calcified spherical conglomerations of collagen bundles formed by leptomeningeal cells *(8)*. Coating the choroid plexus is a layer of cuboidal epithelial cells, which bear microvilli, but few, if any, cilia *(2)*. This epithelium is derived from ependyma, from which it differs, in that ependymal cells are highly ciliated. Furthermore, carbonic anhydrase C is present in the choroid plexus epithelium, but this enzyme is absent from normal ependyma *(9)*. At its junction with the underlying stroma, the choroid plexus epithelium rests on a basement membrane, a feature that is only seen in ependyma where such cells come into contact with blood vessels. Both in rat and in human, resident epiplexus macrophages (Kolmer cells) cling to the surface of the choroid plexus epithelium, anchored to the microvilli *(2)*.

Cerebrospinal fluid is produced from the blood. Capillaries in the stroma of the choroid plexus are permeable to fluid and to proteins; the endothelium is fenestrated with thin attenuated areas of the cell, across which high water-flow can occur *(2)*. It is the epithelium of the choroid plexus that is the blood–CSF barrier *(2)*. Electron microscopy reveals characteristic intercellular junctions (zonulae occludentes) between choroid plexus epithelial cells; these junctions form the barrier between the stroma and CSF in the ventricles. A good demonstration of the blood–CSF barrier is seen in patients who are jaundiced; although the CSF and brain tissue remain relatively free of bile pigments, the choroid plexus stroma is discolored yellow, indicating that bile pigments have leaked into the stroma, but not through the choroid plexus epithelium into the CSF.

In the rat, CSF is secreted at 2.1 μL/min *(2)*, and at the considerably greater rate of 350 μL/min in human *(10)*. Hydration of carbon dioxide by carbonic anhydrase and the action of sodium–potassium pumps play major roles in the formation of CSF; there is a net transport of Na, chloride, and bicarbonate ions into the CSF. Protein levels in the CSF are very low, and the specific gravity is close to that of brain tissue. Functionally, CSF may act partially as a sink for extracellular fluid from the CNS parenchyma *(2)*, but perhaps its major function in human is to act as a buffer against the effects of sudden impact or movement, and to compensate from the changes in intracranial volume induced by blood flow. Of the mean total intracranial volume of CSF in human of 123 mL, some 25 mL is in the cerebral ventricles, and the greater part (98 mL) is in the subarachnoid space over the surface of the brain *(2)*. The amount of intracranial CSF increases with age, reflecting the overall loss of brain tissue.

3. CIRCULATION AND DRAINAGE OF THE CEREBROSPINAL FLUID IN HUMANS

The circulation of CSF in human has been well-characterized by the use of magnetic resonance imaging (MRI) *(11)*. Produced by the choroid plexus, CSF flows through the ventricular system and subarachnoid space to drain into blood through arachnoid villi and granulations in the walls of major venous sinuses and veins.

3.1. *Ventricular System*

CSF produced in the LVs flows through the foramina of Monro, into the third ventricle, through the narrow aqueduct of Sylvius, and into the fourth ventricle. It receives contributions from the choroid plexuses in the third and fourth ventricles. The paired foramina of Luschka, the lateral angles of the fourth ventricle, and the single dorsal foramen of Magendie provide pathways by which CSF escapes into cisternae and the subarachnoid space *(7)*. MRI studies suggest that there is a slow, steady flow of CSF through much of the ventricular system, and that there is rapid flow through restricted passages, such as the aqueduct and foramina of Monro *(11)*. Rapid to-and-fro motion of flow of CSF, caused by cardiac pulsations, is seen in the aqueduct and other narrow portions of the ventricular system, and in the subarachnoid space *(11)*. The cerebral hemispheres expand during cardiac systole, as does the choroid plexus *(11)*. Pulsations also occur at the foramen magnum, but decrease in extent toward the lumbar region.

3.2. Leptomeninges and the Subarachnoid Space

The pia mater separates cerebrospinal fluid in the subarachnoid space from the surface of the brain. As the pia is reflected onto the surface of blood vessels in the subarachnoid space *(12)*, it separates the cerebrospinal fluid from the brain *(5)*. Although the pia mater forms a regulating interface between CSF and brain, it is a delicate membrane, composed of only 2–3 layers of flattened cells joined by desmosomes and nexus junctions *(13)*. It is impermeable to particulate matter; red blood cells, for example, do not pass through the pia mater in significant numbers in patients with subarachnoid hemorrhage (SAH) *(12)*. Leukocytes, however, and especially macrophages, readily pass through the pia *(14)*. As part of its function as a regulating interface, the pia mater contains enzymes, such as glutamate synthetase, which hydrolyze neurotransmitters *(13)*. Pia mater cells are similar in their histological and ultrastructural characteristics to cells of the arachnoid, although the arachnoid mater is a thicker and more compact layer on the outer aspect of the subarachnoid space; it is impermeable to CSF, because of the presence of tight junctions in its outer layers *(15)*.

The structure of the leptomeninges varies in different parts of the neuraxis *(7)*. Arachnoid mater forms a delicate, translucent layer over the surface of the brain and the spinal cord. Trabeculae composed of arachnoid cells traverse the cerebral subarachnoid space, forming suspensory sheets for the arteries and veins in that space *(15)*, and large cisternae containing CSF are present, particularly at the base of the brain, underlying the hypothalamus and the pons. A more complex arrangement of arachnoid exists over the surface of the spinal cord *(16)*, with a highly perforated intermediate layer of arachnoid extending over the dorsal surface of the cord, between the pia mater and the outer coating of arachnoid. It is from this intermediate layer that the dorsal and ventral ligaments of the cord are formed. Pia mater closely invests the surface of the brain and the spinal cord, extending down into fissures and sulci; it is continuous with the arachnoid through the trabeculae that traverse the subarachnoid space. A layer of pia mater is reflected from the surface of the brain to coat leptomeningeal vessels entering and leaving brain tissue, thus separating the subpial and perivascular spaces of the brain from the CSF in the subarachnoid space *(7,12)*.

3.3. Arachnoid Villi and Granulations

In addition to the arachnoid and pia mater, arachnoid cells form a number of specialized structures, such as the stroma of the choroid plexus, as mentioned previously, and the arachnoid granulations and villi *(7,17)*. Arachnoid granulations in human are formed by a protrusion of arachnoid mater through perforations in the dura, into the lumina of venous sinuses, such as the superior sagittal sinus and cavernous sinuses *(7)*. The core of an arachnoid granulation is a trabecular meshwork of collagen bundles coated by arachnoid cells, and is, in effect, an extension of the subarachnoid space. On this core sits a cap of arachnoid cells, through which channels pass to the vascular endothelial coating on the venous surface of the granulation. Although, in human, it is unclear exactly how bulk flow of CSF occurs across the vascular endothelium into the blood, experiments in monkeys suggests that large vacuoles form on the abluminal surface of the endothelium, and subsequently open into the vascular lumen, carrying quanta of CSF by bulk flow through endothelial cells into the blood *(18)*.

3.4. Disorders Affecting the Cerebrospinal Fluid

A number of disorders affect human cerebrospinal fluid. Chief among these are hydrocephalus, in which cerebral ventricles dilate primarily because of obstruction to the drainage of CSF; SAH, caused by rupture of an artery, either an aneurysm or malformation; and meningitis caused by bacteria, fungi, or viruses infecting the cerebrospinal fluid.

3.4.1. Hydrocephalus

Dilatation of the cerebral ventricles results from obstruction to cerebrospinal fluid flow by tumors or other mass lesions within the ventricles, or compressing them from adjacent intracranial tissues *(19)*. A variety of hydrocephalic syndromes, often presenting in fetal or neonatal life, may also result from stenosis of the cerebral aqueduct *(19)*. CSF flow may be impeded by obstruction to the subarachnoid space most frequently as a sequela to meningitis or subarachnoid hemorrhage. In these cases, the acute stages of inflammation or bleeding are followed by fibrous scarring, which obliterates the subarachnoid space, thus, interfering with the flow of CSF. In experimental animals, hydrocephalus has been produced by viral infections, implantation of inflammatory agents such as kaolin, or by occluding the aqueduct with oil or silastic *(19)*. A number of hereditary forms of hydrocephalus also occur in rodents and in human *(19)*. The major pathological effect of acute hydrocephalus is the infusion of CSF into periventricular tissues, producing periventricular edema, particularly of cerebral white matter *(5)*. Such edema is seen as pallor of staining in histological sections of hydrocephalic brain, and is well-demonstrated on computerized tomography as periventricular lucency, or by MRI *(19)*. If the hydrocephalus remains untreated, slow but progressive destruction of axons within the edematous white matter occurs, resulting in loss of nerve fibers from the periventricular white matter and gliosis *(19)*. Grey matter is relatively well preserved in hydrocephalic brains *(5)*.

In adults, hydrocephalus may be characterized by an increase in the size of the cerebral ventricles, but with only intermittently raised CSF pressure. This syndrome is known as "normal-pressure hydrocephalus," to distinguish it from the much more life-threatening acute hydrocephalus, with raised pressure. Nevertheless, a syndrome of dementia, ataxia, and incontinence may result from normal pressure hydrocephalus, and such patients benefit from shunting of the cerebrospinal fluid *(19,20)*. The cerebral ventricles may also become enlarged in demented elderly patients, because of loss of brain tissue from infarction or Alzheimer's disease (AD) *(21)*. This is the condition called "hydrocephalus ex vacuo," which does not necessarily benefit from CSF shunting.

Free communication between cranial and spinal CSF may be obstructed, particularly at the foramen magnum, because of fibrosis of the meninges, or by occlusion of the foramen magnum by elongated cerebellar tonsils. In such cases, cystic expansion of the cervical spinal cord (syringomyelia) may occur *(22)*, with destruction of pain and temperature pathways in the spinal cord and of anterior horn cells. Patients with syringomyelia suffer burns to the fingers, arthritis, and wasting of muscles, particularly in the hands *(23)*.

3.4.2. Subarachnoid Hemorrhage

Bleeding into the subarachnoid space most frequently results from the rupture of saccular aneurysms on major arterial branches of the circle of Willis *(24)*. Arterial blood floods into the CSF, and surrounds arteries, veins, and nerves in the subarachnoid space. Spasm of arteries with cerebral infarction, caused by deprivation of blood supply to brain tissue is a recognized complication in the early period following subarachnoid hemorrhage. Communication between CSF in cranial and spinal subarachnoid spaces is well-demonstrated in cases of subarachnoid hemorrhage. Blood spreads widely through the subarachnoid space and CSF withdrawn from the lumbar sac contains erythrocytes in the acute stages. Organization and fibrosis of remaining blood in the subarachnoid space may result in fibrous adhesions in the meninges, and hydrocephalus may be a late complication of subarachnoid hemorrhage.

3.4.3. Meningitis

A variety of bacteria, viruses, and fungi may cause meningitis, although amoebae and metazoan parasites are much less frequently involved *(25)*. *Escherichia coli, Haemophilus influenzae, Neisseria meningitidis,* and *Streptococcus pneumoniae* are the commonest bacteria that colonize the cerebral and spinal subarachnoid spaces resulting in a polymorphonuclear leukocyte exudate and pus within the CSF *(25)*. Such infections can be devastating, particularly in the very young and in the elderly. Arteries and veins traversing the subarachnoid space are normally bathed by CSF; they become inflamed when surrounded by the pus of meningitis, and thrombosis of the lumina may result in cerebral infarction and subsequent neurological disability. Cranial nerves may also be damaged by inflammation in meningitis. It is unclear how bacteria enter the CSF in the subarachnoid space from the blood, but polymorphonuclear leukocytes do traverse the endothelium and walls of veins, to enter the CSF. In general, inflammation in meningitis is restricted to the subarachnoid space and rarely extends into the glia limitans or underlying brain or spinal cord. Any damage that is caused to underlying central nervous system tissue usually results from venous or arterial infarction that follows inflammation of the vessel wall and thrombotic occlusion of the vessel lumina. Meningitis resolves by the ingestion and removal of dead bacteria and polymorphonuclear leukocytes by macrophages. However, fibrous scarring during this stage may result in adhesions in the subarachnoid space, impedance of CSF flow, and post-meningitic hydrocephalus.

CSF withdrawn from the lumbar region of the spinal canal in cases of meningitis usually reveals a fall in the components of the CSF that are normally in excess of blood levels, and a rise in those components of the serum that are at low levels in the normal CSF. This reflects a change in the blood–CSF barrier, but it is unclear whether this is a change in the choroid plexus secretion or an inflammatory exudate derived from veins in the subarachnoid space. In meningitis, the levels of magnesium and Cl fall; phosphorus, potassium, and protein levels rise. Microscopy of the CSF in the acute stages of bacterial meningitis shows an increase in the number of polymorphonuclear leukocytes; bacteria may also be identified *(25)*. Culture will subsequently identify the organisms. In tuberculous meningitis, the protein level may be very high, and the main cells in the CSF are lymphocytes. *Mycobacterium tuberculosis* may be identified by microscopy, by culture, or by polymerase chain reaction. Viral infections involving the CSF and subarachnoid space are chiefly characterized by a lymphocyte exudate.

3.4.4. CSF Analysis in Other Diseases of Nervous System

Examination of the CSF is most informative in diseases such as meningitis and subarachnoid hemorrhage which directly involve the subarachnoid spaces. Similarly, in

Guillain-Barré syndrome, an autoimmune disease involving spinal nerve roots, protein levels in the CSF also rise.

Metastatic tumor cells invade the subarachnoid space and may spread diffusely in the CSF, resulting in "carcinomatous meningitis," hydrocephalus, and widespread signs and symptoms, particularly involving cranial and spinal nerves. Protein levels rise, and tumor cells may be detected in centrifuged deposits of lumbar CSF. This is a useful diagnostic tool for detecting spread of metastatic carcinomas, lymphomas, and malignant melanoma, to the leptomeninges. However, CSF cytology is less valuable for the diagnosis of primary tumors of the CNS, except for primitive neuroectodermal tumors (e.g., medulloblastomas of the cerebellum), which spread diffusely through the cerebrospinal fluid. Although astrocytic tumor cells may invade the subarachnoid space, they are usually adherent to the underlying CNS tissue, and are not released into the CSF in the same way as carcinoma cells *(26)*.

Analysis of the lumbar CSF is less reliable for disease processes deep within the brain. Nevertheless, diagnostic information in multiple sclerosis with oligoclonal bands of immunoglobulin may help to establish the diagnosis *(27)*. In dementias, such as AD, analysis of β-amyloid peptides (Aβ) suggest that the level of soluble Aβ 1-40 in the CSF falls *(28)*; the level of the more insoluble Aβ 1-42 increases, especially in the presence of cerebral amyloid angiopathy (CAA) *(29)*.

4. CIRCULATION AND LYMPHATIC DRAINAGE OF CSF IN EXPERIMENTAL ANIMALS

A high proportion, at least 50%, of intracranial CSF in the rat appears to drain directly to cervical lymph nodes *(3)*, and CSF from the spinal column drains to para-aortic lymph nodes *(4)*. Tracers injected into the rat CSF via the cisterna magna drain within minutes to cervical lymph nodes *(5)*. Histological studies *(4)*, following injection of Indian ink into the CSF, have revealed direct connections between the subarachnoid space and nasal lymphatics, through the cribriform plate of the ethmoid bone just below the olfactory bulbs. Similar lymphatic drainage pathways of the CSF have been identified in larger mammals. In sheep, radiolabeled serum albumin in the CSF drains almost equally into the lymphatic system and into the blood via the arachnoid villi *(30)*. Although they are present, arachnoid villi in the rat are relatively primitive, and pout into venous sinuses related to the olfactory bulbs *(4)*.

5. DRAINAGE OF INTERSTITIAL FLUID FROM THE BRAIN IN EXPERIMENTAL ANIMALS

Many of the data relating to the extracellular spaces (ECS) of the brain and drainage of interstitial fluid have been gathered from experimental studies, and, with some modification, these have important implications for the human brain *(5)*.

ECSs in the grey matter and white matter differ in their extensibility. As is seen with vasogenic edema and the interstitial edema of hydrocephalus *(5,31)*, myelinated nerve fibers in the white matter can be widely separated by expansion of the ECS. This is not so in the grey matter, in which the narrow, intercellular clefts between glial and neuronal processes are tightly controlled *(32)*. Resistance to bulk flow of interstitial fluid is high in these intercellular clefts, so that the main bulk flow of interstitial fluid appears to be along the wider, low-resistance perivascular channels (Fig. 2; *3*).

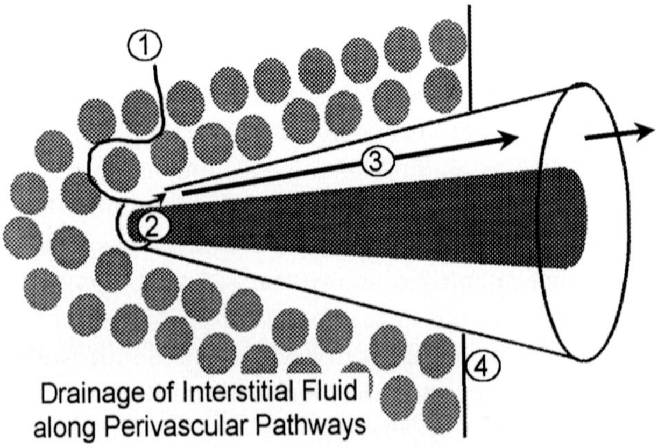

Fig. 2. Drainage of interstitial fluid along perivascular pathways. Anatomical and pathological studies, in the rat and human brain, suggest that interstitial fluid (ISF) and proteins drain from narrow intercellular clefts (1) into the PVSs of capillaries (2), and thence along PVSs of arteries within the brain (3), to the wide PVSs of leptomeningeal arteries. In relation to arteries, the PVS is separated from either brain or CSF by a layer of pia mater (4).

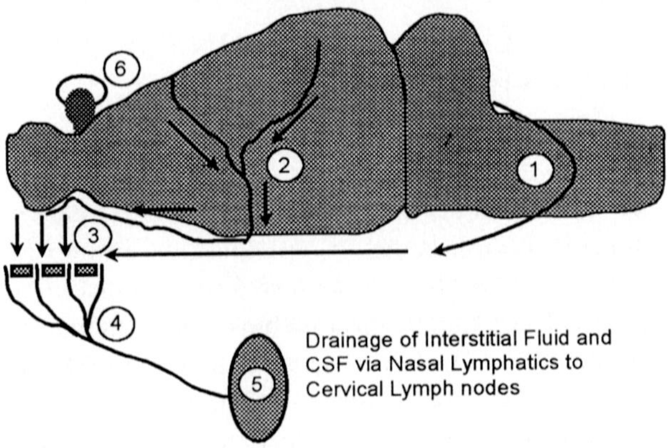

Fig. 3. Drainage of interstitial fluid and CSF via nasal lymphatics to cervical lymph nodes in the rat. CSF from the ventricles drains from the cisternal magna (1) to the subarachnoid space on the under surface of the brain, where it joins interstitial fluid draining along PVSs related to branches of the major cerebral arteries (2) and the circle of Willis. Fluid then passes anteriorly to the region of the olfactory bulbs (3), where it enters lymphatics at the cribriform plate, to drain, via the nasal submucosa (4), to cervical lymph nodes (5). CSF also passes directly back into the blood via primitive arachnoid villi (6).

5.1. Injection Studies

Tracers injected into the grey matter of rat and rabbit brains can be identified in cervical lymph nodes within a few hours *(3,5,33)*. Microscopic studies show how the tracers pass along the PVSs of arteries, both within the brain and within the subarachnoid space to drain by lymphatics through the cribriform plate of the ethmoid bone into the nasal submucosa, and thence to cervical lymph nodes (Fig. 3; *34*).

PVSs in the brain show a number of adaptations to their role as ISF drainage pathways. They are expandable, and they contain a relatively stable population of resident macrophages, the perivascular cells *(34,35)*. In the rat, perivascular cells demonstrate an individual phenotype as they express the ED2 antigen (Ag), which is not expressed by microglia in the parenchyma of the brain *(36)*. Injection of insoluble particulate matter into the brain results in the flow of particles along PVSs, and their ingestion by perivascular cells. Phagocytosis of particulate matter by macrophages is also seen within perivascular pathways around leptomeningeal vessels in the rat *(34)*.

5.2. Immunological Significance of Interstitial Fluid Drainage of CNS

The main biological significance for interstitial fluid drainage from the brain, which has been demonstrated in experimental animals, is related to immunological reactions.

5.2.1. B-Cell-Mediated Immunity in CNS

The most extensively investigated immunological relationships between the brain and the regional lymph nodes are those related to B-cell immune reactions *(3)*. Injection of human serum albumin into the central grey matter of the rat brain results in preferential production of antibodies in cervical lymph nodes; similar antibody production in cervical lymph nodes occurs when Ags are injected into the rat cerebrospinal fluid *(3)*. Removal of the cervical lymph nodes significantly reduces such antibody formation *(3)*, emphasizing the role played by cervical nodes as the regional lymph nodes for the brain.

5.2.2. T-Cell-Mediated Immunity in CNS

T-cell-mediated reactions in the brain appear to be more complicated than B-cell reactions. It has long been known that the brain is an immunologically privileged site, as shown by the long survival of tissue allografts in the brain *(3)*. However, T-cell-mediated immune reactions do occur in the brain, as seen in virus infections and in autoimmune diseases, such as experimental autoimmune encephalomyelitis (EAE) and multiple sclerosis in human. The question, therefore, is how does the brain relate to the immune system regarding T-cell-mediated immunity?

In most organs of the body, there is a defined lymphatic drainage to regional lymph nodes. In a number of organs, such as the gut, lung, and skin, T-cells appear to traffic from the tissue to regional lymph nodes then return to those target organs *(37)*. A certain degree of specificity may exist in the T-cell subsets of the various organs, as demonstrated by the expression of integrins on their surfaces *(37)*. Similar restriction is seen, whereby the majority of T-cells isolated from the CNS express $\alpha 4\beta 1$-integrin *(38)*.

In organs such as the gut, T-lymphocytes are readily identified within the submucosa, and the concept of lymphocyte surveillance applies to these organs. Few, if any, lymphocytes are seen in a normal brain, so any T-cell surveillance is not on the same scale as in organs such as the gut. In autoimmune diseases, such as EAE, it has been proposed *(39)* that activated lymphocytes enter the brain irrespective of the Ag against which they are directed. The entry of activated T-cells may initiate the autoimmune reaction, and, as more lymphocytes are recruited, the majority undergo apoptosis *(40)*; there is no firm evidence that T-cells escape from the brain. The relationship between the brain and the immune system, therefore, differs from that of other organs in one major respect: T-cells do not seem to circulate from the brain to regional lymph nodes, although lymphatic drainage of soluble Ags readily occurs.

Despite immunological privilege in the brain, the results of a number of experiments suggest that there is a firm interrelationship between the brain and regional lymph nodes with respect to T-cell-mediated immunity. Although intracerebral neural allografts survive indefinitely, rapid rejection of the intracerebral graft can be induced by orthotopic skin grafting *(41)*, which suggests that T-cell-mediated immunity in the brain may depend on peripheral immunization, as well as the presence of Ag in the CNS.

The role of cervical lymph nodes as a possible source for lymphocytes targeting the brain, has been studied in EAE. If active EAE is produced by the injection of Ags into the footpad of susceptible animals, such as Lewis rats, clinical signs of spinal cord damage, with paralysis in the lower limbs, develop ~10 d later. Inflammation characterized by T-lymphocyte infiltration and microglial activation is concentrated mainly in the spinal cord in these animals. The cerebral hemispheres are relatively spared at this early stage, and it is only later, at ~20 d postinoculation, that substantial inflammation is seen in the cerebral hemispheres *(42)*. Cerebral EAE can be enhanced by a brain wound induced 8 d after footpad inoculation *(43)*. Such a procedure results in a sixfold increase in inflammation in the cerebral hemispheres 15 d postinoculation. Removal of the cervical lymph nodes at the time of the brain injury reduces the amount of inflammation in the brain at 15 d by 50%, suggesting that the cervical lymph nodes play a major role in inflammation in the cerebral hemispheres, possibly as the origin of T-lymphocytes targeting the brain *(44)*. When lymphocytes, isolated from EAE animals with brain wounds, are injected into naïve recipients, inflammation in the brain is concentrated in the cerebral hemispheres, at the expense of the spinal cord *(45)*. Again, this suggests that drainage of Ags from the brain to cervical lymph nodes may induce targeting of lymphocytes toward the cerebral hemispheres.

From the EAE experiments outlined above, a picture is emerging that T-cell-mediated immunity in the CNS depends upon two major factors *(5)*. First, the presence in the circulation, and in the cervical and para-aortic lymph nodes of activated T-lymphocytes sensitized to CNS components. Second, the drainage of Ags from the brain or spinal cord to the regional lymph nodes; it is this event that may trigger invasion of the CNS by substantial numbers of T-lymphocytes. T-cell-mediated immunity to virus infections may be analogous to that seen in EAE with brain wounds. Entry of viruses into the brain requires initial passage through other tissues and blood. Peripheral immunization, therefore, occurs, and subsequent drainage of viral proteins from the brain to regional lymph nodes may then initiate T-cell-mediated immune reactions within the CNS.

Information regarding the role of brain injury in the enhancement of autoimmune disease in the brain continues to emerge. Despite the pivotal role played by cervical LNs in the enhancement of autoimmune inflammation, it has also been shown that local factors, such as the release of chemokines and stimulation of neuronal pathways, play an important role in determining the site of inflammation in EAE related to brain wounds *(42)*. Nevertheless, the local reactions to trauma are greatly enhanced in the presence of circulating reactive T-cells.

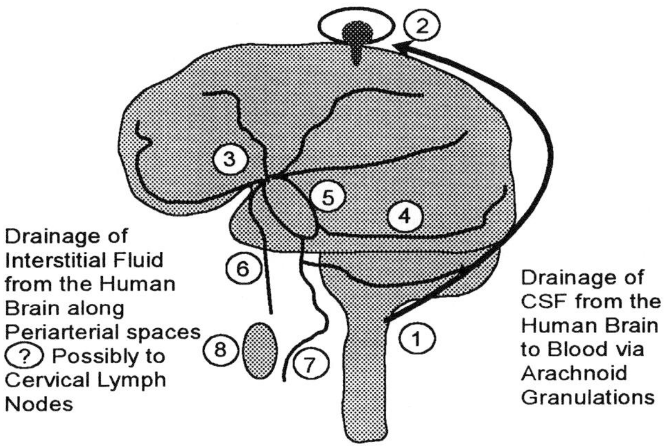

Fig. 4. Drainage of interstitial fluid and CSF from the human brain. CSF from the ventricles drains from the cisternal magna (1) to the subarachnoid space and thence into the blood via arachnoid granulations related to major venous sinuses (2) in the cranial and spinal compartments. Anatomical and pathological studies suggest that interstitial fluid from the brain parenchyma drains along PVSs around major cerebral arteries, e.g., anterior and middle (3) and posterior (4) cerebral arteries related to the circle of Willis (5). It is possible that ISF drains along the PVSs of carotid (6) and vertebral (7) arteries to cervical lymph nodes (8).

6. DRAINAGE OF INTERSTITIAL FLUID FROM THE HUMAN BRAIN

Tracer experiments to examine the drainage of interstitial fluid from the human brain have not been developed, as yet (Fig. 4). Evidence for ISF drainage pathways, therefore, relies on anatomical studies, analogy with experimental animals, and on natural disease processes, such as CAA.

6.1. *Anatomical Pathways*

Anatomical study of PVSs in the brain and leptomeninges is easier in the human brain than in animals, because of the larger size of the structures in human. Electron microscopy has shown that interstitial spaces in grey matter are in direct continuity with the PVSs of capillaries *(32)*. PVSs around capillaries are mostly occupied by basement membrane, and are continuous from capillaries along the walls of cortical arterioles, and ultimately with the PVSs of arteries in the leptomeninges (Fig. 2; *46*). A layer of pia mater surrounds the PVSs of arterioles in the cerebral cortex *(7,46)*. This layer of pia not only separates the periarterial spaces from surrounding brain tissue, but it also ensures continuity of the intracortical PVSs with the wider spaces around leptomeningeal arteries *(46)*. Passing from capillary to leptomeningeal artery, the PVS becomes progressively larger, and may, therefore, offer progressively lower resistance to the drainage of ISF and proteins.

Evidence for functional lymphatic drainage of ISF has not yet been established in human, but pericarotid and perivertebral artery pathways would seem to be likely routes, because of their continuity with the perivascular pathways in the brain parenchyma. The perivascular compartment around the intracranial carotid artery can be traced through the base of the skull, to the region of the deep cervical lymph nodes (unpublished data), and may represent a route for lymphatic drainage of ISF from the brain. There is evidence from early studies by Zwillinger, in 1912 *(47)*, that a small amount of CSF may drain through arachnoid prolongations in the nasal mucosa in human. But, in contrast to the rat, dog, and other species, the main olfactory artery, which is so prominent as a conduit for perivascular drainage of ISF in the rat, no longer exists in human *(4,34)*. It would seem, therefore, that lymphatic drainage of interstitial fluid through the cribriform plate to cervical LNs in a way similar to the rat, is unlikely in human, and that periarterial routes are more probable.

6.2. CAA as Reflection of ISF Drainage Pathways

Data that would support periarterial channels as interstitial fluid drainage pathways in human also come from the study of pathological tissues. For example, edema fluid appears to drain from grey matter along PVSs, with activation of the phagocytic perivascular cells *(36)*. However, much more dramatic visual evidence for perivascular drainage of ISF is seen in CAA.

Amyloid peptides form insoluble 7-nm fibrils, with a β-pleated sheet structure, in a number of different diseases in human. Such peptides vary in their origin, and may consist of Aβ amyloid and its various isoforms in AD, cystatin C in hereditary cystatin-C amyloid angiopathy, or the prion protein typical of sporadic and familial Creutzfeldt-Jakob disease *(6,48)*. Deposition of amyloid occurs not only as plaques within brain parenchyma, but also as CAA caused by the accumulation of amyloid peptides within the walls of cortical and leptomeningeal blood vessels *(21)*.

Amyloid in vessel walls in CAA outlines perivascular interstitial fluid drainage pathways analogous to those in experimental animals *(6)*. Aβ, for example, accumulates in PVSs around capillaries, where it is focally continuous with Aβ in the ECSs of the surrounding brain tissue. Deposition of Aβ is seen within the walls of cortical arterioles, and in leptomeningeal arteries *(6)*. As amyloid deposition increases, smooth muscle cells disappear from the artery walls, the vessel wall is weakened, aneurysms form, and hemorrhage occurs, as a direct complication of CAA. In the meninges, it is mostly the smaller arteries that show deposition of histologically stainable Aβ in the adventitial PVSs, and within the media. Nevertheless, biochemical studies have shown that Aβ is present in the walls of large arteries, such as the middle cerebral and basilar arteries, even in individuals as young as 30 yr of age *(49)*; this supports the concept that perivascular pathways are the natural route for the drainage of Aβ throughout life. As yet, the ultimate destination of interstitial fluid draining from the brain along perivascular pathways, is unknown, and the amount draining to cervical lymph nodes is similarly unquantified.

Further evidence that CAA results from the deposition of amyloid in perivascular interstitial fluid drainage pathways is derived from the study of transgenic mice, in which the transgene is a mutation in the amyloid precursor protein gene, and the promoter is restricted to the brain. In such mice, plaques of amyloid form in the brain parenchyma, as in AD, and there is a heavy involvement of blood vessels, both within the brain and within the leptomeninges by CAA *(50)*.

7. CONCLUSIONS

The physiology of cerebrospinal fluid production and drainage has received much attention for many years, in contrast to the relative lack of interest in the production and drainage of interstitial fluid from the brain. Studies in experimental animal and human brains, nevertheless, reveal mounting evidence for selective interstitial fluid drainage pathways, particularly in the human brain. This has implications both in health and disease. A combination of experimental studies in animals and the pathoanatomical observations in human enhances progress in the study of interstitial fluid drainage pathways, as long as the anatomical, and possibly the physiological, differences between human and smaller mammals are taken into account. Recognition that drainage of interstitial fluid and Ags from the brain to regional lymph nodes in man may play a role in the induction and regulation of immunological diseases, such as multiple sclerosis, may lead to reorientation of therapeutic strategies. Similarly, the relationship between drainage of amyloid peptides along perivascular pathways and the pathogenesis of AD and other dementias, may stimulate research into the enhancement of interstitial fluid drainage for the therapy or prevention of these devastating diseases.

REFERENCES

1. Roitt, I., Brostoff, J., and Male, D. (1998) *Immunology*, 5th ed. Mosby, London.
2. Davson, H., Welch, K., and Segal, M. B. (1987) *Physiology and Pathophysiology of the Cerebrospinal Fluid.* Churchill Livingstone, Edinburgh.
3. Cserr, H. F., Harling-Berg, C. J., and Knopf, P. M. (1992) Drainage of brain extracellular fluid into blood and deep cervical lymph and its immunological significance. *Brain Pathol.* **2**, 269–276.
4. Kida, S., Pantazis, A., and Weller, R. O. (1993) CSF drains directly from the subarachnoid space into nasal lymphatics in the rat. Anatomy, histology and immunological significance. *Neuropathol. Appl. Neurobiol.* **19,** 480–488.
5. Weller, R. O. (1998) Pathology of cerebrospinal fluid and interstitial fluid of the CNS: significance for Alzheimer disease, prion disorders and multiple sclerosis. *J. Neuropathol. Exp. Neurol.* **57,** 885–894.
6. Weller, R. O., Massey, A. P., Newman, T. A., Hutchings, M., Kuo, Y.-M., and Roher, A. (1998) Cerebral amyloid angiopathy: β-amyloid accumulates in putative interstitial fluid drainage pathways in Alzheimer's disease. *Am. J. Pathol.* **153,** 725–734.
7. Weller, R. O. (1995) Fluid compartments and fluid balance in the central nervous system, in *Gray's Anatomy*, 38th ed. (Williams, P. L., ed.), Churchill Livingstone, Edinburgh, pp. 1202–1224.
8. Alcolado, J. C., Moore, I. E., and Weller, R. O. (1986) Calcification in the human choroid plexus meningiomas and pineal gland. *Neuropathol. Appl. Neurobiol.* **12,** 235–250.
9. Weller, R. O., Steart, P. V., and Moore, I. E. (1986) Carbonic anhydrase C as a marker antigen in the diagnosis of choroid plexus papillomas and other tumours. An immunoperoxidase study, in *Biology of Brain Tumour* (Walker, M. D. and Thomas, D. G. T., eds.), Martinus Nighoff and Dr. W. Junk, Dordrecht, pp. 115–120.
10. Cutler, R. W. P., Page, L., Galicich, J., and Watters, G. V. (1968) Formation and absorption of cerebrospinal fluid in man. *Brain* **91,** 707–720.
11. Bradley, W. G. (1990) Flow phenomena, in *MRI Atlas of the Brain* (Bradley, W. G. and Bydder, G., eds.), Martin Dunitz, London, pp. 68–111.
12. Hutchings, M. and Weller, R. O. (1986) Anatomical relationships of the pia mater to cerebral blood vessels in man. *J. Neurosurg.* **63,** 316–325.
13. Feurer, D. J. and Weller, R. O. (1991) Barrier function of the leptomeninges: a study of normal meninges and meningiomas in tissue culture. *Neuropathol. Appl. Neurobiol.* **17,** 391–405.
14. Krahn, V. (1982) The pia mater at the site of entry of blood vessels into the central nervous system. *Anat. Embryol.* **164,** 257–263.

15. Alcolado, R., Weller, R. O., Parrish, E. P., and Garrod, D. (1988) The cranial arachnoid and pia mater in man: anatomical and ultrastructural observations. *Neuropathol. Appl. Neurobiol.* **14,** 1–17.
16. Nicholas, D. S. and Weller, R. O. (1988) The fine structure of the spinal meninges in man. *J. Neurosurg.* **69,** 276–282.
17. Upton, M. L. and Weller, R. O. (1985) The morphology of cerebrospinal fluid drainage pathways in human arachnoid granulations. *J. Neurosurg.* **63,** 867–875.
18. Tripathi, B. J. and Tripathi, R. C. (1974) Vacuolar transcellular channels as a drainage pathway for cerebrospinal fluid. *J. Physiol. (Lond.)* **239,** 195–206.
19. Schurr, P. H. and Polkey, C. E., eds. (1993) *Hydrocephalus*, Oxford University Press, Oxford.
20. Del Bigio, M. R., Cardoso, E. R., and Halliday, W. C. (1997) Neuropathological changes in chronic adult hydrocephalus: cortical biopsies and autopsy findings. *Can. J. Neurol. Sci.* **24,** 121–126.
21. Esiri, M. M., Hyman, B. T., Beyreuther, K., and Masters, C. L. (1997) Ageing and dementia, in *Greenfield's Neuropathology*, 6th ed., vol. II (Graham, D. I. and Lantos, P. L., eds.), Arnold, London, pp. 153–233.
22. Harding, B. and Copp, A. J. (1997) Malformations, in *Greenfield's Neuropathology*, 6th ed. (Graham, D. I. and Lantos, P. L., eds.), Edward Arnold, London, pp. 397–533.
23. Walton, J., ed. (1993) *Brain's Diseases of the Nervous System*, Oxford University Press, Oxford, pp. 56–65.
24. Kalimo, H., Kaste, M., and Haltia, M. (1997) Vascular disease, in *Greenfield's Neuropathology*, 6th ed. (Graham, D. I. and Lantos, P. L., eds.), Edward Arnold, London, pp. 315–396.
25. Lambert, H. P., ed. (1991) *Infections of the Central Nervous System*, B. C. Decker, Philadelphia.
26. Ellison, D. W. and Love, S., eds. (1998) *Neuropathology*, Mosby, London.
27. Thompson, E. J. (1995) Cerebrospinal fluid. *J. Neurol. Neurosurg. Psychol.* **59,** 349–357.
28. Trojanowski, J. Q. and Growdon, J. H. (1998) A new consensus report on biomarkers for the early antemortem diagnosis of Alzheimer Disease: current status, relevance to drug discovery, and recommendations for future research. *J. Neuropathol. Exp. Neurol.* **5,** 634–644.
29. Pitschke, M., Prior, R., Haupt, M., and Riesner, D. (1998) Detection of single amyloid β-protein aggregates in the cerebrospinal fluid of Alzheimer's patients by fluorescence correlation spectroscopy. *Nature Med.* **4,** 832–834.
30. Boulton, M., Young, A., Hay, J., et al. (1996). Drainage of CSF through lymphatic pathways and arachnoid villi in sheep: measurement of ^{125}I-albumin clearance. *Neuropathol. Appl. Neurobiol.* **22,** 325–333.
31. Miller, J. D. and Ironside, J. W. (1997) Raised intracranial pressure, oedema and hydrocephalus, in *Greenfield's Neuropathology*, 6th ed. (Graham, D. I. and Lantos, P. L., eds.), Edward Arnold, London, pp. 157–196.
32. Nicholson, C. and Syková, E. (1998) Extracellular space structure revealed by diffusion analysis. *Trends Neurosci.* **21,** 207–215.
33. Foldi, M., Csillik, B., and Zoltan, O. T. (1968) Lymphatic drainage of the brain. *Experientia* **24,** 1283–1287.
34. Zhang, E. T., Richards, H. F., Kida, S., and Weller, R. O. (1992) Directional and compartmentalised drainage of interstitial fluid and cerebrospinal fluid from the rat brain. *Acta Neuropathol.* **83,** 233–239.
35. Graeber, M. B., Streit, W. J., Buringer, D., Sparks, D. L., and Kreutzberg, G. W. (1992) Ultrastructural location of major histocompatibility complex (MHC) Class II positive perivascular cells in histologically normal human brain. *J. Neuropathol. Exp. Neurol.* **51,** 303–311.

36. Kida, S., Ellison, D. W., Steart, P. V., and Weller, R. O. (1995) Characterisation of perivascular cells in astrocytic tumours and oedematous brain. *Neuropathol. Appl. Neurobiol.* **21,** 121–129.
37. Springer, T. A. (1994) Traffic signals for lymphocyte recirculation and leukocyte emigration: the multiple paradigm. *Cell* **76,** 301–314.
38. Engelhardt, B., MartinSimnet, M. T. G., Rött, C. S., Butcher, E. C., and Michie, S. A. (1998) Adhesion molecule phenotype of T lymphocytes in inflamed central nervous system. *J. Neuroimmunol.* **84,** 92–104.
39. Wekerle, H., Engelhardt, B., and Risau, W. (1991) Interaction of T lymphocytes with cerebral endothelium *in vivo*. *Brain Pathol.* **1,** 107–114.
40. Bauer, J., Bradl, M., Hickey, W. F., et al. (1998) T-cell apoptosis in inflammatory brain lesions. *Am. J. Pathol.* **153,** 715–724.
41 Duan, W. M., Widner, H., Frodl, E. M., and Brundin, P. (1995) Immune reactions following systemic immunisation prior or subsequent to intrastitial transplantation of allogenic mesencephalic tissue in adult rats. *Neuroscience* **64,** 629–641.
42. Sun, D., Tani, M., Newman, T. A., et al. (2000) Role of chemokines and neuronal projections in the enhancement of cerebral EAE following focal brain damage *J. Neuropathol. Exp. Neurol.* **59,** 1031–1043.
43. Phillips, M. J., Weller, R. O., Kida, S., and Iannotti, F. (1995) Focal brain damage enhances experimental allergic encephalomyelitis in brain and spinal cord. *Neuropathol. Appl. Neurobiol.* **21,** 189–200.
44. Phillips, M. J., Needham, M., and Weller, R. O. (1997) Role of cervical lymph nodes in autoimmune encephalomyelitis in the Lewis rat. *J. Pathol.* **182,** 457–464.
45. Lake, J., Weller, R. O., Phillips, M. J., and Needham, M. (1998) Lymphocyte targeting of the brain in adoptive transfer cryolesion: EAE. Implications for multiple sclerosis. *J. Pathol.* **187,** 259–265.
46. Zhang, E. T., Inman, C. B. E., and Weller, R. O. (1990) Interrelationships of the pia mater and perivascular (Virchow Robin) spaces in the human cerebrum. *J. Anat.* **170,** 111–123.
47. Zwillinger, H. (1912) Die Lymphbahnen des oberen Nasalschnittes und deren Beziehungen zu den perimeningealen Lymphraumen. *Arch. F. Laryngol. Und. Rhinol.* **26,** 66–78.
48. Vinters, H. V. (1996) Brain parenchymal and microvascular amyloid in Alzheimer's disease. *Brain Pathol.* **6,** 179–195.
49. Shinkai, Y., Yoshimura, M., Ito, Y., et al. (1995) Amyloid β-proteins 1-40 and 1 42(43) in the soluble fraction of extra- and intracranial blood vessels. *Ann. Neurol.* **38,** 421–428.
50. Calhoun, M. E., Burgermeister, P., Phinney, A. L., et al. (1999) Neuronal overexpression of mutant amyloid precursor protein results in prominent deposition of cerebrovascular amyloid. *Proc. Natl. Acad. Sci. USA* **96,** 14,088–14,093.

11
The Blood–Brain Barrier

Richard F. Keep

1. INTRODUCTION

The main function of the cardiovascular system is the exchange of solutes between blood and tissue interstitial space. Thus, blood is the source of oxygen (O_2) and nutrients such as glucose (GLU) for the tissue, and the mechanism by which waste products, such as carbon dioxide (CO_2), are removed from tissue. The site of such exchange is the capillary. In most tissues of the body, plasma solutes (except plasma proteins) freely diffuse across or between the capillary endothelial cells (ECs). In such tissues, the stability of the tissue microenvironment (the interstitial fluid [ISF]) depends chiefly on the stability of plasma composition. In the case of the brain, stability of the microenvironment is essential for normal brain function, and the cerebral capillaries (with the exception of certain specialized regions: *see* Chapter 12) are structurally and functionally different from other capillaries, forming what is known as the blood–brain barrier (BBB). The cerebral capillary ECs are linked by tight junctions and form a diffusion barrier to the entry of many compounds from blood to brain. However, the cerebral capillaries are not just a passive barrier: There are many different transport processes at the cerebral capillaries to facilitate the movement of nutrients into brain, and to control the brain microenvironment. They also serve as an enzymatic barrier to the movement of compounds between blood and brain via degradation of unwanted compounds.

The cerebral capillaries are not the only tissues involved in regulating the passage of compounds between blood and brain. There is no barrier tissue between cerebrospinal fluid (CSF) and brain, and, therefore, compounds that enter CSF via choroid plexus or arachnoid membrane (the sites of the blood–CSF barrier; *see* Chapter 10) are free to diffuse into brain. Thus, control of the brain microenvironment involves regulation of exchange at both BBB and blood–CSF barrier (*see* Chapter 10). However, unlike for the choroid plexuses and arachnoid membrane, all the cells of the brain are within about 100 µm of a cerebral capillary. This difference in proximity may be important in determining the relative roles of these different tissues in controlling the brain microenvironment.

An understanding of the BBB (in concert with other cell types) is important in providing an insight into how the composition of the brain microenvironment is normally controlled, and how this may be altered in various pathophysiological situations.

From: *The Neuronal Environment: Brain Homeostasis in Health and Disease*
Edited by: W. Walz © Humana Press Inc., Totowa, NJ

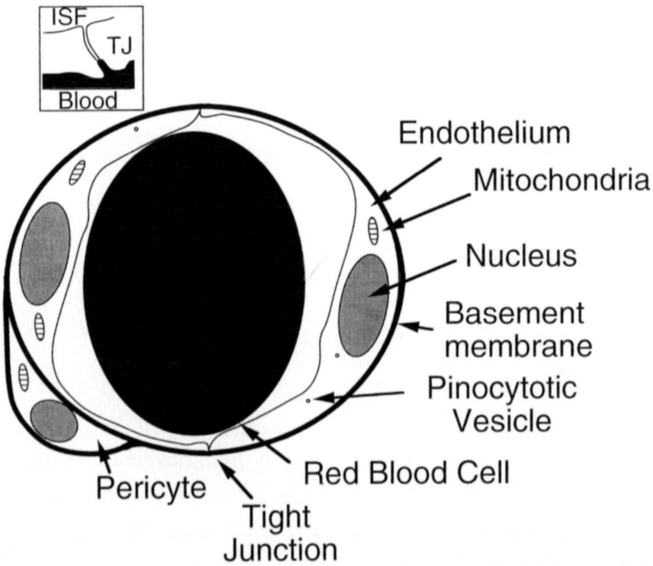

Fig. 1. Schematic of the structural elements of the cerebral capillary. The cerebral capillaries are composed of ECs and pericytes. The latter are enclosed in the same basement membrane as the EC. The ECs are linked by tight junctions (TJ). This, along with the paucity of vesicles and the absence of fenestrations, results in the diffusion barrier to the migration of solutes between blood and brain (the BBB). This is demonstrated in the inset, which depicts the classic electron microscope experiments of Brightman and Reese *(3)*, in which the passage of intravascularly injected lanthanum between two ECs is blocked at the level of the tight junction (the electron-dense lanthanum does not pass into the ISF).

It is also an important therapeutic consideration, since the presence of a BBB often limits the penetration of therapeutic agents into the brain ISF extracellular space (ECS).

The overall purpose of this chapter is to provide insight into the role of the BBB in controlling the composition of the brain ISF, i.e., the neuronal microenvironment. One of the factors limiting understanding is the relative inacessibility of the interstitial space. In this chapter, therefore, inferences are sometimes made concerning the brain ISF from measurements of CSF. There is (at least in the adult) little barrier to diffusion between those two compartments, but there are potential pitfalls in such inferences.

2. BARRIER FUNCTIONS: OVERVIEW

The purpose of this subheading is to give an overview of the barrier functions of the cerebral ECs. More details on each type of barrier function are given in later subheadings.

2.1. *Physical*

The existence of a barrier between blood and brain was first suggested because intravenous administration of vital dyes resulted in staining of most tissues, but not the brain *(1)*. The classic studies of Reese and Karnovsky *(2)* and Brightman and Reese *(3)* demonstrated that this barrier is caused by the ECs of the cerebral capillaries, which, unlike ECs in other tissues, are linked by tight junctions. Those junctions limit the paracellular passage of compounds between blood and brain (Fig. 1). Electron micro-

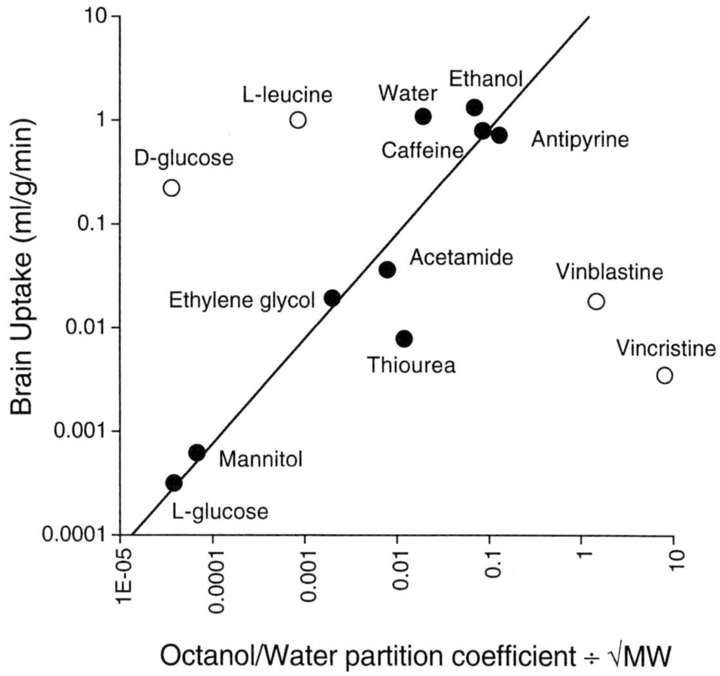

Fig. 2. The relation between lipid solubility and the brain uptake of different compounds. Allowing for differences in molecular weight (MW), there is a good relation between lipid solubility (as assessed by the octanol–water partition coefficient) and brain uptake for a number of compounds (filled circles). However, some compounds either fall above (e.g., D-GLU and L-leucine) or below (e.g., vinblastine and vincristine) this relationship, because of carrier-mediated transport into and out of the brain, respectively. The data are from refs. *144–147.*

scopic studies also indicate that the ECs of the BBB have relatively few pinocytotic vesicles, compared to other endothelia, suggesting that the passage of large-mol-wt compounds by a vesicular pathway is also limited at the BBB *(2)*. The ECs that form the BBB also lack the fenestrations found in high-permeability capillaries.

Around the abluminal membrane of the ECs (and enclosed by the endothelial basement membrane), there are sporadic pericytes (Fig. 1). These, along with the astrocyte endfeet that almost completely encircle the cerebral capillaries, do not form a physical barrier to the diffusion of substances from blood-to-brain ISF, although they have other important functions.

2.2. Diffusion and Transport Across EC

The absence of significant paracellular flux across the BBB raises questions as to how nutrients enter brain from blood, and how waste products generated by the brain are cleared to blood. Some compounds, e.g., O_2 and CO_2, can readily diffuse across the EC. For such compounds, the rate of diffusion is closely linked to their lipid solubility (Fig. 2), with compounds with a high lipid solubility (usually assessed as their octanol:water partition coefficient) having a higher permeability. However, this does not apply to all compounds. There are many compounds with an entry rate much greater than that expected by lipid solubility, including D-GLU and L-amino acids and these compounds have specific transporters mediating blood-to-brain transport (*see* Subheading 4.). There is a second group of compounds, such as vinblastine and vincristine, with an entry rate much less than that expected by lipid solubility. That group, which includes a number of xenobiotics, have specific transporters mediating brain-to-blood (or at least EC–blood) transport (*see* Subheading 8.).

2.3. Metabolic

The ECs that form the BBB may also form a metabolic barrier to the entry of compounds from blood to brain. Thus, the BBB possesses many enzymes that may prevent the entry of compounds via degradation. Alternately, they may chemically modify compounds, so that they are substrates for efflux mechanisms that transport compounds from the endothelium to blood. It should be noted, however, that it is not always evident whether the enzymes involved in these two types of processes are protective for the brain as a whole, or just the EC itself.

3. BARRIER INTEGRITY

3.1. Physiology

As described above, the BBB is formed by the cerebral capillary EC and the tight junctions (*zona occludens*) that link them. The resultant barrier has a high electrical resistance. In vivo measurements of 3000 Ω/cm^2 have been recorded *(4)*; there are indications that the barrier may be even tighter (8000 Ω/cm^2 *[5]*).

Much work has centered on determining the components of tight junctions at the molecular level *(6,7)*. A number of proteins appear to be integral to tight junction function (the claudins, occludin, and perhaps JAM); others appear to have a peripheral role (such as ZO-1, -2, and -3; cinguilin, 7H6, Rab3B, symplekin, AF-6, and ASIP). A model of tight junction structure and the relationship between these proteins, is given in Fig. 3. The complexity of this tight junction structure, as well as their association with signaling molecules (such as the G proteins, $G\alpha_0$ and $G\alpha_{i2}$, and the protein kinase C isotypes, PKCζ and PKCλ *[6]*), suggests that permeability may be regulated by a number of factors. At the BBB, there is evidence that a number of mediators may indeed modulate paracellular permeability in vivo, including histamine, serotonin and bradykinin *(4,8,9)*. The physiological significance of such changes, in terms of fluid or solute flux across the BBB, has not been fully elucidated, although such changes may have a pathophysiological role and have practical importance in terms of drug delivery to the brain (*see* below).

A number of groups have also examined regulation of the permeability of brain microvessel endothelial monolayers in vitro. Culture of brain microvessel ECs alone, or culture of immortalized brain microvessel ECs (e.g., RBE4 cells), results in monolayers with a low resistance (the different in vitro systems are reviewed in ref. *10*). Co-culture of such ECs with astrocytes results in monolayers with higher resistances, although these are still far less than those that occur in vivo *(11,12)*. Given the understanding that these cultures do not fully express all BBB features *(13,14)*, and that they might even express characteristics not found in the BBB in vivo, these cultures have been useful in examining the potential regulators of BBB permeability, and particularly the cellular mechanisms involved. Thus, for example, increasing EC cyclic adenosine monophosphate levels causes them to form a higher electrical resistance

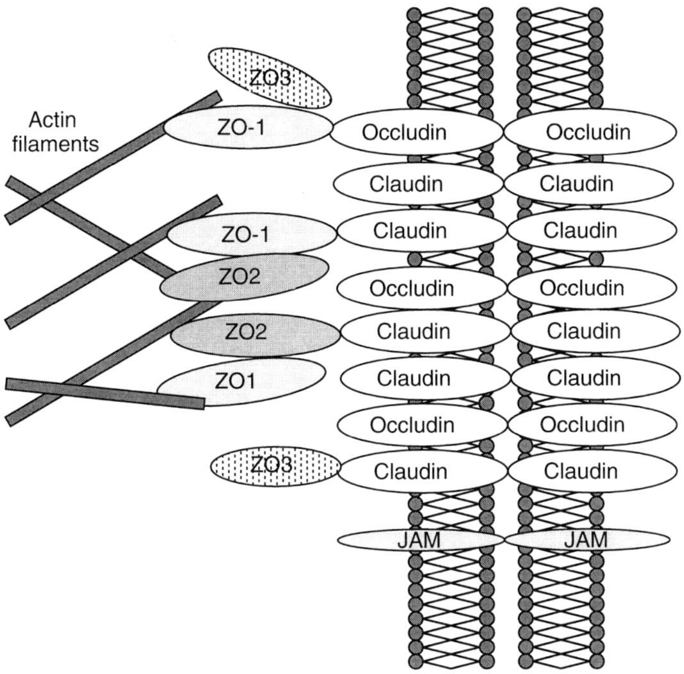

Fig. 3. A model of some of the proteins involved in tight junction structure. Claudins, occludins, and JAM are transmembrane proteins that appear to be integral to tight junction function *(6,7)*. ZO-1, ZO-2, and ZO-3 bind to occludins, and probably claudins. ZO-1 binds to both ZO-2 and ZO-3 to form heterodimers. In addition, ZO-1 and ZO-2 bind to actin filaments, and, thus, for the link between the tight junction and the cytoskeleton. Not shown are a number of proteins associated with tight junctions, and potentially their regulation, such as cinguilin, 7H6, Rab3B, symplekin, AF-6, and ASIP.

monolayer *(15)*; a variety of agents that induce a rise in intracellular calcium (Ca) cause a fall in resistance *(16)*.

3.2. Pathophysiology

Alterations in BBB integrity occur in many disease states, such as brain tumors, traumatic brain injury, cerebral hemorrhage, ischemic brain damage, meningitis, and hypertension *(17)*. As discussed in the overview, the physical barrier to diffusion at the BBB reflects the tight junctions linking the ECs, the absence of open interendothelial clefts, the lack of fenestrations, and the paucity of vesicles. The cause of the decreased BBB integrity differs between the different disease states, and, in the case of brain tumors, can even vary within a single tumor. Tumor vasculature may lack tight junctions or have wide interendothelial clefts, fenestrations, and numerous vesicles, resulting in markedly enhanced penetration of substances from blood to tissue *(18)*. In ischemic brain damage, there may be a more subtle loss of integrity. Thus, the permeability of the BBB to a nontransported amino acid, α-aminoisobutyric acid (AIB), may only increase 2–3-fold following focal cerebral ischemia *(19)*; in rats bearing W256 and C6 gliomas, it increases 8- and 22-fold, respectively *(20)*. In focal cerebral ischemia, BBB disruption appears to occur via alteration in tight junction function *(21)*.

One consequence of a loss of BBB integrity is an increased movement of fluid from blood, so-called "vasogenic" brain edema *(22)*. Because of the brain is enclosed in the skull, such edema can lead to increased intracranial pressure and, potentially, brain herniation. A loss of BBB integrity may also lead to an increased entry of potentially neurotoxic compounds from blood to brain, which may further exacerbate the brain injury. Dexamethasone has been used as a treatment for vasogenic edema formation. Particularly in some brain tumors, it reduces both permeability of the tumor vasculature and edema *(23,24)*. Its mechanism of action is still not well-understood.

Although alterations in BBB integrity are a complication in many disease states, there are situations when a disruption of the BBB is a desired therapeutic outcome, since it may enable a greater entry of therapeutic agents into brain. Two major methods have been used to disrupt the BBB. The first is hyperosmotic opening, in which a bolus of hyperosmotic solution is injected into the carotid artery. This is thought to shrink the ECs of the BBB, causing a disruption in their linking tight junctions *(25)*. The second method is via activation of endothelial receptors that may modify junctional structure. The best-studied of these systems is the bradykinin B_2 receptor, with a number of studies indicating that the agonist, RMP-7 (Cereport), can increase BBB permeability *(26)*. That effect is particularly pronounced in tumor vasculature, but RMP-7 also has effects on the normal BBB.

4. NUTRIENT TRANSPORT

4.1. *Physiology*

Brain function is dependent on the entry of nutrients from blood to the brain ISF. Although some nutrients can cross the BBB by diffusion (e.g., O_2) the physical characteristics of others (e.g., GLU and amino acids) are such that they do not readily diffuse across endothelial plasma membranes and diffusion via the paracellular pathway is precluded by the tight junctions. For these latter compounds, there are an array of transport systems, some of which are listed in Table 1.

There are high rates of entry of GLU, mediated by the GLU transporter, GLUT1 *(27)*. GLU is the primary energy substrate for the brain, and accounts for almost all the O_2 consumption of the brain. GLUT1 is a facilitative transporter, and, therefore, it does not specifically transport GLU in one direction, but merely facilitates the movement of GLU down its concentration gradient. It is present on both the luminal (blood-facing) and abluminal (brain-facing membrane). GLUT1 is stereoselective, transporting D- but not L-GLU (the latter only penetrates the brain very slowly; *see* Fig. 2).

Although GLU is the main energy substrate for the adult brain, lactate and ketone bodies are an important energy source prior to weaning. There is a specific transporter at the BBB for transporting monocarboxylic acids (including ketone bodies, lactate, acetate, and pyruvate). That transporter, MCT1, is present at very high levels in the BBB of neonates, but is still present in substantial amounts at the adult BBB *(28)*. Like GLUT1, MCT1 is present on both the luminal and abluminal membrane of the BBB. However, unlike GLUT1, MCT1 is proton-dependent, and, thus, the direction of monocarboxylate transport is dependent on the hydrogen ion gradient across each membrane *(29)*.

Table 1
Blood-to-Brain Transporters for Metabolic Substrates at BBB

Transport system	Mediator	Typical substrates
Hexose	GLUT1	GLU
Monocarboxylic acid	MCT1	Lactate; ketone bodies
Large neutral amino acid (System L)	LAT1	Phe, leucine
Basic amino acid (System y^+)	CAT1	Lysine, arginine
Acidic amino acid (System X^-)		Glu, aspartate
System N	SN1 (?)	Gln, histidine
β-amino acid		Taurine, β-alanine
Amine		Choline
Purine		Adenine
Nucleoside		Adenosine
Saturated fatty acid		Octanoate

Cloned transporters are listed under Mediator.

Large, neutral amino acids are also transported from blood to brain rapidly via the System L amino acid transporter *(30)*; *see* Fig. 4 summary of amino acid transport at the BBB). This transporter, LAT1, was recently cloned *(31)*, and is selectively expressed at the BBB, compared to other tissues *(32)*. It transports a wide range of substrates (such as phenylalanine [Phe], leucine, tyrosine, isoleucine, valine, tryptophan, methionine, and histidine). These include essential amino acids that cannot be synthesized by the brain. Like GLUT1, it is a facilitative transporter, as is another amino acid transporter, System y^+, which transports cationic amino acids (such as lysine, arginine, and ornithine) between blood and brain. Three cationic amino acid transporters have been cloned (CAT1, CAT2, CAT3), and the BBB System y^+ appears to be CAT1 *(30)*. Recently, it has been shown that a major portion of blood to brain glutamine (Gln) transport is mediated by System N amino acid transport *(33)*. This transporter was named "System N," because its naturally occurring substrates (Gln, histidine, and asparagine) have a nitrogen in their side chains *(34)*. It has recently been cloned (SN1) *(35)*. Unlike System L and y^+, System N amino acid transport is Na^+-dependent. Because of the high concentration of Gln in plasma, System N-mediated transport may represent a substantial portion of total amino acid flux into brain. Compared to System L, System y^+, and System N substrates, many other amino acids have a very low uptake from blood to brain, and are actually the subject of active efflux from brain to blood (e.g., substrates for System A amino acid transport; *see* Subheading 8.).

There are other blood to brain transporters for other metabolic substrates *(36)*. Their transport rate is less than those described above, but they may be vital for brain function. Thus, there is a blood–brain choline transporter. Since choline is not synthesized by the brain, choline transport is hypothesized as a regulator of acetylcholine formation in the brain *(37)*. The BBB also transports purines, nucleosides, and saturated fatty acids from blood to brain. It also transports vitamins and cofactors necessary for metabolism.

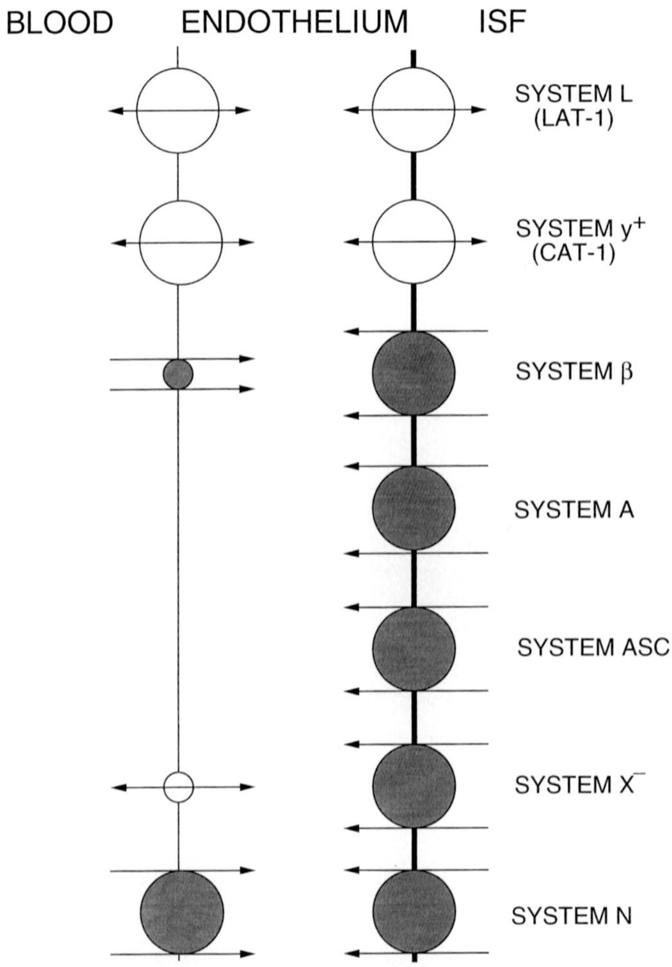

Fig. 4. Polarity in the distribution of amino acid transporters at the BBB. Amino acid transporters are either facilitative (shown with bidirectional arrows, e.g., System L) or Na^+-dependent (filled symbols and unidirectional arrows, e.g., System A). Some transporters are present on both the luminal and abluminal (e.g., System L); others are totally polarized to one membrane (e.g., System A); and yet others may be predominantly on one membrane (e.g., System β). For acidic amino acids, there may be a Na^+-independent transporter on the luminal membrane and a Na^+-dependent transporter on the abluminal membrane. For System N, there is in vivo evidence for a luminal distribution and in vitro evidence for an abluminal distribution *(33,120)*.

4.2. Pathophysiology

As yet, relatively few disease states have been identified that result from changes in BBB nutrient transporters. This may reflect the importance of such systems, which could lead to genetic mutations in such transporters being lethal *in utero*. However, recent efforts in the cloning of different transporters may provide insights into whether subtle modifications in transporter function might lead to alterations in brain function.

Three disease states that are associated with alterations in BBB nutrient transport are GLU transporter protein syndrome, hepatic encephalopathy, and phenylketonuria (PKU). In GLU transporter protein syndrome, there are partial defects in GLUT1-mediated GLU transport, with, as yet, two identified causes: GLUT1 hemizygosity, and a truncated GLUT-1 protein, because of nonsense mutations *(38,39)*. This rare syndrome is associated with infantile seizures, developmental delay, and acquired microcephaly.

One of the effects of liver failure is a deterioration in brain function, known as "hepatic encephalopathy." The probable underlying cause is a rise in plasma and, thus, brain ammonia concentrations *(40)* and a build-up in brain Gln, because of the conversion of ammonia and glutamate (Glu) to Gln *(41)*. Hepatic encephalopathy is associated with marked increases in the brain concentrations of amino acids transported by System L *(42,43)*, and it has been suggested *(40)* that those changes may result in a derangement of neurotransmitter concentrations in the brain.

In PKU, there are high levels of Phe in plasma (primarily because of a partial deficiency in the enzyme, Phe hydroxylase). Phe is one of a number of structurally related amino acids that competes for transport by the large neutral (System L) amino acid transporter and high levels of Phe result in a reduction in the brain uptake of other essential amino acids *(44)*.

The presence of specific nutrient transporters at the BBB has importance for drug delivery to the brain. For example, dopamine does not cross the BBB, but L-DOPA, a precursor of dopamine, is a substrate for the L-system amino acid transporter *(45)*. Thus, to increase brain dopamine concentrations in patients with Parkinson's disease, L-DOPA, rather than dopamine is administered. Investigators have also modified the structure of therapeutic agents, in order to gain affinity for BBB transport systems. Thus, melphalan, an anticancer agent, has limited BBB permeability, but modifications to the side chains of this compound can result in greatly enhanced transport by the L-system amino acid transporter *(46)*. Others have used alternative approaches to target BBB transport systems. For example, antibodies have been raised against the transferrin receptor involved in vesicular-mediated iron transport across the BBB. By linking therapeutic agents to those antibodies, that vesicular system can be used as a vector to transport the therapeutic agent across the BBB *(47)*.

The use of such techniques may not only increase blood–brain transport, but they may also provide some targeting for brain. For, although there may be no transporters that are specific for the BBB alone, it is possible that the BBB may express higher levels of those transporters than other tissues.

5. ION REGULATION AND FLUID MOVEMENT

5.1. Physiology

It has long been known that Ca and potassium (K) concentrations in CSF are tightly regulated, with either marked increases or decreases in plasma concentration having little effect on CSF composition (*48*; Fig. 5). This work has since been extended to demonstrate that both the total brain content and the brain interstitial concentration of these two ions are also tightly regulated (e.g., *48–51*; Fig. 5). This is the case for both acute and chronic changes in plasma composition (e.g., *52*).

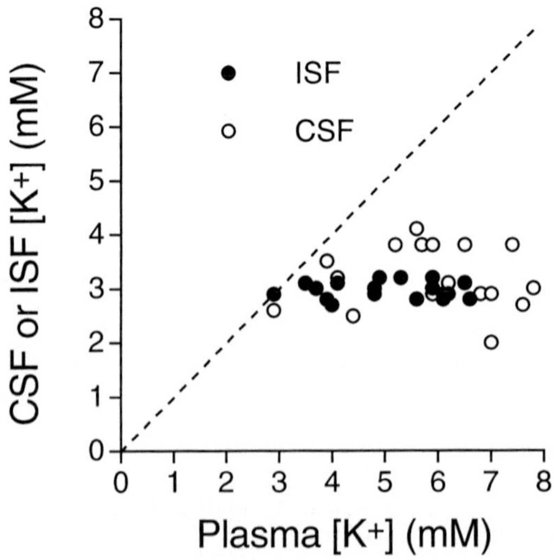

Fig. 5. Relation between plasma and CSF or ISF K concentration ([K^+]) in the rat *(49)*. During acute hyperkalemia, increases in plasma [K^+] do not produce corresponding increases in CSF or ISF plasma [K^+].

Astrocytes have a important role in brain ISF K regulation during neuronal activation, through sequestration of released K^+ and spatial buffering *(53,54)*. However, astrocytes alone do not account for brain K regulation during chronic changes in plasma composition, since there is regulation of total brain K^+ content, as well as ISF K^+ concentration ([K^+]), suggesting that the total amount of K^+ in the astrocytic compartment is not changing.

Before discussing the mechanisms by which K^+ and Ca^{2+} may be regulated by different transporters at the BBB, it should be noted that the permeabilities of these two ions (and Na^+ and Cl^-) at the BBB is low (Table 2). Thus, the turnover of these ions in the brain is very slow, and this means that acute changes in plasma composition take time to be reflected in brain ISF composition, even without compensatory alterations in BBB transport. This is important in situations in which there may be acute changes in plasma composition, which are quickly corrected. For example, there may be short-term elevations in plasma K^+ concentration during strenuous exercise *(55)*.

The absence of changes in total brain K^+ or Ca^{2+} content during chronic perturbations in plasma composition implies that there is regulation of the flux of these two ions. Thus, for example, during plasma hyperkalemia, brain K^+ content could be kept constant either by maintaining a constant influx of K^+ from blood to brain, in the face of an increase in plasma concentration, or by increasing efflux from brain to blood, or a combination of the two. For chronic changes in plasma [K^+] and [Ca^{2+}], there is evidence that there is regulation of influx *(48,52,56)*; for acute changes in plasma [K^+], efflux appears to be the regulated parameter *(51)*.

Na^+/K^+-ATPase has a polarized distribution at the BBB, with a predominant abluminal membrane distribution *(57,58)*. That abluminal Na^+/K^+-ATPase, which transports K^+ from the ISF to the EC, may be involved in the upregulation of K^+ clearance during acute hyperkalemia, or increases in ISF [K^+] *(51,59)*. Some histochemical

Table 2
BBB Permeabilities and Fluxes for Na^+, K^+, Ca^{2+}, and Cl^-

Ion	PS product (µL/g/min)	Flux (nmol/g/min)	Turnover (%/h)
Na^+	1.2–2.5	170–350	~35
K^+	3–12	12–48	~2
Ca^{2+}	0.3–0.6	0.6–1.2	~6
Cl^-	0.8–1.3	80–130	~20

Permeability surface area (PS) product values are from a variety of sources *(5,52,68,80,81,149,150)*. The flux data assume plasma concentrations for Na^+, K^+, Ca^{2+}, and Cl^- of 140, 4, 2, and 100 m*M*, respectively. The turnover rate is the percentage of total brain content that will be replaced by BBB ion transport/h, assuming that the brain ion contents for Na^+, K^+, Ca^{2+}, and Cl^- are 43, 96, 1, and 30 µmol/g wet wt, respectively.

and physiological studies *(58,60,61)* have suggested that there is some luminal Na^+/K^+-ATPase, and findings by Zlokovic et al. *(62)* indicate that multiple isoforms of Na^+/K^+-ATPase are present in isolated cerebral microvessels (α_1, α_2, α_3, β_1, and β_2 subunits all being present). This indicates the possibility that different isoforms might be present on the luminal and abluminal membranes of the BBB. The amount of one of these isoforms (α_3) is decreased in brain microvessels isolated from chronically hyperkalemic rat *(63)*. Such a downregulation could participate in maintaining K^+ influx constant, if α_3 has a luminal distribution. In a similar fashion, the amount of plasma membrane Ca^{2+}-ATPase is increased in brain microvessels isolated from chronically hypocalcemic rats *(63)*. Such upregulation may participate in ISF Ca^{2+} regulation, if it reflects an upregulation in luminal plasma membrane Ca^{2+}-ATPase.

There is less information on the role of the BBB in buffering changes in brain ISF ion concentrations. There is a clearance of K^+ from the brain during ischemic brain damage (*see* below), but the extent to which there might be a clearance of K^+ across the BBB during normal neuronal release is much less certain. It has been suggested *(64)*, however, that spatial buffering of K^+ by astrocytes results in the release of K^+ by glial endfeet that encircle the cerebral capillaries, and that the ECs then transport that K^+ to blood *(64)*.

K and Ca are not the only two ions that are regulated in the brain interstitial space. Both Mg^{2+} and H^+ concentrations are also regulated during fluctuations in plasma composition *(48)*. However, in the case of H^+, that regulation depends on the form of the acidosis/alkalosis. Brain ISF/CSF H^+ concentration is regulated during metabolic acidosis/alkalosis, but there is little regulation during respiratory acidosis/alkalosis, since CO_2 can readily diffuse between blood and brain. Respiratory-induced changes in brain pH affect the respiratory centers of the brainstem that normally correct the respiratory acidosis/alkalosis.

Unlike with the other ions described above, alterations in plasma Na^+ and Cl^- concentration result in almost equal changes in the concentration of these ions in CSF and brain ISF *(65)*. This reflects the fact that the brain appears incapable of maintaining an osmotic gradient between itself and blood. Na and Cl are by far the major osmolytes (~90%) in brain ISF, and maintaining these ion concentrations constant, during changes in plasma composition, would necessitate maintaining an osmotic gradient.

Although changes in plasma osmolality result in similar changes in brain osmolality, there is a time lag before the brain reaches osmotic equilibrium (*65*; Fig. 6). This

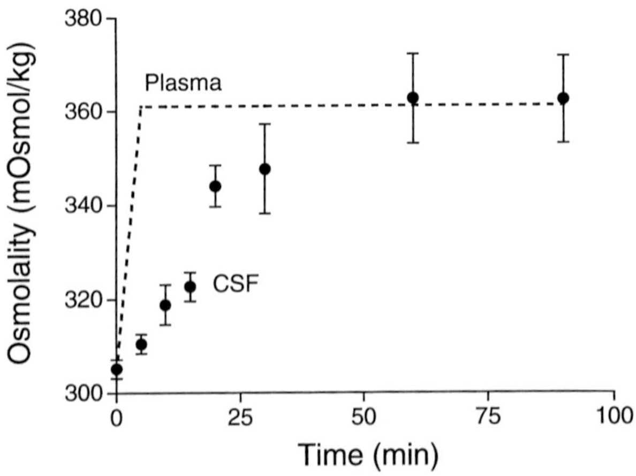

Fig. 6. The effect of acute hyperosmotic stress on CSF osmolality in the rat (Keep, unpublished observations). There is a time lag before CSF comes back into osmotic equilibrium with plasma, after an acute rise in plasma osmolality (*see* dashed line). The hyperosmolality was induced by an intraperitoneal injection of 2 *M* NaCl (2 mL/100 g). Values are means ± SE, $n = 4 - 8$.

reflects the fact that the BBB has a low hydraulic conductivity (the rate at which water moves in response to a hydrostatic or osmotic gradient). Transient changes in blood osmolality may, therefore, not be fully reflected in brain osmolality.

The brain also volume-regulates. Thus, during acute hyperosmotic stress, the brain only loses half the water expected for a perfect osmometer, i.e., a tissue in which osmotic equilibrium is only achieved by the movement of water *(66)*. This acute volume regulation results from a net uptake of ions into brain, primarily Na^+ and Cl^- *(66)*. This uptake comes partly from CSF, as a result of the hydrostatic gradients established during the hyperosmotic stress *(67)*. However, there are also alterations in BBB ion transport with an increase in the unidirectional influx of Na^+ from blood and a marked increase in the unidirectional influx of K^+ *(68)*.

The specific transporters involved in these changes in ion flux during osmotic stress have not been identified, nor has the question of whether these ion fluxes are directly linked to water movement been answered. One transporter that may be involved is Na^+/K^+-ATPase. The polarized distribution of Na^+/K^+-ATPase at the cerebral ECs is similar to the distribution found at the choroid plexus epithelium, where Na^+/K^+-ATPase is only found on the apical (CSF-facing) membrane *(69)*. That epithelial Na^+/K^+-ATPase is thought to be involved in CSF secretion, raising the possibility that the abluminal Na^+/K^+-ATPase of the cerebral capillaries might secrete ISF *(70,71)*, and that this might be increased during hyperosmotic stress. The concept of ISF secretion by the BBB is supported by findings *(72)* that choroid plexus removal reduces, but does not eliminate, CSF production, and that there is a normal flow of fluid from ISF to CSF *(73,74)*.

There is a second phase of brain volume regulation that is involved in returning brain volume toward normal. This generally involves changes in organic osmolytes. Thus, for example, during prolonged hypo-osmotic stress, there is a gradual loss of total brain taurine *(75)*. Because taurine metabolism in the brain is limited *(76)*, this suggests that taurine is being cleared from the brain at either the BBB or blood–CSF barrier, or both. There are taurine transporters at both sites (the β-amino acid trans-

Table 3
Ion Transporters and Channels at BBB

Transport system	Proposed functions
Na^+/K^+-ATPase (multiple isoforms)	Transendothelial Na^+ and K^+ transport, ISF secretion, EC ion regulation, Na^+ gradient for secondary active transport[a,b,d,e]
Na^+/H^+ exchange	Brain and endothelial pH regulation, ISF secretion[a]
Na^+/Ca^{2+} exchange	Transendothelial Ca^{2+} transport, EC Ca^{2+} regulation[b]
$Na^+/K^+/Cl^-$ cotransport	Endothelial and brain volume regulation[b,e]
Ca^{2+}-ATPase	Transendothelial Ca^{2+} transport, EC Ca^{2+} regulation[d,e]
Cl^-/HCO_3^--exchange	Brain and endothelial pH regulation, ISF secretion[a]
H^+-ATPase	pH regulation of EC compartments[e]
Nonselective amiloride-sensitive cation channel	ISF secretion, brain K^+ regulation, endothelial Ca^{2+} regulation[a,c]
Inward rectifying K^+ channel	Brain K^+ regulation, endothelial K^+, and membrane potential regulation[c]
K^+(ATP) channel	K^+ regulation during ischemia[c]
Ca^{2+}-activated K^+ channel(s)	Brain K^+ regulation, endothelial K^+, and membrane potential regulation[c]
Stretch-activated nonselective cation channel	Endothelial and brain volume regulation, endothelial Ca^{2+} regulation[c]
Ca^{2+}/ATP sensitive nonselective cation channel	Ion regulation during ischemia, endothelial Ca^{2+} regulation[c]

Evidence from refs. *151–154*. Evidence is from [a]in vivo transport, [b]in vitro transport, [c]in vitro patch clamp, [d]immunohistochemistry, or [e]Western blots. For transporters/channels shown in italics, there is some conflicting evidence on their presence.

porter), which are important in maintaining the concentration of taurine low in CSF and the brain interstitial space *(77,78)*. During hypo-osmotic stress, brain parenchymal cells lose taurine, resulting in a build-up in the taurine concentration in the ISF *(79)*. Whether the barrier transporters respond passively to those changes in extracellular taurine concentration, or whether there is an active upregulation of taurine transport, has not been examined.

Although total brain volume regulation must involve movements of water and osmolytes between blood and brain, it should be noted that there may be a redistribution of water and osmolytes between the different compartments of the brain. This has already been referred to in terms of movements of CSF into brain during hyperosmotic stress, but there is also a redistribution between the brain interstitial and intracellular spaces, with a disproportionate amount of brain shrinkage during hyperosmotic stress occurring in the interstitial space, i.e., cellular volume is relatively spared *(65)*.

The above description has focused on the role of the BBB in brain ion and volume regulation. Some of the ion transporters may also be involved in EC ion homeostasis. Thus, for example, EC Ca^{2+}-ATPase and Na^+/Ca^{2+} exchange could be involved in brain Ca^{2+} homeostasis, but they may also be involved in maintaining the constant EC Ca^{2+} concentration that is important for the maintenance of BBB integrity *(16)*.

Table 3 summarizes some of the available data on the ion transporters and ion channels at the BBB, and also gives some of their proposed functions. It does not cover the important topic of the transport of trace metals (e.g., zinc and ion; this is reviewed in ref. *80*).

5.2. Pathophysiology

In Subheading 3., vasogenic brain edema formation was discussed in relation to BBB disruption. There are, however, many forms of disease or brain injury that cause brain edema in the absence of BBB disruption. Klatzo *(22)* termed such edema "cytotoxic," reflecting that it occurs in states in which parenchymal cells are damaged. Even in such forms of edema, however, water and osmolytes (since the brain is in osmotic equilibrium with blood) must come from plasma, i.e., across the BBB. This raises the question of how injury to parenchymal cells causes such a net movement of fluid across the intact BBB. The ensuing discussion focuses on this question (which is still being debated), with reference to the brain edema formation that follows focal cerebral ischemia where a large portion of edema formation occurs prior to BBB disruption *(19)*.

During parenchymal cell injury, such as occurs in cerebral ischemia, there is a redistribution of ions between the intracellular space and the ISF. Thus, Na^+ moves into and K^+ out of the intracellular space as normal cellular ion gradients are not be maintained. The former tends to exceed the latter, and the cells also swell. The movement of Na^+ and K^+ results in an increase in extracellular K^+ concentration (concentrations rise from 3 to 30 m*M* in focal cerebral ischemia) *(81)* and, presumably, a decrease in ISF Na^+ concentration. This produces a concentration gradient for Na^+ to move into the brain from blood. There is also a concentration that would move K^+ from brain ISF to blood. However, during focal cerebral ischemia, the movement of Na^+ into brain exceeds the movement of K^+ out of the brain, i.e., there is a net gain in brain cations *(82)*. That gain in cations is roughly matched by a gain in Cl^-, and, since brain and blood are basically in osmotic equilibrium, a gain in brain water (*82*; and *see* an example in Fig. 7). If this scenario is the cause of ischemic brain edema, a fundamental question is, why does the movement of Na^+ into brain exceed the movement of K^+ out. As described above, there are a number of potential ion transporters and channels involved in ion movement at the BBB. This difference in the rate of Na^+ and K^+ movement may reflect how those transporters respond to changes in ISF composition. The gain in brain Na^+ content during focal cerebral ischemia exceeds the K loss by about 3:2, a ratio that has led to the suggestion that the abluminal BBB Na^+/K^+-ATPase may play a crucial role in determining the net movement of these ions *(83)*.

One disease state that results in an increase in brain water content is hydrocephalus, a disease that is almost always caused by an obstruction in the CSF drainage pathway *(84)*. If, as discussed above, ISF is secreted by the BBB, an obstruction to the ISF drainage pathway may lead to edema formation, and it is possible that this may occur during cerebral ischemia, since parenchymal cells swell, resulting in a reduction in the volume of the interstitial space *(85,86)*.

Ion transporters at the BBB have a role in brain ion homeostasis. However, there are, as yet, no known genetic mutations in those ion transporters that alter brain function. Possibly, such mutations would be lethal, or have a more profound effect on another tissue (e.g., kidney) masking an brain effect.

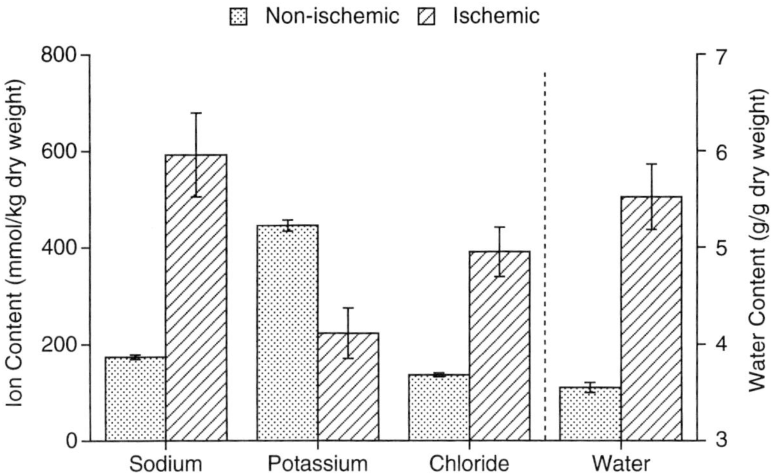

Fig. 7. Brain ion and water contents 24 h after focal cerebral ischemia induced by middle-cerebral artery occlusion in the rat (Keep, Betz, Ren, and Beer, unpublished observations). Focal cerebral ischemia is associated with an increase in brain Na^+ content and a decrease brain K^+ content. The former exceeds the latter, and this difference (along with a concomitant increase in brain Cl) may be the cause of the profound increase in brain water content (edema) that occurs following cerebral ischemia. Values are means ± SE, $n = 6$.

6. HORMONE TRANSPORT

6.1. Physiology

A number of hormones that are secreted in the periphery (e.g., leptin, insulin, and vasopressin) have a site of action in the brain. This subheading concentrates on leptin as an example of such hormone transport from blood to the brain ISF. Leptin is secreted by adipose tissue, but has its main site of action as a satiety signal in the hypothalamus *(87)*. Mutant obese mice (ob/ob) do not produce leptin, and have marked increases in body fat, as well as being hyperglycemic, hyperinsulinemic, and hypothermic *(87)*. Because leptin and other hormones would not generally be expected to cross the BBB by passive diffusion (leptin is a 16-kDa polypeptide), there must be transport systems for those hormones. In the case of leptin, saturable transport from blood has been identified at both the BBB and blood–CSF barrier *(88–90)*. Indeed, a leptin receptor was first cloned from the choroid plexus *(88)*. Unlike the hypothalamic receptor *(91,92)*, the choroid plexus and BBB leptin receptors are not linked to a cytoplasmic signaling cascade, but appear to be solely involved in transport *(89)*. Bypassing the BBB and blood–CSF barrier by direct intracerebroventricular administration of leptin, results in marked reductions in feeding and body wt at much lower doses than are required systemically *(93,94)*.

Leptin transport into brain via the BBB and blood–CSF barrier is regulated. Thus, feeding rats a high-fat diet results in an upregulation in the mRNA for the BBB leptin receptor and the amount of the protein *(95)*. Thus, potentially, obesity may not only increase the amount of circulating leptin, but also the transport of that leptin to brain.

6.2. *Pathophysiology*

The study of the effects of leptin on body wt has been greatly aided by the availability of mice and rats that either have mutations in the gene encoding for leptin (the ob/ob mice *[87]*) or the leptin receptor (e.g., the db/db mice and the Zucker and Koletsky rats *[96–98]*). These animals are overweight. For animals with a mutation in the leptin gene, obesity can be corrected by systemic or intracerebroventricular administration of leptin *(99)*. Most animals lacking the leptin receptors (the different receptors are different splice variants of one gene *[96]*) are unresponsive to leptin administration *(96)*, an exception is the New Zealand Obese (NZO) mouse, which is responsive to intracerebroventricular, but not subcutaneous, administration of leptin *(100)*. This suggests that the obesity in this latter strain of mice results from defective blood-to-brain leptin transport.

In humans, there are few known cases of obesity that are caused by a mutation in the leptin gene, and it appears that the circulating levels of leptin in the obese population are appropriate for their body wt *(101)*. This focuses attention on the concept of leptin resistance; i.e., the failure of an appropriate level of leptin to produce an appropriate response *(101)*. Such resistance could result from alterations in the signaling cascade initiated by the long form of the leptin receptor, or it could result from changes in blood-to-brain leptin transport, an extreme version of which would be the NZO mouse (Fig. 8).

7. CELL TRAFFICKING

This topic is dealt with in Chapter 15 and, therefore, is not covered here, apart from a few comments. Under normal circumstances, there is little trafficking of cells across the BBB. Such trafficking is important, however, in a number of disease states, including multiple sclerosis, and in brain injury, when an influx of neutrophils and monocytes may contribute to the progression of the injury. The cerebral endothelium is not a passive participant in this process. The adhesion of neutrophils to the endothelium requires the expression of adhesion molecules on both the infiltrating cell and the endothelium; migration across the endothelium appears to involve rearrangement of the cytoskeleton of both the infiltrating cell and the EC. Thus, modification of cerebral EC function is a therapeutic target in such diseases.

8. EFFLUX (BRAIN-TO-BLOOD) TRANSPORT MECHANISMS

8.1. *Physiology*

As described in the overview, a number of compounds with high lipid solubility do not have the expected high rate of uptake into brain from blood. Thus, for example, vinblastine and vincristine have permeability surface area products 2–3 orders of magnitude less than that predicted by their lipid solubility (Fig. 2). This reflects the fact that they are P-glycoprotein substrates. P-glycoprotein (GP), the product of the *mdr* (multidrug resistance) gene, is an ATP-dependent pump. In cells that express this transporter, it transports substrates from the cytosol to the extracellular space *(102)*. In the case of the BBB, most, but not all *(103)* studies indicate that it is predominantly expressed on the luminal membrane of the ECs *(104)*, i.e., it is involved in the clearance of compounds from the endothelium to the blood. This concept is supported by

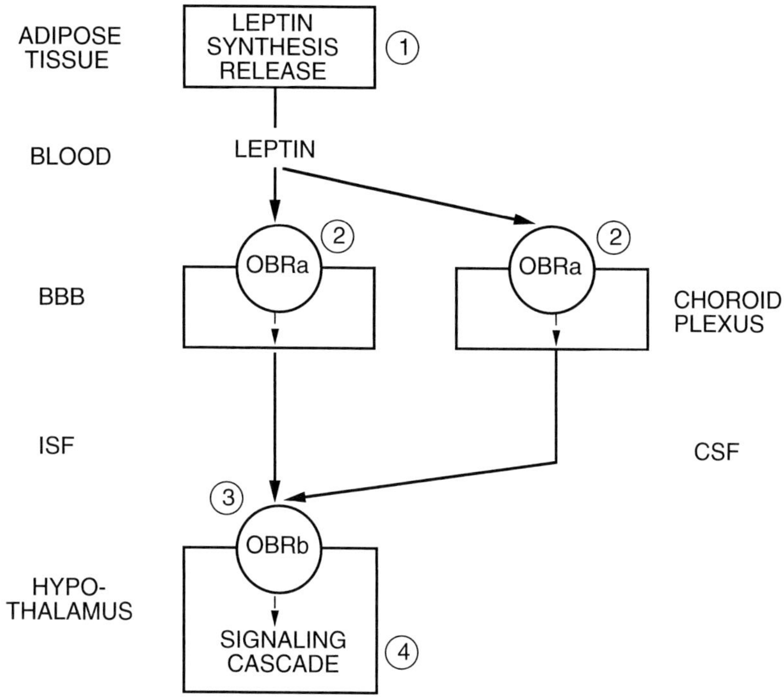

Fig. 8. Leptin and obesity. Leptin is secreted by adipose tissue, and has its primary site of action in the hypothalamus. Thus, it must cross the BBB or blood–CSF barrier, and this occurs via receptor (OBRa)-mediated transcytosis. Leptin controls body wt by decreasing feeding and increasing energy expenditure. A number of defects in the leptin system could cause obesity: (1) An inappropriately low production of leptin (e.g., a mutation in the *Ob* gene, as occurs in the Ob/Ob mouse); (2) an reduction in the OBRa-mediated transport of leptin into brain (as may occur in the NZO mouse); (3) an alteration in the OBRb receptor of the hypothalamus (e.g., there is a mutation in OBRb and OBRa receptors in the db/db mouse); and (4) an alteration in the signaling cascade following OBRb activation.

the finding that mice lacking mdr1a (the BBB form of P-GP) attain much higher brain concentrations of vinblastine, paclitaxel, cyclosporin A, dexamethasone, loperamide, digoxin, and ondansetron, after systemic administration, than do wild-type mice *(105,106)*. The mdr1a knockout mice develop normally, and are fertile, suggesting that the mdr1a gene product is not essential for survival, and that perhaps the main function of BBB mdr1a is exclusion of foreign substrates from the brain, rather than extrusion of substrates that normally occur from metabolism in the animal *(102)*.

Another efflux pump, the multidrug resistance-associated protein (MRP), has been identified at the BBB *(107)*. It is selective for organic anions, as well as glutathione and glucuronide conjugates, and it is probenecid sensitive *(107–109)*. Like P-GP, with which it shares approx 15% amino acid sequence homology *(108)*, MRP, of which there are several isoforms, belongs to a superfamily of ATP-dependent efflux transporter proteins. Like P-GP, MRP may be important in limiting the access of drugs to the brain, although it may also control the concentration of normally occurring metabolites.

There are also probably other organic anion efflux mechanisms at the BBB, although, unlike P-GP and MRP, these are probably involved in transport from the brain interstitial space to the EC, and they may act in concert with P-GP and MRP in clearing substances from EC to blood. For example, the transport of para-aminohippuric acid from brain to blood is much faster than from blood to brain *(110)*. Para-aminohippuric acid is a substrate for the family of organic anion transporters (OATs) (OAT1, OAT2, OAT3, and OAT4), and, although OAT1, OAT2, and OAT4 are predominantly expressed in kidney, liver, and placenta, OAT3 is present in brain, and has been suggested as a possible BBB transporter *(111)*. There is another family of OATs, the organic anion transporting polypeptides (oatp), which are involved in the transport of organic anions into cells. Oatp1 is present on the apical membrane of the choroid plexus where it is probably involved in clearing organic anions from CSF *(112)*. However, not all oatps appear to be involved in clearing organic anions from brain. oatp2 is present on the basolateral membrane of choroid plexus, and may, therefore, be involved in transport from blood to CSF *(113)*, and it is also present at the BBB, and may account for the entry of cardiac glycosides into brain *(113)*.

Another group of transporters involved in clearing organic compounds from the brain ISF are those for specific groups of amino acids (Fig. 4). A number of amino acids are either neurotransmitters (Glu, aspartate, glycine, and γ-aminobutyric acid); others are precursors for neurotransmitters (tryptophan, arginine, and tyrosine are precursors for serotonin, nitric oxide, and catecholamines, respectively). The CSF concentration of many amino acids is substantially lower than that in plasma *(114)*, and this is probably also the case for brain ISF. The finding that the System-A amino acid transporter is present in isolated cerebral microvessels but not on the luminal membrane of the EC, first led to the concept that the ECs of the BBB are polarized, i.e., System A is purely abluminal *(115)*. Unlike System L, the System A amino acid transporter is coupled to Na^+, and is thought to participate in maintaining the low concentration of some amino acids (such as proline, methionine, alanine, and glycine) in the brain ECS.

Like System A, there appear to be a number of Na^+-dependent amino acid transporters that are polarized, with at least a predominantly abluminal distribution, including Systems ASC (a transporter of small neutral amino acids such as alanine, serine, and cysteine), β (a transporter of β-amino acids such as taurine) and X^- (an anionic amino acid transporter) *(30)*. A number of their substrates are either neurotransmitters or neuromodulators (e.g., Glu and taurine). The Na^+ dependency of all these transporters indicates that they are concentrative, and they may contribute to the low extracellular concentration of their substrates. System X^-, for example, transports Glu and aspartate, two excitatory amino acids that have lower CSF concentrations than that found in plasma *(114)*. Glu is released from presynaptic vesicles during neuronal activity. Two predominantly glial Glu transporters, GLT-1 (excitatory amino acid transporter [EAAT2] is the human homolog) and astrocyte specific Glu transporter (EAAT1 is the human homolog), and a predominantly neuronal Glu transporter, EAAC1 (EAAT3 is the human homolog), are involved in clearing Glu from the brain microenvironment *(116)*. This is important, because excessive levels of Glu can cause excitotoxicity *(117)*. The importance, if any, of the BBB System X^- transporter in clearing Glu that may spillover from the synaptic cleft is uncertain. It may serve another purpose, i.e., to regulate the Glu influx that occurs across the BBB. Recently, O'Kane et al. *(118)*

examined whether System X^- corresponds to EAAT1, EAAT2, or EAAT3. In vesicles isolated from brain microvessels, they found evidence for all three types of transporter. However, in such preparations, there is always the concern that the transporters may be present on pericytes or fragments of astrocytic endfeet, rather than endothelium.

Although the BBB distribution of the System A amino acid transporter appears to be totally polarized (solely abluminal), for System β, this is not be the case. Although this System may be predominantly abluminal *(77)*, there appears to be some transport activity on the luminal membrane *(119)*. Taurine, the major substrate for this transporter, is present at high concentrations inside cells, and it may be that the luminal transporter is necessary to allow some influx of taurine into brain from blood, against a plasma–EC concentration gradient.

Most amino acids appear to have a lower concentration in the brain ISF than in plasma, which may reflect that most of the BBB Na^+-dependent amino acid transporters have a predominantly abluminal distribution. One amino acid that is an exception is Gln. Glutamine is by far the most prevalent amino acid in CSF *(114)*, with a concentration close to that found in plasma (500 vs 600 μ*M*). There is a Na^+-dependent transporter, System N, for Gln at the BBB *(33)*, but that transporter is present on the luminal membrane of the BBB in vivo (*33*; *see* Subheading 4.1.), although there is also some evidence for an abluminal distribution *(120)*.

8.2. Pathophysiology

The effect of disease states on BBB efflux transporters has been little studied, partly because of the difficulty in determining the rate of efflux transport in vivo. Disease effects on efflux may be important, because they could prevent the clearance of endogenously produced toxic compounds, and alter the distribution of therapeutic agents between blood and brain. There is evidence in cerebral ischemia that P-GP is inhibited, possibly as a consequence of ATP depletion in the cerebral ECs *(121)*. Similarly, penicillin is normally excluded from the brain by efflux transporters, but it achieves much higher CSF concentrations in patients with meningitis *(48)*. This may reflect barrier disruption, but there is evidence, at least at the blood–CSF barrier, of efflux transport inhibition *(122)*.

9. METABOLIC BARRIER

9.1. Physiology

There are a number of enzymes at the BBB that limit the entry of compounds from blood to brain, via degradation. For example, the brain capillary contains a variety of neurotransmitter-metabolizing enzymes, such as monoamine oxidase (MAO), cholinesterases, GABA transaminase, aminopeptidases, and endopeptidases *(123–125)*. Thus, along with the absence of specific transporters mediating blood-to-brain transport at the BBB, this metabolic barrier serves to protect the brain from neurotransmitters in the circulation. In addition, several drug- and toxin-metabolizing enzymes, typically found in the liver, are also found in brain capillaries *(126,127)*.

The metabolic barrier may potentially act in concert with the brain efflux systems discussed in Subheading 8. Thus, MRP, which are present at the BBB *(107)*, transport glutathione and glucuronide conjugates *(107–109)*, and two enzymes involved in such conjugation reactions, 1-naphthol UDP-glucuronosyltransferase and an α-class glutathione S-transferase (Y_kY_k), appear to be associated with cerebral microvessels *(128,129)*.

A number of detoxification enzymes are present in brain ECs at much higher concentrations than in the brain as a whole. Thus, for example, 1-naphthol UDP-glucuronosyltransferase is present at sevenfold higher concentrations in cerebral microvessels compared to whole brain homogenates *(128)*. A question, therefore, arises as to the importance of BBB-mediated detoxification of compounds in the ISF, compared with that which occurs in the other cell types in the brain. However, the total mass of brain parenchymal cells far outweighs the mass of cerebral endothelium (only about 0.5% of the brain is endothelium).

9.2. *Pathophysiology*

The fact that the cerebral ECs form a metabolic as well as a physical barrier, has a number of important pathophysiological sequelae. Thus, the metabolic barrier may limit the delivery of some therapeutic agents (Fig. 9). For example, L-DOPA is administered to Parkinson's disease patients, because of its affinity for the BBB System-L amino acid transporter, but both aromatic amino acid decarboxylase and MAO, enzymes that metabolize L-DOPA and dopamine, are present at the BBB *(123)*. Thus, high concentrations of L-DOPA must be given to Parkinson's disease patients in order to elevate brain dopamine levels.

Conversely, the enzymes that form the metabolic barrier may be beneficial in some disease states (Fig. 9). Thus, for example, leukotriene C_4 (LTC_4) is produced in the brain, in some forms of injury, as a product of the lipoxygenase branch of the arachidonic acid cascade. This potentially toxic chemical is taken up into the BBB, and is detoxified by γ-glutamyl transpeptidase to LTD_4, and is then further metabolized to LTE_4 by a dipeptidase *(130)*. It appears that LTE_4 is the form by which it is secreted into the blood *(130)*. A loss of γ-glutamyl transpeptidase in brain tumor ECs appears to be the reason why LTC_4 administration enhances blood-to-brain-tumor permeability, but very high doses are required to have any effect on the normal BBB *(130–132)*.

The presence of MAO-B in brain microvessels is also important in determining the toxicity of 1-methyl-4-phenyl-1,2,3,6-tetrahydropyridine (MPTP) *(133)*. This compound is converted into its dihydropyridium form (MPP^+) by mitochondrial MAO-B, and, if present in the brain parenchyma, MPP^+ causes a syndrome similar to Parkinson's disease. Rats and mice have a greater tolerance of MPTP than humans and primates, and this seems to reflect the fact that the cerebral microvessels of these rodents have higher levels of MAO-B. This results in the conversion of MPTP to MPP^+ in the vasculature, rather than the brain parenchyma, and the vascular MPP^+ may not gain access to the parenchyma, because of its charge *(133)*.

10. INTEGRATION WITH OTHER HOMEOSTATIC MECHANISMS

10.1. *Acute and Chronic Homeostasis*

Unless compounds are produced or metabolized within the brain itself, whole-brain homeostasis must rely on the movement of compounds between blood and brain, which can either be across the BBB and/or across the blood–CSF barrier (at the choroid plexuses and arachnoid membrane).

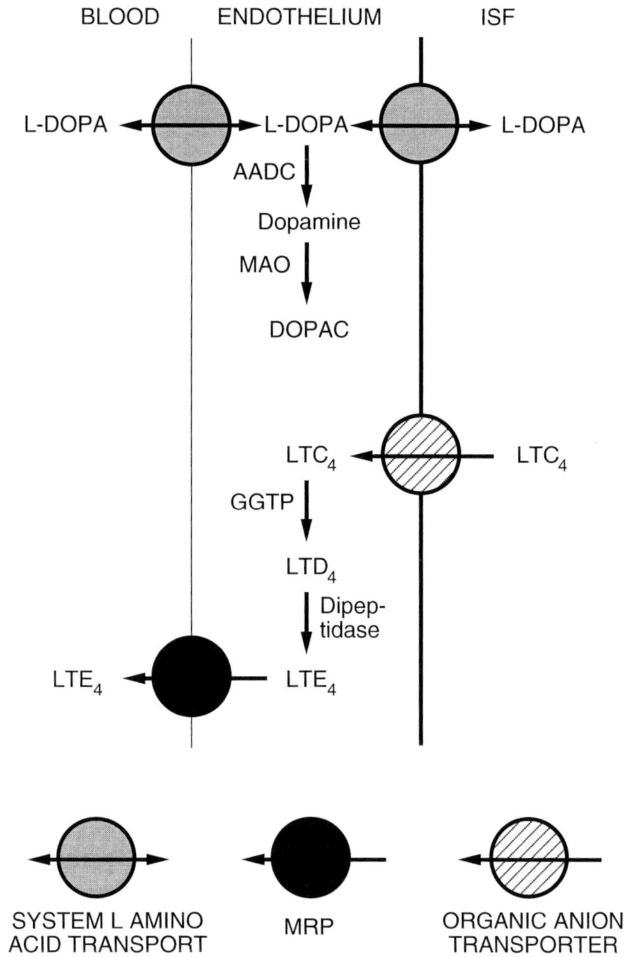

Fig. 9. Examples of the enzymatic BBB: (1) L-DOPA enters the endothelium by the System L amino acid transporter, but its entry to brain is limited by successive degradation to dopamine by aromatic amino acid decarboxylase (AADC) and then to 3,4-dihydroxyphenylacetic acid (DOPAC) and MAO. (2) Potential mechanism for the clearance of LTC_4 from brain ISF, involving an OAT at the abluminal membrane, followed by conversion to LTD_4 by γ-glutamyl transpeptidase (GGTP), and further metabolism to LTE_4 by a dipeptidase. LTE_4 may then be transported to blood, possibly by MRP.

In terms of the brain ISF, however, homeostasis might be achieved by movements of compounds between the ISF and another compartment in the brain (e.g., the astrocytic compartment), which has distinct advantages (e.g., proximity) if the perturbations in the interstitial space are short-term. However, in the long-term, there must, presumably, be a limitation on the capacity of any intraparenchymal sink. If the source of the original ISF space perturbation is another parenchymal compartment, this limitation in sink capacity might be circumvented by recycling the compound back to the original compartment (as occurs with the Glu–Gln cycle between neurons and astrocytes). However, if the source of the compound is from blood, or if the compound is a waste product, clearance to blood is necessary. That is not meant to imply that parenchymal cells are not involved in the regulatory process. Thus, for example, the physical encirclement of the cerebral ECs by astrocyte endfeet may participate in the long-term regulation of ISF [K^+] by delivering K^+ to the BBB *(64)*.

Although this and previous subheadings have discussed the role of the BBB, the blood–CSF barrier, astrocytes, and neurons in regulating the composition of the brain interstitial space, but other cells types also participate. Thus, for example, the previous subheading described the potential role of EC γ-glutamyl transpeptidase in LTC_4 detoxification, but that enzyme is also present in high concentrations in pericytes *(134)*, suggesting that they may also play a detoxification role. Similarly, activated microglia and infiltrating neutrophils may participate in the response of the brain to injury and the removal of injured cells. The release of the cellular contents of damaged cells may effect ISF composition and disrupt the function of nearby uninjured cells.

10.2. Control of BBB Function

As discussed in Subheading 10.1., the BBB acts in concert with other cell types in regulating the brain microenvironment. A question arises as to whether the BBB regulation of that microenvironment is influenced by those other cell types (e.g., neurons or astrocytes) (Fig. 10). In culture, astrocytes modify brain microvessel EC functions *(15,135)*. They greatly enhance the tightness of endothelial monolayers, and they change other structural and physiological functions. This has led to the suggestion that astrocytes are involved in the induction of the BBB phenotype in vivo. Experiments in which toxic agents have been used to destroy astrocytes in vivo have been less compelling *(136,137)*, but it is possible that changes in astrocyte function, e.g., during neural activation might in turn alter BBB function (such as upregulating the transporters clearing metabolic waste products). Alternately, there may be direct neural modulation of BBB function. Such neural regulation has been relatively little studied in vitro or in vivo, although cerebral microvessels are innervated. Thus, for example, there is innervation by noradrenergic neurons from the locus ceruleus *(138)*, which modulate microvessel Na^+/K^+-ATPase activity *(139)*.

Another potentially interesting interaction is that migrating leukocytes may modulate BBB function in ways other than just to facilitate migration across the endothelium. Federici et al. *(140)* examined the cytoplasmic domain of intercellular adhesion molecule-1 (ICAM-1), an endothelial adhesion molecule involved in the trans-endothelial migration of leukocytes, in a rat brain microvessel EC line. They found that ICAM-1 is linked to the glycolytic enzyme, glyceraldehyde-3-phosphate dehydrogenase (GraP-DH). This might be part of a mechanism to regulate the EC microtubule network (and migration), because GraP-DH is a microtubule bundling enzyme, but it also suggests the possibility that ICAM-1 may have wider effects on EC function, by affecting metabolism.

The above interactions describe the effects of cells that are in contact with, or in close proximity to, the cerebral ECs. Another potential form of interaction occurs when a remote sensor releases a hormone to alter BBB function. The BBB has a wide variety of hormone receptors, and a number of groups *(141)* have examined the effects of various hormones on the BBB function, particularly in relation to ion and water homeostasis. There is evidence that hormones, such as vasopressin, angiotensin II, and

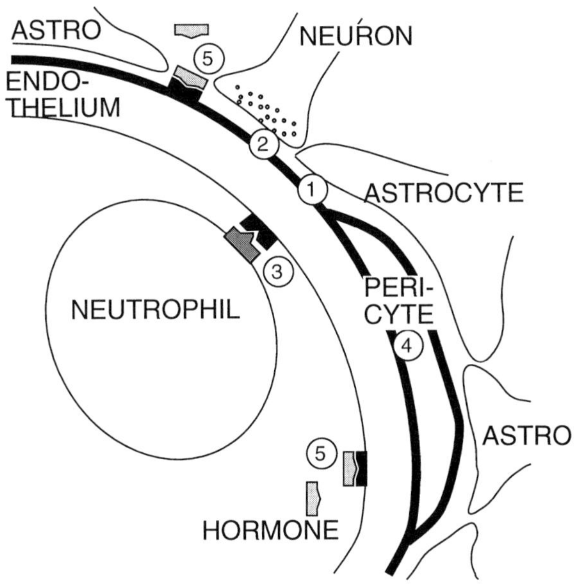

Fig. 10. Schematic showing potential for regulation of BBB function by many different cell types (astro = astrocyte). (1) Factors secreted by astrocytes have been shown to modulate a number of BBB functions *(15,135)*. (2) The BBB is innervated and, for example, noradrenergic neurons from the locus ceruleus affect microvessel Na^+/K^+-ATPase activity *(139)*. (3) Adhering neutrophils alter the BBB, to allow for migration across the barrier, but they may have effects on function. (4) Pericytes are the cell types in closest proximity to the endothelium. Their effect on BBB has been relatively little-studied, although they, for example, produce cytokines that can modulate BBB function *(148)*. (5) Cells distant from the BBB may also affect barrier function via many hormone receptors. Hormones present in the blood or the brain ISF may regulate BBB function, such as ion transport *(141)*.

atrial natriuretic peptide, can modulate BBB ion transport and ion channels *(142,143)*. Hormones may affect the BBB from the blood side of the barrier, but there is also evidence for hormones circulating in the brain ISF and potentially influencing the BBB from the brain side of the barrier.

11. OTHER FUNCTIONS

This chapter has focused on the barrier functions of the cerebral ECs, and their role in controlling the brain microenvironment. However, cerebral ECs possess many of properties of ECs elsewhere in the body, particularly in relation to hemostasis.

12. CONCLUSIONS

The BBB has a whole series of physical, enzymatic, and transport characteristics that enable it too control the movement of compounds between the brain ISF and blood, and, thus, aid in the control of the composition of the fluid that forms the neuronal microenvironment. Indeed, the BBB may also secrete that fluid. The BBB may respond passively to changes in the composition of the ISF, but may also respond dynamically by up- or downregulating permeability, transporters, and enzymes in response to changes in ISF composition or signals from other cell types.

ACKNOWLEDGMENTS

This work was supported by grants from the National Institutes of Health (NS-34709; HL 18575; NS-17760; NS-39866).

REFERENCES

1. Goldmann, E. (1909) Die äussere und innere Sekretion des Gesunden und Kranken organismus im Lichte der "vitalen Färbung. " *Bietr Klin Chirurg* **64,** 192–265.
2. Reese, T. S. and Karnovsky, M. J. (1967) Fine structural localization of a blood-brain barrier to exogenous peroxidase. *J. Cell Biol.* **34,** 207–217.
3. Brightman, M. W. and Reese, T. S. (1969) Junctions between intimately apposed cell membranes in the vertebrate brain. *J. Cell Biol.* **40,** 648–677.
4. Butt, A. M. and Jones, H. C. (1992) Effect of histamine and antagonists on electrical resistance across the blood-brain barrier in rat brain-surface microvessels. *Brain Res.* **569,** 100–105.
5. Smith, Q. R. and Rapoport, S. I. (1986) Cerebrovascular permeability coefficients to sodium, potassium, and chloride. *J. Neurochem.* **46,** 1732–1742.
6. Tsukita, S., Furuse, M., and Itoh, M. (1999) Structural and signaling molecules come together at tight junctions. *Curr. Opin. Cell Biol.* **11,** 628–633.
7. Stevenson, B. R. (1999) Understanding tight junction clinical physiology at the molecular level. *J. Clin. Invest.* **104,** 3–4.
8. Olesen, S.-P. and Crone, C. (1986) Substances that rapidly augment ionic conductance of endothelium in cerebral venules. *Acta Physiol. Scand.* **127,** 233–241.
9. Sarker, M. H., Easton, A. S., and Fraser, P. A. (1998) Regulation of cerebral microvascular permeability by histamine in the anaesthetized rat. *J. Physiol.* **507,** 909–918.
10. de Boer, A. G., Gaillard, P. J., and Breimer, D. D. (1999) The transference of results between blood-brain barrier cell culture systems. *Eur. J. Pharm. Sci.* **8,** 1–4.
11. Raub, T. J. (1996) Signal transduction and glial cell modulation of cultured brain microvessel endothelial cell tight junctions. *Am. J. Physiol.* **271,** C495–503.
12. Ramsohoye, P. V. and Fritz, I. B. (1998) Preliminary characterization of glial-secreted factors responsible for the induction of high electrical resistances across endothelial monolayers in a blood-brain barrier model. *Neurochem. Res.* **23,** 1545–1551.
13. Meyer, J., Mischeck, U., Veyhl, M., Henzel, K., and Galla, H. J. (1990) Blood-brain barrier characteristic enzymatic properties in cultured brain capillary endothelial cells. *Brain Res.* **514,** 305–309.
14. Hemmila, J. M. and Drewes, L. R. (1993) Glucose transporter (GLUT1) expression by canine brain microvessel endothelial cells in culture: an immunohistochemical study. *Adv. Exp. Med. Biol.* **331,** 13–18.
15. Rubin, L. L., Hall, D. E., Porter, S., Barbu, K., Cannon, C., Horner, H. C., et al. (1991) A cell culture model of the blood-brain barrier. *J. Cell Biol.* **115,** 1725–1735.
16. Abbott, N. J. (1998) Role of intracellular calcium in regulation of brain endothelial permeability, in *Introduction to the Blood-Brain Barrier. Methodology, Biology and Pathology* (Pardridge, W. M., ed.), Cambridge University Press, Cambridge, pp. 345–353.
17. Betz, A. L., Iannotti, F., and Hoff, J. T. (1989) Brain edema: a classification based on blood-brain barrier integrity. *Cerebrovasc. Brain Metab. Rev.* **1,** 133–154.
18. Stewart, P. A. and Mikulis, D. (1998) The blood-brain barrier in brain tumors, in *Introduction to the Blood-Brain Barrier. Methodology, Biology and Pathology* (Pardridge, W. M., ed.), Cambridge University Press, Cambridge, pp. 434–440.
19. Menzies, S. A., Betz, A. L., and Hoff, J. T. (1993) Contributions of ions and albumin to the formation and resolution of ischemic brain edema. *J. Neurosurg.* **78,** 257–266.

20. Hiesiger, E. M., Voorhies, R. M., Basler, G. A., Lipschutz, L. E., Posner, J. B., and Shapiro, W. R. (1986) Opening the blood-brain and blood-tumor barriers in experimental rat brain tumors: the effect of intracarotid hyperosmolar mannitol on capillary permeability and blood flow. *Ann. Neurol.* **19,** 50–59.
21. Preston, E. and Foster, D. O. (1997) Evidence for pore-like opening of the blood-brain barrier following forebrain ischemia in rats. *Brain Res.* **761,** 4–10.
22. Klatzo, I. (1967) Neuropathological aspects of brain edema. *J. Neuropathol. Exp. Neurol.* **26,** 1–14.
23. Bodsch, W., Rommel, T., Ophoff, B. G., and Menzel, J. (1987) Factors responsible for the retention of fluid in human tumor edema and the effect of dexamethasone. *J. Neurosurg.* **67,** 250–257.
24. Reichman, H. R., Farrell, C. R., and Del Maestro, R. F. (1986) Effect of steroids and non-steroid anti-inflammatory agents on vascular permeability in a rat glioma model. *J. Neurosurg.* **65,** 233–237.
25. Rapoport, S. I. and Robinson, P. J. (1986) Tight junctional modification as the basis of osmotic opening of the blood-brain barrier. *Ann. NY Acad. Sci.* **481,** 250–266.
26. Bartus, R. T. (1999) The blood-brain barrier as a target for pharmacological manipulation. *Curr. Opin. Drug Disc. Dev.* **2,** 152–167.
27. Drewes, L. R. (1998) Biology of the blood-brain barrier glucose transporter, in *Introduction to the Blood-Brain Barrier. Methodology, Biology and Pathology* (Pardridge, W. M., ed.), Cambridge University Press, Cambridge, pp. 165–174.
28. Gerhart, D. Z., Emerson, B. E., Zhdankina, O. Y., Leino, R. L., and Drewes, L. R. (1997) Expression of monocarboxylate transporter MCT1 by brain endothelium and glia in adult and suckling rats. *Am. J. Physiol.* **273,** E207–E213.
29. Drewes, L. R. (1999) Transport of the brain fuels, glucose and lactate, in *Brain Barrier Systems* (Paulson, O. B., Knudsen, G. M., and Moos, T., eds.), Munksgaard, Copenhagen, pp. 285–295.
30. Smith, Q. R. and Stoll, J. (1998) Blood-brain barrier amino acid transport, in *Introduction to the Blood-Brain Barrier. Methodology, Biology and Pathology* (Pardridge, W. M., ed.), Cambridge University Press, Cambridge, pp. 188–197.
31. Kanai, Y., Segawa, H., Miyamoto, K., Uchino, H., Takeda, E., and Endou, H. (1998) Expression cloning and characterization of a transporter for large neutral amino acids activated by the heavy chain of 4F2 antigen (CD98). *J. Biol. Chem.* **273,** 23,629–23,632.
32. Boado, R. J., Li, J. Y., Nagaya, M., Zhang, C., and Pardridge, W. M. (1999) Selective expression of the large neutral amino acid transporter at the blood-brain barrier. *Proc. Natl. Acad. Sci. USA* **96,** 12,079–12,084.
33. Ennis, S. R., Kawai, N., Ren, X.-D., Abdelkarim, G. E., and Keep, R. F. (1998) Glutamine uptake at the blood-brain barrier is mediated by System-N transport. *J. Neurochem.* **71,** 2565–2573.
34. Kilberg, M. S., Handlogten, M. E., and Christensen, H. N. (1980) Characteristics of an amino acid transport system in rat liver for glutamine, asparagine, histidine and closely related analogs. *J. Biol. Chem.* **255,** 4011–4019.
35. Chaudhry, F. A., Reimer, R. J., Krizaj, D., Barber, D., Storm-Mathisen, J., Copenhagen, D. R., and Edwards, R. H. (1999) Molecular analysis of System N suggests novel physiological roles in nitrogen metabolism and synaptic transmission. *Cell* **99,** 769–780.
36. Laterra, J., Keep, R. F., Betz, A. L., and Goldstein, G. W. (1998) Blood-brain-cerebrospinal fluid barriers, in *Basic Neurochemistry: Molecular Cellular and Medical Aspects* (Siegel, G. J., Agranoff, B. W., Albers, R. W., Fisher, S. K., and Uhler, M. D., eds.), Lippincott-Raven, Philadelphia, pp. 671–689.
37. Cohen, E. and Wurtman, R. J. (1976) Brain acetylcoline synthesis: Control by dietary choline. *Science* **191,** 561–562.

38. Seidner, G., Alvarez, M. G., Yeh, J. I., O'Driscoll, K. R., Klepper, J., Stump, T. S., et al. (1998) GLUT-1 deficiency syndrome caused by haploinsufficiency of the blood-brain barrier hexose carrier. *Nature Genet.* **18,** 188–191.
39. Klepper, J., Wang, D., Fischbarg, J., Vera, J. C., Jarjour, I. T., O'Driscoll, K. R., and De Vivo, D. C. (1999) Defective glucose transport across brain tissue barriers: a newly diagnosed neurological syndrome. *Neurochem. Res.* **24,** 587–594.
40. Pulsinelli, W. A. and Cooper, A. J. L. (1994) Metabolic encephalopathies and coma, in *Basic Neurochemistry* (Siegel, G. J., Agranoff, B. W., Albers, R. W., and Molinoff, P. B., eds.), Raven, New York, pp. 841–857.
41. Hawkins, R. A. and Jessy, J. (1991) Hyperammonemia does not impair brain function in the absence of glutamine synthesis. *Biochem. J.* **277,** 697–703.
42. Rigotti, P., Jonung, T., Peters, J. C., James, J. H., and Fischer, J. E. (1985) Methionine sulfoximine prevents the accumulation of large amino acids in brain of portacaval-shunted rats. *J. Neurochem.* **44,** 929–933.
43. Zanchin, G., Rigotti, P., Dussini, N., Vassanelli, P., and Battistin, L. (1979) Cerebral amino acid levels and uptake in rats after portocaval anastomosis: II. Regional studies in vivo. *J. Neurosci. Res.* **4,** 301–310.
44. Oldendorf, W. H. (1973) Saturation of blood-brain barrier transport of amino acids in phenylketonuria. *Arch Neurol.* **28,** 45–48.
45. Wade, L. A. and Katzman, R. (1975) Synthetic amino acids and the nature of L-DOPA transport at the blood-brain barrier. *J. Neurochem.* **25,** 837–842.
46. Takada, Y., Vistica, D. T., Greig, N. H., Purdon, D., Rapoport, S. I., and Smith, Q. R. (1992) Rapid high-affinity transport of a chemotherapeutic amino acid across the blood-brain barrier. *Cancer Res.* **52,** 2191–2196.
47. Pardridge, W. M. (1995) Blood-brain barrier peptide transport and peptide drug delivery to the brain, in *Peptide-Based Drug Design* (Taylor, M. D. and Amidon, G. L., eds.), pp. 265–296.
48. Bradbury, M. W. B. (1979) *The Concept of a Blood-Brain Barrier.* John Wiley & Sons, Chichester, UK.
49. Jones, H. C. and Keep, R. F. (1987) The control of potassium concentration in the cerebrospinal fluid and brain interstitial fluid of developing rats. *J. Physiol.* **383,** 441–453.
50. Jones, H. C. and Keep, R. F. (1988) Brain fluid calcium concentration and response to acute hypercalcaemia during development in the rat. *J. Physiol.* **402,** 579–593.
51. Stummer, W., Betz, A. L., and Keep, R. F. (1995) Mechanisms of brain ion homeostasis during acute and chronic variations of plasma potassium. *J. Cereb. Blood Flow Metab.* **15,** 336–344.
52. Stummer, W., Keep, R. F., and Betz, A. L. (1994) Rubidium entry into brain and cerebrospinal fluid during acute and chronic alterations in plasma potassium. *Am. J. Physiol.* **266,** H2239–H2246.
53. Walz, W. and Hertz, L. (1983) Functional interactions between neurons and astrocytes. II. Potassium homeostasis at the cellular level. *Progr. Neurobiol.* **20,** 133–183.
54. Kimelberg, H. K. and Norenberg, M. D. (1989) Astrocytes. *Sci. Am.* **260,** 66–76.
55. Harris, P. and Snow, D. H. (1986) Alterations in plasma potassium concentrations during and following short-term strenuous exercise in the horse. *J. Physiol.* **357,** 46P.
56. Murphy, V. A. and Rapoport, S. I. (1988) Increased transfer of ^{45}Ca into brain and cerebrospinal fluid from plasma during chronic hypocalcemia in rats. *Brain Res.* **454,** 315–320.
57. Betz, A. L., Firth, J. A., and Goldstein, G. W. (1980) Polarity of the blood-brain barrier: distribution of enzymes between the luminal and antiluminal membranes of brain capillary endothelial cells. *Brain Res.* **192,** 17–28.
58. Vorbrodt, A. W. (1988) Ultrastructural cytochemistry of blood-brain barrier endothelia. *Progress in Histochemistry and Cytochemistry* **18,** 1–99.

59. Bradbury, M. W. B. and Stulcova, B. (1970) Efflux mechanism contributing to the stability of the potassium concentration in cerebrospinal fluid. *J. Physiol.* **208,** 415–430.
60. Sanchez del Pino, M. M., Hawkins, R. A., and Peterson, D. R. (1995) Biochemical discrimination between luminal and abluminal enzyme and transport activities of the blood-brain barrier. *J. Biol. Chem.* **270,** 14,907–14,912.
61. Ennis, S. R., Keep, R. F., Ren, X.-D., and Betz, A. L. (1997) Potassium transport at the luminal membrane of the blood-brain barrier. *J. Cereb. Blood Flow Metab.* **17(Suppl.),** S515.
62. Zlokovic, B. V., Mackic, J. B., Wang, L., McComb, J. G., and McDonough, A. (1993) Differential expression of Na,K-ATPase a and b subunit isoforms at the blood-brain barrier and the choroid plexus. *J. Cell Biol.* **268,** 8019–8025.
63. Keep, R. F., Ulanski, L. J., Xiang, J., Ennis, S. R., and Betz, A. L. (1999) Blood-brain barrier mechanisms involved in brain calcium and potassium homeostasis. *Brain Res.* **815,** 200–205.
64. Paulson, O. B. and Newman, E. A. (1987) Does the release of potassium from astrocyte endfeet regulate cerebral blood flow. *Science* **237,** 896–898.
65. Cserr, H. F., DePasquale, M., Nicholson, C., Patlak, C. S., Pettigrew, K. D., and Rice, M. E. (1991) Extracellular volume decreases while cell volume is maintained by ion uptake in rat brain during acute hypernatremia. *J. Physiol.* **442,** 277–295.
66. Cserr, H. F., Depasquale, M., and Patlak, C. S. (1987) Regulation of brain water and electrolytes during acute hyperosmolality in rats. *Am. J. Physiol.* **253,** F522–F529.
67. Pullen, R. G. L., Depasquale, M., and Cserr, H. F. (1987) Bulk flow of cerebrospinal fluid into brain in response to acute hyperosmolality. *Am. J. Physiol.* **253,** F538–F545.
68. Cserr, H. F., DePasquale, M., and Patlak, C. S. (1987) Volume regulatory influx of electrolytes from plasma to brain during acute hyperosmolality. *Am. J. Physiol.* **253,** F530–F537.
69. Masuzawa, T., Saito, T., and Sato, F. (1981) Cytochemical study on enzyme activity associated with cerebrospinal fluid secretion in the choroid plexus and ventricular ependyma. *Brain Res.* **222,** 309–322.
70. Betz, A. L. (1986) Transport of ions across the blood-brain barrier. *Fed. Proc.* **45,** 2050–2054.
71. Crone, C. (1986) The blood-brain barrier as a tight epithelium: where is information lacking? *Ann. NY Acad. Sci.* **481,** 174–185.
72. Milhorat, T. H., Hammock, M. K., Fenstermacher, J. D., Rall, D. P., and Levin, V. A. (1971) Cerebrospinal fluid production by the choroid plexus and brain. *Science* **173,** 330–332.
73. Cserr, H. F., Cooper, D. N., Suri, P. K., and Patlak, C. S. (1981) Efflux of radiolabeled polyethylene glycols and albumin from rat brain. *Am. J. Physiol.* **240,** F319–F328.
74. Szentistványi, I., Patlak, C. S., Ellis, R. A., and Cserr, H. F. (1984) Drainage of interstitial fluid from different regions of rat brain. *Am. J. Physiol.* **246,** F835–F844.
75. Thurston, J. H. and Hauhart, R. E. (1987) Brain amino acids decrease in chronic hyponatremia and rapid correction causes brain dehydration: possible clinical significance. *Life Sci.* **40,** 2539–2542.
76. Huxtable, R. J. (1989) Taurine in the central nervous system and the mammalian actions of taurine. *Prog. Neurobiol.* **32,** 471–533.
77. Tayarani, I., Cloez, I., Lefauconnier, J.-M., and Bourre, J.-M. (1989) Sodium-dependent high affinity uptake of taurine by isolated rat brain capillaries. *Biochim. Biophys. Acta* **985,** 168–172.
78. Keep, R. F. and Xiang, J. (1996) Choroid plexus taurine transport. *Brain Res.* **715,** 17–24.
79. Wade, J. V., J. P. O., Samson, F. E., Nelson, S. R., and Pazdernik, T. L. (1988) A possible role for taurine in osmoregulation within the brain. *J. Neurochem.* **51,** 740–745.
80. Smith, Q. R. (1990) Transport of calcium and other metals across the blood-brain barrier: mechanisms and implications for neurodegenerative disorders. *Adv. Neurol.* **51,** 217–222.

81. Schielke, G. P., Moises, H. C., and Betz, A. L. (1991) Blood to brain sodium transport and interstitial fluid potassium concentration during early focal ischemia in the rat. *J. Cereb. Blood Flow Metab.* **11,** 466–471.
82. Betz, A. L., Keep, R. F., Beer, M. E., and Ren, X.-D. (1994) Blood-brain barrier permeability and brain concentration of sodium, potassium, and chloride during focal ischemia. *J. Cereb. Blood Flow Metab.* **14,** 29–37.
83. Betz, A. L., Ennis, S. R., and Schielke, G. P. (1989) Blood-brain barrier sodium transport limits development of brain edema during partial ischemia in gerbils. *Stroke* **20,** 1253–1259.
84. Gjerris, F. and Borgesen, S. E. (2000) Pathophysiology of cerebrospinal fluid circulation, in *Neurosurgery* (Crockard, A., Hayward, R., and Hoff, J. T., eds.), *The Scientific Basis of Clinical Practice.* Blackwell Science, Oxford, pp. 147–168.
85. Hansen, A. J. and Olsen, C. E. (1980) Brain extracellular space during spreading depression and ischemia. *Acta Physiol. Scand.* **108,** 355–365.
86. Syková, E., Svoboda, J., Polák, J., and Chvátal, A. (1994) Extracellular volume fraction and diffusion characteristics during progressive ischemia and terminal anoxia in the spinal cord of the rat. *J. Cereb. Blood Flow Metab.* **14,** 301–311.
87. Zhang, Y., Proenca, R., Maffei, M., Barone, M., Leopold, L., and Friedman, J. M. (1994) Positional cloning of the mouse *obese* gene and its human homologue. *Nature* **372,** 425–432.
88. Tartaglia, L. A., Dembski, M., Weng, X., Deng, N., Culpepper, J., Devos, R., et al. (1995) Identification and expression cloning of a leptin receptor, OB-R. *Cell* **83,** 1263–1271.
89. Golden, P. L., Maccagnan, T. J., and Pardridge, W. M. (1997) Human blood-brain barrier leptin receptor. *J. Clin. Invest.* **99,** 14–18.
90. Banks, W. A., Kastin, A. J., Huang, W., Jaspan, J. B., and Maness, L. M. (1996) Leptin enters the brain by a saturable system independent of insulin. *Peptides* **17,** 305–311.
91. Vaisse, C., Halaas, J. L., Horvarth, C. M., Darnell, J. E., Stoffel, M., and Friedman, J. M. (1996) Leptin activation of Stat3 in the hypothalamus of wild-type and ob/ob mice but not db/db mice. *Nat. Genet.* **14,** 95–97.
92. Ghilardi, N., Ziegler, S., Wiestner, A., Stoffel, R., Heim, M. H., and Skoda, R. C. (1996) Defective STAT signaling by the leptin receptor in diabetic mice. *Proc. Natl. Acad. Sci. USA* **93,** 6231–6235.
93. Campfield, L. A., Smith, F. J., Guisez, Y., Devos, R., and Burn, P. (1995) Recombinant mouse ob protein: evidence for a peripheral signal linking adiposity and central neural networks. *Science* **269,** 546–549.
94. Jacob, R. J., Dziura, J., Medwick, M. B., Leone, P., Caprio, S., During, M., Shulman, G. I., and Sherwin, R. S. (1997) The effect of leptin is enhanced by microinjection into the ventromedial hypothalamus. *Diabetes* **46,** 150–152.
95. Boado, R. J., Golden, P. L., Levin, N., and Pardridge, W. M. (1998) Upregulation of blood-brain barrier short-form leptin receptor gene products in rats fed a high fat diet. *J. Neurochem.* **71,** 1761–1764.
96. Lee, G.-H., Proenca, R., Montez, J. M., Carroll, K. M., Darvishzadeh, J. G., Lee, J. I., and Friedman, J. M. (1996) Abnormal splicing of the leptin receptor in diabetic mice. *Nature* **379,** 632–635.
97. Takaya, K., Ogawa, Y., Isse, N., Okazaki, T., Satoh, N., Masuzaki, H., et al. (1996) Molecular cloning of rat leptin receptor isoform complementary DNAs: identification of a missense mutation in Zucker Fatty (*fa/fa*) rats. *Biochem. Biophys. Res. Commun.* **225,** 75–83.
98. Wu-Peng, X. S., Chua, S. C., Okada, N., Liu, S.-M., Nicholson, M., and Leibel, R. L. (1997) Phenotype of the obese Koletsky (*f*) rat due to Tyr763Stop mutation in the extracellular domain of the leptin receptor (Lepr). *Diabetes* **46,** 513–518.
99. Pelleymounter, M., Cullen, M., Baker, M., et al. (1995) Effects of the obese gene product on body weight regulation in *ob/ob* mice. *Science* **269,** 540–543.

100. Halaas, J. L., Boozer, C., Blair-West, J., Fidahusein, N., Denton, D. A., and Friedman, J. M. (1997) Physiological response to long-term peripheral and central leptin infusion in lean and obese mice. *Proc. Natl. Acad. Sci. USA* **94,** 8878–8883.
101. Considine, R. V., Sinha, M. K., Heiman, M. L., Kriauciunas, A., Tephens, T. W., Nyce, M. R., et al. (1996) Serum immunoreactive-leptin concentrations in normal-weight and obese humans. *N. Engl. J. Med.* **334,** 292–295.
102. Schinkel, A. H. (1997) The physiological function of drug-transporting P-glycoproteins. *Semin. Cancer Biol.* **8,** 161–170.
103. Pardridge, W. M., Golden, P. L., Kang, Y. S., and Bickel, U. (1997) Brain microvascular and astrocyte localization of P-glycoprotein. *J. Neurochem.* **68,** 1275–1285.
104. Borst, P. and Schinkel, A. H. (1998) P-glycoprotein, a guardian of the brain, in *Introduction to the Blood-Brain Barrier. Methodology, Biology and Pathology* (Pardridge, W. M., ed.), Cambridge University Press, Cambridge, pp. 198–206.
105. Schinkel, A. H., Smit, J. J. M., van Tellingen, O., Beijnen, J. H., Wagenaar, E., van Deemter, L., et al. (1994) Disruption of the mouse mdr1a P-glycoprotein gene leads to a deficient in the blood-brain barrier and to increased sensitivity to drugs. *Cell* **77,** 491–502.
106. Schinkel, A. H., Wagenaar, E., Mol, C. A. A. M., and van Deemter, L. (1996) P-glycoprotein in the blood-brain barrier of mice influences the brain penetration and pharmacological activity of many drugs. *J. Clin. Invest.* **97,** 2517–2524.
107. Huai-Yun, H., Secrest, D. T., Mark, K. S., Carney, D., Branquist, C., Elmquist, W. F., and Miller, D. W. (1998) Expression of multidrug resistance-associated protein (MRP) in brain microvessel endothelial cells. *Biochem. Biophys. Res. Commun.* **243,** 816–820.
108. Loe, D. W., Deeley, R. G., and Cole, S. P. (1996) Biology of the multidrug resistance-associated protein, MRP. *Eur. J. Cancer* **32A,** 945–957.
109. Barrand, M. A., Bagrij, T., and Neo, S. Y. (1997) Multidrug resistance-associated protein: a protein distinct from P-glycoprotein involved in cytotoxic drug expulsion. *Gen. Pharmacol.* **28,** 639–645.
110. Kakee, A., Terasaki, T., and Sugiyama, Y. (1997) Selective brain to blood efflux transport of para-aminohippuric acid across the blood-brain barrier: in vivo evidence by use of the brain efflux index method. *J. Pharmacol. Exp. Ther.* **283,** 1018–1025.
111. Kusuhara, H., Seline, T., Utsunomiya-Tate, N., Tsuda, M., Kojima, R., Cha, S. H., et al. (1999) Molecular cloning and characterization of a new multispecific organic anion transporter from rat brain. *J. Biol. Chem.* **274,** 13,675–13,680.
112. Angeletti, R. H., Novikoff, P. M., Juvvadi, S. R., Fritschy, J. M., Meier, P. J., and Wolkoff, A. W. (1997) The choroid plexus epithelium is the site of the organic anion transport protein in the brain. *Proc. Natl. Acad. Sci. USA* **94,** 283–286.
113. Gao, B., Stieger, B., Noe, B., Fritschy, J. M., and Meier, P. J. (1999) Localization of the organic anion transporting polypeptide 2 (Oatp2) in capillary endothelium and choroid plexus epithelium of rat brain. *J. Histochem. Cytochem.* **47,** 1255–1264.
114. Clarke, D. D., Lajtha, A. L., and Maker, H. (1989) Intermediary metabolism, in *Basic Neurochemistry* (Siegel, G. J., Agranoff, B. W., Albers, R. W., and Molinoff, P. B., eds.), Raven, New York, pp. 541–564.
115. Betz, A. L. and Goldstein, G. W. (1978) Polarity of the blood-brain barrier: neutral amino acid transport into isolated brain capillaries. *Science* **202,** 225–227.
116. Dingledine, R. and McBain, C. J. (1998) Glutamate and aspartate, in *Basic Neurochemistry: Molecular Cellular and Medical Aspects* (Siegel, G. J., Agranoff, B. W., Albers, R. W., Fisher, S. K., and Uhler, M. D., eds.), Lippincott-Raven, Philadelphia, pp. 315–333.
117. Choi, D. W. (1997) The excitotoxic concept, in *Primer on Cerebrovascular Diseases* (Welch, K. M. A., Caplan, L. R., Reis, D. J., Siesjo, B. K., and Weir, B., eds.), Academic, San Diego, pp. 187–190.

118. O'Kane, R. L., Martinez-Lopez, I., DeJoseph, M. R., Vina, J. R., and Hawkins, R. A. (1999) Na^+-dependent glutamate transporters (EAAT1, EAAT2, and EAAT3) of the blood-brain barrier. *J. Biol. Chem.* **274,** 31,891–31,895.
119. Benrabh, H., Bourre, J.-M., and Lefauconnier, J.-M. (1995) Taurine transport at the blood-brain barrier: an in vivo brain perfusion study. *Brain Res.* **692,** 57–65.
120. Lee, W.-J., Hawkins, R. A., Vina, J. R., and Peterson, D. R. (1998) Glutamine transport by the blood-brain barrier: a possible mechanism for nitrogen removal. *Am. J. Physiol.* **274,** C1101–C1107.
121. Ohnishi, T., Tamai, I., Sakanaka, K., Sakata, A., Yamashita, J., and Tsuji, A. (1995) In vivo and in vitro evidence for ATP-dependency of P-glycoprotein-mediated efflux of doxorubicin at the blood-brain barrier. *Biochem. Pharmacol.* **49,** 1541–1544.
122. Spector, R. and Lorenzo, A. V. (1974) Inhibition of penicillin transport from cerebrospinal fluid after intracisternal inoculation of bacteria. *J. Clin. Invest.* **54,** 316–325.
123. Hardebo, J. E. and Owman, C. (1979) Barrier mechanisms for neurotransmitter monoamines and their precursors at the blood-brain barrier. *Ann. Neurol.* **8,** 1–31.
124. Hardebo, J. E., Emson, P. C., Falck, B., Owman, C., and Rosengren, E. (1980) Enzymes related to monoamine transmitter metabolism in brain microvessels. *J. Neurochem.* **35,** 1388–1393.
125. Kalaria, R. N. and Harik, S. I. (1987) Blood-brain barrier monoamine oxidase: enzyme characterization in cerebral microvessels and other tissues from six mammalian species, including human. *J. Neurochem.* **49,** 856–864.
126. Minn, A., Ghersi-Egea, J. F., Perrin, R., Leininger, B., and Siest, G. (1991) Drug metabolizing enzymes in the brain and cerebral microvessels. *Brain Res.* **16,** 65–82.
127. Ghersi-Egea, J.-F., Leininger-Muller, B., Cecchelli, R., and Fenstermacher, J. D. (1995) Blood-brain interfaces: relevance to cerebral drug metabolism. *Toxicol. Lett.* **82/83,** 645–653.
128. Ghersi-Egea, J. F., Leninger-Muller, B., Suleman, G., Siest, G., and Minn, A. (1994) Localization of drug-metabolizing enzyme activities to blood-brain interfaces and circumventricular organs. *J. Neurochem.* **62,** 1089–1096.
129. Johnson, J. A., El Barbary, A., Kornguth, S. E., Brugge, J. F., and Siegel, F. L. (1993) Glutathione-S-transferase isoenzymes in rat brain neurons and glia. *J. Neurosci.* **13,** 2013–2023.
130. Frey, A. (1993) Gamma-glutamyl transpeptidase: molecular cloning and structural and functional features of a blood-brain barrier marker protein, in *The Blood-Brain Barrier* (Pardridge, W. M., ed.), Raven, New York, pp. 339–368.
131. Black, K. L., King, W. A., and Ikezaki, K. (1990) Selective opening of the blood-tumor barrier by intracarotid infusion of leukotriene C_4. *J. Neurosurg.* **72,** 912–916.
132. Black, K. L. and Hoff, J. T. (1985) Leukotrienes increase blood-brain barrier permeability following intraparenchymal injections in rats. *Ann. Neurol.* **18,** 349–351.
133. Riachi, N. J., Harik, S. I., Kalaria, R. N., and Sayre, L. M. (1988) On the mechanisms underlying 1-methyl-4-phenyl-1,2,3,6-tetrahydropyridine neurotoxicity. II Susceptibility among mammalian species correlates with the toxin's metabolic patter in brain microvessels and liver. *J. Pharmacol. Exp. Ther.* **244,** 443–448.
134. Risau, W., Dingler, A., Albrecht, U., Dehouck, M.-P., and Cecchelli, R. (1992) Blood-brain barrier pericytes are the main source of γ-glutamyltranspeptidase activity in brain capillaries. *J. Neurochem.* **58,** 667–672.
135. Dehouck, M.-P., Méresse, S., Delorme, P., Fruchart, J.-C., and Cecchelli, R. (1990) An easier, reproducible, and mass-production method to study the blood-brain barrier in vitro. *J. Neurochem.* **54,** 1798–1801.
136. Krum, J. and Rosenstein, J. M. (1993) Effect of astroglial degeneration on the blood-brain barrier to protein in neonatal rats. *Dev. Brain Res.* **74,** 41–50.
137. Krum, J. M. (1994) Experimental gliopathy in the adult rat CNS: effect on the blood-spinal cord barrier. *Glia* **11,** 354–366.

138. Kalaria, R. N., Stockmeier, C. A., and Harik, S. I. (1989) Brain microvessels are innervated by locus ceruleus noradrenergic neurons. *Neurosci. Lett.* **97,** 203–208.
139. Harik, S. I. (1986) Blood-brain barrier sodium/potassium pump: modulation by central noradrenergic innervation. *Proc. Natl. Acad. Sci. USA* **83,** 4067–4070.
140. Frederici, C., Camoin, L., Hattab, M., Strosberg, A. D., and Couraud, P. O. (1996) Association of the cytoplasmic domain of intercellular-adhesion molecule-1 with glyceraldehyde-3-phosphate dehydrogenase and beta-tubulin. *Eur. J. Biochem.* **238,** 173–180.
141. Betz, A. L. (1992) An overview of the multiple functions of the blood-brain barrier. *NIDA Res. Monogr.* **120,** 54–72.
142. Ibaragi, M.-A., Niwa, M., and Ozaki, M. (1989) Atrial natriuretic peptide modulates amiloride-sensitive Na^+ transport across the blood-brain barrier. *J. Neurochem.* **53,** 1802–1806.
143. Hoyer, J., Popp, R., Meyer, J., Galla, H.-J., and Gögelein, H. (1991) Angiotensin II, vasopressin and GTP[γ-S] inhibit inward-rectifying K^+ channels in porcine cerebral capillary endothelial cells. *J. Membrane Biol.* **123,** 55–62.
144. Smith, Q. R., Momma, S., Aoyagi, M., and Rapoport, S. I. (1987) Kinetics of neutral amino acid transport across the blood-brain barrier. *J. Neurochem.* **49,** 1651–1658.
145. Rapoport, S. I. (1992) Drug delivery to the brain: barrier modification and drug modification methodologies, in *Bioavailability of Drugs to the Brain and Blood-Brain Barrier* (Frankenheim, J. and Brown, R. M., eds.), US Department of Health and Human Services, Rockville, MD, pp. 121–137.
146. Fenstermacher, J. D. (1983) Drug transfer across the blood-brain barrier, in *Topics in Pharmaceutical Sciences 1983* (Breimer, D. D. and Speiser, P., eds.), Elsevier Science, Amsterdam, pp. 143–154.
147. Robinson, P. J. and Rapoport, S. I. (1992) Transport of drugs, in *Physiology and Pharmacology of the Blood-Brain Barrier* (Bradbury, M. W. B., ed.), Springer-Verlag, Berlin, pp. 279–300.
148. Balabanov, R. and Dore-Duffy, P. (1998) Cytokines and the blood-brain barrier, in *Introduction to the Blood-Brain Barrier: Methodology, Biology and Pathology* (Pardridge, W. M., ed.), Cambridge University Press, Cambridge, pp. 354–361.
149. Davson, H. and Welch, K. (1971) The permeation of several materials into the fluids of the rabbit's brain. *J. Physiol.* **218,** 337–351.
150. Ennis, S. R., Keep, R. F., Schielke, G. P., and Betz, A. L. (1990) Decrease in perfusion of cerebral capillaries during incomplete ischemia and reperfusion. *J. Cereb. Blood Flow Metab.* **10,** 213–220.
151. Frelin, C. and Vigne, P. (1998) Ion channels in endothelial cells, in *Introduction to the Blood-Brain Barrier: Methodology, Biology and Pathology* (Pardridge, W. M., ed.), Cambridge University Press, Cambridge, pp. 214–220.
152. Keep, R. F., Ennis, S. R., and Betz, A. L. (1998) Blood-brain barrier ion transport, in *Introduction to the Blood-Brain Barrier: Methodology, Biology and Pathology* (Pardridge, W. M., ed.), Cambridge University Press, Cambridge, pp. 207–213.
153. Ennis, S. R., Keep, R. F., Abdelkarim, G. E., and Betz, A. L. (1997) Chloride transport at the luminal membrane of the blood-brain barrier. *J. Cereb. Blood Flow Metab.* **19(Suppl.),** S246.
154. Domotor, E., Abbott, N. J., and Adam-Vizi, V. (1999) Na^+-Ca^{2+} exchange and its implications for calcium homeostasis in primary cultured rat brain microvascular endothelial cells. *J. Physiol.* **515,** 147–155.

12
Circumventricular Organs

James W. Anderson and Alastair V. Ferguson

1. INTRODUCTION

The dominant role of the brain in the hierarchical control of the autonomic nervous system demands that it receive extensive afferent information regarding not only the external environment, but also the internal environment. The assertion for many years has been that the hypothalamus (HT) represents the apex of this hierarchy, and, as such, must represent the primary forebrain site at which such feedback occurs. However, this conclusion ignores the problematic issue that, although this region is highly vascular, HT capillaries demonstrate a normal blood–brain barrier (BBB), so that lipid-insoluble substances cannot gain direct access to HT neurons. Thus, although the BBB acts to protect and insulate the brain from large shifts in these variables, it also precludes the central nervous system (CNS) from monitoring these same important parameters. The absence of clear mechanisms through which such feedback can occur argues for the existence of additional, physiologically significant, sensory centers involved in the monitoring of these variables. The sensory circumventricular organs (CVOs), which are specialized CNS structures that lack a normal BBB, represent a potential CNS window for such autonomic feedback. These structures, namely the subfornical organ (SFO), organum vasculosum of the lamina terminalis (OVLT), and area postrema (AP), consist primarily of neuronal cell bodies, which receive afferent input from the circulation, and communicate with other regions of the CNS through the efferent neural projections of these neurons. In contrast, a second group of secretory CVOs, composed primarily of axon terminals (median eminence [ME], neurohypophysis [NH]) or nonneuronal cell bodies (pineal gland, subcommissural organ and choroid plexus) have primary functions apparently associated with secretions of chemical messengers on one or other specific side of the BBB. This chapter describes the specific features of this unique group of CNS structures, which permit them to play such essential roles at the blood–brain interface.

2. ANATOMICAL FEATURES

2.1. *Location*

The CVOs of the brain have been defined as the SFO, OVLT, ME, NH, SCO, pineal gland, AP, and the choroid plexus. As implied from their names, all are located on the

From: *The Neuronal Environment: Brain Homeostasis in Health and Disease*
Edited by: W. Walz © Humana Press Inc., Totowa, NJ

periphery of the ventricular system of the brain, at the interface between the bulk cerebrospinal fluid (CSF) and the CNS.

The SFO, which bulges down from the roof of the third ventricle, forms a midline structure ventral to the junction of the two fornical columns, and consists of three distinct zones: dorsal stalk, main body, and ventral extension, which merges into the median preoptic nucleus. Similarly, the OVLT is located on the midline in the ventral anterior wall of the third ventricle, but is superior to the optic chiasm. The ME, so well understood for its secretory role as the central hub for neuroendocrine control, sits on the floor of the third ventricle, immediately caudal to the optic chiasm, close to the NH, which is connected to the HT through the pituitary stalk. Another secretory CVO located in the forebrain, the SCO, is positioned on the rostral, ventral surface of the third ventricle, forming the initial part of the roof of the aqueduct of Sylvius, between the third and fourth ventricle. In humans, it is almost absent in adults, although clearly visible in both the fetus and newborn. The pineal gland, another midline CVO, is found on the roof of the third ventricle, and is joined by peduncles to the habenular and posterior commissures.

The AP, the final sensory CVO, is located in the medulla at the level of the obex, and immediately adjacent to the nucleus tractus solitarius (NTS). In the rat, the AP consists of two bulges, together forming a V-shaped structure, from each lateral wall of the fourth ventricle, with the two halves converging at the obex. In other species, such as the rabbit, these two bulges of the AP do not meet, and the structure is truly a bilateral one.

The choroid plexus, located in lateral ventricles, roof of the third ventricle immediately adjacent to SFO and fourth ventricle adjacent to AP, is also considered by some to be a true CVO.

2.2. Specializations

2.2.1. Absence of BBB

The capillaries of all CVOs (except for SCO) lack detectable ZO-1, a tight junction-associated protein; immunoreactivity for ZO-1 forms a continuous line in SCO capillaries and other brain regions, with an intact BBB *(1)*. Despite the absence of a functional BBB in OVLT, this region does show ZO-1 staining comparable to brain regions that have an intact blood barrier *(1)*. In CVOs, the lack of BBB results from the presence of not only type III fenestrated capillaries *(2)*, but also type I capillaries, which are remarkable for the number of vesicles and pits present in endothelial cells. These high numbers of vesicles may result from rapid shuttling of substances across the endothelial cell via receptor-mediated exo- or endocytotic mechanisms *(3,4)*.

2.2.2. Vascularity

Another defining characteristic of the CVOs is their extreme vascularity. In addition to simply a large number of capillaries in each CVO, the arrangement of blood vessels is complex *(5,6)*, leading to a high blood volume per tissue weight, yet a slow transit time across the structure. Some of these specializations include capillary loops extending to the ependymal surface of the structure, and large, perivascular spaces (Virchow-Robin spaces) surrounding the blood vessels *(5)*. Most, if not all, of these complex specializations in the microcirculation are consistent with maximizing not only the time, but also the area for exposure of bloodborne substances to the cellular components of the organ *(6)*.

2.2.3. Ependymal Cells

These cells, found preferentially at the periphery of the CVOs, and referred to as "tanycytes" *(7)*, are irregular in shape, compared to the cuboidal cells of most of the ependymal surface of the ventricles. These elongated, flattened, or columnar cells have few cilia on their luminal surface, are rich in glial elements, and have tight junctions, indicated by continuous line of ZO-1 immunoreactivity *(1)* between cells, forming a CSF–CVO barrier.

2.2.4. Enzyme Activity

Although much of the BBB, or lack of it in CVOs, is dependent on morphological features of the capillary-associated endothelial cells, the presence of high specific enzyme concentrations tends to prevent entry, and to protect the CVOs from both endogenous and exogenous circulating compounds. Some of these enzyme systems include epoxide hydrolase and UDP-glucuronosyltransferase *(8)* and cytochrome P-450 reductase, expressed in the choroid plexus *(9)*, meningeal membranes, pineal, and the NH and median eminence in concentrations much higher than the rest of cortex. These enzymes all presumably act to prevent entry of, and to detoxify, harmful compounds *(8,10,11)*. In fact, enzyme concentrations are so high within choroid plexus that they may represent the major site of drug metabolism in the brain *(10)*. Similarly, monoamine oxidase B, which deaminates neurotransmitters and xenobiotic amines, is highly expressed in CVOs *(12)*. Expression of aminopeptidases, such as AM-dipeptidyl-peptidase IV, γ-glutamyl transpeptidase and dipeptidylpeptidase IV are also strongly expressed in all CVOs *(13)*.

Certain specific transport systems have a CVO-specific expression pattern within the brain. The high-affinity peptide/peptide fragment transporter, PEPT2, responsible for removal of peptide/peptide fragments from brain extracellular fluid (ECF), is highly expressed within not only the choroid epithelial cells, but also the ependymal and subependymal cells lining the ventricles *(14)*. The glucose transporter (GLUT1) is, logically, underexpressed in brain regions with leaky capillary barriers (CVOs) *(15)*. At birth, all capillaries that have tight junctions in adult express GLUT1 *(16)*; brain regions with leaky capillaries, except for AP and SFO, lack GLUT1 at birth, although these CVOs lose GLUT1 soon after birth *(16)*.

The net effect of all these characteristic physical features of CVOs is to permit these specialized structures to act as points of either entry or egress for circulating hormones, peptides, and other molecules between the blood, CNS, and the ventricular system. Measurements of the distribution of serum protein, immunoglobulins G and M, complement C9, and exogenous horseradish peroxidase, show that these large molecules all have potential sites of entry to the CNS at the CVOs, then widespread distribution by entry from blood through CVOs into not only adjacent white and grey matter, but also into the ventricular system *(3,17,18)*.

2.3. Receptor Localization

There is a diverse and prolific literature describing high densities of a startling variety of not only the widely studied peptidergic receptors but also other less-well-appreciated receptors for a wide variety of ligands, including steroids, specific ions, and lipopolysaccharides (LPS) in the CVOs. A complete review of this literature is beyond the scope of this chapter, but the diversity is highlighted by some specific examples.

2.3.1. *Angiotensin II*

Angiotensin II (ANGII), an octapeptide produced by the action of angiotensin-converting enzyme, to convert ANGI to ANGII, underlies most biological actions, although the more recently described components of the system (ANGIII and -IV) also have biological effects *(19)*. The development of specific blockers of ANGII receptors has led to designation of AT1 and AT2 receptor subtypes: saralasin blocks binding to both types *(20)*; losartan only prevents specific binding to the AT1 type *(21)*.

This peptide-receptor system, widely studied in the CVOs, in relation to the neuroendocrine control of both body fluid balance and the cardiovascular system (CVS), has been the subject of numerous recent reviews *(22,23)*. Of the three receptor subtypes present in rat brain *(24)*, in general, the AT1 type is associated with most physiological functions, especially CV, body fluid, and neuroendocrine regulation *(25)*; the AT2 type, because of its higher levels of expression in young and developing animals, has been implicated during development *(26–29)*.

In general, in the CNS, the two subtypes of the AT1 receptor (AT1A and AT1B) and the AT2 receptor are expressed only on neurons, and not on glial elements *(30,31)*. In most, if not all, CVOs studied, mostly in the rat, the presence of either the receptor mRNA and/or the specific, displaceable binding, has indicated a preponderance of AT1, and more specifically the AT1A receptor *(24,32–40)*. In addition, the SFO exhibits weak AT1B receptor binding *(38)*. The CVOs of the rabbit exhibit qualitatively similar receptor distributions, except for less AT1 binding in AP *(41)*, and much less AT2 binding in non-CVO brain regions.

Stimulation of isolated SFO and OVLT neurons by ANGII leads to calcium (Ca) transients because of entry from the ECF, which are blocked by losartan, indicating involvement of AT1 receptor subtype *(42)*; in isolated and cultured AP neurons, stimulation of AT1 receptor leads to increase in intracellular (IC) Ca, independent of extracellular Ca, through a pertussis-toxin-sensitive G-protein mechanism *(43)*.

2.3.1.1. Components of Renin–Angiotensin System

In addition to the receptor system, the CVOs exhibit a high concentration of most, if not all of the other components of the renin–ANG system *(44)*. For example, within the SFO, conversion of ANGI to ANGII is blocked by the ANG-converting enzymes-inhibitor, captopril *(45,46)*, leading to a blockade of physiological responses (drinking, pressor response) to systemic infusion of large doses of ANGI by intracerebroventricular (icv) captopril *(46)*; SFO injection of losartan or captopril blocks both drinking and c-*fos* activation of supraoptic nuclei (SON) and paraventricular nuclei (PVN) following ANGI inj into SFO *(47)*. In rabbits, the highest concentration of ANG-converting enzyme is found in choroid plexus and all sensory CVOs (OVLT, SFO, and AP) *(48)*. Taken together, these facts indicate the importance of the local production of ANGII within the CVO, permitting local production and/or breakdown of the active forms of ANG to regulate/control ANG levels in the CVO separately from either the rest of the body or rest of the brain.

Additionally, immunoreactivity for glutamyl aminopeptidase, which is able to convert ANGII to ANGIII *(49)*, is highest in CVOs and in other brain areas containing ANGII (PVN), and is concentrated along the microvessel adventitium *(49)*; aminopeptidase A, a membrane-bound zinc metalloprotease, which also produces ANGIII from ANGII, is most concentrated in the pituitary, SFO and OVLT, followed by AP, ME, and choroid plexus *(50)*. Aminopeptidase activity is up to 2.5× higher in spontaneously hypertensive rats although the concentrations of ANGI-converting enzyme are not affected. Angiotensinogen cRNA is also expressed within the glial elements of SFO *(31)*.

2.3.1.2. Modulation of AII Receptors

A number of experimental manipulations have been shown to lead to changes in ANGII receptor expression, including induction of experimental hydrocephalus *(51)*, subarachnoid hemorrhage *(52)*, water deprivation *(34)*, stress *(53)*, and depletion of sodium (Na) in the ECF *(54)* and during lactation as compared to diestrus in female rats *(55)*.

2.3.2. Steroid Hormones

Specific receptors, both cytoplasmic and nuclear, for a wide variety of both natural and synthetic steroid compounds, are found throughout the CVOs *(56)*. Cells that express the estrogen receptor (ER) are found on all CVOs in a wide variety of species including human *(57–61)*. Up to 15% of cells from OVLT and SFO, which project to the SON, express ERs *(57)*; ER-positive cells are especially abundant in AP of female sheep *(59)* and are present in primate (human) pineal *(60)* and SFO *(61)*. In the SFO, neurons that express the ER seem to be concentrated in a ring around the central core of the SFO *(58)*. Human male pineal expresses receptors to both gonadal steroids *(60)*. Although there is apparently no circadian or seasonal variation in gonadal steroid receptors *(60)*, the rat pineal expresses receptors for 5-α-dihydrotestosterone that do show a circadian rhythm *(62)*. Type 1 corticosteroid receptors are concentrated in all CVOs *(63)*; expression of aldosterone-selective mineralocorticoid receptors is concentrated in the anteroventral third ventricle (AV3V) region of brain, including SFO and OVLT *(64,65)*.

Most research into the potential physiological effects of these steroid receptors has concentrated on the ability of steroids to modify other receptor and/or enzyme systems. Following treatment with glucocorticoids, the angiotensinogen gene was upregulated in the AP, decreased in the median eminence, and no effect was seen in SFO *(66)*. Exogenous testosterone or castration in mice leads to changes in secretory function of pineal gland *(67)*; treatment of animals with both mineralocorticoid (deoxycorticosterone acetate) and glucocorticoid (dexamethasone) led to increases in AT1 receptor binding of ANGII in AFO and AP *(68)*.

2.3.3. Gonadotrophins

Human male pineal expresses gonadotrophin receptors *(60)*, which have been reported to undergo significant seasonal (gonadotrophic-releasing hormone [GnRH] higher in winter), and circadian (follicle-stimulating hormone [FSH] receptors higher at night) *(60)* variations. Gonadotropin receptors are also present in choroid plexus, AP *(69)*, and OVLT *(70,71)*.

2.3.4. Relaxin

The ovarian hormone, relaxin, produced in high concentrations during the second half of gestation (in rat) or highest in first trimester (human), is produced by the corpus luteum, placenta, endometrium, and, in males, the prostate *(72)*. The best-understood functions during pregnancy include lengthening of the pubic ligament, inhibition of uterine contractions, and softening of the cervix *(73)*, although relaxin also has a direct stimulatory effect (both chronotropic and inotrophic) on isolated rat atria *(74)*. Continuous infusion in SHR and control rats, at levels that mimic increases seen in the second half of pregnancy, lead not only to a decrease in blood pressure in SHR rats (as in pregnancy) *(75)* but also to a blunting of the vasoconstrictor response in isolated mesenteric vasculature to AVP and norepinephrine (not AII) *(76)* in control rats.

Binding sites for relaxin have been described within the SFO and OVLT, but not AP *(77)*. After delivery, the main physiological effect of relaxin is to inhibit milk ejection in lactating rats, an effect secondary to stimulation of an intact SFO *(78)*, with subsequent influences on oxytocin and AVP release from both SON and PVN *(78)*. Central injection of exogenous relaxin raises blood pressure, through a SFO-dependent mechanism, since it is blocked in SFO-lesioned rats *(79)*, and leads to c-*fos* increases in SFO and OVLT, as well as the MnPO- and AVP/OXY-containing neurons in SON and PVN *(80,81)*.

2.3.5. CD14 (Endotoxins)

Endotoxins, the LPS constituents of the outermost part of Gram-negative bacterial cell membranes, when purified and administered to animals, tend to mimic the acute phases of the effect of a bacterial infection, and are used in many animal models to study neuroendocrine responses to both bacterial infection and sepsis. The CD14 receptor, expressed mostly on cells of myeloid origin, when stimulated by LPS, is thought to be a key inducer of septic shock *(82)*, and to lead to the increased expression of various proinflammatory cytokines within the CNS and CVOs in particular *(83–85)*. Circulating levels of LPS lead to a rapid expression of CD14 mRNA within the CVOs, followed by later expression in parenchymal cells on borders of CVOs and subsequent induction of CD14 expression in microglia across whole-brain parenchyma *(86)*. This induced expression of CD14 is an important step in the cascade of steps leading to transcription of proinflammatory cytokines, first in the CVOs, then throughout the whole brain in sepsis *(87–89)*. Similarly, the microglial-derived proinflammatory cytokine, tumor necrosis factor α, which is produced in choroid plexus and all sensory CVOs, following systemic LPS treatment *(89)*, can induce CD14 transcription in CVOs in a paracrine/autocrine short loop manner *(90)*.

2.3.6. Cytokines

Two receptors (Type I and Type II) for interleukin 1 (IL-1) have been cloned, and their molecular biology, physiology, pharmacology, and potential roles in CNS disorders have been recently reviewed *(91,92)*. Type 1 IL receptors, which bind recombinant forms of both IL-1α and IL-1β with similar potencies, are thought to mediate most biological effects of this cytokine *(93)*; the type II receptor binds with less affinity, and does not seem to be central to observed biological effects. A potential involvement with IL Type II receptors has been suggested by the observation that the thermogenic response to IL-1β can be blocked by administration of a monoclonal antibody to block the type II receptor *(94)*.

Type 1 IL-1 receptors, described within the AP, appear to be located not just at blood vessels because of diffuse labeling over the entire structure *(95)*. After intravenous (iv) inj, IL-1α is able to enter CNS in intact form *(96)*, indicating the presence of a saturable transport system able to transport cytokines across BBB *(96,97)*. Similarly, IL-1ra has been reported to cross the BBB via a saturable transport system *(98)*. Recently, use of radioiodinated IL-1α, injected iv, showed that, although the molecule can gain access to regions lacking a BBB (SFO and choroid plexus), the subsequent movement of IL-1α is greatly restricted to the immediate region of the brain *(99)*.

2.3.7. *Ca Receptor*

This G-protein-coupled receptor, which senses Ca directly, and is also sensitive to overall ionic strength of the ECF *(100)*, is expressed in high levels in brain areas associated with body fluid balance, being especially high in SFO *(101–103)*, AP, and ME *(103)*. Interaction of Ca^{2+} with the calcium receptor (CaR) in human astrocytoma cell lines (U87) leads to opening of an outwardly rectifying potassium (K) channel *(102)*, and has been shown to modulate the nonselective cation current present in SFO neurons *(104)*.

A number of other peptide receptors have been identified in one or a combination of the CVOs, including the structurally related vasoactive intestinal peptide/pituitary adenylate cyclase-activating polypeptide *(105–109)*, endothelin *(110–117)*, glucagon-like peptide *(118)*, amylin *(119,120)*, bradykinin *(121,122)*, cholecystokinin *(123)*, orexin/hypocretin *(124)*, thyrotropin-releasing hormone *(125–128)*, leptin *(129–131)*, natriuretic peptides *(132–138)*, somatostatin *(139,140)*, vasopressin (VP) *(141–145)*, oxytocin *(146)*, neuropeptide Y *(147,148)*, and prolactin *(149,150)*. This list is growing on an annual basis; the above list is far from exhaustive and is provided only to emphasize the diversity.

2.4. Anatomical Connections

2.4.1. *SFO*

Utilization of both anterograde and retrograde labeling techniques have permitted the description of the primary efferent projections of the SFO to both the HT and the AV3V region *(151,152)*. Areas of termination within the AV3V include the median preoptic area and the OVLT; HT projections, which include both the SON and PVN, point to the involvement of the SFO in the regulation of both oxytocin and VP secretion. Other efferent projections include the regions of the zona incerta, raphe nuclei *(153)*, infralimbic cortex, and the rostral and ventral parts of the bed nucleus of the stria terminalis *(154,155)*.

Afferents from other regions of the brain, including the midbrain raphe *(154)*, median preoptic nucleus and nucleus reunions of the thalamus *(156)*, outer layer of the lateral division of the parabrachial nucleus *(157)*, and NTS *(158,159)* do provide a relatively sparse input to the region of the SFO.

2.4.2. *Area Postrema*

The anatomical connections of the AP provide a framework to understand the central position that the AP seems to take in influencing the CVS, by providing and receiving input from a variety of autonomic centers in the medulla, pons, and forebrain *(160)*.

Afferent projections from AP include not only the vagus nerve *(161)* and the NTS *(162)*, but also the parabrachial nucleus *(163)* and the PVN *(162)*. In the cat, AP afferents have been described from the carotid sinus *(164–166)*, aortic depressor *(167)*, and glossopharyngeal nerves *(168)*. The main efferent projections arising from AP go to both the NTS *(162,169)* and the PBN *(162,163,169,170)*, with other regions receiving input, including the A1 region of the nucleus ambiguous, the dorsal and dorsolateral tegmental nuclei, and the dorsal motor nucleus of the vagus *(162,169)*.

2.4.3. *OVLT*

Fiber tracts arising from the OVLT project mostly to the region of the SON *(171,172)*, with smaller, less-dense projections to both the stria medullaris and the basal ganglia *(172)*. Efferent fibers, providing input to the region of OVLT, arise most distinctly from the MnPO, brainstem, and SFO *(172)*; additional afferent input arises from a large number of HT regions *(172)*.

2.4.4. *Pineal*

The trigeminal nerve provides a significant proportion of CGRP- or SP-immunoreactive innervation to the pineal gland *(173,174)*. The pineal also receives polysynaptic input from the suprachiasmatic nucleus, presumably transmitting visual information from the retina to pineal *(174,175)*, and other HT regions, including the anterolateral region *(176)* and posterior HT *(177)*. Retrograde tracers injected into the pineal also label neurons in the superior cervical ganglia *(178,179)*, indicative of the important innervation by the sympathetic nervous system.

2.4.5. *SCO*

This CVO, whose main functions are not well-understood, but that involve secretions into the CSF and circulatory system, receive serotoninergic innervation from raphe nucleus *(180)* and innervating fibers from the lateral geniculate nucleus *(181)*.

2.4.6. *Median Eminence and Neurohypophysis*

These widely studied and best-understood of the secretory CVOs are predominantly composed of axon terminals arising from neurons of the SON, PVN, and arcuate regions of the HT *(182)*. These cells play established and well-understood roles as essential regulators of the neuroendocrine axis.

2.4.7. *Choroid Plexus*

This CVO, which receives an abundant sympathetic neural input, contains nerve fibers containing NPY and VIP, found both close to blood vessels and opposed to secretory epithelium of choroid plexus in a variety of species *(183,184)*. The main physiological effects of stimulation of these nerve fibers are effects on choroid plexus blood flow and/or CSF production *(185)*. Local release of a variety of vasoactive substances, such as AVP *(186)*, ANGII *(187,188)* or VIP *(185)*, all seem to affect both blood flow and CSF production; stimulation of the neurons of the AP have been reported to decrease blood flow to the choroid plexus *(189)*.

3. ELECTROPHYSIOLOGY OF CVO NEURONS

Previous reviews *(190,191)* give a comprehensive description of earlier studies utilizing both IC and EC recording techniques both in vivo *(153,163,192)* and in vitro

(193–195) slice preparations. Much of this work has provided information on such aspects of CVO electrophysiology as connectivity between CVOs and other brain nuclei, neuronal responses to exogenous hormones and neurotransmitters, and effects of synaptic input on CVO neuronal activity *(196,197)*.

Recently, the development of the ability to record, using the patch-clamp technique, both cellular activity (current-clamp) and ionic currents (voltage-clamp), in synaptically isolated CVO neurons, has helped to expand knowledge of the membrane properties of CVO neurons through which they fulfil their unique sensory functions *(104,198–205)*. These studies, in addition to confirming the presence of typical fast, TTX-sensitive inward Na currents and outwardly rectifying K currents, already described utilizing slice preparation with either IC or whole-cell patch recording techniques *(194,206)*, have provided further insights into intrinsic mechanisms determining cellular firing patterns.

Exposure of isolated SFO neurons to VP leads to a dose-dependent and reversible inhibition of outwardly rectifying K current *(200)*, with EC_{50} values similar to both in vivo and in vitro slice preparations. Separation of these outward currents, either by changes in holding potential or by the use of TEA, to preferentially block the delayed rectifier current, indicated that, although the majority of the current inhibited by VP was of a delayed-rectifier type, the rapidly activating, rapidly inactivating I_A in SFO neurons was also inhibited. In current-clamp mode, AVP application led to a depolarization and increase in action potential frequency, secondary to the inhibition of these repolarizing K currents. Until recently, the novel extracellular CaR, which is expressed in high levels in brain areas involved in body fluid balance, including the SFO *(101)*, was without known physiological effect in CVOs. Stimulation of this receptor, either by the use of the allosteric agonist of the CaR, NPS R-467, or with increased levels of extracellular Ca in isolated SFO neurons, leads to a depolarization caused by activation of both a nonselective cation conductance *(104)* and the hyperpolarization-activated inward current *(207)*. As would be predicted, activation of the CaR in the SFO in the whole animal also leads to a site-specific, centrally mediated augmentation of blood pressure *(104)*. This CaR-mediated excitation of SFO neurons also leads to an increased tendency for these CVO neurons to fire in a bursting pattern, although the majority of the bursting behavior in SFO neurons is dependent on the actions of a persistent Na current *(199)*. Recently, we have also established that SFO neurons exhibit an intrinsic osmosensitivity, so that increases or decreases in osmolality, secondary to changes in mannitol concentration, lead to activation or inhibition of SFO neurons, respectively *(201)*. The intrinsic mechanism underlying this osmosensitivity is unknown, but it is not secondary to activation of either the nonselective cation current or the swelling-activated chloride conductance *(201)*. Thus, the use of the isolated SFO neuron preparation, combined with the patch-clamp technique, has allowed description of four previously unknown conductances (the NSCC, I_h, I_{NaP}, and the swelling-activated Cl current), and should continue to advance understanding of the potential integrative roles of SFO neurons in a wide variety of physiological processes.

Similar patch recordings from dissociated AP neurons have characterized voltage-gated K *(204)* and Ca *(203)* channels in these cells, as well as describing the effects of specific peptides and steroids in modulating the gating characteristics of these channels *(208)*.

Although the electrophysiology of CVOs, traditionally regarded as primarily secretory in nature, have lagged behind research into the electrophysiology of sensory CVOs, recent reports *(209,210)* have shown that cells within pineal are capable of producing spike potentials, when studied ex vivo, i.e., deprived of neural input. This spiking pattern, which is often associated with, but does not seem to be a requirement for, stimulation of melatonin production and release from the pineal gland *(211)*, is still present in TTX, indicating that Ca^{2+} currents underlie spikes *(210)*. These spiking cells exhibit both regularly firing and rhythmically bursting patterns, which can be regulated by stimulation of β-adrenergic receptors within the pineal *(209)*: Application of norepinepherine reduced firing and/or prevented bursting; phenylepherine had no effect *(209)*. Adjacent cells within the pineal seem to fire in a highly synchronized pattern within a bursting group, and stimulation of one cluster of cell can immediately evoke a synaptically driven response in an adjacent cluster of cells *(212)*. This highly organized pattern of firing has led to the hypothesis that pineal cells form a network to control synthesis and/or release of melatonin *(212)*.

In pinealocytes that produce melatonin, stimulation by noradrenaline leads, via an α-adrenoceptor mechanism, to a biphasic increase in IC Ca^{2+}: the initial spike phase, caused by release of Ca from IC stores; and plateau phase dependent on influx from external sources *(213)*. Acetylcholine and carbachol, via stimulation of muscarinic receptors, also lead to Ca^{2+} increases from IC stores, in contrast to the adrenergic response *(213)*. Acetylcholine responses, mediated through nicotinic receptors that are completely dependent on influx of extracellular Ca, have also been reported *(214)*. One-half of this Ca response is nifedipine-sensitive, indicating L-type Ca channel involvement *(214)*; acetylcholine also leads to a depolarization secondary to a Na influx through the nonselective cation channels *(214)*.

4. FUNCTIONAL ROLES OF CVOS

4.1. *Fluid Balance*

The involvement of CVOs, in general, and the sensory CVOs in the forebrain (SFO and OVLT) in the control of overall body fluid balance, is well-established, and has been the subject of numerous reviews *(215,216)*. In particular, the profound dipsogenic actions of ANGII in the SFO *(217,218)* was perhaps the single observation that served to direct attention to the study of functional roles for the CVOs. Physiologically, body fluid balance is maintained both by behavioral control of salt and water intake and by physiological mechanisms to control both diuresis and naturiesis. Stimuli that can modify one or more of these responses, and can interact at the CVOs, include, but are not limited to, overall body osmolality, specific peptide concentrations (ANGII, AVP, and so on), alterations in specific ion concentration (Na^+), and neural input form other brain regions. Complicating factors, which have prevented complete understanding of the involvement of CVOs in whole body fluid balance, include the fact that the important stimuli are almost never altered in isolation in the in vivo situation, and, presumably because of the absolute requirement that fluid balance be regulated in a relatively narrow range, a great degree of redundancy is built into the system. These redundancies are indicated by studies in sheep, which indicate that only near-complete disruption of the lamina terminalis (OVLT and median preoptic nucleus), combined with destruction of the SFO, were sufficient to block drinking, either after a period of water deprivation or in response to iv infusion of hypertonic saline *(219)*.

Since the earliest studies *(220)*, the osmosensitive elements responsible for the initiation of these physiological mechanisms have been proposed to lie within the cranial cavity, and to receive blood flow directly from the carotid artery. Systemic increases in osmolality via intracarotid infusion of either NaCl or sucrose, which cross both cell membranes and the BBB slowly, compared to water, act as efficient stimuli to both thirst and VP secretion, compared to equivalent infusions of urea *(221)*. The fact that systemic urea, which readily crosses the BBB, is relatively ineffective, compared to either NaCl or sucrose, in stimulating osmotic release of VP, have been taken by most researchers in this field to indicate that the physiologically relevant osmosensor must lie outside the BBB *(221–224)*. Possible central sites, which lie outside the BBB and have been shown to be involved in the physiological responses to altered salt and water balance, include the OVLT and the SFO *(225–228)*. Whole-animal studies, utilizing the activation of c-*fos* as a marker of neuronal activation, have indicated that cells of the AV3V, OVLT, HT nuclei, and SFO are all involved in the neuronal responses to changes in extracellular osmolality *(225–228)*. Similarly, in vivo and in vitro studies, utilizing either recordings or lesions, have indicated that neurons within the AV3V region *(229,230)*, including the OVLT *(196,231–233)* and the SFO, in particular *(234–237)*, are either capable of influencing physiological responses to osmotic stimuli or are directly osmosensitive themselves.

Rats depleted of Na^+ by prior treatment with furosemide, regardless of whether water was available or not, showed strong c-*fos* immunoreactivity in both SFO and OVLT, and drank both water and NaCl solution *(238,239)*. These effects on both salt appetite and c-*fos* activation in the CVOs were blocked by prior treatment with losartan, the ANGII type I receptor antagonist *(238)*. Rats, similarly depleted of Na^+ drank less saline, with lesions of the SFO *(240,241)*; after access to NaCl, c-*fos* in SFO was decreased while cfos in NTS was increased *(239)*.

In studies with acute ECF depletion to induce both salt and water ingestion, electrolytic lesions of SFO were effective, greatly reducing and nearly abolishing the intakes of water and salt, respectively, within the first 2 h of ECF depletion *(242)*. During prolonged (up to 48 h) water deprivation, c-*fos* positive cells were increased in OVLT within 5 h, but were not consistently increased in SFO, until at least 24 h of water deprivation *(243)*.

Of the regions that show c-*fos* activation following water deprivation and dehydration, namely, the SON, PVN, and SFO, only SFO was unaffected by electrolytic lesion of the AV3V region in the rat *(244)*, although lesions of these forebrain regions (SFO, AV3V) did not entirely prevent activation of PVN and SON *(245)*. c-*fos* expression in brainstem centers, following ECF depletion, was not affected by destruction of forebrain circumventricular structures *(245)*.

In addition to the forebrain sensory CVOs, the sensory CVO located in the brainstem, the AP, also has an important part to play in the control of body fluid balance. Stimulation of AP in rabbits leads to inhibition of renal sympathetic nerve activity *(246)*, and, after lesion of AP, rats show a spontaneous increase in the consumption of concentrated NaCl solution *(247,248)*. Combined with this excessive salt appetite, animals with AP lesions also exhibited insufficient thirst and ability to excrete Na^+, leading to impaired osmoregulation *(247)*. Similarly, in response to an injected NaCl load, animals showed these same deficits, combined with a blunted response of the NH hormones to the increased osmotic load *(247)*, provided that Na^+ (compared to mannitol) was the main component of the osmotic load *(249)*. In response to a gastric Na^+ load, activation of c-*fos*, in both SON and PVN, although dependent on either vagal or splanchnic afferents, was significantly reduced by AP lesion *(250)*. A Na^+ not osmotic load to portal vein, which stimulates hepatic nerve afferents, and electrical stimulation of these same afferents, stimulated c-*fos* in AP *(251,252)*. Thus, the AP is involved in all aspects of central regulation of body fluid balance, and may be most important in whole-body responses to specific alteration in Na^+ concentrations.

4.2. Cardiovascular Regulation

The involvement of CVOs in the central control of cardiovascular (CV) function is well-established, and has been the subject of recent reviews *(253,254)*. Although the central effects of CVO stimulation on CV regulation are complex *(111,255–257)*, most can be generally described as resulting from the ability of CVO neurons to respond to appropriate vascular cues, and to induce appropriate autonomic adjustments through the anatomical connections of these CVOs with PVN, NTS, the rostral ventrolateral medulla, and the intermediolateral cell column of the spinal cord.

Sensory CVOs are certainly involved in CV responses to acutely induced changes in blood pressure. Hypertension induced by phenylepherine, or hypotension induced by Na nitroprusside for 60 min, both lead to c-*fos* expression in AP, along with regions of NTS and venterolateral medulla *(258)*; hypotension secondary to hemorrhage resulted in c-*fos* expression in NTS, VLM, and AP, an effect that, in AP, was attenuated by prior treatment with an ANGII antagonist *(259)*. Two-thirds of SFO neurons, identified as receiving projections from NTS, were excited by both electrical stimulation of NTS and by hemorrhage *(159)*, through an α-adrenergic-REC-mediated mechanism *(159)*. Both the increase in blood pressure and increase in urinary Na excretion, seen after icv inj of hypertonic saline, are blocked by SFO inj of losartan *(260)*.

A potential role of CVOs in normal blood pressure regulation is indicated by the fact that losartan, the AT1 receptor antagonist, decreases arterial pressure in Na-replete rats although the response is attenuated in rats with lesions to the AP *(261)*. The augmentation of blood pressure and HR recorded in response to muscle contraction, known as the "exercise pressor reflex," has recently been shown to be inhibited in animals with an intact AP *(262)*. This reflex, which opposes the arterial baroreflex, and maintains or augments blood pressure during exercise, is augmented in anesthetized cats, following either chemical or thermal lesions of AP *(262)*.

SHR rats show modified BP responses to microinjection of VP into AP, compared to age-matched WKY controls *(263)*. In addition, baroreceptor reflex gain control is improved in the SHR by iv AT1 receptor blockade, an effect that is not observed in APX rats, indicating that the resetting of the baroreceptor gain in SHR must be taking place via/through the AP *(264)*.

4.3. Reproductive Function

One indication of the importance of the CVOs to reproductive function is the fact that the SFO, OVLT, SCO, and AP all have significant concentrations (4–14 ng/mg protein) of luteinizing hormone (LH)-releasing hormone, with levels in the OVLT reaching 50× higher than immediately adjacent regions *(265)*. The OVLT has been implicated as part of the negative feedback loop controlling the secretion of LH, because GnRH inhibits OVLT neurons *(70)*, although a stimulatory effect has also been reported *(71)*. Lesions to the OVLT, in normally cycling female rats, leads to failure to cycle normally, and to a diminished or delayed LH and FSH surge at ovulation, in the few animals that still showed a normal cycle *(266)*. The SCO has a role to play in the cyclic, but not in tonic release of gonadotrophs *(267)*, since lesions of the SCO led to failure to cycle in 50% of animals, but gonadotroph levels were maintained like d 2 of diestrus. In animals still cycling, a delayed LH surge occurred on the day of pro-estrus, but the FSH surge was absent *(267)*. Similarly, lesions to the SFO lead to a disruption of the estrus cycle, with the majority of animals entering a period of prolonged diestrus *(268)*, with a normal LH surge occurring at ovulation, but any stimulation of FSH was absent *(268)*. Hypoglycemic inhibition of pulsatile release of LH is completely inhibited by prior removal of AP, indicating that the hypoglycemic signal to the HT GnRH pulse generator may be mediated via the AP *(269)*.

During pregnancy, rapid changes in overall body fluid balance lead to a drop in plasma osmolality and extracellular Na^+ during pregnancy *(270)*, secondary to resetting of threshold for AVP release to a lower osmolality. A similar resetting of the threshold for AVP release has been observed in ovariectomized rats, following iv relaxin *(271)*, to mimic the levels of this peptide hormone seen during normal pregnancy. The importance of ANGII and the CVOs in reproduction is indicated by the rise in brain ANGII, independent of whole-body ANGII, immediately prior to LH surge *(272)*, and by an increase in ANGII binding in preoptic area during lactation, compared to diestrus *(55)*. Lesions to SFO on d 12 in rat led to earlier delivery, although lesions later in pregnancy (d 19) had no effect *(273)*. Relaxin, which normally falls on d 20, just prior to delivery, was able to prolong pregnancy in control and sham-lesioned rats, but did not prolong pregnancy in SFOX rats *(273)*, following iv infusion to maintain relaxin levels as high as in pregnancy.

4.3.1. *Pineal and Melatonin*

Although the sites of action are not well established, the photoperiod effect on melatonin appears to be involved in the photoperiod control of reproduction in a number of species *(274,275)*, including rodents *(276–278)* and human *(279–281)*. Findings in humans include abnormal melatonin secretion in reproductive disorders and clinical abnormalities of the reproductive hormones, in cases with abnormalities of the pineal gland *(279–281)*. In sheep, either timed, daily administration, or continuous administration via implant of melatonin can mimic short daylength effect on breeding *(275)*. In the naked mole-rat, like other equatorial species of rodents, both the pineal gland and SCO are atrophied, but circadian rhythm and seasonal cycles in reproduction are maintained by other cues *(282)*.

4.4. *Immune-Nervous System Regulation*

Recent reviews of possible involvement of both inflammatory and anti-inflammatory cytokines on normal brain function, including regulation of sleep, synaptic plasticity, the hypothalamic-pituitary-adrenal axis, and Ca^{2+} signaling *(283,284)*, have focused attention on understanding the mechanisms underlying these modulatory actions. In addition to their production in circulating white blood cells, genes for the production of various cytokines and their receptors are expressed on both neurons and glial cells in different brain regions, in both a region- and cell-type-specific manner. Although the importance of various cytokines for both the production and regulation of fever and activation of the hypothalamic-pituitary-adrenal axis is generally appreciated *(284,285)*, the specific roles of various mediators, and the importance of the CVOs to the process, remain as areas of active research.

4.4.1. CVOs as Access Point

Given the large molecular size of cytokines (8–65 kDa), and their generally hydrophilic nature, the CVOs must represent a possible site to permit entry of peripherally derived factors into the brain. Following iv administration of IL-1β, c-*fos* activation is first noted in less than 30 min within nonneuronal cells characterized as being in the barrier regions of the CNS (outer meninges [arachnoid]), blood vessels and choroid plexus), but is gone within 1 h) *(286)*. A second region of activation within the CNS, evident by 3 h, was noted close to the cells showing the initial activation, including OVLT, SFO, and AP *(286)*, with what was described as a spread of c-*fos* activation into the CNS, away from CVOs with time *(286)*. Similarly, icv and iv inj of IL-1α both led to a concentration of radiolabel in CVOs *(99,287)*. This movement of the cytokines mostly resulted from leakage and diffusion across the compromised BBB of the CVOs following icv administration; the transport of radiolabel into the CVOs, following iv administration, was proportionately more dependent on a saturable transport system across the blood vessels of the brain *(287)*. However, following entry into the CVO, the subsequent movement of cytokines seems to be relatively restricted to the immediate region of the CVO *(99)*. Finally, septic doses of LPS lead to expression and production of the proinflammatory cytokine, IL-1β, globally throughout the brain of rats; subseptic doses of LPS lead to expression of IL-1β in only choroid plexus and at CVOs *(288)*. Following iv LPS in rats, marked increase in c-*fos* in not only OVLT, but also in AP and SFO *(289)* at 3 h, low doses of LPS in rats only induced c-*fos* in sensory CVOs; higher doses led to a more widespread activation of not only c-*fos*, but also immediate early genes and nerve growth factor-inducible gene B *(290)*, prior inhibition of prostaglandins by iv indomethacin; but blocking activation of many brain centers, including OVLT and PVN, had no effect on the induction of c-*fos* in SFO, ME, or the choroid plexus following LPS administration *(290)*, again indicating the importance of these brain regions, which lack a BBB as the initiation point for responses to circulating pyrogens. Recently, production of double-knockout mice, deficient in genes for either IL-1α or IL-1β, have provided evidence that IL-1β is crucial for the normal febrile response seen following the inflammatory response associated with subcutaneous inj of turpentine *(291)*.

4.4.2. Development of Fever

The OVLT is the CVO most studied and best understood in relation to the development and regulation of fever *(292)*. In rabbit, OVLT is one site of production of IL-1 in response to iv injection of endotoxin, causing fever *(83)*, lesion of OVLT attenuates fever response to iv IL-1β *(293)* and LPS *(294)*, but not to intracerebroventricular IL-1β *(293)*. Injection of LPS causes increase in IL-1β in OVLT, SFO, and AP within 1 h but the same CVO production of IL-1β is not seen in endotoxin tolerant rabbits *(295)*, made resistant to exogenous LPS administration by daily (5) inj of LPS. Ca, channels must be involved in OVLT-mediated fever production, since injection of verapamil, the Ca^{2+} channel blocker, either intravenously or, at 1/250 the dose, directly into OVLT, attenuates fever produced by iv injection of endogenous pyrogen into OVLT *(296)*. The SFO is also involved in the fever response to LPS, since the increase in body temperature, noted following ip injection of LPS (maximal at 3 h), was significantly reduced by injection of IL-1ra into either PVN or SFO, but had no effect when injected into either OVLT or ventromedial HT *(297)*. Fever following iv LPS is significantly reduced following specific lesion of the SFO, with no effect on fever generation, following destruction of either AP or OVLT *(298)*.

4.4.3. Interleukins and the HPA Axis

The involvement of cytokines in both physiological and pathophysiological activation and regulation of the HPA has recently been extensively reviewed *(284)*, and involves many systems, in addition to the CVOs. However, the importance of CVOs is clear, since doses of IL-1β, when injected either ip or icv, both lead to activation, as measured by c-*fos* expression in periventricular regions, including the CVOs *(299)*, and prior removal of AP, compared to sham-lesioned rats, abolishes not only the increase in plasma ACTH and corticosterone in response to iv IL-1β, but also prevents increases in c-*fos* mRNA in the NTS and PVN *(300)*.

4.4.4. CVOs and Synthesis of Cytokines

Following ip LPS, IL-1β has been reported to be produced (as measured by presence of immunoreactive protein *(85)*, or by presence of IL-1β mRNA *[87]*) within what are described as ramified microglia within AP, SFO, OVLT, pineal, and ME *(301)*, within 2 h, with little or no c-*fos* activation. Later on, at the time of the rise in body temperature, increases in IL-1β production and c-*fos* activation in CVOs *(85)* were noted to be followed by the appearance of small IL-1β mRNA-positive cells (glia) throughout the whole of the brain parenchyma *(87,301)*. The functional importance of these changes remains to be established, although they clearly suggest important broad-based signaling roles for the cytokines within the CNS, as part of the integrated host response.

5. CONCLUDING REMARKS

Until recently, primary interest in the CVOs focused on their established roles in the control of body fluid balance. The diversity of REC systems concentrated within these structures, combined with an emerging literature suggesting roles for the CVOs in reproduction, and immune-neural regulation, is now beginning to force a broader assessment of the functional importance of these specialized CNS structures. Combined use of molecular, cellular, and more traditional integrative physiological techniques promise major advances in understanding of roles of the CVOs in the complex integrated control of the autonomic nervous system.

REFERENCES

1. Petrov, T., Howarth, A. G., Krukoff, T. L., and Stevenson, B. R. (1994) Distribution of the tight junction-associated protein ZO-1 in circumventricular organs of the CNS. *Mol. Brain Res.* **21,** 235–246.
2. Gross, P. M. (1992) Circumventricular organ capillaries. *Progr. Brain Res.* **91,** 219–233.
3. Balin, B. J. and Broadwell, R. D. (1988) Transcytosis of protein through the mammalian cerebral epithelium and endothelium. I. Choroid plexus and the blood-cerebrospinal fluid barrier. *J. Neurocytol.* **17,** 809–826.
4. Ghitescu, L., Fixman, A., Simionescu, M., and Simionescu, N. (1986) Specific binding sites for albumin restricted to plasmalemmal vesicles of continuous capillary endothelium: receptor-mediated transcytosis. *J. Cell. Biol.* **102,** 1304–1311.
5. Gross, P. M. (1992) Circumventricular organ capillaries. *Progr. Brain Res.* **91,** 219–233.
6. Gross, P. M. (1991) Morphology and physiology of capillary systems in subregions of the subfornical organ and area postrema. *Can. J. Physiol. Pharmacol.* 69, 1010–1025.
7. Lindberg, L. A., Sukura, A., and Talanti, S. (1991) Morphology of the ependymal cells of the bovine area postrema. *Anat. Histol. Embryol.* **20,** 97–100.
8. Ghersi-Egea, J. F., Leininger-Muller, B., Cecchelli, R., and Fenstermacher, J. D. (1995) Blood-brain interfaces: relevance to cerebral drug metabolism. *Toxicol. Lett.* **82/83,** 645–653.
9. Dey, A., Jones, J. E., and Nebert, D. W. (1999) Tissue- and cell type-specific expression of cytochrome P450 1A1 and cytochrome P450 1A2 mRNA in the mouse localized *in situ* hybridization. *Biochem. Pharmacol.* **58,** 525–537.
10. El Bacha, R. S. and Minn, A. (1999) Drug metabolizing enzymes in cerebrovascular endothelial cells afford a metabolic protection to the brain. *Cell. Mol. Biol.* **45,** 15–23.
11. Morse, D. C., Stein, A. P., Thomas, P. E., and Lowndes, H. E. (1998) Distribution and induction of cytochrome P450 1A1 and 1A2 in rat brain. *Toxicol. Applied Pharmacol.* **152,** 232–239.
12. Luque, J. M., Kwan, S. W., Abell, C. W., Da Prada, M., and Richards, J. G. (1995) Cellular expression of mRNAs encoding monoamine oxidases A and B in the rat central nervous system. *J. Comp. Neurol.* **363,** 665–680.
13. Schnabel, R., Bernstein, H. G., Luppa, H., Lojda, Z., and Barth, A. (1992) Aminopeptidases in the circumventricular organs of the mouse brain: a histochemical study. *Neuroscience* **47,** 431–438.
14. Broadwell, R. D. (1992) Pathways into, through, and around the fluid-brain barriers. *NIDA Res. Monogr.* **120,** 230–258.
15. Rahner-Welsch, S., Vogel, J., and Kuschinsky, W. (1995) Regional congruence and divergence of glucose transporters (GLUT1) and capillaries in rat brains. *J. Cereb. Blood Flow Metab.* **15,** 681–686.
16. Zeller, K., Vogel, J., and Kuschinsky, W. (1996) Postnatal distribution of Glut1 glucose transporter and relative capillary density in blood-brain barrier structures and circumventricular organs during development. *Brain Res. Dev. Brain Res.* **91,** 200–208.
17. Hashimoto, P. H. (1988) Tracer in cisternal cerebrospinal fluid is soon detected in choroid plexus capillaries. *Brain Res.* **440,** 149–152.
18. Broadwell, R. D. and Sofroniew, M. V. (1996) Serum proteins bypass the blood-brain fluid barriers for extracellular entry to the central nervous system. *Exp. Neurol.* **120,** 245–263.
19. Wright, J. W. and Harding, J. W. (1992) Regulatory role of brain angiotensins in the control of physiological and behavioural responses. *Brain Res. Rev.* **17,** 227–262.
20. Brunner, H. R., Nussberger, J., Burnier, M., and Waeber, B. (1993) Angiotensin II antagonists. *Clin. Exp. Hypertens.* **15,** 1221–1238.
21. Brunner, H. R., Nussberger, J., and Waeber, B. (1993) Inhibitors of the renin-angiotensin system. *Arzneimittelforschung* **43,** 274–278.

22. Hohle, S., Blume, A., Lebrun, C., Culman, J., and Unger, T. (1995) Angiotensin receptors in the brain. *Pharmacol. Toxicol.* **77,** 306–315.
23. Lenkei, Z., Corvol, P., and Llorens-Cortes, C. (1995) Comparative expression of vasopressin and angiotensin type-1 receptor mRNA in rat hypothalamic nuclei: a double *in situ* hybridization study. *Mol. Brain Res.* **34,** 135–142.
24. Johren, O., Inagami, T., and Saavedra, J. M. (1995) AT1A, AT1B, and AT2 angiotensin II receptor subtype gene expression in rat brain. *Neuroreport* **6,** 2549–2552.
25. Lenkei, Z., Corvol, P., and Llorens-Cortes, C. (1994) The angiotensin receptor subtype AT1A predominates in rat forebrain areas involved in blood pressure, body fluid homeostasis and neuroendocrine control. *Brain Res. Mol. Brain Res.* **30,** 53–60.
26. Allen, A. M., Zhuo, J., and Mendelsohn, F. A. (1999) Localization of angiotensin AT1 and AT2 receptors. *J. Am. Soc. Nephrol.* **10(Suppl. 11),** S23–S29.
27. Millan, M. A., Jacobowitz, D. M., Aguilera, G., and Catt, K. J. (1991) Differential distribution of AT1 and AT2 angiotensin II receptor subtypes in the rat brain during development. *Proc. Natl. Acad. Sci. USA* **88,** 11,440–11,444.
28. Millan, M. A., Kiss, A., and Aguilera, G. (1991) Developmental changes in brain angiotensin II receptors in the rat. *Peptides* **12,** 723–737.
29. Nuyt, A. M., Lenkei, Z., Palkovits, M., Corvol, P., and Llorens-Cortes, C. (1999) Ontogeny of angiotensin II type 2 receptor mRNA expression in fetal and neonatal rat brain. *J. Comp. Neurol.* **407,** 193–206.
30. Saavedra, J. M. (1999) Emerging features of brain angiotensin receptors. *Regul. Pept.* **85,** 31–45.
31. Lippoldt, A., Bunnemann, B., Iwai, N., Metzger, R., Inagami, T., Fuxe, K., and Ganten, D. (1993) Cellular localization of angiotensin type 1 receptor and angiotensinogen mRNAs in the subfornical organ of the rat brain. *Neurosci. Lett.* **150,** 153–158.
32. Andersson, B., Eriksson, S., and Rundgren, M. (1995) Angiotensin and the brain. *Acta Physiol. Scand.* **155,** 117–125.
33. Leikei, Z., Palkovits, M., Corvol, P., and Llorens-Cortes, C. (1998) Distribution of angiotensin type-I receptor messenger RNA expression in the adult rat brain. *Neuroscience* **82,** 827–841.
34. Sanvitto, G. L., Johren, O., Hauser, W., and Saavedra, J. M. (1997) Water deprivation upregulates ANG II AT1 binding and mRNA in rat subfornical organ and anterior pituitary. *Am. J. Physiol.* **273,** E156–E163.
35. Bunnemann, B., Iwai, N., Metzger, R., Fuxe, K., Inagami, T., and Ganten, D. (1992) The distribution of angiotensin II AT 1 receptor subtype mRNA in the rat brain. *Neurosci. Lett.* **142,** 155–158.
36. Rowe, B. P., Saylor, D. L., and Speth, R. C. (1992) Analysis of angiotensin II receptor subtypes in individual rat brain nuclei. *Neuroendocrinology* **55,** 563–573.
37. Obermuller, N., Unger, T., Culman, J., Gohlke, P., de Gasparo, M., and Bottari, S. (1991) Distribution of angiotensin II receptor subtypes in rat brain nuclei. *Neurosci. Lett.* **132,** 11–15.
38. Lenkei, Z., Palkovits, M., Corvol, P., and LlorensCortes, C. (1998) Distribution of angiotensin TYPE-I receptor messenger RNA expression in the adult rat brain. *Neuroscience* **82,** 827–841.
39. Johren, O. and Saavedra, J. M. (1996) Expression of AT(1A) and AT(1B) angiotensin II receptor messenger rna in forebrain of 2-wk-old rats. *Am. J. Physiol.* **34,** E:104–E112.
40. Song, K., Allen, A. M., Paxinos, G., and Mendelsohn, F. A. O. (1992) Mapping of angiotensin II receptor subtype heterogeneity in rat brain. *J. Comp. Neurol.* **316,** 467–484.
41. Aldred, G. P., Chai, S. Y., Song, K., Zhuo, J., MacGregor, D. P., and Mendelsohn, F. A. O. (1993) Distribution of angiotensin II receptor subtypes in the rabbit brain. *Regul. Pept.* **44,** 119–130.

42. Gebke, E., Muller, A. R., Jurzak, M., and Gerstberger, R. (1998) Angiotensin II-Induced calcium signalling in neurons and astrocytes of rat circumventricular organs. *Neuroscience* **85,** 509–520.
43. Consolim-Colombo, F. M., Hay, M., Smith, T. C., Elizondo-Fournier, M., and Bishop, V. S. (1996) Subcellular mechanisms of angiotensin II and arginine vasopressin activation of area postrema neurons. *Am. J. Physiol.* **271,** R34–R41.
44. Chai, S. Y., McKinley, M. J., and Mendelsohn, F. A. (1987) Distribution of angiotensin converting enzyme in sheep hypothalamus and medulla oblongata visualized by in vitro autoradiography. *Clin. Exp. Hypertens. [A]* **9,** 449–460.
45. Rauch, M. and Schmid, H. A. (1999) Functional evidence for subfornical organ-intrinsic conversion of angiotensin I to angiotensin II. *Am. J. Physiol.* **276,** R1630–R1638.
46. Robinson, M. M., McLennan, G. P., Thunhorst, R. L., and Johnson, A. K. (1999) Interactions of the systemic and brain renin-angiotensin systems in the control of drinking and the central mediation of pressor responses. *Brain Res.* **842,** 55–61.
47. Xu, Z. and Xinghong, J. (1999) Drinking and Fos-immunoreactivity in rat brain induced by local injection of angiotensin I into the subfornical organ. *Brain Res.* **817,** 67–74.
48. Rogerson, F. M., Schlawe, I., Paxinos, G., Chai, S. Y., McKinley, L. J., and Mendelsohn, F. A. (1995) Localization of angiotensin converting enzyme by in vitro autoradiography in the rabbit brain. *J. Chem. Neuroanat.* **8,** 227–243.
49. Healy, D. P. and Wilk, S. (1993) Localization of immunoreactive glutamyl aminopeptidase in rat brain. II. Distribution and correlation with angiotensin II. *Brain Res.* **606,** 295–303.
50. Zini, S., Masdehors, P., Lenkei, Z., FournieZaluski, M. C., Roques, B. P., Corvol, P., and LlorensCortes, C. (1997) Aminopeptidase A: distribution in rat brain nuclei and increased activity in spontaneously hypertensive rats. *Neuroscience* **78,** 1187–1193.
51. Acikgoz, B., Akpinar, G., Bingol, N., and Usseli, I. (1999) Angiotensin II receptor content within the circumventricular organs increases after experimental hydrocephalus in rats. *Acta Neurochir. (Wien)* **141,** 1095–1099.
52. Acikgoz, B., Ozgen, T., Ozdogan, F., Sungur, A., and Tekkok, I. H. (1996) Angiotensin II receptor content within the subfornical organ and organum vasculosum lamina terminalis increases after experimental subarachnoid haemorrhage in rats. *Acta Neurochir.* **138,** 460–465.
53. Castren, E. and Saavedra, J. M. (1988) Repeated stress increases the density of angiotensin II binding sites in rat paraventricular nucleus and subfornical organ. *Endocrinology* **122,** 370–372.
54. Ray, P. E., Ruley, E. J., and Saavedra, J. M. (1990) Down-regulation of angiotensin II receptors in subfornical organ of young male rats by chronic dietary sodium depletion. *Brain Res.* **510,** 303–38.
55. Speth, R. C., Barry, W. T., Smith, M. S., and Grove, K. L. (1999) A comparison of brain angiotensin II receptors during lactation and diestrus of the estrous cycle in the rat. *Am. J. Physiol.* **277,** R904–R909.
56. Stumpf, W. E., Bidmon, H. J., and Ruhle, H. J. (1992) Steroid hormones and circumventricular organs. *Progr. Brain Res.* **91,** 271–277.
57. Voisin, D. L., Simonian, S. X., and Herbison, A. E. (1997) Identification of estrogen receptor-containing neurons projecting to the rat supraoptic nucleus. *Neuroscience* **78,** 215–228.
58. Rosas-Arellano, M. P., Solano-Flores, L. P., and Ciriello, J. (1999) Co-localization of estrogen and angiotensin receptors within subfornical organ neurons. *Brain Res.* **837,** 254–262.

59. Scott, C. J., Rawson, J. A., Pereira, A. M., and Clarke, I. J. (1998) The distribution of estrogen receptors in the brainstem of female sheep. *Neurosci. Lett.* **241,** 29–32.
60. Luboshitzky, R., Dharan, M., Goldman, D., Herer, P., Hiss, Y., and Lavie, P. (1997) Seasonal variation of gonadotropins and gonadal steroids receptors in the human pineal gland. *Brain Res. Bull.* **44,** 665–670.
61. Blurton-Jones, M. M., Roberts, J. A., and Tuszynski, M. H. (1999) Estrogen receptor immunoreactivity in the adult primate brain: neuronal distribution and association with p75, trkA, and choline acetyltransferase. *J. Comp. Neurol.* **405,** 529–542.
62. Gupta, D., Haldar, C., Coeleveld, M., and Roth, J. (1993) Ontogeny, circadian rhythm pattern, and hormonal modulation of 5 alpha-dihydrotestosterone receptors in the rat pineal. *Neuroendocrinology* **57,** 45–53.
63. Ahima, R., Krozowski, Z., and Harlan, R. (1991) Type I corticosteroid receptor-like immunoreactivity in the rat CNS: distribution and regulation by corticosteroids. *J. Comp. Neurol.* **313,** 522–538.
64. Reul, J. M., Sutanto, W., van Eekelen, J. A., Rothuizen, J., and de Kloet, E. R. (1990) Central action of adrenal steroids during stress and adaptation. *Adv. Exp. Med. Biol.* **274,** 243–256.
65. de Kloet, E. R., Reul, J. M., de Ronde, F. S., Bloemers, M., and Ratka, A. (1986) Function and plasticity of brain corticosteroid receptor systems: action of neuropeptides. *J. Steroid Biochem.* **25,** 723–731.
66. Bunnemann, B., Lippoldt, A., Aguirre, J. A., Cintra, A., and Metzger, R. (1993) Glucocorticoid regulation of angiotensinogen gene expression in discrete areas of the male rat brain. An *in situ* hybridization study. *Neuroendocrinology* **57,** 856–862.
67. Redins, C. A., Novaes, J. C., and Torres, K. B. (1999) The effects of testosterone on the mice pinealocytes: a quantitative study. *Tissue Cell* **31,** 233–239.
68. Shelat, S. G., King, J. L., Flanagan-Cato, L. M., and Fluharty, S. J. (1999) Mineralocorticoids and glucocorticoids cooperatively increase salt intake and angiotensin II receptor binding in rat brain. *Neuroendocrinology* **69,** 339–351.
69. Lei, Z. M., Rao, C. V., Kornyei, J. L., Licht, P., and Hiatt, E. S. (1993) Novel expression of human chorionic gonadotropin/luteinizing hormone receptor gene in brain. *Endocrinology* **132,** 2262–2270.
70. Felix, D. and Phillips, M. I. (1979) Inhibitory effects of luteinizing hormone releasing hormone (LH-RH) on neurons in the organum vasculosum lamina terminalis (OVLT). *Brain Res.* **169,** 204–208.
71. Sayer, R. J., Hubbard, J. I., and Sirett, N. E. (1984) Rat organum vasculosum laminae terminalis in vitro: responses to transmitters. *Am. J. Physiol.* **247,** R374–R379.
72. Kakouris, H., Eddie, L. W., and Summers, R. J. (1993) Relaxin: more than just a hormone of pregnancy. *Trends Pharmacol. Sci.* **14,** 4–6.
73. Kemp, B. E. and Niall, H. D. (1984) *Relaxin. Vitam. Horm.* **41,** 79–115.
74. Kakouris, H., Eddie, L. W., and Summers, R. J. (1992) Cardiac effects of relaxin in rats. *Lancet* **339,** 1076–1078.
75. St-Louis, J. and Massicotte, G. (1985) Chronic decrease of blood pressure by rat relaxin in spontaneously hypertensive rats. *Life Sci.* **37,** 1351–1357.
76. Massicotte, G., Parent, A., and St-Louis, J. (1989) Blunted responses to vasoconstrictors in mesenteric vasculature but not in portal vein of spontaneously hypertensive rats treated with relaxin. *Proc. Soc. Exp. Biol. Med.* **190,** 254–259.
77. Osheroff, P. L. and Phillips, H. S. (1991) Autoradiographic localization of relaxin binding sites in rat brain. *Neurobiology* **88,** 6413–6417.
78. Summerlee, A. J., O'Byrne, K. T., Jones, S. A., and Eltringham, L. (1987) The subfornical organ and relaxin-induced inhibition of reflex milk ejection in lactating rats. *J. Endocrinol.* **115,** 347–353.

79. Mumford, A. D., Parry, L. J., and Summerlee, A. J. (1989) Lesion of the subfornical organ affects the haemotensive response to centrally administered relaxin in anaesthetized rats. *J. Endocrinol.* **122,** 747–755.
80. McKinley, M. J., Burns, P., Colvill, L. M., Oldfield, B. J., Wade, J. D., Weisinger, R. S., and Tregear, G. W. (1997) Distribution of Fos immunoreactivity in the lamina terminalis and hypothalamus induced by centrally administered relaxin in conscious rats. *J. Neuroendocrinol.* **9,** 431–437.
81. Heine, P. A., Di, S., Ross, L. R., Anderson, L. L., and Jacobson, C. D. (1997) Relaxin-induced expression of Fos in the forebrain of the late pregnant rat. *Neuroendocrinology* **66,** 38–46.
82. Pugin, J., Heumann, I. D., Tomasz, A., Kravchenko, V. V., Akamatsu, Y., Nishijima, M., et al. (1994) CD14 is a pattern recognition receptor. *Immunity* **1,** 509–516.
83. Nakamori, T., Morimoto, A., Yamaguchi, K., Watanabe, T., Long, N. C., and Murakami, N. (1993) Organum vasculosum laminae terminalis (OVLT) is a brain site to produce interleukin-1 beta during fever. *Brain Res.* **618,** 155–159.
84. Nakamori, T., Morimoto, A., Yamaguchi, K., Watanabe, T., and Murakami, N. (1994) Interleukin-1 beta production in the rabbit brain during endotoxin-induced fever. *J. Physiol.* **476,** 177–186.
85. Konsman, J. P., Kelley, K., and Dantzer, R. (1999) Temporal and spatial relationships between lipopolysaccharide-induced expression of Fos, interleukin-1beta and inducible nitric oxide synthase in rat brain. *Neuroscience* **89,** 535–548.
86. Lacroix, S., Feinstein, D., and Rivest, S. (1998) The bacterial endotoxin lipopolysaccharide has the ability to target the brain in upregulating its membrane CD14 receptor within specific cellular populations. *Brain Pathol.* **8,** 625–640.
87. Quan, N., Whiteside, M., and Herkenham, M. (1998) Time course and localization patterns of interleukin-1 beta messenger RNA expression in brain and pituitary after peripheral administration of lipopolysaccharide. *Neuroscience* **83,** 281–293.
88. Vallieres, L. and Rivest, S. (1997) Regulation of the genes encoding interleukin-6, its receptor, and gp130 in the rat brain in response to the immune activator lipopolysaccharide and the proinflammatory cytokine interleukin-1 beta. *J. Neurochem.* **69,** 1668–1683.
89. Nadeau, S. and Rivest, S. (1999) Regulation of the gene encoding tumor necrosis factor alpha (TNF-alpha) in the rat brain and pituitary in response in different models of systemic immune challenge. *J. Neuropathol. Exp. Neurol.* **58,** 61–77.
90. Nadeau, S. and Rivest, S. (2000) Role of microglial-derived tumor necrosis factor in mediating CD14 transcription and nuclear factor kappa B activity in the brain during endotoxemia. *J. Neurosci.* **20,** 3456–3468.
91. Loddick, S. A., Liu, C., Takao, T., Hashimoto, K., and De Souza, E. B. (1998) Interleukin-1 receptors: cloning studies and role in central nervous system disorders. *Brain Res. Brain Res. Rev.* **26,** 306–319.
92. Rothwell, N. J., Luheshi, G., and Toulmond, S. (1996) Cytokines and their receptors in the central nervous system: physiology, pharmacology, and pathology. *Pharmacol. Ther.* **69,** 85–95.
93. Anforth, H. R., Bluthe, R. M., Bristow, A., Hopkins, S., Lenczowski, M. J., Luheshi, G., et al. (1998) Biological activity and brain actions of recombinant rat interleukin-1alpha and interleukin-1beta. *Eur. Cytokine Network* **9,** 279–288.
94. Luheshi, G., Hopkins, S. J., Lefeuvre, R. A., Dascombe, M. J., Ghiara, P., and Rothwell, N. J. (1993) Importance of brain IL-1 type II receptors in fever and thermogenesis in the rat. *Am. J. Physiol.* **265,** E585–E591.
95. Ericsson, A., Liu, C., Hart, R. P., and Sawchenko, P. E. (1995) Type 1 interleukin-1 receptor in the rat brain: distribution, regulation, and relationship to sites of IL-1-induced cellular activation. *J. Comp. Neurol.* **361,** 681–698.

96. Banks, W. A., Kastin, A. J., and Gutierrez, E. G. (1993) Interleukin-1 alpha in blood has direct access to cortical brain cells. *Neurosci. Lett.* **163,** 41–44.
97. Banks, W. A., Kastin, A. J., and Broadwell, R. D. (1995) Passage of cytokines across the blood-brain barrier. *Neuroimmunomodulation* **2,** 241–248.
98. Gutierrez, E. G., Banks, W. A., and Kastin, A. J. (1994) Blood-borne interleukin-1 receptor antagonist crosses the blood-brain barrier. *J. Neuroimmunol.* **55,** 153–160.
99. Maness, L. M., Kastin, A. J., and Banks, W. A. (1998) Relative contributions of a CVO and the microvascular bed to delivery of blood-borne IL-1alpha to the brain. *Am. J. Physiol.* **275,** E207–E212.
100. Quinn, S. J., Kifor, O., Trivedi, S., Diaz, R., Vassilev, P., and Brown, E. M. (1998) Sodium and ionic strength sensing by the calcium receptor. *J. Biol. Chem.* **273,** 19,579–19,586.
101. Rogers, K. V., Dunn, C. K., Hebert, S. C., and Brown, E. M. (1997) Localization of calcium receptor mRNA in the adult rat central nervous system by in situ hybridization. *Brain Res.* **744,** 47–56.
102. Chattopadhyay, N., Ye, C. P., Yamaguchi, T., Vassilev, P. M., and Brown, E. M. (1999) Evidence for extracellular calcium-sensing receptor mediated opening of an outward K^+ channel in a human astrocytoma cell line (U87). *Glia* **26,** 64–72.
103. Ferry, S., Traiffort, E., Stinnakre, J., and Ruat, M. (2000) Developmental and adult expression of rat calcium-sensing receptor transcripts in neurons and oligodendrocytes. *Eur. J. Neurosci.* **12,** 872–884.
104. Washburn, D. L. S., Smith, P. M., and Ferguson, A. V. (1999) Control of neuronal excitability by an ion sensing receptor. *Eur. J. Neurosci.* **11,** 1947–1954.
105. Simonneaux, V., Kienlen-Campard, P., Loeffler, J. P., Basille, M., Gonzalez, B. J., Vaudry, H., Robberecht, P., and Pevet, P. (1998) Pharmacological, molecular and functional characterization of vasoactive intestinal polypeptide/pituitary adenylate cyclase-activating polypeptide receptors in the rat pineal gland. *Neuroscience* **85,** 887–896.
106. Vertongen, P., Schiffmann, S. N., Gourlet, P., and Robberecht, P. (1998) Autoradiographic visualization of the receptor subclasses for vasoactive intestinal polypeptide (VIP) in rat brain. *Ann. NY Acad. Sci.* **865,** 412–415.
107. Shioda, S., Shuto, Y., Somogyvari-Vigh, A., Legradi, G., Onda, H., Coy, D. H., Nakajo, S., and Arimura, A. (1997) Localization and gene expression of the receptor for pituitary adenylate cyclase-activating polypeptide in the rat brain. *Neurosci. Res.* **28,** 345–354.
108. Olcese, J., McArdle, C., Mikkelsen, J., and Hannibal, J. (1996) PACAP and type I PACAP receptors in the pineal gland. *Ann. NY Acad. Sci.* **805,** 595–600.
109. Masuo, Y., Ohtaki, T., Masuda, Y., Tsuda, M., and Fujino, M. (1992) Binding sites for pituitary adenylate cyclase activating polypeptide (PACAP): comparison with vasoactive intestinal polypeptide (VIP) binding site localization in rat brain sections. *Brain Res.* **575,** 113–123.
110. Gebke, E., Muller, A. R., Pehl, U., and Gerstberger, R. (2000) Astrocytes in sensory circumventricular organs of the rat brain express functional binding sites for endothelin. *Neuroscience* **97,** 371–381.
111. Ferguson, A. V. and Smith, P. (1990) Cardiovascular responses induced by endothelin microinjection into area postrema. *Regul. Pept.* **27,** 75–85.
112. Koseki, C., Imai, M., Hirata, Y., Yanagisawa, M., and Masaki, T. (1989) Autoradiographic localization of [^{125}I]-endothelin-1 binding sites in rat brain. *Neurosci. Res.* **6,** 581–585.
113. Niwa, M., Kawaguchi, T., Fujimoto, M., Kataoka, Y., and Taniyama, K. (1991) Receptors for endothelin in the central nervous system. *J. Cardiovasc. Pharmacol.* **17,** S137–S139.
114. Kohzuki, M., Chai, S. Y., Paxinos, G., Karavas, A., Casley, D. J., Johnston, C. I., and Mendelsohn, F. A. O. (1991) Localization and characterization of endothelin receptor

binding sites in the rat brain visualized by in vitro autoradiography. *Neuroscience* **42,** 245–260.
115. Jones, C. R., Hiley, C. R., Pelton, J. T., and Mohr, M. (1989) Autoradiographic visualization of the binding sites for [^{125}I] endothelin in rat and human brain. *Neurosci. Lett.* **97,** 276–279.
116. Hemsen, A. and Lundberg, J. M. (1991) Presence of endothelin-1 and endothelin-3 in peripheral tissues and central nervous system of the pig. *Regul. Pept.* **36,** 71–83.
117. Garrido, M. D. and Israel, A. (1999) Endothelin-1 stimulates phosphoinositide hydrolysis in the rat pineal gland. *Arch. Physiol. Biochem.* **107,** 138–143.
118. Orskov, C., Poulsen, S. S., Moller, M., and Holst, J. J. (1996) Glucagon-like peptide I receptors in the subfornical organ and the area postrema are accessible to circulating glucagon-like peptide I. *Diabetes* **45,** 832–835.
119. Sexton, P. M., Paxinos, G., Kenney, M. A., Wookey, P. J., and Beaumont, K. (1994) In vitro autoradiographic localization of amylin binding sites in rat brain. *Neuroscience* **62,** 553–567.
120. Christopoulos, G., Paxinos, G., Huang, X. F., Beaumont, K., Toga, A. W., and Sexton, P. M. (1995) Comparative distribution of receptors for amylin and the related peptides calcitonin gene related peptide and calcitonin in rat and monkey brain. *Can. J. Physiol. Pharmacol.* **73,** 1037–1041.
121. Murone, C., Paxinos, G., McKinley, M. J., Oldfield, B. J., Muller-Esterl, W., Mendelsohn, F. A., and Chai, S. Y. (1997) Distribution of bradykinin B2 receptors in sheep brain and spinal cord visualized by in vitro autoradiography. *J. Comp. Neurol.* **381,** 203–218.
122. Shi, B., Bhat, G., Mahesh, V. B., Brotto, M., Nosek, T. M., and Brann, D. W. (1999) Bradykinin receptor localization and cell signaling pathways used by bradykinin in the regulation of gonadotropin-releasing hormone secretion. *Endocrinology* **140,** 4669–4676.
123. Kubota, Y., Inagaki, S., Shiosaka, S., Cho, H. J., Tateishi, K., Hashimura, E., Hamaoka, T., and Tohyama, M. (1983) The distribution of cholecystokinin octapeptide-like structures in the lower brain stem of the rat: an immunohistochemical analysis. *Neuroscience* **9,** 587–604.
124. Date, Y., Mondal, M. S., Matsukura, S., Ueta, Y., Yamashita, H., Kaiya, H., Kangawa, K., and Nakazato, M. (2000) Distribution of orexin/hypocretin in the rat median eminence and pituitary. *Brain Res. Mol. Brain Res.* **76,** 1–6.
125. Eskay, R. L., Long, R. T., and Palkovits, M. (1983) Localization of immunoreactive thyrotropin releasing hormone in the lower brainstem of the rat. *Brain Res.* **277,** 159–162.
126. Iwase, M., Homma, I., Shioda, S., and Nakai, Y. (1988) Thyrotropin-releasing hormone-like immunoreactive neurons in rabbit medulla oblongata. *Neurosci. Lett.* **92,** 30–33.
127. Iwase, M., Shioda, S., Nakai, Y., and Homma, I. (1991) Immunocytochemistry of thyrotropin-releasing hormone in the rabbit medulla oblongata. *Brain Res. Bull.* **26,** 49–57.
128. Merchenthaler, I. and Liposits, Z. (1994) Mapping of thyrotropin-releasing hormone (TRH) neuronal systems of rat forebrain projecting to the median eminence and the OVLT. Immunocytochemistry combined with retrograde labeling at the light and electron microscopic levels. *Acta Biol. Hung.* **45,** 361–374.
129. Couce, M. E., Burguera, B., Parisi, J. E., Jensen, M. D., and Lloyd, R. V. (1997) Localization of leptin receptor in the human brain. *Neuroendocrinology* **66,** 145–150.
130. Zamorano, P. L., Mahesh, V. B., De Sevilla, L. M., Chorich, L. P., Bhat, G. K., and Brann, D. W. (1997) Expression and localization of the leptin receptor in endocrine and neuroendocrine tissues of the rat. *Neuroendocrinology* **65,** 223–228.
131. Hoggard, N., Mercer, J. G., Rayner, D. V., Moar, K., Trayhurn, P., and Williams, L. M. (1997) Localization of leptin receptor mRNA splice variants in murine peripheral tissues by RT-PCR and *in situ* hybridization. *Biochem. Biophys. Res. Comm.* **232,** 383–387.

132. van Leeuwen, F. W., Caffe, A. R., and De Vries, G. J. (1985) Vasopressin cells in the bed nucleus of the stria terminalis of the rat: sex differences and the influence of androgens. *Brain Res.* **325,** 391–394.
133. Saavedra, J. M., Correa, F. M. A., Plunkett, L. M., Israel, A., Kurihara, M., and Shigematsu, K. (1986) Binding of angiotensin and atrial natriuretic peptide in brain of hypertensive rats. *Nature* **320,** 758–760.
134. Himeno, A., Niwa, M., Nakao, K., Suga, S., Yamashita, K., Kataoka, Y., et al. (1992) C-type natriuretic peptide-22 differentiates between natriuretic peptide receptors in rat choroid plexus and subfornical organ. *Eur. J. Pharmacol.* **215,** 337–440.
135. Saavedra, J. M. (1988) Alterations in atrial natriuretic peptide receptors in rat brain nuclei during hypertension and dehydration. *Can. J. Physiol. Pharmacol.* **66,** 288–294.
136. Niwa, M., Shigematsu, K., Kurihara, M., Kataoka, Y., Maeda, T., Nakao, K., et al. (1988) Receptor autoradiographic evidence of specific brain natriuretic peptide binding sites in the porcine subfornical organ. *Neurosci. Lett.* **95,** 113–118.
137. Konrad, E. M., Thibault, G., and Schiffrin, E. L. (1992) Atrial natriuretic factor binding sites in rat area postrema: autoradiographic study. *Am. J. Physiol.* **263,** R747–R755.
138. Bianchi, C., Gutkowska, J., Ballak, M., Thibault, G., Garcia, R., Genest, J., and Cantin, M. (1986) Radioautographic localization of ^{125}I-atrial natriuretic factor binding sites in the brain. *Neuroendocrinology* **44,** 365–372.
139. Krisch, B. (1992) Somatostatin-binding sites on structures of circumventricular organs. *Progr. Brain Res.* **91,** 247–250.
140. Fodor, M., Slama, A., Guillaume, V., Videau, C., Csaba, Z., Oliver, C., and Epelbaum, J. (1997) Distribution and pharmacological characterization of somatostatin receptor binding sites in the sheep brain. *J. Chem. Neuroanat.* **12,** 175–182.
141. Schoots, O., Hernando, F., Knoers, N. V., and Burbach, J. P. (1999) Vasopressin receptors: structural functional relationships and role in neural and endocrine regulation. *Results Probl. Cell Differ.* **26,** 107–133.
142. Raggenbass, M., Tribollet, E., Dubois-Dauphin, M., and Dreifuss, J. J. (1989) Vasopressin receptors of the vasopressor (V1) type in the nucleus of the solitary tract of the rat mediate direct neuronal excitation. *J. Neurosci.* **9,** 3929–3936.
143. Ostrowski, N. L., Lolait, S. J., and Young, W. S. (1994) Cellular localization of vasopressin V1a receptor messenger ribonucleic acid in adult male rat brain, pineal, and brain vasculature. *Endocrinology* **135,** 1511–1528.
144. Phillips, P. A., Abrahams, J. M., Kelly, J., Paxinos, G., Grzonka, Z., Mendelsohn, F. A. O., and Johnston, C. I. (1988) Localization of vasopressin binding sites in rat brain by in vitro autoradiography using a radioiodinated V_1 receptor antagonist. *Neuroscience* **27,** 749–761.
145. Tribollet, E., Raufaste, D., Maffrand, J., and Serradeil-Le Gal, C. (1999) Binding of the non-peptide vasopressin V1a receptor antagonist SR-49059 in the rat brain: an in vitro and in vivo autoradiographic study. *Neuroendocrinology* **69,** 113–120.
146. de Kloet, E. R., Voorhuis, D. A. M., Boschma, Y., and Elands, J. (1986) Estradiol modulates density of putative 'oxytocin receptors' in discrete rat brain regions. *Neuroendocrinology* **44,** 415–421.
147. Nakajima, T., Yashima, Y., and Nakamura, K. (1986) Quantitative autoradiographic localization of neuropeptide Y receptors in the rat lower brainstem. *Brain Res.* **380,** 144–150.
148. Nilsson, S. F. (1991) Neuropeptide Y (NPY): a vasoconstrictor in the eye, brain and other tissues in the rabbit. *Acta Physiol. Scand.* **141,** 455–467.
149. Roky, R., Paut-Pagano, L., Goffin, V., Kitahama, K., Valatx, J. L., Kelly, P. A., and Jouvet, M. (1996) Distribution of prolactin receptors in the rat forebrain. Immunohistochemical study. *Neuroendocrinology* **63,** 422–429.
150. Mangurian, L. P., Jurjus, A. R., and Walsh, R. J. (1999) Prolactin receptor localization to the area postrema. *Brain Res.* **836,** 218–220.

151. Lind, R. W. (1987) A review of the neural connections of the subfornical organ, in *Circumventricular Organs and Body Fluids* (Gross, P., ed.), CRC, Boca Raton, FL, pp. 27–43.
152. Lind, R. W., Van Hoesen, G. W., and Johnson, A. K. (1982) An HRP study of the connections of the subfornical organ of the rat. *J. Comp. Neurol.* **210,** 265–277.
153. Tanaka, J., Ushigome, A., Hori, K., and Nomura, M. (1998) Responses of raphe nucleus projecting subfornical organ neurons to angiotensin II in rats. *Brain Res. Bull.* **45,** 315–318.
154. Lind, R. W. (1986) Bi-directional, chemically specified neural connections between the subfornical organ and the midbrain raphe system. *Brain Res.* **384,** 250–261.
155. Swanson, L. W. and Lind, R. W. (1986) Neural projections subserving the initiation of a specific motivated behavior in the rat: new projections from the subfornical organ. *Brain Res.* **379,** 399–403.
156. Lind, R. W., Swanson, L. W., and Ganten, D. (1984) Angiotensin II immunoreactivity in the neural afferents and efferents of the subfornical organ of the rat. *Brain Res.* **321,** 209–215.
157. Gu, G. B. and Ju, G. (1995) The parabrachio-subfornical organ projection in the rat. *Brain Res. Bull.* **38,** 41–47.
158. Zardetto-Smith, A. M. and Gray, T. S. (1987) A direct neural projection from the nucleus of the solitary tract to the subfornical organ in the rat. *Neurosci. Lett.* **80,** 163–166.
159. Tanaka, J., Hayashi, Y., Shimamune, S., and Nomura, M. (1997) Ascending pathways from the nucleus of the solitary tract to the subfornical organ in the rat. *Brain Res.* **777,** 237–241.
160. Leslie, R. A. and Gwyn, D. G. (1984) Neuronal connections of the area postrema. *Fed. Proc.* **43,** 2941–2943.
161. Contreras, R. J., Beckstead, R. M., and Norgren, R. (1982) The central projections of the trigeminal, facial, glossopharyngeal and vagus nerves: an autoradiographic study in the rat. *J. Auton. Nerv. Syst.* **6,** 303–322.
162. van der Kooy, D. and Koda, L. Y. (1983) Organization of the projections of a circumventricular organ: the area postrema in the rat. *J. Comp. Neurol.* **219,** 328–338.
163. Papas, S. and Ferguson, A. V. (1990) Electrophysiological characterization of reciprocal connections between the parabrachial nucleus and the area postrema in the rat. *Brain Res. Bull.* **24,** 577–582.
164. Davies, R. O. and Kalia, M. (1981) Carotid sinus nerve projections to the brain stem in the cat. *Brain Res. Bull.* **6,** 531–541.
165. Ruiz-Pesini, P., Tome, E., Balaguer, L., Romano, J., and Yllera, M. (1995) The projections to the medulla of neurons innervating the carotid sinus in the dog. *Brain Res. Bull.* **37,** 41–47.
166. Papas, S. and Ferguson, A. V. (1991) Electrophysiological evidence of baroreceptor input to area postrema. *Am. J. Physiol.* **261,** R9–R13.
167. Kalia, M. and Mesulam, M.-M. (1980) Brain stem projections of sensory and motor components of the vagus complex in the cat: II. Laryngeal, tracheobronchial, pulmonary, cardiac, and gastrointestinal branches. *J. Comp. Neurol.* **193,** 467–508.
168. Ciriello, J., Hrycyshyn, A. W., and Calaresu, F. R. (1981) Glossopharyngeal and vagal afferent projections to the brain stem of the cat: a horseradish peroxidase study. *J. Auton. Nerv. Syst.* **4,** 63–79.
169. Shapiro, R. E. and Miselis, R. R. (1995) The central neural connections of the area postrema of the rat. *J. Comp. Neurol.* **234,** 344–364.
170. Cedarbaum, J. M. and Aghajanian, G. K. (1978) Afferent projections to the rat locus coeruleus as determined by a retrograde tracing technique. *J. Comp. Neurol.* **178,** 1–16.
171. Lind, R. W. (1988) Angiotensin and the lamina terminalis: illustrations of a complex unity. *Clin. Exp. Hypertens.* **10,** 79–105.

172. Phillips, M. I. and Camache, A. (1987) Neural connections of the organum vasculosum of the lamina terminalis, in *Circumventricular Organs and Body Fluids* (Gross, P., ed.), CRC, Boca Raton, FL, pp. 157–169.
173. Reuss, S. (1999) Trigeminal innervation of the mammalian pineal gland. *Microsc. Res. Tech.* **46,** 305–309.
174. Moller, M. and Liu, W. (1999) Innervation of the rat pineal gland by nerve fibres originating in the sphenopalatine, otic and trigeminal ganglia. A retrograde in vivo neuronal tracing study. *Reprod. Nutr. Dev.* **39,** 345–353.
175. Teclemariam-Mesbah, R., ter Horst, G. J., Postema, F., Wortel, J., and Buijs, R. M. (1999) Anatomical demonstration of the suprachiasmatic nucleus-pineal pathway. *J. Comp. Neurol.* **406,** 171–182.
176. Fink-Jensen, A. and Moller, M. (1990) Direct projections from the anterior and tuberal regions of the lateral hypothalamus to the rostral part of the pineal complex of the rat. An anterograde neuron-tracing study by using Phaseolus vulgaris leucoagglutinin. *Brain Res.* **522,** 337–341.
177. Mikkelsen, J. D., Panula, P., and Moller, M. (1992) Histamine-immunoreactive nerve fibers in the rat pineal gland: evidence for a histaminergic central innervation. *Brain Res.* **597,** 200–208.
178. Bowers, C. W., Dahm, L. M., and Zigmond, R. E. (1984) The number and distribution of sympathetic neurons that innervate the rat pineal gland. *Neuroscience* **13,** 87–96.
179. Stanley, L. C., Horikawa, K., and Powell, E. W. (1987) Innervation of the superficial pineal of the rat using retrograde tracing methods. *Am. J. Anat.* **180,** 249–254.
180. Mikkelsen, J. D., Hay-Schmidt, A., and Larsen, P. J. (1997) Central innervation of the rat ependyma and subcommissural organ with special reference to ascending serotoninergic projections from the raphe nuclei. *J. Comp. Neurol.* **384,** 556–568.
181. Mikkelsen, J. D. (1994) Analysis of the efferent projections of the lateral geniculate nucleus with special reference to the innervation of the subcommissural organ and related areas. *Cell Tissue Res.* **277,** 437–445.
182. Alonso, G. and Assenmacher, I. (1981) Radioautographic studies on the neurohypophysical projections of the supraoptic and paraventricular nuclei in the rat. *Cell Tissue Res.* **219,** 525–534.
183. Nilsson, C., Kannisto, P., Lindvall-Axelsson, M., Owman, C., and Rosengren, E. (1990) The neuropeptides vasoactive intestinal polypeptide, peptide histidine isoleucine and neuropeptide Y modulate [3H]noradrenaline release from sympathetic nerves in the choroid plexus. *Eur. J. Pharmacol.* **181,** 247–252.
184. Nilsson, C., Ekman, R., Lindvall-Axelsson, M., and Owman, C. (1990) Distribution of peptidergic nerves in the choroid plexus, focusing on coexistence of neuropeptide Y, vasoactive intestinal polypeptide and peptide histidine isoleucine. *Regul. Pept.* **27,** 11–26.
185. Nilsson, C., Lindvall-Axelsson, M., and Owman, C. (1992) Effects of vasoactive intestinal polypeptide on choroid plexus blood flow and cerebrospinal fluid production. *Progr. Brain Res.* **91,** 445–449.
186. Segal, M. B., Chodobski, A., Szmydynger-Chodobska, J., and Cammish, H. (1992) Effect of arginine vasopressin on blood vessels of the perfused choroid plexus of the sheep. *Progr. Brain Res.* **91,** 451–453.
187. Chodobski, A., Szmydynger-Chodobska, J., Epstein, M. H., and Johanson, C. E. (1995) The role of angiotensin II in the regulation of blood flow to choroid plexuses and cerebrospinal fluid formation in the rat. *J. Cereb. Blood Flow Metab.* **15,** 143–151.
188. Maktabi, M. A., Heistad, D. D., and Faraci, F. M. (1990) Effects of angiotensin II on blood flow to choroid plexus. *Am. J. Physiol.* **258,** H414–H418.
189. Williams, J. L., Thebert, M. M., Schalk, K. A., and Heistad, D. D. (1991) Stimulation of area postrema decreases blood flow to choroid plexus. *Am. J. Physiol.* **260,** H902–H908.

190. Ferguson, A. V. (1992) Neurophysiological analysis of mechanisms for subfornical organ and area postrema involvement in autonomic control. *Progress in Brain Research* **91,** 413–421.
191. Ferguson, A. V. and Bains, J. S. (1996) Electrophysiology of the circumventricular organs. *Front. Neuroendocrinol.* **17,** 440–475.
192. Donevan, S. D. and Ferguson, A. V. (1988) Subfornical organ connections with septal neurons projecting to the median eminence. *Neuroendocrinology* **48,** 67–71.
193. Lowes, V. L., Sun, K., Li, Z., and Ferguson, A. V. (1995) Vasopressin actions on area postrema neurons in vitro. *Am. J. Physiol.* **269,** R463–R468.
194. Sun, K. and Ferguson, A. V. (1996) Angiotensin II and glutamate influence area postrema neurons in rat brain slices. *Regul. Pept.* **63,** 91–98.
195. Sun, K. and Ferguson, A. V. (1997) Cholecystokinin activates area postrema neurons in rat brain slices. *Am. J. Physiol.* **272,** R1625–R1630.
196. Richard, D. and Bourque, C. W. (1992) Synaptic activation of rat supraoptic neurons by osmotic stimulation of the organum vasculosum lamina terminalis. *Neuroendocrinology* 609–611.
197. Tanaka, J., Kaba, H., Saito, H., and Seto, K. (1985) Electrophysiological evidence that circulating angiotensin II sensitive neurons in the subfornical organ alter the activity of hypothalamic paraventricular neurohypophyseal neurons in the rat. *Brain Res.* **342,** 361–365.
198. Ferguson, A. V., Bicknell, R. J., Carew, M. A., and Mason, W. T. (1997) Dissociated adult rat subfornical organ neurons maintain membrane properties and angiotensin responsiveness for up to 6 days. *Neuroendocrinology* **66,** 409–415.
199. Washburn, D. L. S., Anderson, J. W., and Ferguson, A. V. (2000) A subthreshold persistent sodium current mediates bursting in rat subfornical organ neurones. *J. Physiol.* **529,** 359–371.
200. Washburn, D. L. S., Beedle, A. M., and Ferguson, A. V. (1999) Inhibition of subfornical organ neuronal potassium channels by vasopressin. *Neuroscience* **93,** 349–359.
201. Anderson, J. W., Washburn, D. L. S., and Ferguson, A. V. (2000) Intrinsic osmosensitivity of subfornical organ neurons. *Neuroscience* **100,** 539–547.
202. Hay, M. and Lindsley, K. A. (1999) AMPA receptor activation of area postrema neurons. *Am. J. Physiol.* **276,** R586–R590.
203. Hay, M., Hasser, E. M., and Lindsley, K. A. (1996) Area postrema voltage-activated calcium currents. *J. Neurophysiol.* **75,** 133–141.
204. Hay, M. and Lindsley, K. A. (1995) Membrane properties of area postrema neurons. *Brain Res.* **705,** 199–208.
205. Johnson, R. F., Beltz, T. G., Jurzak, M., Wachtel, R. E., and Johnson, A. K. (1999) Characterization of ionic currents of cells of the subfornical organ that project to the supraoptic nuclei. *Brain Res.* **817,** 226–231.
206. Sun, K. and Ferguson, A. V. (1997) Cholecystokinin activates area postrema neurons in rat brain slices. *Am. J. Physiol.* **272,** R1625–R1630.
207. Washburn, D. L. S., Anderson, J. W., and Ferguson, A. V. (2000) The calcium receptor modulates the hyperpolarization-activated current in subfornical organ neurons. *Neuroreport* **11,** 3231–3235.
208. Hay, M. and Lindsley, K. A. (1999) AMPA receptor activation of area postrema neurons. *Am. J. Physiol.* **276,** R586–R590.
209. Schenda, J. and Vollrath, L. (1998) Demonstration of action-potential-producing cells in the rat pineal gland in vitro and their regulation by norepinephrine and nitric oxide. *J. Comp. Physiol. [A]* **183,** 573–581.
210. Freschi, J. E. and Parfitt, A. G. (1986) Intracellular recordings from pineal cells in tissue culture: membrane properties and response to norepinephrine. *Brain Res.* **368,** 366–370.
211. McCance, I., Parkington, H. C., and Coleman, H. A. (1996) The association between melatonin production and electrophysiology of the guinea pig pineal gland. *J. Pineal Res.* **21,** 79–90.

212. Schenda, J. and Vollrath, L. (1999) An intrinsic neuronal-like network in the rat pineal gland. *Brain Res.* **823,** 231–233.
213. Marin, A., Urena, J., and Tabares, L. (1996) Intracellular calcium release mediated by noradrenaline and acetylcholine in mammalian pineal cells. *J. Pineal Res.* **21,** 15–28.
214. Letz, B., Schomerus, C., Maronde, E., Korf, H. W., and Korbmacher, C. (1997) Stimulation of a nicotinic ACh receptor causes depolarization and activation of L-type Ca^{2+} channels in rat pinealocytes. *J. Physiol.* **499,** 329–340.
215. McKinley, M. J., Pennington, G. L., and Oldfield, B. J. (1996) Anteroventral wall of the third ventricle and dorsal lamina terminalis: headquarters for control of body fluid homeostasis? *Clin. Exp. Pharmacol. Physiol.* **23,** 271–281.
216. McKinley, M. J., Bicknell, R. J., Hards, D., McAllen, R. M., Vivas, L., Weisinger, R. S., and Oldfield, B. J. (1992) Efferent neural pathways of the lamina terminalis subserving osmoregulation. *Progr. Brain Res.* 395–402.
217. Simpson, J. B. and Routtenberg, J. B. (1975) Subfornical organ lesions reduce intravenous angiotensin-induced drinking. *Brain Res.* **88,** 154–161.
218. Simpson, J. B. and Routtenberg, A. (1978) Subfornical organ: a dipsogenic site of action of angiotensin II. *Science* **201,** 379–381.
219. McKinley, M. J., Mathai, M. L., Pennington, G., Rundgren, M., and Vivas, L. (1999) Effect of individual or combined ablation of the nuclear groups of the lamina terminalis on water drinking in sheep. *Am. J. Physiol.* **276,** R673–R683.
220. Verney, E. B. (1947) The antidiuretic hormone and the factors which determine its release. *Proc. R. Soc. Lond.* **135,** 25–106.
221. McKinley, M. J., Denton, D. A., and Weisinger, R. S. (1978) Sensors for antidiuresis and thirst: osmoreceptors or CSF sodium detectors? *Brain Res.* **141,** 89–103.
222. Thrasher, T. N., Brown, C. J., Keil, L. C., and Ramsay, D. J. (1980) Thirst and vasopressin release in the dog: an osmoreceptor or sodium receptor mechanism? *Am. J. Physiol.* **238,** R333–R339.
223. Bourque, C. W. and Oliet, S. H. (1997) Osmoreceptors in the central nervous system. *Annu. Rev. Physiol.* **59,** 601–619.
224. Bourque, C. W., Oliet, S. H. R., and Richard, D. (1994) Osmoreceptors, osmoreception, and osmoregulation. *Front. Neuroendocrinol.* **15,** 231–274.
225. Bisley, J. W., Rees, S. M., McKinley, M. J., Hards, D. K., and Oldfield, B. J. (1996) Identification of osmoresponsive neurons in the forebrain of the rat: a fos study at the ultrastructural level. *Brain Res.* **720,** 25–34.
226. Oldfield, B. J., Badoer, E., Hards, D. K., and McKinley, M. J. (1994) Fos production in retrogradely labelled neurons of the lamina terminalis following intravenous infusion of either hypertonic saline or angiotensin II. *Neuroscience* **60,** 255–262.
227. Giovannelli, L. and Bloom, F. E. (1992) c-Fos protein expression in the rat subfornical organ following osmotic stimulation. *Neurosci. Lett.* **139,** 1–6.
228. Caston-Balderrama, A., Nijland, M. J. M., McDonald, T. J., and Ross, M. G. (1999) Central Fos expression in fetal and adult sheep after intraperitoneal hypertonic saline. *Am. J. Physiol.* **276,** H725–H735.
229. Honda, K., Negoro, H., Higuchi, T., and Tadokoro, Y. (1989) The role of the anteroventral 3rd ventricle area in the osmotic control of paraventricular neurosecretory cells. *Exp. Brain Res.* **76,** 497–502.
230. Negoro, H., Higuchi, T., Tadokoro, Y., and Honda, K. (1988) Osmoreceptor mechanism for oxytocin release in the rat. *Jpn. J. Physiol.* **38,** 19–31.
231. Johnson, A. K., Cunningham, J. T., and Thunhorst, R. L. (1996) Integrative role of the lamina terminalis in the regulation of cardiovascular and body fluid homeostasis. *Clin. Exp. Pharmacol. Physiol.* **23,** 183–191.
232. Thrasher, T. N., Keil, L. C., and Ramsay, D. J. (1982) Lesions of the organum vasculosum of the lamina terminalis (OVLT) attenuate osmotically-induced drinking and vasopressin secretion in the dog. *Endocrinology* **110,** 1837–1839.

233. Thrasher, T. N. and Keil, L. C. (1987) Regulation of drinking and vasopressin secretion: role of organum vasculosum laminae terminalis. *Am. J. Physiol.* **253,** R108–R120.
234. Lind, R. W., Thunhorst, R. L., and Johnson, A. K. (1984) The subfornical organ and the integration of multiple factors in thirst. *Physiol. Behav.* **32,** 69–74.
235. Mangiapane, M. L., Thrasher, T. N., Keil, L. C., Simpson, J. B., and Ganong, W. F. (1984) Role of the subfornical organ in vasopressin release. *Brain Res. Bull.* **13,** 43–47.
236. Sibbald, J. R., Hubbard, J. I., and Sirett, N. E. (1988) Responses from osmosensitive neurons of the rat subfornical organ in vitro. *Brain Res.* **461,** 205–214.
237. Tanaka, J., Saito, H., and Yagyu, K. (1989) Impaired responsiveness of paraventricular neurosecretory neurons to osmotic stimulation in rats after local anesthesia of the subfornical organ. *Neurosci. Lett.* **98,** 51–56.
238. Rowland, N. E., Fregly, M. J., Han, L., and Smith, G. (1996) Expression of Fos in rat brain in relation to sodium appetite: furosemide and cerebroventricular renin. *Brain Res.* **728,** 90–96.
239. Houpt, T. A., Smith, G. P., Joh, T. H., and Frankmann, S. P. (1998) c-fos-like immunoreactivity in the subfornical organ and nucleus of the solitary tract following salt intake by sodium-depleted rats. *Physiol. Behav.* **63,** 505–510.
240. Starbuck, E. M., Lane, J. R., and Fitts, D. A. (1997) Interaction of hydration and subfornical organ lesions in sodium- depletion induced salt appetite. *Behav. Neurosci.* **111,** 206–213.
241. Starbuck, E. M. and Fitts, D. A. (1998) Influence of salt intake, ANG II synthesis and SFO lesion on thirst and blood pressure during sodium depletion. Subfornical organ. *Appetite* **31,** 309–331.
242. Thunhorst, R. L., Beltz, T. G., and Johnson, A. K. (1999) Effects of subfornical organ lesions on acutely induced thirst and salt appetite. *Am. J. Physiol.* **277,** R56–R65.
243. Morien, A., Garrard, L., and Rowland, N. E. (1999) Expression of Fos immunoreactivity in rat brain during dehydration: effect of duration and timing of water deprivation. *Brain Res.* **816,** 1–7.
244. Xu, Z. C. and Herbert, J. (1996) Effects of unilateral or bilateral lesions within the anteroventral third ventricular region on c-fos expression induced by dehydration or angiotensin ii in the supraoptic and paraventricular nuclei of the hypothalamus. *Brain Res.* **713,** 36–43.
245. Hochstenbach, S. L. and Cirello, J. (1996) Effect of lesions of forebrain circumventricular organs on c-fos expression in the central nervous system to plasma hypernatremia. *Brain Res.* **713,** 17–28.
246. Hasser, E. M., Nelson, D. O., Haywood, J. R., and Bishop, V. S. (1987) Inhibition of renal sympathetic nervous activity by area postrema stimulation in rabbits. *Am. J. Physiol.* **253,** H91–H99.
247. Curtis, K. S., Huang, W., Sved, A. F., Verbalis, J. G., and Stricker, E. M. (1999) Impaired osmoregulatory responses in rats with area postrema lesions. *Am. J. Physiol.* **277,** R209–R219.
248. Edwards, G. L., Beltz, T. G., Power, J. D., and Johnson, A. K. (1993) Rapid-onset "need-free" sodium appetite after lesions of the dorsomedial medulla. *Am. J. Physiol.* **264,** R1242–R1247.
249. Huang, W., Sved, A. F., and Stricker, E. M. (2000) Vasopressin and oxytocin release evoked by NaCl loads are selectively blunted by area postrema lesions. *Am. J. Physiol.* **278,** R732–R740.
250. Carlson, S. H., Collister, J. P., and Osborn, J. W. (1998) The area postrema modulates hypothalamic fos responses to intragastric hypertonic saline in conscious rats. *Am. J. Physiol.* **275,** R1921–R1927.
251. Morita, H., Yamashita, Y., Nishida, Y., Tokuda, M., Hatase, O., and Hosomi, H. (1997) Fos induction in rat brain neurons after stimulation of the hepatoportal Na-sensitive mechanism. *Am. J. Physiol.* **272,** R913–R923.

252. Carlson, S. H., Beitz, A., and Osborn, J. W. (1997) Intragastric hypertonic saline increases vasopressin and central Fos immunoreactivity in conscious rats. *Am. J. Physiol.* **272,** R750–R758.
253. DiBona, G. F. (1999) Central sympathoexcitatory actions of angiotensin II: role of type 1 angiotensin II receptors. *J. Am. Soc. Nephrol.* **10(Suppl. 11),** S90–S94.
254. Ferguson, A. V., Bains, J. S., and Lowes, V. L. (1994) Circumventricular organs and cardiovascular homeostasis, in *Central Neural Mechanisms in Cardiovascular Regulation* (Kunos, G. and Ciriello, J., eds.), Birkhäuser, New York, pp. 80–101.
255. Ferguson, A. V., Day T. A., and Renaud, L. P. (1984) Influence of subfornical organ stimulation on the excitability of hypothalamic paraventricular neurons projecting to the dorsal medulla. *Am. J. Physiol.* **247,** R1088–R1092.
256. Renaud, L. P., Ferguson, A. V., Day, T. A., Bourque, C. W., and Sgro, S. (1985) Electrophysiology of the subfornical organ and its hypothalamic connections: an in-vivo study in the rat. *Brain Res. Bull.* **15,** 83–86.
257. Ferguson, A. V. and Marcus, P. (1988) Area postrema stimulation induced cardiovascular changes in the rat. *Am. J. Physiol.* **255,** R855–R860.
258. Li, Y. W. and Dampney, R. A. (1994) Expression of Fos-like protein in brain following sustained hypertension and hypotension in conscious rabbits. *Neuroscience* **61,** 613–634.
259. Chan, R. K. W. and Sawchenko, P. E. (1994) Spatially and temporally differentiated patterns of c fos expression in brain stem catecholaminergic cell groups induced by cardiovascular challenges in the rat. *J. Comp. Neurol.* **348,** 433–460.
260. Rohmeiss, P., Beyer, C., Hocher, B., Qadri, F., Gretz, N., Strauch, M., and Unger, T. (1995) Osmotically induced natriuresis and blood pressure response involves angiotensin AT1 receptors in the subfornical organ. *J. Hypertens.* **13,** 1399–1404.
261. Gatto, E. M., Carreras, M. C., Pargament, G. A., Riobo, N. A., Reides, C., Repetto, M., et al. (1996) Neutrophil function, nitric oxide, and blood oxidative stress in parkinsons disease. *Movement Disorders* **11,** 261–267.
262. Bonigut, S., Bonham, A. C., and Stebbins, C. L. (1997) Area postrema-induced inhibition of the exercise pressor reflex. *Am. J. Physiol.* **272,** H1650–H1655.
263. Lowes, V. L. and Ferguson, A. V. (1994) Modified cardiovascular sensitivity of the area postrema to vasopressin in spontaneously hypertensive rats. *Brain Res.* **636,** 165–168.
264. Matsumura, K., Averill, D. B., and Ferrario, C. M. (1999) Role of AT1 receptors in area postrema on baroreceptor reflex in spontaneously hypertensive rats. *Brain Res.* **850,** 166–172.
265. Kizer, J. S., Palkovits, M., and Brownstein, M. J. (1976) Releasing factors in the circumventricular organs of the rat brain. *Endocrinology* **98,** 311–317.
266. Piva, F., Limonta, P., and Martini, L. (1982) Role of the organum vasculosum laminae terminalis in the control of gonadotrophin secretion in rats. *J. Endocrinol.* **93,** 355–364.
267. Limonta, P., Maggi, R., Martini, L., and Piva, F. (1982) Role of the subcommissural organ in the control of gonadotrophin secretion in the female rat. *J. Endocrinol.* **95,** 207–213.
268. Limonta, P., Maggi, R., Giudici, D., Martini, L., and Piva, F. (1981) Role of the subfornical organ (SFO) in the control of gonadotropin secretion. *Brain Res.* **229,** 75–84.
269. Cates, P. S. and O'Byrne, K. T. (2000) The area postrema mediates insulin hypoglycaemia-induced suppression of pulsatile LH secretion in the female rat. *Brain Res.* **853,** 151–155.
270. Lindheimer, M. D. and Davison, J. M. (1995) Osmoregulation, the secretion of arginine vasopressin and its metabolism during pregnancy. *Eur. J. Endocrinol.* **132,** 133–143.
271. Weisinger, R. S., Burns, P., Eddie, L. W., and Wintour, E. M. (1993) Relaxin alters the plasma osmolality–arginine vasopressin relationship in the rat. *J. Endocrinol.* **137,** 505–510.
272. Phillips, M. I., Wang, H., Kimura, B., Speth, R. C., and Ghazi, N. (1995) Brain angiotensin and the female reproductive cycle. *Adv. Exp. Med. Biol.* **377,** 357–370.
273. Summerlee, A. J. and Wilson, B. C. (1994) Role of the subfornical organ in the relaxin-induced prolongation of gestation in the rat. *Endocrinology* **134,** 2115–2120.

274. Lincoln, G. (1999) Melatonin modulation of prolactin and gonadotrophin secretion. Systems ancient and modern. *Adv. Exp. Med. Biol.* **460,** 137–153.
275. Kennaway, D. J. (1988) Short- and long-term effects of manipulation of the pineal/melatonin axis in ewes. *Reprod. Nutr. Dev.* **28,** 399–408.
276. Pevet, P. (1988) The role of the pineal gland in the photoperiodic control of reproduction in different hamster species. *Reprod. Nutr. Dev.* **28,** 443–458.
277. Dardes, R. C., Baracat, E. C., and Simoes, M. J. (2000) Modulation of estrous cycle and LH, FSH and melatonin levels by pinealectomy and sham-pinealectomy in female rats. *Prog. Neuropsychopharmacol. Biol. Psychiatry* **24,** 441–453.
278. Schwartz, S. M. (1982) Effects of constant bright illumination on reproductive processes in the female rat. *Neurosci. Biobehav. Rev.* **6,** 391–406.
279. Luboshitzky, R. and Lavie, P. (1999) Melatonin and sex hormone interrelationships: a review. *J. Pediatr. Endocrinol. Metab.* **12,** 355–362.
280. Sandyk, R. (1992) The pineal gland and the menstrual cycle. *Int. J. Neurosci.* **63,** 197–204.
281. Aleandri, V., Spina, V., and Morini, A. (1996) The pineal gland and reproduction. *Hum. Reprod. Update* **2,** 225–235.
282. Quay, W. B. (1981) Pineal atrophy and other neuroendocrine and circumventricular features of the naked mole-rat, Heterocephalus glaber (Ruppell), a fossorial, equatorial rodent. *J. Neural. Transm.* **52,** 107–115.
283. Vitkovic, L., Bockaert, J., and Jacque, C. (2000) "Inflammatory" cytokines: neuromodulators in normal brain? *J. Neurochem.* **74,** 457–471.
284. Turnbull, A. V. and Rivier, C. L. (1999) Regulation of the hypothalamic-pituitary-adrenal axis by cytokines: actions and mechanisms of action. *Physiol. Rev.* **79,** 1–71.
285. Netea, M. G., Kullberg, B. J., and Van Der Meer, J. W. (1999) Do only circulating pyrogenic cytokines act as mediators in the febrile response? A hypothesis. *Eur. J. Clin. Invest.* **29,** 351–356.
286. Herkenham, M., Lee, H. Y., and Baker, R. A. (1998) Temporal and spatial patterns of c-fos mRNA induced by intravenous interleukin-1: a cascade of non-neuronal cellular activation at the blood-brain barrier. *J. Comp. Neurol.* **400,** 175–196.
287. Plotkin, S. R., Banks, W. A., and Kastin, A. J. (1996) Comparison of saturable transport and extracellular pathways in the passage of interleukin-1 alpha across the blood-brain barrier. *J. Neuroimmunol.* **67,** 41–47.
288. Quan, N., Stern, E. L., Whiteside, M. B., and Herkenham, M. (1999) Induction of pro-inflammatory cytokine mRNAs in the brain after peripheral injection of subseptic doses of lipopolysaccharide in the rat. *J. Neuroimmunol.* **93,** 72–80.
289. Hare, A. S., Clarke, G., and Tolchard, S. (1995) Bacterial lipopolysaccharide-induced changes in FOS protein expression in the rat brain: correlation with thermoregulatory changes and plasma corticosterone. *J. Neuroendocrinol.* **7,** 791–799.
290. Lacroix, S. and Rivest, S. (1997) Functional circuitry in the brain of immune-challenged rats: partial involvement of prostaglandins. *J. Comp. Neurol.* **387,** 307–324.
291. Horai, R., Asano, M., Sudo, K., Kanuka, H., Suzuki, M., Nishihara, M., Takahashi, M., and Iwakura, Y. (1998) Production of mice deficient in genes for interleukin (IL)-1alpha, IL-1beta, IL-1alpha/beta, and IL-1 receptor antagonist shows that IL-1beta is crucial in turpentine-induced fever development and glucocorticoid secretion. *J. Exp. Med.* **187,** 1463–1475.
292. Zeisberger, E. and Merker, G. (1992) The role of OVLT in fever and antipyresis. *Progr. Brain Res.* **91,** 403–408.
293. Hashimoto, M., Ueno, T., and Iriki, M. (1994) What roles does the organum vasculosum laminae terminalis play in fever in rabbits? *Pflugers Arch.* **429,** 50–57.
294. Blatteis, C. M., Bealer, S. L., Hunter, W. S., Llanos-Q., J., Ahokas, R. A., and Mashburn, T. A., Jr. (1983) Suppression of fever after lesions of the anteroventral third ventricle in guinea pigs. *Brain Res. Bull.* **11,** 519–526.

295. Nakamori, T., Sakata, Y., Watanabe, T., Morimoto, A., Nakamura, S., and Murakami, N. (1995) Suppression of interleukin-1β production in the circumventricular organs in endotoxin-tolerant rabbits. *Brain Res.* **675,** 103–109.
296. Stitt, J. T. and Shimada, S. G. (1991) Site of action of calcium channel blockers in inhibiting endogenous pyrogen fever in rats. *J. Appl. Physiol.* **71,** 956–960.
297. Cartmell, T., Luheshi, G. N., and Rothwell, N. J. (1999) Brain sites of action of endogenous interleukin-1 in the febrile response to localized inflammation in the rat. *J. Physiol.* **518,** 585–594.
298. Takahashi, Y., Smith, P. M., Ferguson, A. V., and Pittman, Q. J. (1997) Circumventricular organs and fever. *Am. J. Physiol.* **273,** R1690–R1695.
299. Day, H. E. and Akil, H. (1996) Differential pattern of c-fos mRNA in rat brain following central and systemic administration of interleukin-1-beta: implications for mechanism of action. *Neuroendocrinology* **63,** 207–218.
300. Lee, H. Y., Whiteside, M. B., and Herkenham, M. (1998) Area postrema removal abolishes stimulatory effects of intravenous interleukin-1beta on hypothalamic-pituitary-adrenal axis activity and c-fos mRNA in the hypothalamic paraventricular nucleus. *Brain Res. Bull.* **46,** 495–503.
301. Wong, M. L., Bongiorno, P. B., Rettori, V., McCann, S. M., and Licinio, J. (1997) Interleukin (IL) 1beta, IL-1 receptor antagonist, IL-10, and IL-13 gene expression in the central nervous system and anterior pituitary during systemic inflammation: pathophysiological implications. *Proc. Natl. Acad. Sci. USA* **94,** 227–232.

13
Glial Linings of the Brain

Marc R. Del Bigio

1. INTRODUCTION

Neurons are exquisitely sensitive to conditions in their external environment, disturbances of which can impair neuronal function. For this reason, the nervous system is equipped with regulatory barriers that ensure a stable environment. Most people are aware of the blood–brain barrier (BBB) (*see* Chapter 11) and of the blood–cerebrospinal fluid (CSF) barrier (*see* Chapter 10). Completing the boundary, albeit perhaps less efficiently, are the brain–CSF barriers. On the outer surface of the brain, the glia limitans and pia mater serve this function; on the inner surface of the brain, i.e., the ventricular system, the ependyma offers some barrier capability. This chapter discusses the features of these two boundary layers, focusing on the mammalian brain.

2. GLIA LIMITANS

Surrounding the brain are several layers of "coating," alternating with fluid-filled compartments of various dimensions. The outermost layer of brain is the glia limitans, a specialized astroglial layer associated with a basement membrane *(1)*. Next is a narrow subpial compartment, followed by the pia mater, the subarachnoid compartment filled with CSF, the arachnoid, and the dura mater. The pia, arachnoid, and dura are mesenchyme-derived layers *(2,3)*. The intimate relationship between the glial cell layer and the mesenchymal cells that form the pia make it necessary to discuss the two in conjunction. Although these layers are thicker in animals with larger brains, the structural features, molecular composition, and functional attributes are probably similar. The intimate contact of these two layers sometimes makes distinction between the two difficult, when defining protein or mRNA expression by immunohistochemistry or *in situ* hybridization respectively. The glia limitans, being a specialized form of astroglia, exhibits many of the same properties as other astroglia, although there are local enhancements of function and/or content that are probably related to the specific role of this layer. The reader is referred to published reviews for general information concerning astrocyte development *(4)*, properties *(5,6)*, and reactions *(7,8)*.

2.1. *Development of Outer Boundary Layer of Central Nervous System*

2.1.1. Glia Limitans Development

Development of the glia limitans has been studied extensively in rodent brain, and considerable morphological information is available from studies of human brains.

From: *The Neuronal Environment: Brain Homeostasis in Health and Disease*
Edited by: W. Walz © Humana Press Inc., Totowa, NJ

Radial glia span the early neural tube, from the lumen to the pial surface. Expanded endfeet of these cells abut on mesenchyme surrounding the neural tube. These are associated with a continuous extracellular (EC) basement membrane by postconception d 10–11, around the brain and spinal cord of mice *(9–12)*; by postconception d 12, around the brain of rats *(13)*; and by postconception wk 6, around the human embryo brain *(14,15)*. Soon thereafter, in rodents, glial cell processes separate from the ventricular surface, and the glia limitans becomes defined *(13,16)*. Expression of glial fibrillary acidic protein (GFAP), the major intermediate filament of astrocytes, begins first in the glia limitans; this occurs at about postconception d 14 in mice and rats *(9,17)*, and is clear by postconception wk 12 in humans *(18)*. Expression of S100, a calcium (Ca)-binding protein, begins at postconception d 16 in rats *(19)*. With maturation, the glia limitans continues to thicken, particularly at sites of blood vessel penetration *(20–22)*. In rats, at postconception d 14, glia limitans cells develop transient "strap junctions," which act as sites of diffusional restriction, by postconception d 16 *(23)*. Gap junctions between glial cells are not apparent at birth; they appear in the early postnatal period in rodents *(11)*, which corresponds to the late fetal period in humans.

2.1.2. Pia Mater Development

The developing neural tube is surrounded by a loose collection of mesenchymal cells with some collagen. On postconception d 13–14 of mouse spinal cord *(12)*, and d 14–16 around brain *(10)*, the fibroblast-like mesenchymal cells condense to form an inner (pia mater) and outer layer (arachnoid) surrounding a subarachnoid space. In rat, the subarachnoid space becomes defined slightly later, at postconception d 19 *(24)*, and, in humans, cavitation of the space occurs at ~44 d postconception *(25)*. During early development, the pia mater forms a discontinuous layer around the neural tube *(10)*. There is no clear distinction between pial and arachnoidal cells, with the minor exception of greater glycogen content in the latter *(26)*. The arachnoid remains in intimate contact with the more densely collagenous dura mater: There is no space between the two *(3)*. In the mouse, the pia and arachnoid acquire a mature character by 21 d after birth *(27)*.

2.1.3. Pia Mater Influences on Glia Limitans

Contact between mesenchymal cells and the neural tube appears to induce development of the glia limitans. When astrocytes are co-cultured with meningeal cells, they produce more basement membrane *(28,29)*, with more tenascin and chondroitin sulfate but less laminin, which reduces their ability to support growing neurites *(30)*. Gap junctional communication is increased *(31)*. Injection of meningeal cells into rat brain induces glia limitans-like specializations at contact sites with astrocytes *(32)*.

Similarly, an interesting series of experiments, conducted primarily by Sievers et al., showed that, in the absence of pial cells, the cerebellar (cerebellum) glia limitans fails to develop *(33)*. They injected the neurotoxin, 6-hydroxydopamine, into the cisterna magna of newborn hamsters and rats, and observed destruction of meningeal cells, and disrupted cerebellum development *(34–40)*. Those authors postulated that the primary lesion was meningeal cell loss over the external granular layer of the cerebellum. This was thought to cause migrational abnormalities among neural progenitors, which could no longer interact with a basal lamina *(41)*. In vitro, meningeal cells can induce a radial glia phenotype, and exert a chemotactic effect over cerebellum cells *(42,43)*. However, it hard to ignore the fact that 6-hydroxydopamine can act as directly as a neurotoxin on dopaminergic and noradrenergic neurons *(44,45)*. Destruction of these can disturb cerebellum development *(46)*. Therefore, conclusions concerning the direct effect of pial cells on cerebellum neuron migration are not necessarily correct; they may represent an epiphenomenon.

2.2. *Mature Structure of the Outer Boundary Layer*

2.2.1. *Glia Limitans*

The mature glia limitans is composed of a continuous sheet of specialized astrocytes that have extensively interwoven, layered (up to 25 layers) cell processes running parallel to the external surface of the brain and spinal cord. There are few nuclei. The total thickness of the layer varies from 1 to 20 μm regionally, and between species *(47,48)*. In primates, it tends to become thicker with aging *(49)*. Like astroglia elsewhere in the nervous system, cellular processes of the glia limitans are packed with intermediate filaments, the majority of which are composed of GFAP and, to a lesser extent, vimentin *(50)*. Mice with targeted deletions of GFAP or vimentin have deficient glia limitans *(51)*, as do mice with deletion of the receptor for epidermal growth factor (EGF), which is involved in glial development *(52)*. The cytoskeletal protein, plectin, is highly expressed in the glia limitans, and in astrocytes at the BBB *(53)*.

Glial cell processes are joined to each other by focal membrane condensations and many gap junctions *(48)*. Hemidesmosomes connect the cells to the basement membrane (*see* below) *(54)*. Junctional proteins, R- and B-cadherin, as well as CD-81, are expressed in the glia limitans *(55,56)*. Freeze-fracture preparations readily demonstrate the abundant gap junctions between cell processes *(57,58)*. These are composed mostly of connexin 43 *(59)* and, to a lesser extent, connexin 30 *(60)*. They allow movement of molecules between cells. In addition, orthogonal assemblies of particles are observed in freeze-fracture studies *(61–67)*. These particles have recently been shown to be composed, at least in part, of the water channel protein, aquaporin 4 (AQP4) *(68,69)*. Rearrangement following disruption of microtubules or actin suggest that they are anchored to the cytoskeleton *(70)*. In rats, these assemblies are infrequent at the time of birth, and reach adult quantities by 10 d *(71)*. Coincidentally, this is the period when the water content of the brain rapidly diminishes. They are also abundant in astrocytes at BBB interfaces *(72)*.

A feature that distinguishes astroglial cells in the glia limitans from those elsewhere (with the exception of those at BBB capillary interfaces) is the basement membrane. This distinct coating of EC material, 60–80 nm in thickness, is apparently bound to the glial cells by hemidesmosomes, but not to the opposing pial cells. It is composed of collagens I, II, IV, and V *(73,74)*, tenascin *(75)*, thrombospondin *(76)*, laminin, fibronectin, and heparan sulfate proteoglycans *(77,78)*. These are all potentially produced by the astrocytes *(79,80)*, although some authors suggest that meningeal cells are the source *(33)*. Glia limitans astrocytes and pia mater cells are linked to laminin and fibronectin in the basement membrane through the 67 kDa receptor and integrins *(78)*.

2.2.2. *Pia Mater*

Cells of the human and rodent pia form a thin sheet, usually only one cell layer in thickness, with rare overlap of processes *(81)*. In humans, there are few gaps between the cells, although, in rats, fenestrations appear to be common *(81,82)*. The cells are joined by gap junctions and desmosomes *(83)*. Between the pia mater and the glial basement membrane are delicate collagen fibers and small blood vessels *(48,84)*. Pial cells are almost identical to those of the arachnoid *(85)*. The cells contain intermediate filaments composed of vimentin and, rarely, of cytokeratin *(84,86)*. Gap junctions of arachnoidal and pial cells are composed of connexin 30 *(60)*.

2.2.3. Exit Sites of Nerves and Blood Vessels

Coverings of the central nervous system must contend with the fact that the surface of the brain and spinal cord is interrupted by blood vessels and nerves that communicate with the periphery. At these entry and exit sites the nervous system has further specialization of the boundary cells. During development, growing axons pierce the already-formed glia limitans individually, and are effectively segregated from each other by fine glial processes. During olfactory nerve development, the glia limitans becomes fragmented, then disappears, as axons grow outward *(87)*. In rat spinal cord the glia limitans is similarly loosened, and the definitive intact astrocytic central nervous system–peripheral nervous system barrier is re-established only after birth *(88)*. Some axons run parallel to, or within, the glial layer, prior to exiting *(89)*. At sites of nerve exit and entry, the glia limitans is thickened and raised in a funnel-like configuration *(90,91)*. In the mature animal, abutting on the continuous glia limitans is the inner nerve-root sheath, which is composed of perineurial cells that continue the length of the nerve. The outer root sheath is composed of pial cells. Where the nerve pierces the dura, the outer root sheath ends. Outside of the dura, the dura itself is reflected onto the peripheral nerve, to continue in the form of the epineurium *(92–94)*. The olfactory nerve exit site differs. Instead of astrocytes, the glia limitans is formed by special glial-ensheathing cells that have features of Schwann cells and astrocytes. These occupy the spaces between adjacent olfactory fascicles, and facilitate the continuous regrowth of olfactory axons *(95,96)*.

Blood vessels passing through the subarachnoid space of human *(83,97,98)* and rat *(2)* brains are surrounded by a sheath of pial cells that is continuous with the pia mater. It is also connected to the arachnoid membrane via the fine cellular trabeculae (perhaps fibroblasts *[3]*), spanning the subarachnoid space. The narrow space between the pial sheath and the walls of veins is continuous with the subpial compartment, because the pia at the brain surface reflects onto the vessel surface. However, the space around artery walls is sealed by a cylinder of pial cells, which continues down the periarterial spaces. Perforations in this layer appear at the arteriolar level, and the pial investment disappears at the capillary level. The funnel-shaped perivascular (Virchow-Robin) spaces are bounded by the glia limitans, whose basal lamina becomes continuous with that of capillaries at sites of the BBB *(81)*. Even at locations where the pia leaves the brain surface to surround a blood vessel, the basement membrane of the glia limitans persists. The functional implication of this anatomical arrangement is that the subarachnoid compartment is not in direct communication with perivascular spaces that penetrate the brain *(2)*; hence, the movement of tracer molecules and cells is partially restricted *(99–101)*. Some authors suggest that the pial sheath surrounding brain surface blood vessels is fenestrated *(82)*; others argue that this is an artifact *(2)*. Nonetheless, macromolecules do quickly access blood vessel walls and the perivascular spaces from the subarachnoid compartment *(102–104)*, indicating that bulk transfer is possible. Furthermore, tracer molecules can move from the perivascular or perineural spaces through the dura, then into lymphatic channels *(2,105–109)*.

2.4. Barrier Function of Glia Limitans/Pia Mater

In one of the earliest electron microscopic studies of rat meninges and brain surface, Pease and Schultz *(81)* observed that surface astrocytes were "specialized in a manner that presumably forms a protective layer." Indeed, there is restriction of movement of albumin from the subarachnoid space into the external surface of rat brain, beginning at, or prior to, gestational d 16 *(23,110)*. Although ruthenium red can eventually diffuse through the glia limitans of rat brain, the complex interdigitations serve to somewhat restrict bulk flow *(58)*.

Several substances related to water flux are expressed in or affect the glia limitans. Vasopressin induces water uptake by astroglial cells, including those of the glia limitans, which causes a reduction in the amount of water in brain interstitial fluid *(111)*. Taurine, which is involved in cell volume regulation, is especially abundant in the glia limitans *(112)*. Atrial natriuretic peptide, which can control water and salt fluxes, is produced by cells in the glia limitans *(113,114)*.

It is difficult to directly evaluate barrier function at the brain surface in vivo, and it is especially difficult to separate the functions of the glia limitans and the pia mater *(115)*. Cultured leptomeningeal cells exhibit endocytotic activity and an abundance of glutamine synthetase and catechol-*O*-methyltransferase, which can degrade the neurotransmitters, glutamate and norepinephrine *(116)*. The Na^+/Cl^--dependent plasma membrane γ-aminobutyric acid (GABA) transporter, GAT-2, is expressed in the pia and glia limitans, and probably contributes to regulation of the neurotransmitter, GABA *(117)*. The glutamate transporter, GLAST, but not GLT-1, is also localized to the meningeal layer *(118)*. The outer surface of the rat brain blocks the passage of the heavy metal, cadmium, from CSF into brain *(119)*.

2.5. Glia Limitans in Pathological Conditions

Following cold injury to the brain surface, the orthogonal arrays (of AQP4) in the glia limitans are reorganized within minutes *(70)*, and soon after are increased in quantity *(72)*. Systemic hyponatremia leads to a rapid increase in AQP4 immunoreactivity over most of the brain surface, apparently as the result of posttranslational modification of the protein. These changes may accelerate water fluxes to avoid brain edema *(120)*. During dehydration, the glia limitans adjacent to the hypothalamic supraoptic nucleus, which produces the water-balance-regulating peptide, vasopressin, becomes thinner, less intensely immunoreactive for GFAP, and has reduced tenascin immunolabeling in the basement membrane. The change is reversed upon rehydration, but the significance of this is not understood *(121–125)*. Other subtle changes have been documented in a variety of conditions. Subjecting the immature brain to experimental hypoxia causes upregulation of c-*jun*, an immediate early gene, indicating perhaps response of the glia limitans to changes in CSF composition *(126)*. With aging, astrocytes of the glial limitans can accumulate abundant corpora amylacea, which are carbohydrate-rich "pearls" surrounding proteins that cannot be degraded *(127)*. Other inclusions have been observed in hamster brain *(128)*. To some extent, this could reflect ingestion of proteins by the cells from the CSF.

The surface of the brain is frequently interrupted by physical trauma. Glia limitans astrocytes produce large amounts of tissue factor, which is the major cellular initiator of the coagulation protease cascades, and which may help to limit bleeding at the brain surface *(129)*. Injury to the adult rodent brain surface is associated with expression of insulin-like growth factor (IGF) type II, its receptor, and type-5 binding protein in the glia limitans, after 7 d *(130)*. This is followed by reconstitution of laminin and collagen in the basal lamina by 14 d, and later by re-establishment of the glia limitans layer and meningothelial cell regeneration *(73,131)*. Topical steroid application after stab wound impairs the glial reorganization and laminin deposition, and is associated with meningeal cell infiltration into brain tissue *(132)*. Similarly, irradiation of the immature rat spinal cord is associated with opening of the glia limitans and entry of Schwann cells into the dorsal (but not ventral) spinal cord *(133–135)*. Evidently, failure of the glia limitans to seal the brain surface can allow foreign cells in. After damage to fetal mouse brain, the superficial astrocytes upregulate the proinflammatory cytokine, interleukin-1 β (IL-1β); blockade of the IL-1 receptor with an antagonist prevents reconstruction of the glia limitans at the injury site *(136)*. Intravenous administration of IL-1β causes activation of c-*fos*, an immediate early gene, in the arachnoid within 30 min, followed by c-*fos* activation in the glia limitans at ~3 h, suggesting that the former causes secondary changes in the latter *(137)*. Pial cells are immunoreactive for trkB, which is the receptor for brain-derived neurotrophic factor *(138)*. This trophic factor may allow pial cells to respond to injury in adjacent brain tissue *(139)*. Repair of the glia limitans can be enhanced by transplantation of cultured astrocytes *(140)*.

Puncturing of the developing brain surface in developing brains is associated with formation of ectopic neuron collections in the subarachnoid compartment and focal disorganization of the cortex, which suggests a role for the glia limitans in restricting outward migration of neuronal precursors *(15,141–144)*. The condition known as "Fukuyama congenital muscular dystrophy" is associated with heterotopic neuroglial tissue and cortical dysplasia *(145)*. Breaches in the glia limitans basement membranes may be the primary cause of micropolygyria in this genetic disorder *(146–148)*. Reeler mice lack reelin, an EC glycoprotein enriched in outer cortical layers during development that functions as an instructive signal in the regulation of cell patterning *(149)*. These mice exhibit abnormal contacts between immature neurons and the basal lamina of the glia limitans *(150)*, supporting the concept of a role for the glia limitans in regulation of neuronal precursor migration. Glia limitans astrocytes produce neurotrophin-3, which might influence neuronal growth, particularly after injury *(151)*.

3. EPENDYMA

The ependyma is a terminally differentiated, single-layered, cuboidal-to-columnar, ciliated glial epithelium that covers the surface of the ventricles of the brain and central canal of the spinal cord. The choroid plexus and other circumventricular organs (*see* Chapter 12) are covered by highly specialized ependymal cells. Among the latter are tanycytes, and have basal processes directed toward neurons or blood vessels *(152)*. The topic of ependymal cells has been reviewed in prior publications *(153–163)*, to which the reader is referred for a broader discussion of their development, structure, and biology. Ependymal cells appear to have several functions, including support and guidance of other cells during development (and seemingly during regeneration of spinal cord in amphibians), secretory activity particularly during development, and (perhaps overlapping with tanycytes) neuroendocrine/transport function. Data showing that the ependyma exhibits only minimal regenerative capacity have led to the widespread opinion that the lining of the human cerebral ventricles is only vestigial. More recently, it has become apparent that these cells have the metabolic machinery to subserve barrier functions, at least in principle *(156)*. These functions are the focus in this chapter.

Because ependymal cells are organized into a simple epithelium, they are easy to identify in whole-brain studies, and therefore localization of proteins or mRNAs by immunohistochemistry or *in situ* hybridization, respectively, is unambiguous. Cell isolation and tissue culture techniques have also been applied to ependymal cells, in order to study their metabolic characteristics *(164–166)*. Isolation of ependymal cells, by stripping from the ventricle surface, with subsequent culturing, yields in a stable, nonproliferating epithelium that is not overgrown by other cell types, and the composition of which resembles that found *in situ (157,167,168)*.

3.1. Ependymal Development

3.1.1. Differentiation During Brain Development

Ependymal cells arise from the germinal layer that surrounds the ventricles and central canal in the developing neural tube *(169)*. The precise timing is species- and site-specific *(154,155)*. Electron microscopic analysis of developing chick brain shows that cells at the ventricle surface develop microvilli, apical interdigitations, and junctions, on postconception d 6. These features mature during the following week, even during subsequent mitoses *(170)*. In the postconception-wk-8 human embryo, ependymal cells have well-developed junctions, but not microvilli or cilia *(171)*. Transient junctions anchor neighboring cells to each other, while the germinal tissue is active *(172)*. It is likely that mature ciliated cells differentiate only after the ventricular zone proliferative layer is exhausted *(173)*. Fewer than 1% of stem cell clones in vivo are destined for ependymal differentiation *(174)*, corresponding well to the frequency of ependymal cell differentiation in vitro from dissociated brain tissue *(175)*. These observations indicate that a subpopulation of cells, adjacent to the CSF, are committed very early to forming an epithelial barrier between brain and CSF.

3.1.2. Postmature Proliferation and Growth Factors

In normal animals, mitoses occasionally occur among mature ependymal cells *(176)*. Terminen, a marker of terminally differentiated cells, is not expressed by ependyma *(177)*. However, there is widespread expression of the nuclear protein, statin, which is found in cells that are irreversibly arrested in the G_0 phase of the cell cycle *(178)*. Ependymal proliferation occurs more frequently after various brain and spinal injuries *(179–183)*, but the reaction is insufficient to regenerate the ventricle surface. Repair of ventricle wall lesions is largely a reaction of subependymal cells and astrocytes *(184)*. A great deal of excitement was generated recently by a report *(185)* that ependymal cells themselves could dedifferentiate to yield precursor cells in vivo, which could in turn generate new neurons *(186–188)*. This experiment was, however, quickly criticized, with suggestions that the methods used could not discriminate between ependymal cells and residual subependymal stem cells *(189)*.

Ependymal cells possess receptors for basic fibroblast growth factor (bFGF), EGF, and brain-derived neurotrophic factor (trkB) *(138,190–193)*. Cultured axolotl ependymal cells proliferate and migrate, when exposed to EGF *(194)*. Ependymal cells also possess the receptor for endothelial differentiation gene (*EDG-5*), which may be related to trophic support *(195)*. The lack of a substantial proliferative response by ependyma *in situ* suggests that growth factors may act predominantly as survival factors. Rodent ependymal cells also produce growth factors, including bFGF, whose mRNA is upregulated after brain injury *(196,197)* and acidic FGF, which appears to be released into CSF after feeding *(198)*. Glial cell line-derived neurotrophic factor is also produced by ependyma in the early postnatal period of rats *(199)*. Rodent ependymal cells are strongly immunoreactive for connective tissue growth factor, whose significance is unknown *(200)*. They also produce insulin-like growth factor-binding proteins *(201,202)*, which can modulate the function IGF released, for example, from the choroid plexus *(203)*. Neurotrophic factors from ependymal cells may contribute to growth and attraction of supraependymal axons *(204)*, although a broader effect, consequent to diffusion through CSF or release into the adjacent tissues, cannot be excluded.

3.2. Structural and Functional Specializations

3.2.1. Apical Specializations

Generally speaking, ependymal cells are morphologically similar across a broad range of species. There are small height and width variations at different sites in the ventricular system. The most striking feature of ependymal cells is the central clusters of cilia that project into the CSF. They beat in a coordinated manner, tending to sweep foreign particles in the same direction as bulk CSF flow *(205)*. The CSF is also stirred at the ependymal surface, probably to encourage metabolic interaction between ependyma and the CSF contents *(206)*. Cytoskeletal fibers interconnect basal bodies of the cilia within and between cells. This is the likely mechanism through which coordinated ciliary activity is possible *(207)*. Orientation of the basal body may be determined during development, by expression of forkhead factor, HFH-4, in ependyma and other ciliated epithelia *(208)*. Cilia of the ependyma are longer and beat faster than those of the respiratory epithelium *(209,210)*. Their activity can be inhibited by colchicine *(211)*, and by pneumolysin from *Pneumococcus* bacteria *(212–214)*.

The apical surface of ependymal cells is covered by microvilli that have a glycocalyx coating, the composition of which has been revealed in a number of studies *(215–232)*. Although the specifics are not yet understood, the sugar component of glycoproteins is known to be an important component of chemical recognition and retention, information transfer, and signal transduction mechanisms at the cell surface *(233)*. Thus, the surface of ependymal cells is specialized to interact chemically with a representative sample of CSF, at all times.

3.2.2. Supraependymal Axons

Axons lying on the surface of ependymal cells have been identified in most species, including humans *(234)*. These axons arise in the raphe nuclei of the midbrain, specifically, the dorsomedian part of the dorsal raphe nucleus, immediately under the caudal aqueduct *(235)*, and extend to most parts of the ventricular system *(236)*. A specific affinity of the axons for ependyma, perhaps mediated by the chemical structure of the glycocalyx, is suggested by observations that certain neurons preferentially grow on the apical surface of ependymal cells *(237,238)*. Varicosities with serotonin- and GABA-containing vesicles are widespread *(204)*. Some of the axons contain glutamate *(239)* and others can produce nitric oxide *(240)*. Any of these neurotransmitters might be released into adjacent ependymal cells, although synapse-like formations, between serotonergic endings and ependymal cells, have been observed only on the subcommissural organ and in the cerebral aqueduct *(241,242)*. Destruction of supraependymal axons can influence the shape of ependymal cells, under some circumstances *(243–245)*. Mathew *(246)* has suggested that the axons might modulate the function of cilia, although direct evidence has not been provided. Hamster ependymal cells contain kynurenine, a metabolite of serotonin, suggesting that they take up and process serotonin from these axons *(247)*. This population of axons is to be distinguished from another important group of neurons, whose terminals protrude locally through the ependymal layer, where they sense CSF composition or release substances into the CSF *(248)*.

3.2.3. *Cytoskeleton and Intercellular Junctions*

Like other macroglial cells (astrocytes and oligodendrocytes), at some stage in their development, ependymal cells contain the intermediate filament protein, GFAP, which is generally downregulated following maturation *(154,155,162,169,249)*. The intermediate filament protein, vimentin, is uniformly present at all ages in vertebrate ependyma *(250)*. Some cytokeratin subtypes can be detected in human fetal ependymal cells *(251)*. Rat ependymal cells exhibit strong immunoreactivity for plectin, an intermediate filament-associated protein *(252)*. Actin and myosin, near the basal bodies, are probably related to movement of cilia *(253)*.

Ependymal cells are bound to their neighbors near the apical surface by zonula adherens-type junctions *(254,255)*. Tight junctions are only found between ependymal cells covering the specialized circumventricular organs, including the choroid plexus, wherein the capillaries lack tight junctions *(254)*. Junction-associated proteins localized to ependymal membranes include B-cadherins, αβ-catenin, occludin, and ZO-1 *(256–259)*. CD81 on the cell surface stabilizes contacts between all glial cells, including ependyma *(56)*. Intraventricular administration of an antibody to N-cadherin disrupts the ependymal junctions in developing chicken brain *(260)*. Injection of phorbol esters into rat ventricle disrupts adherens junctions of ependyma *(257)*. Similarly, administration of mannose-containing glycoproteins bind to endogenous lectins at the intercellular junctions, and make the ependymal membrane more permeable to proteins *(261,262)*. If permeability can be modulated in vivo, it may regulate interaction between CSF and the brain extracellular space.

Freeze-fracture studies reveal abundant gap junctions and orthogonal arrays of particles *(263)*, which represent membrane-embedded AQP4 molecules (*see* Subheading 3.3.1.) between ependymal cells, and occasionally between ependymal cells and neighboring astrocytes *(264,265)*. Immunohistochemical studies of gap junction proteins reveal connexins 26 and 43 in primitive neuroepithelium and in mature ependymal cells *(266,267)*. Ependymal cells express the mRNA for connexin 32, although the protein has not been demonstrated *(268)*. After 3 wk age, rat ependymal cells also express connexin 30 *(60)*. Gap junctions allow transfer of substances, including exogenous tracers, and electrical currents between ependymal cells *(269,270)*. This coupling serves to integrate ependymal cell function (perhaps also with underlying astroglia) through intercellular communication *(265)*.

3.3. Ependymal Features Relevant to Barrier Function

3.3.1. Movement of Water Across Ependyma

The ependyma is postulated to regulate bulk movement of water between the extracellular compartment and CSF *(271)*, although it should be noted that it offers minimal resistance to diffusion of water, and there is minimal evidence for unidirectional flow, as, for example, in choroid plexus. Aside from the fact that they lack many of the active ion pumps found in choroid plexus, little is known about ion and water transport mechanisms in ependymal cells *(272–276)*. The cystic fibrosis transmembrane conductance regulator is present *(277)*, as is the inwardly rectifying potassium channel, K_{ir} 2.2, which is found on endothelia but not other glial cells *(278)*. It has recently been shown that ependymal cells have an abundance of the water transport channel, AQP4 *(57,68,69)*. This protein is localized to the orthogonal arrays of membrane particles long ago identified on ependymal cells *(263,279)*. It represents a low-resistance pathway for water movement.

Vasopressin, whose concentration is increased in CSF when intracranial pressure is raised, increases ependymal permeability to water *(111,280,281)*. The probable sources of vasopressin are circumventricular organs or vasopressinergic axons that terminate in the ependymal layer *(282,283)*. There are vasopressin-activated Ca-mobilizing receptors and dual angiotensin II–vasopressin receptors on ependymal cells *(284,285)*, which mediate the signaling of these two peptides, known to be important in regulation of water movements. Angiotensin-converting enzyme in ependymal cells *(286,287)* is capable of producing angiotensin II from angiotensinogen, the probable source of which is a subpopulation of astrocytes *(288)*. Therefore, the angiotensin system could play an autocrine role in ependymal water regulation. To what extent water flux needs to be regulated is unclear. Systemic administration of mannitol increases the permeability of ependyma to large molecules, allowing them to move from CSF into brain, which indicates that bulk flow is flexible *(289)*.

3.3.2. Uptake of Neurotransmitters and Waste Products

A variety of transmitters and neuroactive peptides are transported via the CSF to their site of action *(290)*. Waste products are also dumped into the extracellular space of the nervous system, and they subsequently make their way to the CSF. The ependymal acts as a selectively permeable sieve that allows only slow access of CSF proteins (or exogenous tracers, such as ferritin and horseradish peroxidase) to the brain extracellular space *(291–293)*. Ependymal cells are in a position to modulate the availability of these substances, by uptake and degradation. Many receptors have been identified on ependymal cells in a range of mammalian species. Those which bind and transport neuroactive substances are listed in Table 1. Pinocytotic vesicles in ependymal cells can incorporate macromolecules into lysosomes *(292,294,295)*. Degradative enzymes identified in ependymal cells are listed in Table 2. Inactivation of substances by the ependyma probably serves to protect the brain from re-entry of these substances, which could affect neuronal function *(296,297)*. Some peptidases present in ependymal cells

Table 1
Ependymal Receptors and Transporters Possibly Relevant to Barrier Function

Name	Ref.
Peptide receptors	
Oxytocin receptor	*(381)*
β-endorphin receptor (subpopulation of cells only)	*(382)*
High-affinity peptide transporter, PEPT2 (binds a variety of neuropeptide fragments)	*(383)*
Megalin (low-density lipoprotein receptor/peptide uptake mediator)	*(374)*
Leptin receptor OB-Rb	*(384,385)*
Estrogen receptor ERα (developing rat spinal cord only)	*(386)*
Thyroid hormone receptors α and β	*(387)*
Small molecule receptors/transporters	
Glucose transporters, GLUT1 and -2 (cilia), GLUT4 (plasma membrane)	*(388,389)*
Adenosine receptor	*(390)*
Glutamate receptor, GLAST (but not EAAC1 or GLT1)	*(391)*
GABA transporter, GAT-2	*(117)*
Urea transporter UT3 (? regulation of waste product)	*(392)*
Norepinephrine transporter	*(393)*
σ 1 receptor (postsynaptic protein)	*(394)*
γ-GTP amino acid transporter	*(395)*
Bradykinin receptor B2	*(396)*
Dopaminergic D2 receptor	*(397)*
Peripheral-type benzodiazepine-binding sites	*(398)*
Glutathione-S-transferases (binding of hormones and drugs)	*(399)*
Transthyretin (carrier protein of thyroxine produced by choroid plexus and endocytosed by ependymal cells)	*(400)*

Table 2
Enzymes Found in Ependymal Cells

Name	Ref.
Degradative enzymes	
Aminopeptidase N (EC 3.4.11.2)	*(401)*
Dipeptidyl-peptidase IV (EC 3.4.14.15)	*(402)*
Carboxypeptidase H (EC 3.4.17.10)	*(403)*
Enkephalinase (endopeptidase 24.11; EC 3.4.24.11)	*(404)*
Endo-oligopeptidase (EC 3.4.24.15)	*(405)*
Carboxypeptidase E (enkephalin convertase) (EC 3.4.17.10)	*(406)*
Glutamine synthetase (EC 6.3.1.2) (metabolism of glutamate)	*(407)*
Tripeptidyl peptidase II (EC 3.4.14.10) (inactivation of cholecystokinin)	*(408)*
Acid phosphatase (EC 3.1.3.2) (lysomal, broad specificity)	*(409,410)*
Cathepsins B, L, and S (EC 3.4.22.1) (lysomal, broad specificity)	*(411,412)*
Enzymes with possible peptide-activating function	
Peptidyl glycine α amidating mono-oxygenase (PAM; EC 1.14.17.3)	*(413)*
γ-glutamyl transferase (EC 2.3.2.2)	*(167)*
Furin (EC 3.4.21.75)	*(414)*
Tissue kallikrein (EC 3.4.21.35)	*(415)*
Peptidyl-dipeptidase A (angiotensin-converting enzyme, ACE; EC 3.4.15.1)	*(286,287)*

are implicated in the synthesis or activation of neuropeptides that circulate in CSF. These are also listed in Table 2. It is conceivable that the products are released into the CSF or brain ECS, where they could act diffusely.

3.3.3. Antioxidant Agents and Proteins that Bind Ions and Metals

Potentially harmful toxins or waste products can enter the CSF from the choroid plexus and brain substance, respectively. Ependymal cells are in a position to prevent their (re)entry into the brain. Metallothionein is a low molecular weight protein that binds copper, zinc, and some heavy metals. Choroid plexus accumulates heavy metals from the blood, despite its lack of metallothionein. Nevertheless, some neurotoxic heavy metals ultimately enter the CSF *(298)*. Metallothionein types I and II are present in the ependyma of rodents, sheep, and humans *(299–302)*, and their expression can be induced by parenteral administration of Cd or endotoxin *(303,304)*. Ependymal cells effectively block the movement of Cd from CSF to brain *(119)*. Histochemical methods show that Zn, Cu, iron, and the iron-binding/transfer protein, transferrin (but not the iron storage protein, ferritin) are abundant in ependymal cells *(305–307)*.

Because Fe and Cu ions are involved in the generation of oxygen radicals, ependymal uptake of reactive metal ions from CSF may protect the brain parenchyma from oxidative damage *(308)*. Ependymal cells also have the antioxidant enzymes, peroxiredoxin-I *(309)* and glutaredoxin (thioltransferase) *(310)*, in abundance. Heme oxygenase (EC 1.14.99.3), also known as "heat shock protein 32", converts heme from blood breakdown into bile pigments (which are potent antioxidants). This enzyme is present in normal rat brain at low levels, but, following heat shock, the mRNA increases dramatically in ependymal cells, and less so in neurons *(311)*. Another oxygen-radical scavenger, superoxide dismutase (EC 1.15.1.1), has been detected by immunological methods in rat and human ependymal cells *(312,313)*.

Ependymal cells in fetal and adult humans and animals uniformly contain the protein and mRNA of the Ca-binding protein, S100β *(19,249,314)* and, to a lesser extent, S100A6 (calcylin) *(315)*. Another Ca-binding protein found in rodent ependyma is calbindin D28k *(316–318)*. Calcyphosine, a Ca-binding protein phosphorylated by a cyclic adenosine monophosphate-dependent process, is found in dog ependyma *(319)*. These proteins probably serve as Ca ion buffers *(320)*. A Ca-ion-activated adenosine triphosphatase (EC 3.6.1.3) present on the endoplasmic reticulum of ependymal cells *(321)* presumably acts to sequester Ca that has not been adequately buffered. The Ca pump present at the apical membrane of mammalian choroid plexus is not present on ependymal cells *(322)*. The presence of these proteins in ependymal (as well as astroglial) cells may help to prevent exposure of neurons to high concentrations of Ca ion, which can activate a range of proteolytic enzymes in pathologic situations.

3.4. Ependymal Changes in Pathological Conditions

The expression of c-FOS-like immunoreactivity and upregulation of c-*fos* mRNA is seen in ependymal cells within a few hours of brain ischemia or trauma *(323–325)*, or following iv administration of urocortin, a corticotrophin agonist that mediates autonomic responses to stress *(326)*. The c-*fos* gene encodes for FOS, which is a nuclear protein that binds DNA and modulates gene transcription. Other so-called "immediate early proteins" expressed by ependymal cells include JUN and KROX 24 *(327,328)*.

Like astrocytes, in response to injury or metabolic stress, ependymal cells increase their expression of several protective proteins, as well as the cytoskeletal protein, GFAP *(329)*. Immediate early genes may provide a long-term "memory" of stimuli; heat shock proteins mediate a cell survival response following injury, to protect cells from subsequent acute injury *(330,331)*. Details of the signaling cascades are not known in this cell population.

Perhaps because the ependyma is exposed to a relatively harsh environment, with considerable potential for oxidative and metabolic injury, there is a much higher incidence of DNA strand breaks and oxidative DNA damage than in other glial cell populations *(332,333)*. This may explain the high prevalence of "false-positive" labeling when terminal deoxynucleotidyl transferase-mediated deoxyuridine triphosphate-biotin nick-end labeling (TUNEL) assays are done to detect dying and apoptotic cells (unpublished observations). Perhaps in compensation for frequent damage, but minimal proliferative capacity, ependymal cells in most mammalian species express the anti-apoptotic proteins, Bcl-2 *(334,335)* and Bag-1 *(336)*, to help prevent cell loss.

A common finding in the lateral ventricles of animals with large brains, including humans, is focal loss of ependymal cells from the ventricle wall. These are often associated with undulations, small diverticulae, and thickening of the astroglial layer. In humans, focal loss is first evident in the third-trimester fetus, and progresses into adulthood *(337–340)*. In humans (and small mammals), the central canal usually becomes occluded by 20 yr of age. This is postulated to be the result of a mild injury, perhaps secondary to viral infections *(341–343)*. Ependymal cells can be damaged by a variety of common viruses to which humans are susceptible, including influenza, herpes simplex, varicella zoster, adeno-associated virus, and mumps *(344–348)*. Ependymal cells are the only cells in the mature nervous system that express the adenovirus- and coxsackie virus-binding epitope *(349)*. In rodents ependymal cells are infected by, but not necessarily killed by, neurovirulent Sindbis virus and lymphocytic choriomeningitis virus *(350,351)*.

During viral infection, ependymal cells upregulate expression of major histocompatibility class (MHC) I antigens, perhaps as part of their survival strategy to avoid attack by lymphocytes *(352)*. They also upregulate MHC II antigens, generally reserved for antigen-presenting cells, following intracerebral or iv injection of interferon-γ and systemic infection with *Toxoplasma gondii (353–356)*. Resting ependymal cells in rat, express mRNAs for the proinflammatory cytokine, IL-6, as well as the IL-6 receptor *(357,358)*. Mice with targeted overexpression of IL-6 in glial cells also have high levels of complement C3 in ependymal cells, which may contribute to inflammatory processes *(359)*. Physical, infectious, and chemical injuries to rodent brains induce ependymal production of several cytokines, including tumor necrosis factor α, IL-1α, and transforming growth factor (TGF-α and TGF-β_1) *(360–363)*. Ependymal cells also have receptors for IL-1 (IL-1RI and IL-1RII) *(364)*. It is not clear how these cytokines play a role in the response of ependyma to infection and injury. The expression of cytokines may support the invasion of intraventricular inflammatory cells, or be a mechanism by which other cells in the brain are signaled to the presence of ependymal injury *(365)*. The intercellular adhesion molecules, ICAM-1 and VCAM-1, are expressed on microvilli of ependymal cells, moreso following infection, probably to mediate adhesion of inflammatory cells to the ependymal surface *(353,366)*.

Ependymal cell loss follows a variety of other insults, including dilatation of the cerebral ventricles (hydrocephalus), cerebral ischemia, irradiation, toxin exposure, nutritional disturbances, disorders of neuronal migration, hyperbaric oxygen environment, and "sudden infant death syndrome" *(154,156,161,367–369)*. Sites of ependymal denudation are associated with chronic astroglial changes around the cerebral ventricles, including thickening of the layer and more abundant corpora amylacea (unpublished observations), which are reactive astroglial inclusions that contain undigestible proteins *(127)*. In aged human brains, ependymal cells may contain fibrillar β-amyloid-immunoreactive inclusions that resemble the neurofibrillary tangles found in neurons of Alzheimer's disease *(370–372)*. Amyloid precursor protein *(373)*, the low-density lipoprotein receptor, megalin (which can mediate uptake of apolipoprotein J, a binding protein for the amyloid peptide) *(374)*, and the carboxypeptidase capable of digesting the protein *(375)* are all highly expressed in ependymal cells.

4. SUMMARY AND CONCLUSIONS

Cerebrospinal fluid functions as a sink for some water-soluble waste substances from the brain extracellular space and as a conduit into the lymphatic system *(290,376)*. The glia limitans/pia mater at the outer surface of the brain, and the ependymal lining of the ventricles, are morphologically designed to retard bulk movement of CSF into brain tissue. Furthermore, they are equipped with the metabolic machinery necessary to remove and process undesirable, potentially harmful agents from the CSF. These barrier functions are established very early in the development of the nervous system *(377,378)*. Astrocyte specializations of the glia limitans are similar to those at the BBB, and many of the same enzymes are also found in the glia limitans/pia mater, BBB, and ependyma *(379)*. This implicates them in an important functional role at brain–CSF interfaces. Mercier and Hatton *(380)* proposed that, through gap junctions, the ependyma, astrocytes, and glia limitans form a continuous network of communication throughout the brain. Through AQP channels, water too can be rapidly moved along intracellular pathways. This has obvious advantages for the redistribution of noxious substances away from focal brain disturbances. Hence, the barriers at the brain surfaces are potentially capable of acting in concert with those at blood interfaces, to maintain a stable environment for neuronal development and function.

ACKNOWLEDGMENTS

Some aspects of this chapter are related to work funded by the Canadian Institutes of Health Research.

REFERENCES

1. Held, H. (1909) Ueber die Neuroglia marginalis der menschlichen Grosshirnrinde. *Monat. Psychiatr. Neurol.* **26,** 360–416.
2. Krisch, B., Leonhardt, H., and Oksche, A. (1984) Compartments and perivascular arrangement of the meninges covering the cerebral cortex of the rat. *Cell Tissue Res.* **238,** 459–474.
3. Haines, D. E., Harkey, H. L., and al-Mefty, O. (1993) The "subdural" space: a new look at an outdated concept. *Neurosurgery* **32,** 111–120.
4. Lee, J. C., Mayer Proschel, M., and Rao, M. S. (2000) Gliogenesis in the central nervous system. *Glia* **30,** 105–121.

5. Aschner, M. (1998) Astrocytic functions and physiological reactions to injury: the potential to induce and/or exacerbate neuronal dysfunction: a forum position paper. *Neuro-Toxicology* **19,** 7–17.
6. Walz, W. (1989) Role of glial cells in the regulation of the brain ion microenvironment. *Progr. Neurobiol.* **33,** 309–333.
7. Norenberg, M. D. (1994) Astrocyte responses to CNS injury. *J. Neuropathol. Exp. Neurol.* **53,** 213–220.
8. Eddleston, M. and Mucke, L. (1993) Molecular profile of reactive astrocytes: implications for their role in neurologic disease. *Neuroscience* **54,** 15–36.
9. Choi, B. H. (1988) Prenatal gliogenesis in the developing cerebrum of the mouse. *Glia* **1,** 308–316.
10. McLone, D. G. and Bondareff, W. (1975) Developmental morphology of the subarachnoid space and contiguous structures in the mouse. *Am. J. Anat.* **142,** 273–293.
11. McLone, D. G. (1980) Development of the limiting glial membrane of the brain. *Child's Brain* **6,** 150–162.
12. Sturrock, R. R. (1989) An ultrastructural study of the development of the leptomeninx of the mouse spinal cord. *Anat. Anz.* **169,** 321–327.
13. Rickmann, M. and Wolff, J. R. (1985) Prenatal gliogenesis in the neopallium of the rat. *Adv. Anat. Embryol. Cell Biol.* **93,** 1–104.
14. Choi, B. H. (1988) Developmental events during the early stages of cerebral cortical neurogenesis in man. A correlative light, electron microscopic, immunohistochemical and Golgi study. *Acta Neuropathol.* **75,** 441–447.
15. Choi, B. H. (1994) Role of the basement membrane in neurogenesis and repair of injury in the central nervous system. *Microsc. Res. Technol.* **28,** 193–203.
16. Marin-Padilla, M. (1995) Prenatal development of fibrous (white matter), protoplasmic (gray matter), and layer I astrocytes in the human cerebral cortex: a Golgi study. *J. Comp. Neurol.* **357,** 554–572.
17. Landry, C. F., Ivy, G. O., and Brown, I. R. (1990) Developmental expression of glial fibrillary acidic protein mRNA in the rat brain analyzed by *in situ* hybridization. *J. Neurosci. Res.* **25,** 194–203.
18. Borit, A. and McIntosh, G. C. (1981) Myelin basic protein and glial fibrillary acidic protein in human fetal brain. *Neuropathol. Appl. Neurobiol.* **7,** 279–287.
19. Landry, C. F., Ivy, G. O., Dunn, R. J., Marks, A., and Brown, I. R. (1989) Expression of the gene encoding the β-subunit of S-100 protein in the developing rat brain analyzed by in situ hybridization. *Mol. Brain Res.* **6,** 251–262.
20. Antalikova, L. (1969) Pia mater and membrana limitans gliae superficialis in human ontogenesis. *Folia Morphol.* **17,** 400–407.
21. Astrom, K. E. and Webster, H. D. (1991) The early development of the neopallial wall and area choroidea in fetal rats. A light and electron microscopic study. *Adv. Anat. Embryol. Cell Biol.* **123,** 1–76.
22. Klika, E. and Antalikova, L. (1969) The membrana limitans gliae superficialis in vertebrates. *Folia Morphol.* **17,** 342–347.
23. Balslev, Y., Saunders, N. R., and Mollgard, K. (1997) Ontogenetic development of diffusional restriction to protein at the pial surface of the rat brain: an electron microscopical study. *J. Neurocytol.* **26,** 133–148.
24. Angelov, D. N. and Vasilev, V. A. (1989) Morphogenesis of the rat cranial meninges. A light- and electron-microscopic study. *Cell Tissue Res.* **257,** 207–216.
25. O'Rahilly, R. and Muller, F. (1986) The meninges in human development. *J. Neuropathol. Exp. Neurol.* **45,** 588–608.
26. Sturrock, R. R. (1990) An ultrastructural study of development of the leptomeninx of the rabbit spinal cord. *Anat. Anz.* **170,** 73–77.

27. McLone, D. G. (1980) The subarachnoid space: a review. *Child's Brain* **6,** 113–130.
28. Struckhoff, G. (1995) Cocultures of meningeal and astrocytic cells: a model for the formation of the glial-limiting membrane. *Int. J. Dev. Neurosci.* **13,** 595–606.
29. Lyser, K. M. (1972) The differentiation of glial cells and glia limitans in organ cultures of chick spinal cord. *In Vitro* **8,** 77–84.
30. Ness, R. and David, S. (1997) Leptomeningeal cells modulate the neurite growth promoting properties of astrocytes in vitro. *Glia* **19,** 47–57.
31. Anders, J. J. and Salopek, M. (1989) Meningeal cells increase in vitro astrocytic gap junctional communication as measured by fluorescence recovery after laser photobleaching. *J. Neurocytol.* **18,** 257–264.
32. Abnet, K., Fawcett, J. W., and Dunnett, S. B. (1991) Interactions between meningeal cells and astrocytes in vivo and in vitro. *Dev. Brain Res.* **59,** 187–196.
33. Sievers, J., Pehlemann, F. W., Gude, S., and Berry, M. (1994) Meningeal cells organize the superficial glia limitans of the cerebellum and produce components of both the interstitial matrix and the basement membrane. *J. Neurocytol.* **23,** 135–149.
34. Hartmann, D., Sievers, J., Pehlemann, F. W., and Berry, M. (1992) Destruction of meningeal cells over the medial cerebral hemisphere of newborn hamsters prevents the formation of the infrapyramidal blade of the dentate gyrus. *J. Comp. Neurol.* **320,** 33–61.
35. Pehlemann, F. W., Sievers, J., and Berry, M. (1985) Meningeal cells are involved in foliation, lamination, and neurogenesis of the cerebellum: evidence from 6-hydroxydopamine-induced destruction of meningeal cells. *Dev. Biol.* **110,** 136–146.
36. Sievers, J., Pehlemann, F. W., Gude, S., and Berry, M. (1994) A time course study of the alterations in the development of the hamster cerebellar cortex after destruction of the overlying meningeal cells with 6-hydroxydopamine on the day of birth. *J. Neurocytol.* **23,** 117–134.
37. Sievers, J. and Pehlemann, F. W. (1986) Influences of meningeal cells on brain development. Findings and hypothesis. *Naturwissenschaften* **73,** 188–194.
38. Sievers, J., von Knebel Doeberitz, C., Pehlemann, F. W., and Berry, M. (1986) Meningeal cells influence cerebellar development over a critical period. *Anat. Embryol.* **175,** 91–100.
39. Sievers, J., Pehlemann, F. W., Baumgarten, H. G., and Berry, M. (1985) Selective destruction of meningeal cells by 6-hydroxydopamine: a tool to study meningeal- neuroepithelial interaction in brain development. *Dev. Biol.* **110,** 127–135.
40. von Knebel Doeberitz, C., Sievers, J., Sadler, M., Pehlemann, F. W., Berry, M., and Halliwell, P. (1986) Destruction of meningeal cells over the newborn hamster cerebellum with 6-hydroxydopamine prevents foliation and lamination in the rostral cerebellum. *Neuroscience* **17,** 409–426.
41. Hausmann, B. and Sievers, J. (1985) Cerebellar external granule cells are attached to the basal lamina from the onset of migration up to the end of their proliferative activity. *J. Comp. Neurol.* **241,** 50–62.
42. Hartmann, D., Schulze, M., and Sievers, J. (1998) Meningeal cells stimulate and direct the migration of cerebellar external granule cells in vitro. *J. Neurocytol.* **27,** 395–409.
43. Hartmann, D., Ziegenhagen, M. W., and Sievers, J. (1998) Meningeal cells stimulate neuronal migration and the formation of radial glial fascicles from the cerebellar external granular layer. *Neurosci. Lett.* **244,** 129–132.
44. Breese, G. R., Criswell, H. E., Johnson, K. B., O'Callaghan, J. P., Duncan, G. E., Jensen, K. F., Simson, P. E., and Mueller, R. A. (1994) Neonatal destruction of dopaminergic neurons. *Neurotoxicology* **15,** 149–159.
45. Sachs, C., Pycock, C., and Jonsson, G. (1974) Altered development of central noradrenaline neurons during ontogeny by 6-hydroxydopamine. *Med. Biol.* **52,** 55–65.
46. Lau, C., Cameron, A., Antolick, L., and Slotkin, T. A. (1990) Trophic control of the ornithine decarboxylase/polyamine system in neonatal rat brain regions: lesions caused by 6-hydroxydopamine produce effects selective for cerebellum. *Dev. Brain Res.* **52,** 167–173.

47. Williams, V. (1975) Intercellular relationships in the external glial limiting membrane of the neocortex of the cat and rat. *Am. J. Anat.* **144,** 421–431.
48. Lopes, C. A. and Mair, W. G. (1974) Ultrastructure of the outer cortex and the pia mater in man. *Acta Neuropathol.* **28,** 79–86.
49. Peters, A., Josephson, K., and Vincent, S. L. (1991) Effects of aging on the neuroglial cells and pericytes within area 17 of the rhesus monkey cerebral cortex. *Anat. Rec.* **229,** 384–398.
50. Oudega, M. and Marani, E. (1991) Expression of vimentin and glial fibrillary acidic protein in the developing rat spinal cord: an immunocytochemical study of the spinal cord glial system. *J. Anat.* **179,** 97–114.
51. Gimenez, Y. R. M., Langa, F., Menet, V., and Privat, A. (2000) Comparative anatomy of the cerebellar cortex in mice lacking vimentin, GFAP, and both vimentin and GFAP. *Glia* **31,** 69–83.
52. Kornblum, H. I., Hussain, R., Wiesen, J., Miettinen, P., Zurcher, S. D., Chow, K., Derynck, R., and Werb, Z. (1998) Abnormal astrocyte development and neuronal death in mice lacking the epidermal growth factor receptor. *J. Neurosci. Res.* **53,** 697–717.
53. Lie, A. A., Schroder, R., Blumcke, I., Magin, T. M., Wiestler, O. D., and Elger, C. E. (1998) Plectin in the human central nervous system: predominant expression at pia/glia and endothelia/glia interfaces. *Acta Neuropathol.* **96,** 215–221.
54. Bondareff, W. and McLone, D. G. (1973) The external glial limiting membrane in Macaca: ultrastructure of a laminated glioepithelium. *Am. J. Anat.* **136,** 277–295.
55. Gerhardt, H., Rascher, G., Schuck, J., Weigold, U., Redies, C., and Wolburg, H. (2000) R- and B-cadherin expression defines subpopulations of glial cells involved in axonal guidance in the optic nerve head of the chicken. *Glia* **31,** 131–143.
56. Sullivan, C. D. and Geisert, E. E., Jr. (1998) Expression of rat target of the antiproliferative antibody (TAPA) in the developing brain. *J. Comp. Neurol.* **396,** 366–380.
57. Rash, J. E., Duffy, H. S., Dudek, F. E., Bilhartz, B. L., Whalen, L. R., and Yasumura, T. (1997) Grid-mapped freeze-fracture analysis of gap junctions in gray and white matter of adult rat central nervous system, with evidence for a "panglial syncytium" that is not coupled to neurons. *J. Comp. Neurol.* **388,** 265–292.
58. Wagner, H. J., Barthel, J., and Pilgrim, C. (1983) Permeability of the external glial limiting membrane of rat parietal cortex. *Anat. Embryol.* **166,** 427–437.
59. Ochalski, P. A., Frankenstein, U. N., Hertzberg, E. L., and Nagy, J. I. (1997) Connexin-43 in rat spinal cord: localization in astrocytes and identification of heterotypic astro-oligodendrocytic gap junctions. *Neuroscience* **76,** 931–945.
60. Kunzelmann, P., Schroder, W., Traub, O., Steinhauser, C., Dermietzel, R., and Willecke, K. (1999) Late onset and increasing expression of the gap junction protein connexin30 in adult murine brain and long-term cultured astrocytes. *Glia* **25,** 111–119.
61. Black, J. A. and Waxman, S. G. (1985) Specialization of astrocytic membrane at glia limitans in rat optic nerve: freeze-fracture observations. *Neurosci. Lett.* **55,** 371–378.
62. Dermietzel, R. (1973) Visualization by freeze-fracturing of regular structures in glial cell membranes. *Naturwissenschaften* **60,** 208.
63. Gotow, T. (1984) Cytochemical characteristics of astrocytic plasma membranes specialized with numerous orthogonal arrays. *J. Neurocytol.* **13,** 431–448.
64. Gotow, T. and Hashimoto, P. H. (1988) Deep-etch structure of astrocytes at the superficial glia limitans, with special emphasis on the internal and external organization of their plasma membranes. *J. Neurocytol.* **17,** 399–413.
65. Landis, D. M. and Reese, T. S. (1982) Regional organization of astrocytic membranes in cerebellar cortex. *Neuroscience* **7,** 937–950.
66. Nabeshima, S., Reese, T. S., Landis, D. M., and Brightman, M. W. (1975) Junctions in the meninges and marginal glia. *J. Comp. Neurol.* **164,** 127–169.

67. Sandri, C., Van Buren, J. M., and Akert, K. (1977) Membrane morphology of the vertebrate nervous system. A study with freeze-etch technique. *Progr. Brain Res.* **46,** 1–384.
68. Rash, J. E., Yasumura, T., Hudson, C. S., Agre, P., and Nielsen, S. (1998) Direct immunogold labeling of aquaporin-4 in square arrays of astrocyte and ependymocyte plasma membranes in rat brain and spinal cord. *Proc. Natl. Acad. Sci. USA* **95,** 11,981–11,986.
69. Rash, J. E. and Yasumura, T. (1999) Direct immunogold labeling of connexins and aquaporin-4 in freeze-fracture replicas of liver, brain, and spinal cord: factors limiting quantitative analysis. *Cell Tissue Res.* **296,** 307–321.
70. Anders, J. J. and Brightman, M. W. (1982) Particle assemblies in astrocytic plasma membranes are rearranged by various agents in vitro and cold injury in vivo. *J. Neurocytol.* **11,** 1009–1029.
71. Gotow, T. and Hashimoto, P. H. (1989) Developmental alterations in membrane organization of rat subpial astrocytes. *J. Neurocytol.* **18,** 731–747.
72. Anders, J. J. and Brightman, M. W. (1979) Assemblies of particles in the cell membranes of developing, mature and reactive astrocytes. *J. Neurocytol.* **8,** 777–795.
73. Maxwell, W. L., Duance, V. C., Lehto, M., Ashurst, D. E., and Berry, M. (1984) The distribution of types I, III, IV and V collagens in penetrant lesions of the central nervous system of the rat. *Histochem. J.* **16,** 1215–1229.
74. Wolf, J. (1970) Anchorage of collagenous fibrils in brain surface wrapping basal membrane (membrana limitans gliae superficialis). *Folia Morphol.* **18,** 322–329.
75. Tucker, R. P., Hagios, C., and Chiquet Ehrismann, R. (1999) Tenascin-Y in the developing and adult avian nervous system. *Dev. Neurosci.* **21,** 126–133.
76. Hoffman, J. R. and O'Shea, K. S. (1999) Thrombospondin expression in nerve regeneration II. Comparison of optic nerve crush in the mouse and goldfish. *Brain Res. Bull* **48,** 421–427.
77. Rutka, J. T., Giblin, J. R., Balkissoon, R., Wen, D., Myatt, C. A., McCulloch, J. R., and Rosenblum, M. L. (1987) Characterization of fetal human brain cultures. Development of a potential model for selectively purifying human glial cells in culture. *Dev. Neurosci.* **9,** 154–173.
78. Rutka, J. T., Apodaca, G., Stern, R., and Rosenblum, M. (1988) The extracellular matrix of the central and peripheral nervous systems: structure and function. *J. Neurosurg.* **69,** 155–170.
79. Johnson-Green, P. C., Dow, K. E., and Riopelle, R. J. (1991) Characterization of glycosaminoglycans produced by primary astrocytes in vitro. *Glia* **4,** 314–321.
80. Bernstein, J. J., Getz, R., Jefferson, M., and Keleman, M. (1985) Astrocytes secrete basal lamina after hemisection of rat spinal cord. *Brain Res.* **327,** 135–141.
81. Pease, D. C. and Schultz, R. L. (1958) Electron microscopy of rat cranial meninges. *Am. J. Anat.* **102,** 301–321.
82. Morse, D. E. and Low, F. N. (1972) The fine structure of the pia mater of the rat. *Am. J. Anat.* **133,** 349–367.
83. Zhang, E. T., Inman, C. B., and Weller, R. O. (1990) Interrelationships of the pia mater and the perivascular (Virchow-Robin) spaces in the human cerebrum. *J. Anat.* **170,** 111–123.
84. Alcolado, R., Weller, R. O., Parrish, E. P., and Garrod, D. (1988) The cranial arachnoid and pia mater in man: anatomical and ultrastructural observations. *Neuropathol. Appl. Neurobiol.* **14,** 1–17.
85. Peters, A., Palay, S. L., and Webster, H. def. (1976) *The Fine Structure of the Nervous System*, W. B. Saunders, Philadelphia, pp. 264–279.
86. Achtstatter, T., Fouquet, B., Rungger-Brandle, E., and Franke, W. W. (1989) Cytokeratin filaments and desmosomes in the epithelioid cells of the perineurial and arachnoidal sheaths of some vertebrate species. *Differentiation* **40,** 129–149.

87. Doucette, R. (1989) Development of the nerve fiber layer in the olfactory bulb of mouse embryos. *J. Comp. Neurol.* **285,** 514–527.
88. Fraher, J. P. (1997) Axon-glial relationships in early CNS-PNS transitional zone development: an ultrastructural study. *J. Neurocytol.* **26,** 41–52.
89. Nugent, S. G., O'Sullivan, V. R., Fraher, J. P., and Rea, B. B. (1991) Central-peripheral transitional zone of the spinal accessory nerve in the rat. *J. Anat.* **175,** 19–25.
90. Fraher, J. P., Smiddy, P. F., and O'Sullivan, V. R. (1988) The central-peripheral transitional regions of cranial nerves. Oculomotor nerve. *J. Anat.* **161,** 103–113.
91. Hildebrand, C., Remahl, S., and Waxman, S. G. (1985) Axo-glial relations in the retina-optic nerve junction of the adult rat: electron-microscopic observations. *J. Neurocytol.* **14,** 597–617.
92. Haller, F. R. and Low, F. N. (1971) The fine structure of the peripheral nerve root sheath in the subarachnoid space in the rat and other laboratory animals. *Am. J. Anat.* **131,** 1–19.
93. Haller, F. R., Haller, C., and Low, F. N. (1972) The fine structure of cellular layers and connective tissue space at spinal nerve root attachments in the rat. *Am. J. Anat.* **133,** 109–123.
94. Malloy, J. J. and Low, F. N. (1974) Scanning electron microscopy of the subarachnoid space in the dog. II. Spinal nerve exits. *J. Comp. Neurol.* **157,** 87–107.
95. Doucette, R. (1993) Glial cells in the nerve fiber layer of the main olfactory bulb of embryonic and adult mammals. *Microsc. Res. Tech.* **24,** 113–130.
96. Doucette, R. (1991) PNS-CNS transitional zone of the first cranial nerve. *J. Comp. Neurol.* **312,** 451–466.
97. Hutchings, M. and Weller, R. O. (1986) Anatomical relationships of the pia mater to cerebral blood vessels in man. *J. Neurosurg.* **65,** 316–325.
98. Krahn, V. (1982) The pia mater at the site of the entry of blood vessels into the central nervous system. *Anat. Embryol.* **164,** 257–263.
99. Angelov, D. N. (1990) Ultrastructural investigation of the meningeal compartment of the blood–cerebrospinal fluid-barrier in rats and cats. A horseradish peroxidase study. *Z Mikrosk Anat. Forsch.* **104,** 1–16.
100. Cardoso, E. R., Peterson, E. W., and Hendelman, W. (1985) Sub-pial infiltration of blood products following experimental subarachnoid haemorrhage. *Acta Neurochir (Wien)* **76,** 140–144.
101. Ichimura, T., Fraser, P. A., and Cserr, H. F. (1991) Distribution of extracellular tracers in perivascular spaces of the rat brain. *Brain Res.* **545,** 103–113.
102. Dunker, R. O., Harris, A. B., and Jenkins, D. P. (1976) Kinetics of horseradish peroxidase migration through cerebral cortex. *Brain Res.* **118,** 199–217.
103. Levin, E. and Sisson, W. B. (1972) The penetration of radiolabeled substances into rabbit brain from subarachnoid space. *Brain Res.* **41,** 145–153.
104. Rennels, M. L., Gregory, T. F., Blaumanis, O. R., Fujimoto, K., and Grady, P. A. (1985) Evidence for a "paravascular" fluid circulation in the mammalian central nervous system, provided by the rapid distribution of tracer protein throughout the brain from the subarachnoid space. *Brain Res.* **326,** 47–63.
105. Erlich, S. S., McComb, J. G., Hyman, S., and Weiss, M. H. (1986) Ultrastructural morphology of the olfactory pathway for cerebrospinal fluid drainage in the rabbit. *J. Neurosurg.* **64,** 466–473.
106. Erlich, S. S., McComb, J. G., Hyman, S., and Weiss, M. H. (1989) Ultrastructure of the orbital pathway for cerebrospinal fluid drainage in rabbits. *J. Neurosurg.* **70,** 926–931.
107. Levin, E., Sepulveda, F. V., and Yudilevich, D. L. (1974) Pial vessels transport of substances from cerebrospinal fluid to blood. *Nature* **249,** 266–268.
108. McComb, J. G., Davson, H., and Hollingsworth, J. R. (1975) Further studies on the difference between ventricular and subarachnoid perfusion. *Brain Res.* **89,** 81–91.

109. Shen, J. Y., Kelly, D. E., Hyman, S., and McComb, J. G. (1985) Intraorbital cerebrospinal fluid outflow and the posterior uveal compartment of the hamster eye. *Cell Tissue Res.* **240,** 77–87.
110. Dziegielewska, K. M., Knott, G. W., and Saunders, N. R. (2000) The nature and composition of the internal environment of the developing brain. *Cell. Mol. Neurobiol.* **20,** 41–56.
111. Chen, Y., McNeill, J. R., Hajek, I., and Hertz, L. (1992) Effect of vasopressin on brain swelling at the cellular level: do astrocytes exhibit a furosemide-vasopressin-sensitive mechanism for volume regulation? *Can. J. Physiol. Pharmacol.* **70(Suppl.),** S367–S373.
112. Lake, N., Verdone-Smith, C., and Brownstein, S. (1992) Immunocytochemical localization of taurine and glial fibrillary acidic protein in human optic nerve. *Vis. Neurosci.* **8,** 251–255.
113. McKenzie, J. C. (1992) Atrial natriuretic peptide-like immunoreactivity in astrocytes of parenchyma and glia limitans of the canine brain. *J. Histochem. Cytochem.* **40,** 1211–1222.
114. McKenzie, J. C., Berman, N. E., Thomas, C. R., Young, J. K., Compton, L. Y., Cothran, L. N., Liu, W. L., and Klein, R. M. (1994) Atrial natriuretic peptide-like (ANP-LIR) and ANP prohormone immunoreactive astrocytes and neurons of human cerebral cortex. *Glia* **12,** 228–243.
115. Wright, P. M., Nogueira, G. J., and Levin, E. (1971) Role of the pia mater in the transfer of substances in and out of the cerebrospinal fluid. *Exp. Brain Res.* **113,** 294–305.
116. Feurer, D. J. and Weller, R. O. (1991) Barrier functions of the leptomeninges: a study of normal meninges and meningiomas in tissue culture. *Neuropathol. Appl. Neurobiol.* **17,** 391–405.
117. Conti, F., Zuccarello, L. V., Barbaresi, P., Minelli, A., Brecha, N. C., and Melone, M. (1999) Neuronal, glial, and epithelial localization of gamma-aminobutyric acid transporter 2, a high-affinity gamma-aminobutyric acid plasma membrane transporter, in the cerebral cortex and neighboring structures. *J. Comp. Neurol.* **409,** 482–494.
118. Berger, U. V. and Hediger, M. A. (2000) Distribution of the glutamate transporters GLAST and GLT-1 in rat circumventricular organs, meninges, and dorsal root ganglia. *J. Comp. Neurol.* **421,** 385–399.
119. Takeda, A., Takefuta, S., Ijiro, H., Okada, S., and Oku, N. (1999) ^{109}Cd transport in rat brain. *Brain Res.* Bull **49,** 453–457.
120. Vajda, Z., Promeneur, D., Doczi, T., Sulyok, E., Frokiaer, J., Ottersen, O. P., and Nielsen, S. (2000) Increased aquaporin-4 immunoreactivity in rat brain in response to systemic hyponatremia. *Biochem. Biophys. Res. Commun.* **270,** 495–503.
121. Bobak, J. B. and Salm, A. K. (1996) Plasticity of astrocytes of the ventral glial limitans subjacent to the supraoptic nucleus. *J. Comp. Neurol.* **376,** 188–197.
122. Hawrylak, N., Fleming, J. C., and Salm, A. K. (1998) Dehydration and rehydration selectively and reversibly alter glial fibrillary acidic protein immunoreactivity in the rat supraoptic nucleus and subjacent glial limitans. *Glia* **22,** 260–271.
123. Hawrylak, N., Boone, D., and Salm, A. K. (1999) The surface density of glial fibrillary acidic protein immunopositive astrocytic processes in the rat supraoptic nucleus is reversibly altered by dehydration and rehydration. *Neurosci. Lett.* **277,** 57–60.
124. Salm, A. K. and Bobak, J. B. (1999) Dehydration-associated changes in the ventral glial limitans subjacent to the supraoptic nucleus include a reduction in the extent of the basal lamina but not astrocytic process shrinkage. *Exp. Neurol.* **160,** 425–432.
125. Singleton, P. A. and Salm, A. K. (1996) Differential expression of tenascin by astrocytes associated with the supraoptic nucleus (SON) of hydrated and dehydrated adult rats. *J. Comp. Neurol.* **373,** 186–199.
126. Wickstrom, H. R., Holgert, H., Hokfelt, T., and Lagercrantz, H. (1999) Birth-related expression of c-fos, c-jun and substance P mRNAs in the rat brainstem and pia mater: possible relationship to changes in central chemosensitivity. *Dev. Brain Res.* **112,** 255–266.

127. Cavanagh, J. B. (1999) Corpora-amylacea and the family of polyglucosan diseases. *Brain Res. Rev.* **29,** 265–295.
128. Gianonatti, C., Gianonatti, M., Larana, A., and Bujan, J. (1988) Inclusion bodies in subpial astrocytes of hamster cerebellum (Cricetus cricetus). Ultrastructural study. *J. Hirnforsch.* **29,** 315–318.
129. Eddleston, M., de la Torre, J. C., Oldstone, M. B., Loskutoff, D. J., Edgington, T. S., and Mackman, N. (1993) Astrocytes are the primary source of tissue factor in the murine central nervous system. A role for astrocytes in cerebral hemostasis. *J. Clin. Invest.* **92,** 349–358.
130. Walter, H. J., Berry, M., Hill, D. J., and Logan, A. (1997) Spatial and temporal changes in the insulin-like growth factor (IGF) axis indicate autocrine/paracrine actions of IGF-I within wounds of the rat brain. *Endocrinology* **138,** 3024–3034.
131. Carbonell, A. L. and Boya, J. (1988) Ultrastructural study on meningeal regeneration and meningo-glial relationships after cerebral stab wound in the adult rat. *Brain Res.* **439,** 337–344.
132. Li, M. S. and David, S. (1996) Topical glucocorticoids modulate the lesion interface after cerebral cortical stab wounds in adult rats. *Glia* **18,** 306–318.
133. Sims, T. J. and Gilmore, S. A. (1983) Interactions between intraspinal Schwann cells and the cellular constituents normally occurring in the spinal cord: an ultrastructural study in the irradiated rat. *Brain Res.* **276,** 17–30.
134. Sims, T. J., Gilmore, S. A., Waxman, S. G., and Klinge, E. (1985) Dorsal-ventral differences in the glia limitans of the spinal cord: an ultrastructural study in developing normal and irradiated rats. *J. Neuropathol. Exp. Neurol.* **44,** 415–429.
135. Sims, T. J., Durgun, M. B., and Gilmore, S. A. (1998) Schwann cell invasion of ventral spinal cord: the effect of irradiation on astrocyte barriers. *J. Neuropathol. Exp. Neurol.* **57,** 866–873.
136. Scripter, J. L., Ko, J., Kow, K., Arimura, A., and Ide, C. F. (1997) Regulation by interleukin-1beta of formation of a line of delimiting astrocytes following prenatal trauma to the brain of the mouse. *Exp. Neurol.* **145,** 329–341.
137. Herkenham, M., Lee, H. Y., and Baker, R. A. (1998) Temporal and spatial patterns of c-fos mRNA induced by intravenous interleukin-1: a cascade of non-neuronal cellular activation at the blood-brain barrier. *J. Comp. Neurol.* **400,** 175–196.
138. Yan, Q., Radeke, M. J., Matheson, C. R., Talvenheimo, J., Welcher, A. A., and Feinstein, S. C. (1997) Immunocytochemical localization of TrkB in the central nervous system of the adult rat. *J. Comp. Neurol.* **378,** 135–157.
139. Zhao, L. R., Mattsson, B., and Johansson, B. B. (2000) Environmental influence on brain-derived neurotrophic factor messenger RNA expression after middle cerebral artery occlusion in spontaneously hypertensive rats. *Neuroscience* **97,** 177–184.
140. Franklin, R. J. and Blakemore, W. F. (1995) Reconstruction of the glia limitans by subarachnoid transplantation of astrocyte-enriched cultures. *Microsc. Res. Tech.* **32,** 295–301.
141. Hrabowska, M. (1978) Morphogenesis of cerebral matrix ectopies in human fetus and newborn. *J. Hirnforsch.* **19,** 485–495.
142. Rosen, G. D., Sherman, G. F., Richman, J. M., Stone, L. V., and Galaburda, A. M. (1992) Induction of molecular layer ectopias by puncture wounds in newborn rats and mice. *Dev. Brain Res.* **67,** 285–291.
143. Suzuki, M. and Choi, B. H. (1990) The behavior of the extracellular matrix and the basal lamina during the repair of cryogenic injury in the adult rat cerebral cortex. *Acta Neuropathol.* **80,** 355–361.
144. Choi, B. H. and Matthias, S. C. (1987) Cortical dysplasia associated with massive ectopia of neurons and glial cells within the subarachnoid space. *Acta Neuropathol.* **73,** 105–109.
145. Takada, K., Nakamura, H., Suzumori, K., Ishikawa, T., and Sugiyama, N. (1987) Cortical dysplasia in a 23-week fetus with Fukuyama congenital muscular dystrophy (FCMD). *Acta Neuropathol.* **74,** 300–306.

146. Yamamoto, T., Armstrong, D., Shibata, N., Kanazawa, M., and Kobayashi, M. (1999) Immature astrocytes in Fukuyama congenital muscular dystrophy: an immunohistochemical study. *Pediatr. Neurol.* **20,** 31–37.
147. Saito, Y., Murayama, S., Kawai, M., and Nakano, I. (1999) Breached cerebral glia limitans-basal lamina complex in Fukuyama-type congenital muscular dystrophy. *Acta Neuropathol.* **98,** 330–336.
148. Nakano, I., Funahashi, M., Takada, K., and Toda, T. (1996) Are breaches in the glia limitans the primary cause of the micropolygyria in Fukuyama-type congenital muscular dystrophy (FCMD)? Pathological study of the cerebral cortex of an FCMD fetus. *Acta Neuropathol.* **91,** 313–321.
149. Curran, T. and D'Arcangelo, G. (1998) Role of reelin in the control of brain development. *Brain Res. Rev.* 26, 285–294.
150. Derer, P. (1979) Evidence for the occurrence of early modifications in the 'glia limitans' layer of the neocortex of the reeler mutant mouse. *Neurosci. Lett.* **13,** 195–202.
151. Burette, A., Belliot, G., Albuisson, E., and Romand, R. (1998) Localization of neurotrophin-3-like immunoreactivity in the rat cochlear nucleus. *Microsc. Res. Tech.* **41,** 224–233.
152. Cameron, R. S. and Rakic, P. (1991) Glial cell lineage in the cerebral cortex: a review and synthesis. *Glia* **4,** 124–137.
153. Agduhr, E. (1932) Choroid plexus and ependyma, in *Cytology and Cellular Pathology of the Nervous System*, vol. 2 (Penfield, W., ed.), Paul B. Hoeber, New York, pp. 535–573.
154. Bruni, J. E., Del Bigio, M. R., and Clattenburg, R. E. (1985) Ependyma: normal and pathological. A review of the literature. *Brain Res.* Rev **9,** 1–19.
155. Bruni, J. E. (1998) Ependymal development, proliferation, and functions: a review. *Microsc. Res. Tech.* **41,** 2–13.
156. Del Bigio, M. R. (1995) The ependyma: a protective barrier between brain and cerebrospinal fluid. *Glia* **14,** 1–13.
157. Gabrion, J. B., Herbute, S., Bouille, C., Maurel, D., Kuchler Bopp, S., Laabich, A., and Delaunoy, J. P. (1998) Ependymal and choroidal cells in culture: characterization and functional differentiation. *Microsc. Res. Tech.* **41,** 124–157.
158. Leonhardt, H. (1980) Ependym und circumventriculare Organe, in *Neuroglia I* (Oksche, A., ed.), Springer Verlag, Berlin, pp. 177–544.
159. Mitchell, J. A. (1980) Scanning electron microscopy of brain ventricular surfaces: a bibliography. *Scanning Electron Microsc.* **III,** 475–484.
160. Sarnat, H. B. (1992) Role of the human fetal ependyma. *Pediatr. Neurol.* **8,** 163–178.
161. Sarnat, H. B. (1995) Ependymal reactions to injury. A review. *J. Neuropathol. Exp. Neurol.* **54,** 1–15.
162. Sarnat, H. B. (1998) Histochemistry and immunocytochemistry of the developing ependyma and choroid plexus. *Microsc. Res. Tech.* **41,** 14–28.
163. Scott, D. E., Kozlowski, G. P., and Sheridan, M. N. (1974) Scanning electron microscopy in the ultrastructural analysis of the mammalian cerebral ventricular system. *Int. Rev. Cytol.* **37,** 349–388.
164. Majocha, R. E., Schmidt, R., and Shashoua, V. E. (1982) Cultures of zona ependyma cells of goldfish brain: an immunological study of the synthesis and release of ependymins. *J. Neurosci. Res.* **8,** 331–342.
165. Manthorpe, C. M., Wilkin, G. B., and Wilson, J. E. (1977) Purification of viable ciliated cuboidal ependymal cells from rat brain. *Brain Res.* **134,** 407–415.
166. Weibel, M., Pettmann, B., Artault, J. C., Sensenbrenner, M., and Labourdette, G. (1986) Primary culture of rat ependymal cells in serum-free defined medium. *Dev. Brain Res.* **25,** 199–210.
167. Gee, P., Rhodes, C. H., Fricker, L. D., and Angeletti, R. H. (1993) Expression of neuropeptide processing enzymes and neurosecretory proteins in ependyma and choroid plexus epithelium. *Brain Res.* **617,** 238–248.

168. Perez-Martin, M., Grondona, J. M., Cifuentes, M., Perez-Figares, J. M., Jimenez, J. A., and Fernandez-Llebrez, P. (2000) Ependymal explants from the lateral ventricle of the adult bovine brain: a model system for morphological and functional studies of the ependyma. *Cell Tissue Res.* **300,** 11–19.
169. Gould, S. J., Howard, S., and Papadaki, L. (1990) The development of ependyma in the human fetal brain: an immunohistochemical and electron microscopic study. *Dev. Brain Res.* **55,** 255–267.
170. Wechsler, W. and Meller, K. (1967) Electron microscopy of neuronal and glial differentiation in the developing brain of the chick. *Progr. Brain Res.* **26,** 93–144.
171. Malinsky, J. (1968) Fine structure of ependyma in lateral ventricles of human brain. *Acta Univ Palack Olomucensis* **48,** 65–72.
172. Mollgard, K., Balshev, Y., Lauritzen, B., and Saunders, N. R. (1987) Cell junctions and membrane specializations in the ventricular zone (germinal matrix) of the developing sheep brain: a CSF-brain barrier. *J. Neurocytol.* **16,** 433–444.
173. Takahashi, T., Nowakowski, R. S., and Caviness, V. S. (1993) Cell cycle parameters and patterns of nuclear movement in the neocortical proliferative zone of the fetal mouse. *J. Neurosci.* **13,** 820–833.
174. Grove, E. A., Williams, B. P., Li, D. Q., Hajihosseini, M., Friedrich, A., and Price, J. (1993) Multiple restricted lineages in the embryonic rat cerebral cortex. *Development* **117,** 553–561.
175. Walker, A. G., Chapman, J., Bruce, C. B., and Rumsby, M. G. (1984) Immunocytochemical characterisation of cell cultures grown from dissociated 1-2-day post-natal rat cerebral tissue. A developmental study. *J. Neuroimmunol.* **7,** 1–20.
176. Bernocchi, G., Scherini, E., Giacometti, S., and Mares, V. (1990) Premitotic DNA synthesis in the brain of the adult frog (Rana esculenta L.): an autoradiographic ^{3}H-thymidine study. *Anat. Rec.* **228,** 461–470.
177. Schipper, H. M., Yang, G., and Wang, E. (1994) Expression of terminen, a senescence-related cytoplasmic protein, in the aging rat brain. *Brain Res.* **635,** 224–230.
178. Schipper, H. M. and Wang, E. (1990) Expression of statin, a non-proliferation-dependent nuclear protein, in the postnatal rat brain: evidence for substantial retention of neuroglial proliferative capacity with aging. *Brain Res.* **528,** 250–258.
179. Aikawa, H. and Suzuki, K. (1986) Ultrastructural evidence of mitotic ependymal cells in 6-aminonicotinamide-treated suckling mice. *Acta Neuropathol.* **70,** 71–74.
180. Bernstein, J. J. (1986) Ependyma formation in adult rat spinal cord after fetal cerebral cortex homografts. *J. Neurosci. Res.* **15,** 481–490.
181. Del Bigio, M. R. and Bruni, J. E. (1986) Reaction of rabbit lateral periventricular tissue to shunt tubing implants. *J. Neurosurg.* **64,** 932–940.
182. Del Bigio, M. R. and Bruni, J. E. (1988) Periventricular pathology in hydrocephalic rabbits before and after shunting. *Acta Neuropathol.* **77,** 186–195.
183. Namiki, J. and Tator, C. H. (1999) Cell proliferation and nestin expression in the ependyma of the adult rat spinal cord after injury. *J. Neuropathol. Exp. Neurol.* **58,** 489–498.
184. Collins, P. and Fairman, S. (1990) Repair of the ependyma in hydrocephalic brains. *Neuropathol. Appl. Neurobiol.* **16,** 45–56.
185. Johansson, C. B., Momma, S., Clarke, D. L., Risling, M., Lendahl, U., and Frisen, J. (1999) Identification of a neural stem cell in the adult mammalian central nervous system. *Cell* **96,** 25–34.
186. Barres, B. A. (1999) A new role for glia: generation of neurons! *Cell* **97,** 667–670.
187. Kuhn, H. G. and Svendsen, C. N. (1999) Origins, functions, and potential of adult neural stem cells. *Bioessays* **21,** 625–630.
188. Peschanski, M. (1999) Ependymocyte: a well hidden neural progenitor. *M S Med. Sci.* **15,** 751–753.

189. Chiasson, B. J., Tropepe, V., Morshead, C. M., and van der Kooy, D. (1999) Adult mammalian forebrain ependymal and subependymal cells demonstrate proliferative potential, but only subependymal cells have neural stem cell characteristics. *J. Neurosci.* **19,** 4462–4471.
190. Birecree, E., King, L. E., and Nanney, L. B. (1991) Epidermal growth factor and its receptor in the developing human nervous system. *Dev. Brain Res.* **60,** 145–154.
191. Matsuo, A., Tooyama, I., Isobe, S., Oomura, Y., Akiguchi, I., Hanai, K., Kimura, J., and Kimura, H. (1994) Immunohistochemical localization in the rat brain of an epitope corresponding to the fibroblast growth factor receptor-1. *Neuroscience* **60,** 49–66.
192. Mudo, G., Persson, H., Timmusk, T., Funakoshi, H., Bindoni, M., and Belluardo, N. (1993) Increased expression of trkB and trkC messenger RNAs in the rat forebrain after focal mechanical injury. *Neuroscience* **57,** 901–912.
193. Takami, K., Kiyota, Y., Iwane, M., Miyamoto, M., Tsukuda, R., Igarashi, K., et al. (1993) Upregulation of fibroblast growth factor-receptor messenger RNA in rat brain following transient forebrain ischemia. *Exp. Brain Res.* **97,** 185–194.
194. O'Hara, C. M. and Chernoff, E. A. G. (1994) Growth factor modulation on injury-reactive ependymal cell proliferation and migration. *Tissue Cell* **26,** 599–611.
195. Beer, M. S., Stanton, J. A., Salim, K., Rigby, M., Heavens, R. P., Smith, D., and McAllister, G. (2000) EDG receptors as a therapeutic target in the nervous system. *Ann. NY Acad. Sci.* **905,** 118–131.
196. Cuevas, P., Gimenez-Gallego, G., Martinez-Murillo, R., and Carcellar, F. (1991) Immunohistochemical localization of basic fibroblast growth factor in ependymal cells of the rat lateral and third ventricles. *Acta Anat.* **141,** 307–310.
197. Frautschy, S. A., Walicke, P. A., and Baird, A. (1991) Localization of basic fibroblast growth factor and its mRNA after CNS injury. *Brain Res.* **553,** 291–299.
198. Oomura, Y., Sasaki, K., and Li, A. J. (1993) Memory facilitation educed by food intake. *Physiol. Behav.* **54,** 493–498.
199. Ikeda, T., Xia, X. Y., Xia, Y. X., Ikenoue, T., and Choi, B. H. (1999) Expression of glial cell line-derived neurotrophic factor in the brain and cerebrospinal fluid of the developing rat. *Int. J. Dev. Neurosci.* **17,** 681–691.
200. Kondo, Y., Nakanishi, T., Takigawa, M., and Ogawa, N. (1999) Immunohistochemical localization of connective tissue growth factor in the rat central nervous system. *Brain Res.* **834,** 146–151.
201. Stenvers, K. L., Zimmermann, E. M., Gallagher, M., and Lund, P. K. (1994) Expression of insulin-like growth factor binding protein-4 and -5 mRNAs in adult rat forebrain. *J. Comp. Neurol.* **339,** 91–105.
202. Cardona-Gomez, G. P., Chowen, J. A., and Garcia-Segura, L. M. (2000) Estradiol and progesterone regulate the expression of insulin-like growth factor-I receptor and insulin-like growth factor binding protein-2 in the hypothalamus of adult female rats. *J. Neurobiol.* **43,** 269–281.
203. Walter, H. J., Berry, M., Hill, D. J., Cwyfan-Hughes, S., Holly, J. M., and Logan, A. (1999) Distinct sites of insulin-like growth factor (IGF)-II expression and localization in lesioned rat brain: possible roles of IGF binding proteins (IGFBPs) in the mediation of IGF-II activity. *Endocrinology* **140,** 520–532.
204. Didier-Bazes, M., Voutsinos, B., Aguera, M., Peyron, C., Akaoka, H., and Belin, M. F. (1997) Specific potentialities of embryonic rat serotonergic neurons to innervate different periventricular targets in the adult brain. *J. Comp. Neurol.* **382,** 29–45.
205. Yamadori, T. and Nara, K. (1979) The directions of ciliary beat on the wall of the lateral ventricle and the currents of the cerebrospinal fluid in the brain ventricles. *Scanning Electron Microsc.* **III,** 335–340.
206. Roth, Y., Kimhi, Y., Edery, H., Aharonson, E., and Priel, Z. (1985) Ciliary motility in brain ventricular system and trachea of hamsters. *Brain Res.* **330,** 291–297.

207. Gonzalez Santander, R., Martinez Cuandrado, G., Rubio Saez, M., Bujan Varela, J., and Laraxa Sole, A. (1984) Ultrastructural basis of the coordination of ciliary movement. Contractile function of the interconnecting cross-striated roots in ependymal epithelium. *Acta Anat.* **118,** 82–90.
208. Brody, S. L., Yan, X. H., Wuerffel, M. K., Song, S. K., and Shapiro, S. D. (2000) Ciliogenesis and left-right axis defects in forkhead factor HFH-4-null mice. *Am. J. Respir. Cell Mol. Biol.* **23,** 45–51.
209. O'Callaghan, C., Achaval, M., Forsythe, I., and Barry, P. W. (1995) Brain and respiratory cilia: the effect of temperature. *Biol. Neonate* **68,** 394–397.
210. O'Callaghan, C., Sikand, K., and Rutman, A. (1999) Respiratory and brain ependymal ciliary function. *Pediatr. Res.* **46,** 704–707.
211. Nakamura, Y. and Sato, K. (1993) Role of disturbance of ependymal ciliary movement in development of hydrocephalus in rats. *Child's Nerv. Syst.* **9,** 65–71.
212. Hirst, R. A., Rutman, A., Sikand, K., Andrew, P. W., Mitchell, T. J., et al. (2000) Effect of pneumolysin on rat brain ciliary function: comparison of brain slices with cultured ependymal cells. *Pediatr. Res.* **47,** 381–384.
213. Hirst, R. A., Sikand, K. S., Rutman, A., Mitchell, T. J., Andrew, P. W., and O'Callaghan, C. (2000) Relative roles of pneumolysin and hydrogen peroxide from *Streptococcus pneumoniae* in inhibition of ependymal ciliary beat frequency. *Infect Immun.* **68,** 1557–1562.
214. Mohammed, B. J., Mitchell, T. J., Andrew, P. W., Hirst, R. A., et al. (1999) The effect of the pneumococcal toxin, pneumolysin on brain ependymal cilia. *Microb. Pathol.* **27,** 303–309.
215. Deissler, H., Blass-Kampmann, S., Kindler-Rohrborn, A., Meyer, H. E., and Rajewsky, M. F. (1996) Characterization of rat NCA/CD9 cell surface antigen and its expression by normal and malignant neural cells. *J. Neurosci. Res.* **43,** 664–674.
216. Blass-Kampmann, S., Kindler-Rohrborn, A., Deissler, H., D'Urso, D., and Rajewsky, M. F. (1997) In vitro differentiation of neural progenitor cells from prenatal rat brain: common cell surface glycoprotein on three glial cell subsets. *J. Neurosci. Res.* **48,** 95–111.
217. Kent, M. N., Alvarez, F. J., Ng, A. K., and Rote, N. S. (2000) Ultrastructural localization of monoclonal antiphospholipid antibody binding to rat brain. *Exp. Neurol.* **163,** 173–179.
218. Scholl, F. G., Gamallo, C., Vilaro, S., and Quintanilla, M. (1999) Identification of PA2. 26 antigen as a novel cell-surface mucin-type glycoprotein that induces plasma membrane extensions and increased motility in keratinocytes. *J. Cell Sci.* **112,** 4601–4613.
219. Shinohara, H., Asano, T., Kato, K., Kameshima, T., and Semba, R. (1998) Localization of a G protein G(i2) in the cilia of rat ependyma, oviduct and trachea. *Eur. J. Neurosci.* **10,** 699–707.
220. Acarin, L., Vela, J. M., Gonzalez, B., and Castellano, B. (1994) Demonstration of poly-N-acetyl lactosamine residues in ameboid and ramified microglial cells in rat brain by tomato lectin binding. *J. Histochem. Cytochem.* **42,** 1033–1041.
221. Adam, E., Dziegielewska, K. M., Saunders, N. R., and Schumacher, U. I. (1993) Neuraminic acid specific lectins as markers of early cortical plate neurons. *Int. J. Dev. Neurosci.* **11,** 451–460.
222. Bartlett, P. F., Noble, M. D., Pruss, R. M., Raff, M. C., Raltray, S., and Williams, C. A. (1981) Rat neural antigen-2 (Ran-2): a cell surface antigen on astrocytes, ependymal cells, Muller cells and leptomeninges defined by a monoclonal antibody. *Brain Res.* **204,** 339–351.
223. Damjanov, J. and Black, P. (1987) Lectin binding sites on the luminal surface of ependymal cells of the rat spinal cord: implications for neuropathological investigation. *Neurosurgery* **20,** 722–725.
224. Debbage, P. L. (1996) A systematic histochemical investigation in mammals of the dense glycocalyx glycosylations common to all cells bordering the interstitial fluid compartment of the brain. *Acta Histochem.* **98,** 9–28.

225. Kaneko, Y., Iwaki, T., Matsushima, T., and Fukui, M. (1991) Comprehensive lectin histochemistry of normal and neoplastic human choroid plexus cells: alternation of lectin-binding patterns through neoplastic transformation. *Acta Neuropathol.* **82,** 127–133.
226. Korte, G. E. and Rosenbluth, J. (1982) Anionic sites on the surface of frog ependymal astrocytes and mouse ependymal cells. *Anat Rec.* **204,** 95–100.
227. Kuratsu, J., Sueyoshi, N., Mihara, Y., and Ushio, Y. (1990) Localization and significance of peanut agglutinin-binding sites on ependymoma cells. *Acta Neuropathol.* **79,** 634–639.
228. Mai, J. K. and Reifenberger, G. (1988) Distribution of the carbohydrate epitope 3-fucosyl-N-acetyl-lactosamine (FAL) in the human brain. *J. Chem. Neuroanat.* **1,** 255–285.
229. Momoi, T., Momoi, M. Y., and Kurata, T. (1986) Peanut agglutinin receptor is a marker of myelin in rat brain. Developmental changes in its distribution. *J. Neurochem.* **46,** 229–234.
230. Rhodes, R. H. (1987) Ultrastructure of complex carbohydrates of rodent and monkey ependymal glycocalyx and meninges. *Am. J. Anat.* **179,** 369–384.
231. Roth, J. (1993) Cellular sialoglycoconjugates: a histochemical perspective. *Histochem. J.* **25,** 687–710.
232. Uematsu, Y., Rojas-Corona, R. R., Llena, J. F., and Hirano, A. (1989) Distribution of epithelial membrane antigen in normal and neoplastic human ependyma. *Acta Neuropathol.* **78,** 325–328.
233. Paulson, J. C. (1989) Glycoproteins: what are the sugar chains for? *Trends Biochem. Sci.* **14,** 272–276.
234. Richards, J. G., Lorez, H. P., Columbo, V. E., Guggenheim, R., and Kiss, D. (1980) Supraependymal nerve fibres in human brain: correlative transmission and scanning electron microscopical and fluorescence histochemical studies. *Neuroscience* **5,** 1489–1502.
235. Mikkelsen, J. D., Hay-Schmidt, A., and Larsen, P. J. (1997) Central innervation of the rat ependyma and subcommissural organ with special reference to ascending serotoninergic projections from the raphe nuclei. *J. Comp. Neurol.* **384,** 556–568.
236. Ugrumov, M. V., Taxi, J., Mitskevich, M. S., Aluison, M., and Tramu, G. (1985) Immunocytochemical and radioautographic study of serotonin projections to cerebral ventricles of perinatal rats. *Dev. Brain Res.* **18,** 225–230.
237. Voutsinos, B., Chouaf, L., Mertens, P., Ruiz-Flandes, P., Joubert, Y., Belin, M. F., and Didier-Bazes, M. (1994) Tropism of serotonergic neurons towards glial targets in the rat ependyma. *Neuroscience* **59,** 663–672.
238. Goldman, S. A., Lemman, V., and Chin, S. S. M. (1993) Migration of newly generated neurons upon ependymally derived radial guide cells in explant cultures of the adult songbird forebrain. *Glia* **8,** 150–160.
239. Robinson, S. R., Noone, D. F., and Odowd, B. S. (1996) Ependymocytes and supraependymal axons in rat brain contain glutamate. *Glia* **17,** 345–348.
240. Rodrigo, J., Riveros-Moreno, V., Bentura, M. L., Uttenthal, L. O., Higgs, E. A., Fernandez, A. P., et al. (1997) Subcellular localization of nitric oxide synthase in the cerebral ventricular system, subfornical organ, area postrema, and blood vessels of the rat brain. *J. Comp. Neurol.* **378,** 522–534.
241. Meller, S. T. and Dennis, B. J (1993) A scanning and transmission electron microscopic analysis of the cerebral aqueduct in the rabbit. *Anat. Rec.* **237,** 124–140.
242. Didier-Bazes, M., Voutsinos, B., Chouaf, L., Joubert, Y., and Belin, M. F. (1993) Tropism between serotonergic neurons and their glial targets in a graft model of the third periventricular region. *J. Neurochem.* **61(Suppl.),** S171.
243. Bjugn, R., Haugland, H. K., and Flood, P. R. (1988) Ultrastructure of the mouse spinal cord ependyma. *J. Anat.* **160,** 117–125.
244. Cupedo, R. N. J. and de Weerd, H. (1988) Morphology of ventricular wall surfaces following neurotoxin induced degeneration of supraependymal afferents. *Acta Morphol. Neerl. Scand.* **26,** 17–32.

245. Ribas, J. L. (1977) Morphological evidence for a possible functional role of supraependymal nerves on ependyma. *Brain Res.* **125,** 362–368.
246. Mathew, T. C. (1999) Association between supraependymal nerve fibres and the ependymal cilia of the mammalian brain. *Anat. Histol. Embryol. J. Vet. Med. C* **28,** 193–197.
247. Keith, I. M. and Brownfield, M. S. (1985) Evidence for presence of kynurenine in lung and brain of neonate hamsters. *Histochemistry* **83,** 465–471.
248. Vigh, B. and Vigh-Teichmann, I. (1998) Actual problems of the cerebrospinal fluid-contacting neurons. *Microsc. Res. Tech.* **41,** 57–83.
249. Sarnat, H. B. (1992) Regional differentiation of the human fetal ependyma: immunocytochemical markers. *J. Neuropathol. Exp. Neurol.* **51,** 58–75.
250. Lauro, G. M., Fonti, R., and Margotta, V. (1991) Phylogenetic evolution of intermediate filament associated proteins in ependymal cells of several adult poikilotherm vertebrates. *J. Hirnforsch.* **32,** 257–261.
251. Kasper, M., Perry, G., and Stosiek, P. (1991) Cytokeratin expression in human spinal meninges and ependymal cells. *J. Hirnforsch.* **32,** 19–26.
252. Errante, L. D., Wiche, G., and Shaw, G. (1994) Distribution of plectin, an intermediate filament-associated protein, in the adult rat central nervous system. *J. Neurosci. Res.* **37,** 515–528.
253. Groschel-Stewart, U., Unsicker, K., and Leonhardt, H. (1977) Immunohistochemical demonstration of contractile proteins in astrocytes, marginal glial and ependymal cells. *Cell Tissue Res.* **180,** 133–137.
254. Brightman, M. W. and Reese, T. S. (1969) Junctions between intimately apposed cell membranes in the vertebrate brain. *J. Cell Biol.* **40,** 648–677.
255. Kawamata, S., Stumpf, W. E., and Bidmon, H. J. (1995) Adhesion and fusion of ependyma in rat brain. *Acta Anat.* **152,** 205–214.
256. Petrov, T., Howarth, A. G., Krukoff, T. L., and Stevenson, B. R. (1994) Distribution of the tight junction-associated protein ZO-1 in circumventricular organs of the CNS. *Mol. Brain Res.* **21,** 235–246.
257. Lippoldt, A., Jansson, A., Kniesel, U., Andbjer, B., Andersson, A., Wolburg, H., Fuxe, K., and Haller, H. (2000) Phorbol ester induced changes in tight and adherens junctions in the choroid plexus epithelium and in the ependyma. *Brain Res.* **854,** 197–206.
258. Murphy-Erdosh, C., Napolitano, E. W., and Reichardt, L. F. (1994) The expression of B-cadherin during embryonic chick development. *Dev. Biol.* **161,** 107–125.
259. Smith, G. M. and Shine, H. D. (1992) Immunofluorescent labeling of tight junctions in the rat brain and spinal cord. *Int. J. Dev. Neurosci.* **10,** 387–392.
260. Gerhardt, H., Wolburg, H., and Redies, C. (2000) N-cadherin mediates pericytic-endothelial interaction during brain angiogenesis in the chicken. *Dev. Dyn.* **218,** 472–479.
261. Kuchler, S., Graff, M. N., Gobaille, S., Vincendon, G., Roche, A. C., Delaunoy, J. P., Monsigny, M., and Zanetta, J. P. (1994) Mannose dependent tightening of the rat ependymal cell barrier. In vivo and in vitro study using neoglycoproteins. *Neurochem. Int.* **24,** 43–55.
262. Perraud, F., Kuchler, S., Gobaille, S., Labourdette, G., Vincendon, G., and Zanetta, J. P. (1988) Endogenous lectin CSL is present on the membrane of cilia of rat brain ependymal cells. *J. Neurocytol.* **17,** 745–751.
263. Brightman, M. W. and Reese, T. S. (1975) Membrane specializations of ependymal cells and astrocytes, in *The Nervous System, vol. I. The Basic Neurosciences* (Tower, D. B., ed.), Raven, New York, pp. 267–271.
264. Mack, A., Neuhaus, J., and Wolburg, H. (1987) Relationship between orthogonal arrays of particles and tight junctions as demonstrated in cells of the ventricular wall of the rat brain. *Cell Tissue Res.* **248,** 619–625.
265. Mugnaini, E. (1986) Cell junctions of astrocytes, ependyma, and related cells in the mammalian central nervous system, with emphasis on the hypothesis of a generalized func-

tional syncytium of supporting cells, in *Astrocytes* (Fedoroff, S. and Vernadakis, A., eds.), Academic, Orlando, pp. 329–371.

266. Dermietzel, R., Traub, O., Hwang, T. K., Beyer, E., Bennet, M. V. L., Spray, D. C., and Willecke, K. (1989) Differential expression of three gap junction proteins in developing and mature brain tissues. *Proc. Natl. Acad. Sci. USA* **86,** 10,148–10,152.
267. Yamamoto, T., Vukelic, J., Hertzberg, E. L., and Nagy, J. I. (1992) Differential anatomical and cellular patterns of connexin 43 expression during postnatal development of rat brain. *Dev. Brain Res.* **66,** 165–180.
268. Matsumoto, A., Arai, Y., Urano, A., and Hyodo, S. (1991) Cellular localization of gap junction messenger RNA in the neonatal rat brain. *Neurosci. Lett.* **124,** 225–228.
269. Bouille, C., Mesnil, M., Barriere, H., and Gabrion, J. (1991) Gap junctional intercellular communication between cultured ependymal cells, revealed by Lucifer yellow CH transfer and freeze-fracture. *Glia* **4,** 25-36.
270. Connors, B. W. and Ransom, B. R. (1987) Electrophysiological properties of ependymal cells (radial glia) in dorsal cortex of the turtle, Pseudemys scripta. *J. Physiol.* **385,** 287–306.
271. Pollay, M. and Curl, F. (1967) Secretion of cerebrospinal fluid by the ventricular ependyma of the rabbit. *Am. J. Physiol.* **213,** 1031–1038.
272. Barres, B. A., Chun, L. L. Y., and Corey, D. P. (1989) Glial and neuronal forms of the voltage-dependent sodium channel: characteristics and cell-type distribution. *Neuron* **2,** 1375–1388.
273. Cardy, J. and Firth, J. A. (1993) Adenosine triphosphate-lead histochemical reactions in ependymal epithelia of murine brains do not represent calcium transport adenosine triphosphate. *Histochem. J.* **25,** 319–324.
274. Nielson, S., Smith, B. L., Christensen, E. I., and Agre, P. (1993) Distribution of the aquaporin CHIP in secretory and resorptive epithelia and capillary endothelia. *Proc. Natl. Acad. Sci. USA* **90,** 7275–7279.
275. Wenzel, J., Kunde, D., David, E., and Hecht, A. (1970) Enzymhistochemische befunde am Subkommissuralorgan, am Ependym und an den Plexus Choroidei des Meerschweinchengehirnes (Cavia porcellus L.). *Z Mikrosk Anat. Forsch.* **82,** 242–263.
276. Zalc, B., Collet, A., Monge, M., Ollier-Hartmann, M. P., Jacque, C., Hartmann, L., and Baumann, N. A. (1984) Tamm-Horsfall protein, a kidney marker, is expressed on brain sulfogalactosylceramide-positive astroglial structures. *Brain Res.* **291,** 182–187.
277. Hincke, M. T., Nairn, A. C., and Staines, W. A. (1995) Cystic fibrosis transmembrane conductance regulator is found within brain ventricular epithelium and choroid plexus. *J. Neurochem.* **64,** 1662–1668.
278. Stonehouse, A. H., Pringle, J. H., Norman, R. I., Stanfield, P. R., Conley, E. C., and Brammar, W. J. (1999) Characterisation of Kir2. 0 proteins in the rat cerebellum and hippocampus by polyclonal antibodies. *Histochem. Cell Biol.* **112,** 457–465.
279. Brightman, M. W., Shivers, R. R., and Prescott, L. (1975) Morphology of the walls around fluid compartments in nervous tissue, in *Fluid Environment of the Brain* (Cserr, H. F., Fenstermacher, J. D., and Fencl, V., eds.), Academic, New York, pp. 3–29.
280. Rosenberg, G. A., Kyner, W. T., Fenstermacher, J. D., and Patlak, C. S. (1986) Effect of vasopressin on ependymal and capillary permeability to tritiated water in cat. *Am. J. Physiol.* **251,** F485–F489.
281. Sorensen, P. S. (1986) Studies of vasopressin in the human cerebrospinal fluid. *Acta Neurol. Scand.* **74,** 81–102.
282. Dubois-Dauphin, M., Pevet, P., Tribollet, E., and Dreifuss, J. J. (1990) Vasopressin in the brain of the golden hamster: the distribution of vasopressin binding sites and of immunoreactivity to the vasopressin-related glycopeptide. *J. Comp. Neurol.* **300,** 535–548.
283. Lepetit, P., Ferre-Montagne, M., Gay, N., Belin, M. F., and Bobillier, P. (1993) Vasopressin mRNA in the cerebellum and circumventricular organs: a quantitative in situ hybridization study. *Neurosci. Lett.* **159,** 171–174.

284. Gehlert, D. R., Gackenheimer, S. L., and Schober, D. A. (1991) Autoradiographic localization of subtypes of angiotensin II antagonist binding in the rat brain. *Neuroscience* **44,** 501–514.
285. Hurbin, A., Orcel, H., Ferraz, C., Moos, F. C., and Rabie, A. (2000) Expression of the genes encoding the vasopressin-activated calcium-mobilizing receptor and the dual angiotensin II/vasopressin receptor in the rat central nervous system. *J. Neuroendocrinol.* **12,** 677–684.
286. Chai, S. Y., Mendelsohn, F. A. O., and Paxinos, G. (1987) Angiotensin converting enzyme in rat brain visualized by quantitative in vitro autoradiography. *Neuroscience* **20,** 615–627.
287. Defendini, R., Zimmerman, E. A., Weare, J. A., Alhenc-Gelas, F., and Erdos, E. G. (1983) Angiotensin-converting enzyme in epithelial and neuroepithelial cells. *Neuroendocrinology* **37,** 32–40.
288. Stornetta, R. L., Hawela-Johnson, C. L., Guyenet, P. G., and Lynch, K. R. (1988) Astrocytes synthesize angiotensinogen in brain. *Science* **242,** 1444–1446.
289. Ghodsi, A., Stein, C., Derksen, T., Martins, I., Anderson, R. D., and Davidson, B. L. (1999) Systemic hyperosmolality improves beta-glucuronidase distribution and pathology in murine MPS VII brain following intraventricular gene transfer. *Exp. Neurol.* **160,** 109–116.
290. Fishman, R. A. (1992) *Cerebrospinal Fluid in Diseases of the Nervous System*, 2nd ed., W. B. Saunders, Philadelphia, pp. 40–42.
291. Aird, R. B. (1984) A study of intrathecal, cerebrospinal fluid-to-brain exchange. *Exp. Neurol.* **86,** 342–358.
292. Brightman, M. W. (1965) The distribution within the brain of ferritin injected into cerebrospinal fluid compartments. I. Ependymal distribution. *J. Cell Biol.* **26,** 99–123.
293. Fossan, G., Cavanaugh, M. E., Evans, C. A. N., Malinowska, D. H., Mollgard, K., Reynolds, M. L., and Saunders, N. R. (1985) CSF-brain permeability in the immature sheep fetus: a CSF-brain barrier. *Dev. Brain Res.* **18,** 113–124.
294. Broadwell, R. D. and Sofroniew, M. V. (1993) Serum proteins bypass the blood-brain fluid barriers for extracellular entry to the central nervous system. *Exp. Neurol.* **120,** 245–263.
295. Graff, M. N., Laabich, A., Dunel-Erb, S., Sensenbrenner, M., and Delaunoy, J. P. (1993) Protein endocytosis by rat astrocytes and ependymal cells in primary culture. *Circulat. Metabol. Cerveau* **10,** 11–24.
296. Bach-y-Rita, P. (1993) Neurotransmission in the brain by diffusion through the extracellular fluid: a review. *NeuroReport* **4,** 343–350.
297. Begley, D. J. and Chain, D. G. (1992) Mechanisms regulating peptide levels in the cerebrospinal fluid, in *Barriers and Fluids of the Eye and Brain* (Segal, M. B., ed.), CRC, Boca Raton, FL, pp. 82–105.
298. Zheng, W., Perry, D. F., Nelson, D. L., and Aposhian, H. V. (1991) Choroid plexus protects cerebrospinal fluid against toxic metals. *FASEB J.* **5,** 2188–2193.
299. Blaauwgeers, H. G. T., Smitt, P. A. E. S., De Jong, J. M. B. V., and Troost, D. (1993) Distribution of metallothionein in the human central nervous system. *Glia* **8,** 62–70.
300. Dincer, Z., Haywood, S., and Jasani, B. (1999) Immunocytochemical detection of metallothionein (MT1 and MT2) in copper- enhanced sheep brains. *J. Comp. Pathol.* **120,** 29–37.
301. Gong, Y. H. and Elliott, J. L. (2000) Metallothionein expression is altered in a transgenic murine model of familial amyotrophic lateral sclerosis. *Exp. Neurol.* **162,** 27–36.
302. Suzuki, K., Nakajima, K., Otaki, N., Kimura, M., Kawaharada, U., Uehara, K., et al. (1994) Localization of metallothionein in aged human brain. *Pathol. Int.* **44,** 20–26.
303. Itano, Y., Noji, S., Koyama, E., Taniguchi, S., Taga, N., Takahashi, T., Ono, K., and Kosaka, F. (1991) Bacterial endotoxin-induced expression of metallothionein genes in rat brain as revealed by *in situ* hybridization. *Neurosci. Lett.* **124,** 13–16.

304. Nishimura, N., Nishimura, H., Ghaffer, A., and Tohyama, C. (1992) Localization of metallothionein in the brain of rat and mouse. *J. Histochem. Cytochem.* **40,** 309–315.
305. Benkovic, S. A. and Connor, J. R. (1993) Ferritin, transferrin, and iron in selected regions of the adult and aged rat brain. *J. Comp. Neurol.* **338,** 97–113.
306. Connor, J. R. and Benkovic, S. A. (1992) Iron regulation in the brain: histochemical, biochemical, and molecular considerations. *Ann. Neurol.* **32(Suppl.),** S51–S61.
307. Moos, T. and Mollgard, K. (1993) A sensitive post-DAB enhancement technique for demonstration of iron in the central nervous system. *Histochemistry* **99,** 471–475.
308. Halliwell, B. (1992) Reactive oxygen species and the central nervous system. *J. Neurochem.* **59,** 1609–1623.
309. Sarafian, T. A., Verity, M. A., Vinters, H. V., Shih, C. C., Shi, L., Ji, X. D., Dong, L., and Shau, H. (1999) Differential expression of peroxiredoxin subtypes in human brain cell types. *J. Neurosci. Res.* **56,** 206–212.
310. Takagi, Y., Nakamura, T., Nishiyama, A., Nozaki, K., Tanaka, T., Hashimoto, N., and Yodoi, J. (1999) Localization of glutaredoxin (thioltransferase) in the rat brain and possible functional implications during focal ischemia. *Biochem. Biophys. Res. Commun.* **258,** 390–394.
311. Ewing, J. F., Haber, S. N., and Maines, M. D. (1992) Normal and heat-induced patterns of expression of heme oxygenase-1 (HSP 32) in rat brain: hyperthermia causes rapid induction of mRNA and protein. *J. Neurochem.* **58,** 1140–1149.
312. Mori, K., Miyake, H., Kurisaka, M., and Sakamoto, T. (1993) Immunohistochemical localization of superoxide dismutase in congenital hydrocephalic rat brain. *Child's Nerv. Syst.* **9,** 136–141.
313. Zhang, P., Anglade, P., Hirsch, E. C., Javoy-Agid, F., and Agid, Y. (1994) Distribution of manganese-dependent superoxide dismutase in the human brain. *Neuroscience* **61,** 317–330.
314. Didier, M., Harandi, M., Aguera, M., Bancel, B., Tardy, M., Fages, C., et al. (1986) Differential immunocytochemical staining for glial fibrillary acidic (GFA) protein, S100 protein and glutamine synthetase in the rat subcommissural organ, non-specialized ventricular ependyma and adjacent neuropil. *Cell Tissue Res.* **245,** 343–351.
315. Yamashita, N., Ilg, E. C., Schafer, B. W., Heizmann, C. W., and Kosaka, T. (1999) Distribution of a specific calcium-binding protein of the S100 protein family, S100A6 (calcyclin), in subpopulations of neurons and glial cells of the adult rat nervous system. *J. Comp. Neurol.* **404,** 235–257.
316. Frantz, G. D. and Tobin, A. J. (1994) Cellular distribution of calbindin D28k mRNAs in the adult mouse brain. *J. Neurosci. Res.* **37,** 287–302.
317. Rogers, J. H. (1992) Immunohistochemical markers in rat brain: colocalization of calretinin and calbindin-D28K with tyrosine hydroxylase. *Brain Res.* **587,** 203–210.
318. Sequier, J. M., Hunziker, W., Andressen, C., and Celio, M. R. (1990) Calbindin D-28k protein and mRNA localization in the rat brain. *Eur. J. Neurosci.* **2,** 1118–1126.
319. Halleux, P., Schurmans, S., Schiffman, S. N., Lecocq, R., Conreur, J. L., Dumont, J., and Vanderhaeghen, J. J. (1998) Calcium binding protein calcyphosine in dog central astrocytes and ependymal cells and in peripheral neurons. *J. Chem. Neuroanat.* **15,** 239–250.
320. Seto-Ohshima, A. (1994) Calcium-binding proteins in the central nervous system. *Acta Histochem. Cytochem.* **27,** 93–106.
321. Miller, K. K., Verma, A., Snyder, S. H., and Ross, C. A. (1991) Localization of an endoplasmic reticulum calcium ATPase mRNA in rat brain by *in situ* hybridization. *Neuroscience* **43,** 1–9.
322. Borke, J. L., Caride, A. J., Yaksh, T. L., Penniston, J. T., and Kumar, R. (1989) Cerebrospinal fluid calcium homeostasis: evidence for a plasma membrane Ca^{2+} pump in mammalian choroid plexus. *Brain Res.* **489,** 355–360.
323. Dragunow, M. and Faull, R. L. M. (1989) Rolipram induces c-fos protein-like immunoreactivity in ependymal and glial-like cells in adult rat brain. *Brain Res.* **501,** 382–388.

324. Dragunow, M., Gouldring, M., Faull, R. L. M., Ralph, R., Mee, E., and Frith, R. (1990) Induction of c-fos mRNA and proteins in neurons and glia after traumatic brain injury: pharmacological characterization. *Exp. Neurol.* **107,** 236–248.
325. Wessel, T. C., Joh, T. H., and Volpe, B. R. (1991) *In situ* hybridization analysis of c-fos and c-jun expression in the rat brain following transient forebrain ischemia. *Brain Res.* **567,** 231–240.
326. Wang, L., Martinez, V., Vale, W., and Tache, Y. (2000) Fos induction in selective hypothalamic neuroendocrine and medullary nuclei by intravenous injection of urocortin and corticotropin-releasing factor in rats. *Brain Res.* **855,** 47–57.
327. Dragunow, M. and Hughes, P. (1993) Differential expression of immediate-early proteins in non-nerve cells after focal brain injury. *Int. J. Dev. Neurosci.* **11,** 249–255.
328. Munell, F., Burke, R. E., Bandele, A., and Gubits, R. M. (1994) Localization of c-fos, c-jun, and hsp70 mRNA expression in brain after neonatal hypoxia-ischemia. *Dev. Brain Res.* **77,** 111–121.
329. Sanchez, F., Hernandez, G., Rubio, M., Santos, M., Carretero, J., Riesco, J., Juanes, J., and Vazquez, R. (1995) Adrenalectomy increases the glial fibrillary acidic immunoreactive-elements in the ventricular ependyma and adjacent neuropil of the rat third ventricle. *Acta Histochem.* **97,** 141–149.
330. Kiessling, M. and Gass, P. (1994) Stimulus-transcription coupling in focal cerebral ischemia. *Brain Pathol.* **4,** 77–83.
331. Nowak, T. S. and Jacewicz, M. (1994) The heat shock/stress response in focal cerebral ischemia. *Brain Pathol.* **4,** 67–76.
332. Nakae, D., Akai, H., Kishida, H., Kusuoka, O., Tsutsumi, M., and Konishi, Y. (2000) Age and organ dependent spontaneous generation of nuclear 8-hydroxydeoxyguanosine in male Fischer 344 rats. *Lab Invest.* **80,** 249–261.
333. Pieper, A. A., Blackshaw, S., Clements, E. E., Brat, D. J., Krug, D. K., White, A. J., et al. (2000) Poly(ADP-ribosyl)ation basally activated by DNA strand breaks reflects glutamate-nitric oxide neurotransmission. *Proc. Natl. Acad. Sci. USA* **97,** 1845–1850.
334. Castren, E., Ohga, Y., Berzaghi, M. P., Tzimagiorgis, G., Thoenen, H., and Lindholm, D. (1994) bcl-2 messenger RNA is localized in neurons of the developing and adult rat brain. *Neuroscience* **61,** 165–177.
335. Shin, D. H., Lee, H. Y., Lee, K. H., Kim, H. J., Lee, W. J., Hwang, D. H., Baik, S. H., and Cho, S. S. (2000) Localization of bcl-2 mRNA in the rabbit central nervous system. *Neurosci. Lett.* **278,** 73–76.
336. Hayashi, T., Sakai, K., Sasaki, C., Itoyama, Y., and Abe, K. (2000) Loss of bag-1 immunoreactivity in rat brain after transient middle cerebral artery occlusion. *Brain Res.* **852,** 496–500.
337. Allen, D. J. and Low, F. N. (1973) The ependymal surface of the lateral ventricle of the dog as revealed by scanning electron microscopy. *Am. J. Anat.* **137,** 483–489.
338. Altschul, R. (1941) The growth of ependyma. *J. Comp. Neurol.* **75,** 245–253.
339. Schimrigk, K. (1966) Uber die Wandstruktur der Seitenventrikel und des dritten Ventrikels beim Menschen. *Z Zellforsch Mikrosk Anat.* **70,** 1–20.
340. Dooling, E. C., Chi, J. G., and Gilles, F. H. (1983) Developmental changes in ventricular epithelia, in *The Developing Human Brain* (Gilles, F. H., Leviton, A., and Dooling, E. C., eds.), John Wright, Boston, pp. 113–116.
341. Isomura, G., Kozasa, T., and Tanaka, S. (1986) Absence of the central canal and its obliterative process in the house shrew spinal cord (Suncus murinus). *Anat. Anz.* **161,** 285–296.
342. Milhorat, T. H., Kotzen, R. M., and Anzil, A. P. (1994) Stenosis of the central canal of spinal cord in man: incidence and pathological findings in 232 autopsy cases. *J. Neurosurg.* **80,** 716–722.
343. Milhorat, T. H. and Kotzen, R. M. (1994) Stenosis of the central canal of the spinal cord following inoculation of suckling hamsters with reovirus type I. *J. Neurosurg.* **81,** 103–106.

344. Boerman, R. H., Mitro, A., Bloem, B. R., Arnoldus, E. P. J., Raap, A. K., Peters, A. C. B., and van der Ploeg, M. (1991) Detection of herpes simplex virus in the ependyma of experimentally infected mice. *Acta Virol.* **35,** 450–457.
345. Davidson, B. L., Stein, C. S., Heth, J. A., Martins, I., Kotin, R. M., Derksen, T. A., et al. (2000) Recombinant adeno-associated virus type 2, 4, and 5 vectors: transduction of variant cell types and regions in the mammalian central nervous system. *Proc. Natl. Acad. Sci. USA* **97,** 3428–3432.
346. Chretien, F., Gray, F., Lescs, M. C., Geny, C., Dubreuil-Lemaire, M. L., Ricolfi, F., et al. (1993) Acute varicella-zoster virus ventriculitis and meningo-myelo-radiculitis in acquired immunodeficiency syndrome. *Acta Neuropathol.* **86,** 659–665.
347. Duffy, P. E., Wolf, A., Harter, D. H., Gamboa, E. T., and Hsu, K. C. (1973) Murine influenza virus encephalomyelitis. II. Electron microscopic observations. *J. Neuropathol. Exp. Neurol.* **32,** 72–91.
348. Takano, T., Mekata, Y., Yamano, T., and Shimada, M. (1993) Early ependymal changes in experimental hydrocephalus after mumps virus inoculation in hamsters. *Acta Neuropathol.* **85,** 521–525.
349. Tomko, R. P., Johansson, C. B., Totrov, M., Abagyan, R., Frisen, J., and Philipson, L. (2000) Expression of the adenovirus receptor and its interaction with the fiber knob. *Exp. Cell Res.* **255,** 47–55.
350. Thach, D. C., Kimura, T., and Griffin, D. E. (2000) Differences between C57BL/6 and BALB/cBy mice in mortality and virus replication after intranasal infection with neuroadapted sindbis virus. *J. Virol.* **74,** 6156–6161.
351. Mucke, L. and Oldstone, M. B. (1992) The expression of major histocompatibility complex (MHC) class I antigens in the brain differs markedly in acute and persistent infections with lymphocytic choriomeningitis virus (LCMV). *J. Neuroimmunol.* **36,** 193–198.
352. Kimura, T. and Griffin, D. E. (2000) The role of $CD8^+$ T cells and major histocompatibility complex class I expression in the central nervous system of mice infected with neurovirulent sindbis virus. *J. Virol.* **74,** 6117–6125.
353. Deckert-Schluter, M., Schluter, D., Hof, H., Wiestler, O. D., and Lassmann, H. (1994) Differential expression of ICAM-1, VCAM-1 and their ligands LFA-1, Mac-1, CD43, VLA-4, and MHC class II antigens in murine Toxoplasma encephalitis: a light and ultrastructural immunohistochemical study. *J. Neuropathol. Exp. Neurol.* **53,** 457–468.
354. Sethna, M. P. and Lampson, L. A. (1991) Immune modulation within the brain: recruitment of inflammatory cells and increased major histocompatibility antigen expression following intracerebral injection of interferon-gamma. *J. Neuroimmunol.* **34,** 121–132.
355. Steiniger, B. and van der Meide, P. H. (1988) Rat ependyma and microglia cells express class II MHC antigens after intravenous infusion of recombinant gamma interferon. *J. Neuroimmunol.* **19,** 111–118.
356. Vass, K. and Lassmann, H. (1990) Intrathecal application of interferon gamma: progressive appearance of MHC antigens within the rat nervous system. *Am. J. Pathol.* **137,** 789–800.
357. LeVine, S. M. and Brown, D. C. (1997) IL-6 and TNFalpha expression in brains of twitcher, quaking and normal mice. *J. Neuroimmunol.* **73,** 47–56.
358. Schobitz, B., de Kloet, E. R., Sutano, W., and Holsboer, F. (1993) Cellular localization of interleukin 6 mRNA and interleukin 6 receptor mRNA in rat brain. *Eur. J. Neurosci.* **5,** 1426–1435.
359. Barnum, S. R., Jones, J. L., Muller-Ladner, U., Samimi, A., and Campbell, I. L. (1996) Chronic complement C3 gene expression in the CNS of transgenic mice with astrocyte-targeted interleukin-6 expression. *Glia* **18,** 107–117.
360. Diaz-Ruiz, C., Perez-Tomas, R., Domingo, J., and Ferrer, I. (1993) Immunohistochemical localization of transforming growth factor-α in choroid plexus of the rat and chicken. *Neurosci. Lett.* **164,** 44–46.

361. Logan, A., Berry, M., Gonzalez, A. M., Frautschy, S. A., Sporn, M. B., and Baird, A. (1994) Effects of transforming growth factor β1 on scar production in the injured central nervous system of the rat. *Eur. J. Neurosci.* **6,** 355–363.
362. Tarlow, M. J., Jenkins, R., Camis, S. D., Osborne, M. P., Stephens, S., Stanley, P., and Crocker, J. (1993) Ependymal cells of the choroid plexus express tumour necrosis factor-α. *Neuropathol. Appl. Neurobiol.* **19,** 324–328.
363. Tchelingerian, J. L., Quinonero, J., Booss, J., and Jacque, C. (1993) Localization of TNFa and IL-1A immunoreactivities in striatal neurons after surgical injury to the hippocampus. *Neuron* **10,** 213–224.
364. French, R. A., VanHoy, R. W., Chizzonite, R., Zachary, J. F., Dantzer, R., Parnet, P., Bluthe, R. M., and Kelley, K. W. (1999) Expression and localization of p80 and p68 interleukin-1 receptor proteins in the brain of adult mice. *J. Neuroimmunol.* **93,** 194–202.
365. Patterson, P. H. (1993) Cytokines and the function of the mature nervous system. *C R Acad. Sci. Paris* **316,** 1150–1157.
366. Ling, E. A., Kaur, C., and Lu, J. (1998) Origin, nature, and some functional considerations of intraventricular macrophages, with special reference to the epiplexus cells. *Microsc. Res. Tech.* **41,** 43–56.
367. Del Bigio, M. R. (1993) Neuropathological changes caused by hydrocephalus. *Acta Neuropathol.* **85,** 573–585.
368. Lucena, J. and Cruz-Sanchez, F. F. (1996) Ependymal changes in sudden infant death syndrome. *J. Neuropathol. Exp. Neurol.* **55,** 348–356.
369. Sarnat, H. B., Darwish, H. Z., Barth, P. G., Trevenen, C. L., Pinto, A., Kotagal, S., et al. (1993) Ependymal abnormalities in lissencephaly/pachygyria. *J. Neuropathol. Exp. Neurol.* **52,** 525–541.
370. del Rio-Hortega, P. (1929) Sobre las formaciones fibrilares del epitelio ependimaro. *Mem. Soc. Espan. Hist. Nat.* **15,** 515–526.
371. Miklossy, J., Kraftsik, R., Pillevuit, O., Lepori, D., Genton, C., and Bosman, F. T. (1998) Curly fiber and tangle-like inclusions in the ependyma and choroid plexus: a pathogenetic relationship with the cortical Alzheimer-type changes? *J. Neuropathol. Exp. Neurol.* **57,** 1202–1212.
372. Wen, G. Y., Rudelli, R. D., Kim, K. S., and Wisniewski, H. M. (1988) Tangles of ependyma-choroid plexus contain β-amyloid protein epitopes and represent a new form of amyloid fiber. *Arch Neurol.* **45,** 1298–1299.
373. Ohta, M., Kitamoto, T., Iwaki, T., Ohgami, T., Fukui, M., and Tateishi, J. (1993) Immunohistochemical distribution of amyloid precursor protein during normal rat development. *Dev. Brain Res.* **75,** 151–161.
374. Gliemann, J. (1998) Receptors of the low density lipoprotein (LDL) receptor family in man. Multiple functions of the large family members via interaction with complex ligands. *Biol. Chem.* **379,** 951–964.
375. Matsumoto, A., Itoh, K., and Matsumoto, R. (2000) A novel carboxypeptidase B that processes native beta-amyloid precursor protein is present in human hippocampus. *Eur. J. Neurosci.* **12,** 227–238.
376. Cserr, H. F., Harling-Berg, C. J., and Knopf, P. M. (1992) Drainage of brain extracellular fluid into blood and deep cervical lymph and its immunological significance. *Brain Pathol.* **2,** 269–276.
377. Saunders, N. R., Knott, G. W., and Dziegielewska, K. M. (2000) Barriers in the immature brain. *Cell. Mol. Neurobiol.* **20,** 29–40.
378. Saunders, N. R., Habgood, M. D., and Dziegielewska, K. M. (1999) Barrier mechanisms in the brain. II. Immature brain. *Clin. Exp. Pharmacol. Physiol.* **26,** 85–91.
379. Brownlees, J. and Williams, C. H. (1993) Peptidases, peptides, and the mammalian blood-brain barrier. *J. Neurochem.* **60,** 793–803.

380. Mercier, F. and Hatton, G. I. (2000) Immunocytochemical basis for a meningeo-glial network. *J. Comp. Neurol.* **420,** 445–465.
381. van Leeuwen, F. W., van Heerikhuize, J., van der Meulen, G., and Wolters, P. (1985) Light microscopic autoradiographic localization of (^{3}H) oxytocin binding sites in the rat brain, pituitary and mammary gland. *Brain Res.* **359,** 320–325.
382. Bjelke, B. and Fuxe, K. (1993) Intraventricular β-endorphin accumulates in DARPP-32 immunoreactive tanycytes. *NeuroReport* **5,** 265–268.
383. Berger, U. V. and Hediger, M. A. (1999) Distribution of peptide transporter PEPT2 mRNA in the rat nervous system. *Anat. Embryol. (Berl.)* **199,** 439–449.
384. Burguera, B., Couce, M. E., Long, J., Lamsam, J., Laakso, K., Jensen, M. D., Parisi, J. E., and Lloyd, R. V. (2000) The long form of the leptin receptor (OB-Rb) is widely expressed in the human brain. *Neuroendocrinology* **71,** 187–195.
385. Couce, M. E., Burguera, B., Parisi, J. E., Jensen, M. D., and Lloyd, R. V. (1997) Localization of leptin receptor in the human brain. *Neuroendocrinology* **66,** 145–150.
386. Burke, K. A., Schroeder, D. M., Abel, R. A., Richardson, S. C., Bigsby, R. M., and Nephew, K. P. (2000) Immunohistochemical detection of estrogen receptor alpha in male rat spinal cord during development. *J. Neurosci. Res.* **61,** 329–337.
387. Graff, M. N., Baas, D., Puymart, J., Sarlieve, L. L., and Delaunoy, J. P. (1993) The α and β thyroid receptors are expressed by cultured ependymal cells. Correlation with the effect of L-3. 5. 3'-triiodothyronine on glutamine synthetase mRNAs. *Neurosci. Lett.* **150,** 174–178.
388. Maekawa, F., Toyoda, Y., Torii, N., Miwa, I., Thompson, R. C., Foster, D. L., et al. (2000) Localization of glucokinase-like immunoreactivity in the rat lower brain stem: for possible location of brain glucose-sensing mechanisms. *Endocrinology* **141,** 375–384.
389. Vannucci, S. J. and Simpson, I. A. (1998) Glucose transporter expression in brain: Relationship to cerebral glucose utilization. *Dev. Neurosci.* **20,** 369–379.
390. Jones-Humble, S. A. and Morgan, P. F. (1994) High expression of nitrobenzylthionosine-insensitive dipyridamole binding sites in post mortem human ependymal tissue. *Eur. J. Pharmacol.* **257,** 311–314.
391. Berger, U. V. and Hediger, M. A. (1998) Comparative analysis of glutamate transporter expression in rat brain using differential double *in situ* hybridization. *Anat. Embryol. (Berl.)* **198,** 13–30.
392. Berger, U. V., Tsukaguchi, H., and Hediger, M. A. (1998) Distribution of mRNA for the facilitated urea transporter UT3 in the rat nervous system. *Anat. Embryol.* **197,** 405–414.
393. Schroeter, S., Apparsundaram, S., Wiley, R. G., Miner, L. H., Sesack, S. R., and Blakely, R. D. (2000) Immunolocalization of the cocaine- and antidepressant-sensitive l-norepinephrine transporter. *J. Comp. Neurol.* **420,** 211–232.
394. Alonso, G., Phan, V. L., Guillemain, I., Saunier, M., Legrand, A., Anoal, M., and Maurice, T. (2000) Immunocytochemical localization of the sigma(1) receptor in the adult rat central nervous system. *Neuroscience* **97,** 155–170.
395. Ogawa, M., Shiozawa, M., Hiraoka, Y., Takeuchi, Y., and Aiso, S. (1998) Immunohistochemical study of localization of gamma-glutamyl transpeptidase in the rat brain. *Tissue Cell* **30,** 597–601.
396. Raidoo, D. M. and Bhoola, K. D. (1997) Kinin receptors on human neurones. *J. Neuroimmunol.* **77,** 39–44.
397. Howard, S., Landry, C., Fisher, R., Bezouglaia, O., Handley, V., and Campagnoni, A. (1998) Postnatal localization and morphogenesis of cells expressing the dopaminergic D2 receptor gene in rat brain: expression in non-neuronal cells. *J. Comp. Neurol.* **391,** 87–98.
398. Benavides, J., Quarteronet, D., Imbault, F., Malgouris, C., Uzan, A., Renault, C., et al. (1983) Labelling of "peripheral-type" benzodiazepine binding sites in rat brain by using [^{3}H] PK 11195, an isoquinoline carboxamide derivative: kinetic studies and autoradiographic localization. *J. Neurochem.* **41,** 1744–1750.

399. Abramovitz, M., Homma, H., Ishigaki, S., Tansey, F., Cammer, W., and Listowsky, I. (1988) Characterization and localization of glutathione-S-transferases in rat brain and binding of hormones, neurotransmitters, and drugs. *J. Neurochem.* **50,** 50–57.
400. Kuchler Bopp, S., Ittel, M. E., Dietrich, J. B., Reeber, A., Zaepfel, M., and Delaunoy, J. P. (1998) The presence of transthyretin in rat ependymal cells is due to endocytosis and not synthesis. *Brain Res.* **793,** 219–230.
401. Schwartz, J. C., Bouthenet, M. L., Giros, B., Gros, C., Llorens-Cortes, C., and Pollard, H. (1991) Neuropeptidases and neuropeptide inactivation in the brain, in *Volume Transmission in the Brain* (Fuxe, K. and Agnati, L. F., eds.), Raven, New York, pp. 381–394.
402. Bourne, A., Barnes, K., Taylor, B. A., Turner, A. J., and Kenny, A. J. (1989) Membrane peptidases in the pig choroid plexus and on other cell surfaces in contact with the cerebrospinal fluid. *Biochem. J.* **259,** 69–80.
403. Birch, N. P., Rodriguez, C., Dixon, J. E., and Mezey, E. (1990) Distribution of carboxypeptidase H messenger RNA in rat brain using *in situ* hybridization histochemistry: implications for neuropeptide synthesis. *Mol. Brain Res.* **7,** 53–59.
404. Matsas, R., Kenny, A. J., and Turner, A. J. (1986) An immunohistochemical study of endopeptidase-24. 11 ("enkephalinase") in the pig nervous system. *Neuroscience* **18,** 991–1012.
405. Oliveira, E. S., Leite, P. E. P., Spillantini, M. G., Camargo, A. C. M., and Hunt, S. P. (1990) Localization of endo-oligopeptidase (EC 3. 4. 22. 19) in the rat nervous tissue. *J. Neurochem.* **55,** 1114–1121.
406. MacCumber, M. W., Snyder, S. H., and Ross, C. A. (1990) Carboxypeptidase E (enkephalin convertase): mRNA distribution in rat brain by *in situ* hybridization. *J. Neurosci.* **10,** 2850–2860.
407. Akimoto, J., Itoh, H., Miwa, T., and Ikeda, K. (1993) Immunohistochemical study of glutamine synthetase expression in early glial development. *Dev. Brain Res.* **72,** 9–14.
408. Facchinetti, P., Rose, C., Rostaing, P., Triller, A., and Schwartz, J. C. (1999) Immunolocalization of tripeptidyl peptidase II, a cholecystokinin-inactivating enzyme, in rat brain. *Neuroscience* **88,** 1225–1240.
409. Anderson, P. J. and Song, S. K. (1962) Acid phosphatase in the nervous system. *J. Neuropathol. Exp. Neurol.* **21,** 263–283.
410. Shuttleworth, E. C., Jr. and Allen, N. (1966) Acid hydrolases in a pia-arachnoid and ependyma of rat brain. *Neurology* **16,** 979–985.
411. Petanceska, S., Burke, S., Watson, S. J., and Devi, L. (1994) Differential distribution of messenger RNAs for cathepsins B, L, and S in adult rat brain: an *in situ* hybridization study. *Neuroscience* **59,** 729–738.
412. Bernstein, H. G., Kirschke, H., Kloss, P., Wiederanders, B., Rinne, A., and Frohlich, J. (1990) Cathepsin B during early human brain development. *Acta Histochem.* **39**(Suppl.), S473–S475.
413. Eipper, B. A., Stoffers, D. A., and Mains, R. E. (1992) The biosynthesis of neuropeptides: peptide α-amidation. *Annu. Rev. Neurosci.* **15,** 57–85.
414. Day, R., Schafer, M. K. H., Cullinan, W. E., Watson, S. J., Cretien, M., and Seidah, N. G. (1993) Region specific expression of furin mRNA in the rat brain. *Neurosci. Lett.* **149,** 27–30.
415. Simson, J. A. V., Dom, R., Chao, J., Woodley, C., Chao, L., and Margolis, H. S. (1985) Immunocytochemical localization of tissue kallikrein in brain ventricular epithelium and hypothalamic cell bodies. *J. Histochem. Cytochem.* **33,** 951–953.

IV

Immune System-Neuron Interactions

14
Microglia in the CNS

Sophie Chabot and V. Wee Yong

1. INTRODUCTION

Microglia, the principal immune effector cells of the central nervous system (CNS), were originally described by del Rio-Horgeta *(1)*. Microglia constitute 5–15% of the total glial cell population present in the adult CNS, and it is estimated *(2)* that there are as many microglia cells as neurons, which emphasizes their importance in the CNS. The origin of microglia is still a controversial issue, but substantial evidence supports the theory that microglia are of mesodermal origin, e.g., that they are related to cells of the monocyte/macrophage lineage, which derived from bone marrow cells. Indeed, microglia are frequently referred to as the resident macrophages of the CNS, since they share many phenotypic markers of hematogeneous macrophages, and many of their functions are similar to those of macrophages. Microglia activation is a graded response, through which microglia acquire various functions, such as antigen (Ag) presentation and phagocytosis. Microglia activation is also accompanied by an alteration of gene expression, which results in cellular responses such as proliferation, and in the release of reactive oxygen (O_2) intermediates, inflammatory cytokines, and nitric oxide (NO). Mechanisms of microglia activation are not completely understood, and few have been described. Activated microglia play an important role in the pathophysiology of CNS disorders, including cerebral ischemia, brain abscesses, stab wounds, Alzheimer's disease (AD), multiple sclerosis (MS), and acquired immune deficiency syndrome (AIDS) dementia complex.

2. PHENOTYPIC PROPERTIES OF MICROGLIA

Microglia cells are distributed throughout the CNS, although their density varies between CNS regions. For example, in the mouse brain, the highest microglia densities are encountered in regions such as the hippocampus, the olfactory telecephalon, portions of the basal ganglia, and the substantia nigra *(3)*. Typically, the density of microglia is such that cytoplasmic processes of neighboring cells do not make noticeable contact with each other *(4)*.

The functional plasticity of microglia is shown by their capacity to adopt various morphologies, depending on their activation state. Resting microglia appear as branched ramified glial cells; activated microglia have an amoeboid shape, and closely resemble macrophages with a large, rounded cell body with retracted or shortened

From: *The Neuronal Environment: Brain Homeostasis in Health and Disease*
Edited by: W. Walz © Humana Press Inc., Totowa, NJ

processes *(4)*. Ramified-like microglia represent a relatively permanent population, with little turnover in the adult nervous system, and are uniformly dispersed throughout the adult CNS, more common in gray than white matter *(5)*. In contrast, amoeboid-like microglia are abundant in the developing brain, and are believed to phagocytose debris from naturally occurring cell death in late embryonic and postnatal stages of development *(6,7)*. Microglia can also be found in a hyperramified state, which represent an intermediate stage between ramified and amoeboid forms *(8)*. It is believed that the hyperramification of microglia results from hypertrophy, an early stage of microglia activation.

In the normal adult CNS, two distinct populations of microglial cells have been described, namely, parenchymal and perivascular microglia. In the normal brain, these two cell types are readily distinguished by their morphology and surface immunophenotype. Parenchymal microglia are located in the CNS parenchyma, and are in a resting state, as shown by their ramified morphology. On the other hand, perivascular microglia are found in the vicinity of blood vessels enclosed within the basal lamina, and are distinguished from parenchymal microglia by their amoeboid shape, suggesting that they are activated cells. Perivascular cells express high level of major histocompatibility complex class II (MHC-II) in contrast to parenchymal microglia *(9)*. Recent experiments *(10)* demonstrated that the intracerebral injection of the neuronal tracer substance, fluorogold, results in the labeling of perivascular microglial cells, but not in the labeling of parenchymal microglia. Therefore, because of this capacity to take up foreign particles, it is believed that perivascular microglia represent the only population of constitutive phagocytes in the CNS *(11,12)*.

Immunocytochemically, microglia can be detected using antibodies (Abs) that recognize numerous molecules. Those include MHC-I and -II, CD4, CD11a (LFA-1), CD11b (CR3 complement receptor/Mac-1), CD11c (CR4, p150.95), CD14 (lipopolysaccharide [LPS] receptor), CD45, CD64 (FcγRI), CD68, and F4/80 *(4,13)*. MHC-II is highly expressed on activated microglia under pathological conditions; low or undetectable levels are expressed on resting microglia *(14,15)*. It is difficult to distinguish the intrinsic population of microglia from invading monocytes/macrophages since both cell population can express those molecules. One criteria that has been proposed relates to multiple surface markers. Sedgwick et al. *(16)* demonstrated, by flow cytometry, that murine microglia express constitutively low levels of CD45 and negligible CD11b, designated CD11b–/CD45^{low}, and that monocytes/macrophages express high levels of CD11b and CD45 (CD11b^{+}/CD45^{high}).

3. ORIGIN OF MICROGLIA

The origin of microglia is still a controversial issue, but a substantial body of evidence supports the theory that microglia are of mesodermal origin, and that they are related to cells of the monocyte/macrophage lineage, which derived from bone marrow cells. Another theory argues that microglia arise from neuroectodermal-derived glioblast progenitors *(17–19)*.

The hypothesis that microglia are myelomonocytic cells, of mesodermal origin, was first proposed by del Rio-Hortega *(1)*, who postulated that microglia arise from bloodborne cells of leukocytic morphology, which invade the brain from the meninges *(13)*. Many studies have now reconfirmed this hypothesis, and further demonstrated that parenchymal microglia settle the CNS antenatally, being derived from bone marrow precursor cells that colonize the CNS early during embryogenesis. It was shown *(13)* that the colonization of the CNS by microglia coincides with vascularization, formation of radial glia, neuronal migration, and myelination, primarily in mo 4–5 of gestation in humans and beyond. Once in the CNS, microglia influx generally conforms to a route following white matter tracts to gray areas *(13,20)*. As microglia migrate through the CNS parenchyma, they have an amoeboid shape. After reaching their definitive location, it appears that amoeboid microglia differentiate into ramified microglia, which are distributed more or less homogeneously throughout the entire CNS parenchyma. Moreover, studies of the developing retina provided insights about the origin of microglia, which supports the monocytic origin *(7,21,22)*.

After embryogenesis, perivascular microglia, but not parenchymal microglia, derive from circulating monocytes, as shown using bone marrow chimeras *(23)*. In these studies, rats of strain A are lethally irradiated, then reconstituted with bone marrow taken from strain A/strain B hybrids; Abs against MHC-I are then used to detect the donor origin. Strain-B-specific MHC-I-positive cells belong to the pool of transplanted cells that have invaded the CNS following bone marrow transplantation. These studies have shown that, under normal conditions, perivascular microglia are regularly replaced from the bone marrow in adult animals *(24–27)*; parenchymal microglia undergo very little or no turnover with bone marrow-derived cells *(24,27–29)*.

The view that at least some microglial cells are of neuroectodermal origin is supported by several studies. Autoradiographic analyses of the genesis of microglia within mouse hippocampus showed the presence of glioblasts, which differentiate into both microglia and astrocytes *(30)*. Hao et al. *(17)* provided evidence that microglial cells and astrocytes may even derive from the same progenitor cell. They showed that clones, derived from a single newborn mouse brain cell, contained both astrocytes and microglial cells. These results also raise the possibility that astrocytes, like microglial cells, may be derived from hematopoietic cells that have entered into the CNS, which agrees with the finding by Eglitis and Mezey *(29)* that astrocytes of donor origin appear in mice subjected to bone marrow grafting.

4. FUNCTIONS OF MICROGLIA

Functions of resting microglia are still mostly unknown, as are the factors that contribute to the maintenance of the quiescent state of microglia. All known functions of microglia are performed by activated microglia, which include phagocytosis, processing/presentation of Ags, and the production and release of several inflammatory mediators, including cytokines, chemokines, reactive O_2 intermediates, NO, complement proteins, coagulation proteins, and proteases *(31)*.

The phagocytic function of microglia has long been known. Penfield *(32)* noted that microglia migrate to the lesion site induced in newborn animals, where they phagocytose debris, as shown by their foamy appearance and the intracellular formation of lipoid and iron granule bodies. Like other professional phagocytes, activated microglia possess a respiratory burst system that can generate prodigious quantities of free radicals from O_2 molecules, such as hydroxyl radicals, singlet O_2 species, and hydrogen peroxide *(33)*, and they possess many phagocytic-related enzymes, such as nucleoside diphosphatase and acid phosphatase *(34–38)*. All groups of Fc receptors (FcRI, FcRII, and FcRIII), which mediate Ab-dependent phagocytosis by binding to the Fc portion of immunoglobulins, have been shown on the surface of both reactive and perivascular microglia *(39)*. In vitro, microglial phagocytic activity is modified by the presence of astrocytes and cytokines. For example, co-culturing microglia with astrocytes was shown to suppress phagocytosis *(40)*. Preincubation of cultured microglia with granulocyte/macrophage colony-stimulating factor (GM-CSF), interferon γ (IFN-γ), and tumor necrosis factor α (TNF-α) enhanced their phagocytic activity; transforming growth factor-β1 and interleukin 4 (IL-4) inhibited this function *(13,41)*. Adult human derived microglia phagocytose Ab-coated targets, as shown by increased reduced nicotinamide adenine dinucleotide phosphate (NADPH) oxidase activity, because of their ability to express Fcγ receptors *(39)*.

A role for microglia as antigen-presenting cells has been proposed, because of their ability to express MHC-II. *In situ*, human resting microglia of ramified morphology express low levels of MHC-II molecules. Under normal conditions, perivascular microglia, expressing high levels of MHC-II, function as effective antigen presenting *(42)*; parenchymal microglia fail to perform this function *(43,44)*. However, under pathological conditions, the expression of MHC-II is widely upregulated on parenchymal microglia, as they become activated. Mechanisms responsible for the upregulation of MHC-II on activated microglia are not entirely understood, but, in culture, it is inducible by cytokines, such as IFN-γ *(31,45)*. Microglia can process Ag in vitro *(46,47)*. Sedgwick et al. *(48)* demonstrated that immediately ex vivo microglia, isolated from the CNS of healthy Brown Norway rat, do not express MHC-II, and that they fail to activate MBP-reactive $CD4^+$ T-cells. However, under inflammatory conditions, MHC-II is expressed on the surface of microglia isolated from the spinal cord of experimental autoimmune encephalitis (EAE) animals *(42)*. The Ag-presenting capacity of microglia was also demonstrated in the murine and human systems. Human adult microglia, derived from brain biopsy specimen from patients who underwent surgical resection as a mean to treat intractable epilepsy or brain tumors, express MHC-II and activate T-cells in vitro *(49,50)*. In culture, IFN-γ is known to induce MHC-II expression on human adult microglia *(49)*. Microglia also express the important co-stimulatory molecules required for adequate Ag presentation. In vitro and *in situ* expression of B7-1 (CD80) and B7-2 (CD86), by activated microglia, has been shown *(51–54)*, and blocking B7 interaction results in the reduced APCs capacity of microglia *(55)*. Recent evidence indicates that the activation state of murine microglia determines whether microglia induce MBP-specific T-cell anergy or T-cell activation. Microglia become professional APC only after a multistep activation process involving both stimulation through cytokines, such as GM-CSF and IFN-γ, and cognate interactions, such as B7/CD28 and CD40/CD40L *(56)*.

Activated microglia are important sources of molecules that are involved in CNS inflammation. For example, in vitro experiments demonstrated that activated microglia can produce proinflammatory cytokines, which include TNF-α, IL-1β, IL-6, IL-12, IL-15, and IFN-γ *(57)*. Cytokines have multiple effects in the CNS, which can be both beneficial and detrimental, depending on their concentrations.

Cytokine functions in the CNS include the control of neuronal and glial differentiation, proliferation, differentiation, and survival. Given those functions, cytokines have the potential to influence neuronal and glial plasticity, degeneration, and development and regeneration of the nervous system *(58)*. Other products of activated microglia are secreted proteases, such as elastase, urokinase plasminogen activator, matrix metalloproteinase-2 (MMP-2), and MMP-9. For example, Cuzner et al. *(59)* demonstrated, by *in situ* hybridization, that MMP-2 and MMP-9 mRNAs are predominantly expressed in microglia throughout normal white matter. MMPs have the potential to degrade basement membrane and other matrix components, and to catalyze the release of membrane-bound inflammatory cytokines, such as TNF-α *(60,61)*. Because of their ability to produce neurotoxins, it is believed that microglia may have neurotoxic functions.

In vitro, microglia produce a large amount of glutamate and aspartate *(62)*, and sustained high extracellular levels of these amino acids was shown to cause *N*-methyl-D-asparate receptor-mediated neuronal injury in vivo *(63)*. Complement proteins, which provide signals for scavenger activation, opsonization, and membrane-attack-complex-mediated direct lysis of cells, are also produced by activated microglia *(64)*. For instance, it was shown that cultured microglia expressed enhanced levels of C2, C3, C4, and C1q mRNAs, when treated with LPS and IFNγ *(65)*. Platelet activating factor (PAF) is a lipid molecule shown to be involved in inflammatory processes and in cell–cell communication. In the CNS, there is growing evidence that PAF is a mediator of neuro-injury caused by stroke and trauma by neuronal cells. PAF production by human fetal microglia was shown to be induced by both TNF-α and LPS, in a concentration-dependent manner *(66)*. Finally, microglia produce and secrete chemokines, which are small proteins (8–10 kDa) that induce chemotaxis, tissue extravasation, and functional modulation of a wide variety of leukocytes during inflammation. Chemokines are also thought to be pivotal regulatory molecules implicated in cellular communication, and mediate their biological activities through G-protein-coupled cell-surface receptors. The binding of chemokines to their receptors results in the activation of MAPK-associated and CREB-associated signal transduction pathways *(67)*. Evidence suggests that chemokines and their receptors are important in CNS development and regeneration *(13)*. Fractalkine (neurotactin), a CX3C chemokine, and its receptor are highly expressed by microglia in rat brains *(68)*.

5. MICROGLIA IN DISEASES

Microglia respond to virtually any, even minor, pathological events in the CNS. In most pathological conditions, microglia are aided by infiltrating hematogenous macrophages. Microglia activation appears to play a central role in the pathogenesis of MS and other CNS pathologies, such as cerebral ischemia, brain abscesses, traumatic brain injury, AD, experimental globoid cell dystrophy, AIDS dementia complex, and cerebral malaria.

In MS, a large number of microglia accumulate in and around lesions of demyelinating areas in brains of MS patients *(69)*. Immunocytochemically, microglia activation markers can be used to characterize the distribution of these cells in MS plaques, including MHC-II, Mac-1, and LFA-1 *(14,69,70)*. MHC-II-positive cells are present in active and chronic demyelinated MS lesions. At the edge, and at the center, of the demyelinating region of MS brains, neuropathological analyses demonstrate that the morphology of microglia appears to be elongated and amoeboid, respectively, characteristic of their activated state *(71)*; resting microglia in noninflamed brains have a ramified shape. In MS brains, it was recently shown that microglia proliferate *(72)*.

Microglia are the first CNS cell type to respond to several types of CNS injury. Upon activation, microglia express cellular adhesion molecules, such as ICAM-1, and VCAM-1, involved in the regulation of immune responses. It has been well-documented that enhanced levels of ICAM and VCAM-1 are present in active MS lesions *(73)*. In MS brains, it was recently shown that microglia proliferate *(72)*.

An active role of microglia in demyelination has been proposed. In MS, the presence of phagocytic cells, with a foamy appearance, are found primarily in new lesions, where active myelin phagocytosis is proceeding, and are seen frequently in active MS plaques *(74–76)*. In vitro, cultured microglia have the ability to phagocytose myelin *(76)*. For example, it was shown that the short incubation of ^{14}C-lipid-labeled myelin with microglia results in the rapid intracellular metabolic conversion of myelin into cholesterol ester and triglyceride *(77)*. Mechanisms of myelin phagocytosis by microglia in vitro appear to involve membrane attack complex 2, a galactoside-specific lectin *(55,78)*. There is also in vivo evidence that microglia phagocytose myelin. By the use of bone marrow chimeras in Lewis rats with EAE, Rinner et al. *(79)* demonstrated that resident microglia contained myelin-degradation products. Prineas et al. reported *(80,81)* that myelin is attached to coated pits at sites where receptors are involved in phagocytosis, suggesting that myelin is ingested by receptor-mediated phagocytosis. Receptors implicated in phagocytosis include Fc receptors (FcRI, FcRII, and FcRIII), complement receptors (CR3), and scavenger receptors. FcR and CR3 can act as a ligand for myelin, and are thus thought to be important mechanisms of myelin destruction in vivo. Enhanced levels of Fc receptor and CR3 expression on microglia in MS lesions have been reported *(39,82)*.

Activated microglia play a central role in regulating inflammatory responses of MS lesions. For example, it is believed that activated microglia plays a role in the initiation of the immune response by presenting myelin Ags to infiltrating myelin-reactive T-lymphocytes. As a result, myelin-reactive T-cells become activated, differentiate, and secrete cytokines that are pathogenic in MS, such as TNF-α *(47)*. By producing a large number of inflammatory molecules, such as cytokines, chemokines, and proteases, possibly through their contact with Ag nonspecific T-cells, activated microglia augment CNS local immune responses, which contribute to the progression of MS *(83–85)*. Finally, under certain conditions, it was shown that activated microglia also have the capacity to downregulate T-cell responses *(86)*. Therefore, it appears that activated microglia regulate the initiation, progression, and termination of CNS inflammation of MS, by dictating the fate of the immune response.

Evidence suggests that TNF-α, which is mostly produced by microglia in the CNS, is involved in the pathogenesis of CNS diseases, including meningococcal meningitis, human immunodeficiency virus infections, AD, brain ischemia, and MS *(87)*. For example, in the case of MS, the level of TNF-α is elevated in the serum, cerebrospinal fluid, and the brain lesions of MS patients, and this correlates with disease activity *(73,88–90)*. Moreover, peripheral blood mononuclear cells, isolated from peripheral blood of MS patients at time of relapse, produce significantly increased levels of TNF-α, compared to peripheral blood mononuclear cells from controls *(91)*. In mice, studies, using transgenic animals overexpressing TNF-α in the CNS, demonstrate a role for TNF-α in CNS demyelinating disease. Indeed, it was demonstrated *(92,93)* that these mice develop a spontaneous inflammatory demyelinating disease similar to

MS. In EAE, the administration of soluble TNF-α receptors or TNF-α Abs, as a mean to attenuate TNF-α biologic effects, was shown to prevent the transfer of EAE and to abrogate autoimmune demyelination *(89,94–96)*. It is believed that the pathogenic role of TNF-α in MS/EAE is associated with its ability to damage the myelin/oligodendrocyte complex, and to enhance leukocyte migration through the direct upregulation of adhesion molecules. In vitro studies demonstrated that TNF-α causes apoptotic death of oligodendrocytes *(97–99)*, and demyelination of mouse optic nerve axons results from the intravitreal injection of TNF-α in vivo *(100)*. In EAE-sensitized animals treated with TNF-binding protein, a polyethylene glycol-linked form of TNFR1, a reduction in VCAM-1 and VLA-4 staining was observed, which corresponded to inhibition of CNS inflammation and the prevention of clinical signs of EAE *(101)*.

AD is a progressive neurodegenerative disorder, which causes memory loss and dementia. Neuropathologically, AD is characterized by the presence of numerous senile plaques in the brain tissue, particularly in the hippocampus and cerebral cortex *(102)*. Senile plaques are extracellular deposits composed principally of insoluble aggregates of β-amyloid (Aβ), a protein derived from a larger membrane-spanning glycoprotein called "amyloid precursor protein" (APP). The progressive cerebral accumulation of Aβ protein is widely believed to be an early and necessary feature of AD pathology. There is increasing evidence that activated microglia are involved in plaque formation. It has long been known *(15)* that activated microglia, expressing MHC-II (HLA-DR), cluster around senile plaques in AD. In vitro, activated microglia synthesize and secrete Aβ proteins, and the activation of murine microglia can be induced by Aβ proteins, resulting in an the production and secretion of NO and TNF-α *(103,104)*. In addition, treatment of rat microglia with Aβ peptide induces the release of glutamate, a potential neurotoxin, by activating the Na^+-dependent glutamate transporter *(105)*. Using microaggregates of labeled Aβ peptide, microglial cells were demonstrated *(106)* to phagocytose Aβ-protein via scavenger receptor, suggesting that microglia also play a role in the clearance of Aβ plaques. On the basis of these observations, Aβ–protein/microglia interaction is believed to promote the progression of inflammatory and neurodegenerative changes in senile plaques in AD.

Evidence suggests that microglia play a central role in the pathophysiology of human AIDS dementia complex, or HIV dementia (HIVD). The presence of HIV-infected microglia has been demonstrated in brain and spinal cord sections of AIDS-demented patients. Infection by HIV-1 appears to be selective for microglia: Very limited infection of astrocytes and endothelial cells has been observed *(107–109)*. Viral replication appears to be restricted to microglia only. Human microglia isolated from adult or fetal tissue can be infected with many stains of HIV-1 *(110–112)*. HIV-1 enters cells through a multistep process that requires at least two receptors, namely, CD4 and a chemokine receptor (such as CCR5 or CXCR4). The presence of these receptors on microglia has been shown *(111,113)*, and blocking Abs against those receptors can inhibit HIV-1 infection of cultured microglia *(111)*. HIV-infected microglia produce secretory products. For example, TNF-α is released by HIV-infected microglia in vitro *(114)*, and TNF-α mRNA levels are higher in the subcortical regions of the CNS in patients with HIVD than control patients *(115)*. NO is another substance produced by microglia infected with HIV. Adamson et al. *(116)* demonstrated that gp120 can stimulate low-level production of NO by macrophages, and that high levels of inducible NO synthase, the enzyme responsible for the production of NO, are found in brains of patients with HIVD. NO production by microglia can also be induced in vitro, using the HIV-1 regulatory protein, Tat, possibly by activating the transcription factor, nuclear factor-κB *(117)*.

6. MECHANISMS OF MICROGLIA ACTIVATION

Microglia activation is a multistep process characterized by changes in cellular morphology, cell size, cell number, and in cell surface molecule expression. Mechanisms of microglia activation are not completely understood, but few have been described. Clearly, microglia activation is accompanied by an alteration in gene expression, which also results in the synthesis of inflammatory mediators, such as reactive O_2 and nitrogen intermediates, proteolytic enzymes, arachidonic acid metabolites, and proinflammatory cytokines.

Microglia undergo morphological transformations, from a ramified to an amoeboid form, through mechanisms that remain unclear. However, the ramification of amoeboid microglia, at least in vitro, is dependent on a Ca^{2+}-adenosine triphosphatase (ATPase), as shown by using the specific inhibitor, thapsigargin *(118)*. A commonly described cellular morphological change of activated microglia is hypertrophy. Hypertrophic microglia develop enlarged cell processes, which give the cells a bushy appearance. Hypertrophy is usually apparent by 24 h after CNS injury, which correlates with an upregulation of the complement receptor CR3 *(9)*. Hypertrophy of microglia is thought to be a hallmark of their proliferation. Microglial proliferation has been shown in vivo in various models of CNS injuries *(3,119,120)*. Following 2–3 d of CNS injury, it was shown that microglia begin proliferating, and their numbers reach maximal levels after 4–7 d *(8)*. In vitro, the proliferation of microglia cells can be regulated by soluble factors, including colony-stimulating factors, such as M-CSF and GM-CSF, primarily produced by astrocytes *(121)* and neurotrophins *(122)*. Accordingly, it is possible that microglia proliferation may be regulated indirectly via neuronal–astroglial signaling. Moreover, evidence suggests that molecules produced by hematogeneous macrophages are also involved. After transection of a CNS fiber tract, microglia are insufficiently activated, and this correlates with a lack of hematogenous macrophages infiltration into the degenerating nerve stump *(123)*. Moreover, CNS injuries, where there is the breakdown of the blood–brain barrier, such as cerebral ischemia, brain abscesses, and stab wounds, elicit prompt microglia activation associated with macrophage recruitment *(123)*.

Activators of microglia include the bacterial LPS and IFN-γ. In vitro, treatment with a combination of LPS and IFN-γ induces the production of reactive nitrogen oxides (NO, $ONOO^-$), superoxide, reactive oxygen intermediates (O_2^-, H_2O_2), and enzymes, such as lysozyme, cathepsin B/L, and acid hydrolases *(13)*. This has also been shown in in vivo experiments. For instance, after injecting a mixture of LPS and IFN-γ in the rat hippocampus, microglia are activated, as shown by morphological changes and by an increase in IL-1β and iNOS immunostaining *(124)*.

Direct contact between T-lymphocytes and microglia is another mechanism involved in microglia activation. We have demonstrated that microglia–T-cell interactions results in significant production of cytokines, such as TNF-α and IL-10 *(125,126)*. Some selectivity of ligand–receptor pairs in microglia–T-cell interactions was revealed in

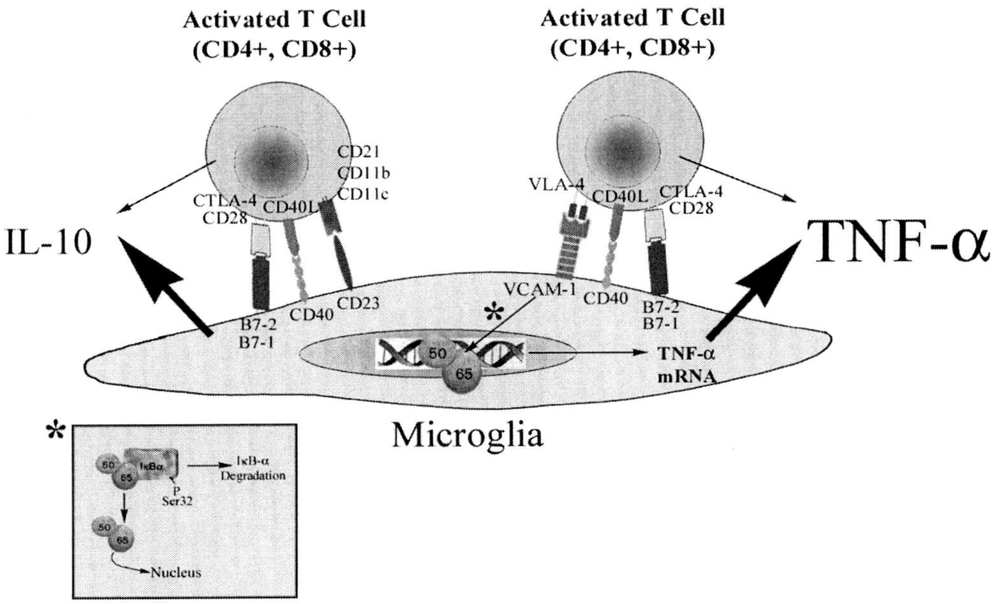

Fig. 1. Molecules that regulate the production of IL-10 and TNF-α in T cell–microglia interaction. The CD40:CD40L and B7:CD28/CTLA-4 pathways regulate the production of both cytokines while VLA-4:VCAM-1 interaction controls TNF-α exclusively. Similarly, CD23: CD21/CD11 interaction selectively regulates IL-10 but not TNF-α. Please refer to refs. *125–127* for details.

this study. Although the CD40/CD40L and CD28-CTLA-4/B7 pathways regulate both IL-10 and TNF-α, the VLA-4–VCAM-1 interaction was specific for TNF-α; in contrast, the CD23 system affected IL-10, but not TNF-α *(125,126*; Fig. 1*)*. Similarly, the ligation of microglial CD40 by CD40L was shown to induce the production of TNF-α *(127)*. Crosslinking of VCAM-1 on microglia was shown to induce TNF-α transcription *(125)* through a mechanism that probably involves the activation of nuclear factor-κB *(128)*.

Microglia cultured from rat, mouse, and human fetal nervous system express an inward rectifying K^+ channel, but no outward currents, which makes them more sensitive to changes in extracellular K than any other cell type in the brain *(13,129,130)*. In the cortical-spreading depression, microglia respond to neuronal depolarizations, which are associated with increased K fluxes across membranes, in the absence of neuronal damage, suggesting that inward-rectifying K^+ channel may play a role in macrophage activation *(131)*. Other ion channels have been implicated in microglia activation, including Cl^- channels. Indeed, it was shown that Cl channel blockers, and not blockers of Na^+ or K^+ channels, inhibit microglia activation, by preventing their ramification from an amoeboid form *(132)*, and by blocking the production of NO *(133)*.

Results obtained from in vitro studies suggest that the co-transmitter, ATP, released from neurons during CNS injury, is involved in microglia activation. Microglia express P2-purinoreceptors, which bind to ATP. The extracellular application of ATP, to cultured microglia from mouse brains, induces complex membrane currents, resulting in the depolarization of microglia *(129,134–136)*. It is not clear whether ATP is a direct activator of microglia, or whether it acts more as a modulator of microglia activation.

Table 1
Activators of Microglia

Activators of microglia	Inducing effects on microglia	Refs.
LPS	Cytokine production	*(148)*
	(TNF-α, IL-1β, IL-6,	*(124)*
	IL-10, IL-12)	*(149)*
	NO production	*(150)*
	Superoxide production	*(151)*
	Phagocytosis	*(75)*
	K^+ channel expression	*(152)*
	trk C, NT-3, and CCR5	*(145)*
	expression	*(143)*
	B7-1 expression	*(153)*
	MMP-9 production	*(154)*
	Immunoproteosome subunits	*(155)*
	production	
IFN-γ	TNF-α and IL-6 production	*(46)*
	Phagocytosis	*(75)*
	B7-1 and B7-2 expression	*(145)*
	Morphological transformation	*(156)*
	Inward-rectifying K^+ current	*(157)*
	MCP-1 production	*(155)*
		(103,158)
		(159)
Contact with activated T-cells	Cytokine production	*(125,126,128)*
	(TNF-α, IL-10, IL-12, IL-1β)	
Inward rectifying K^+ channels	Membrane depolarization	*(131)*
Cl^- channels	Morphological transformation	*(132)*
	NO production	*(133)*
Adenosine A2a receptor	K^+ channel expression	*(160)*
ATP/Purinergic receptors	Membrane depolarization	*(129)*
	IL-1β production	*(134)*
	Ca^{2+} signaling	*(135)*
		(136)
		(137)
		(161)
Gram-positive	Outward rectifying K^+ channel	*(162)*
Streptococcus pneumoniae	production	
	TNF-α, IL-6 and IL-12	
	production	
	MIP-1 and -2 production	
β-amyloid protein (APP)	Cytokine production	*(103,158,163)*
	(TNF-α, IL-1β)	
	NO production	*(164)*
	Ca^{2+} signaling	*(165)*
	Phagocytosis	*(166)*
	NADPH oxidase activation	*(167)*
	MAPK signaling	*(168,169)*

Table 1 *(continued)*

Activators of microglia	Inducing effects on microglia	Refs.
	MIP-1α and MCP-1 production	*(104)*
	CD40 expression	*(159)*
	Glutamate release	*(170)*
		(171)
		(172)
		(105)
		(173)
Tissue plasminogen activator	TNF-α production	*(174)*
Complement fragments (C5a and C3a)	Ca^{2+} signaling	*(138)*
PAF	Ca^{2+} signaling	*(139)*
Gangliosides (GM1, GD1a, GT1b)	NO production	*(140)*
	TNF-α production	
	COX-2 production	
Prion protein	Ca^{2+} signaling	*(175)*
		(164)
HIV-1 Tat protein	NO production	*(117)*
α-crystallin	NO production	*(141)*
	TNF-α production	
Manganese	NO production	*(176)*

It has been demonstrated *(137)* that LPS-induced released of IL-1β is modulated in microglia cells lines via purinergic receptors.

Other activators of microglia include the complement fragments, PAF, gangliosides, and the stress-inducible small heat shock protein, α-crystallin (Table 1). Complement fragments, such as C5a and C3a, which are generated from complement activation, activate microglia by inducing Ca^{2+} signaling *(138)*. In the case of the inflammatory mediator, PAF, it was shown to induce an increase of Ca^{2+} influx in microglia, by causing the activation of store-operated-Ca^{2+} channels *(139)*. Gangliosides, such as GM1, GD1a, GD1b, GT1b, and GQ1b, are glycosphingolipid-containing sialic acid residues that are located in mammalian cell membranes. GT1b was shown to induce the production of NO, TNF-α, and cyclo-oxygenase-2 on rat microglia; GM1 and GD1a induce the production of NO only *(140)*. Finally, α-crystallin induces in vitro activation of microglia, by inducing the production of NO and iNOS *(141)*. α-crystallin also stimulates the synthesis of the proinflammatory cytokine, TNF-α *(141)*.

Deactivators of microglia have also been identified (Table 2). The best-described deactivators of microglia are the anti-inflammatory cytokines, IL-10 and TGF-β. For example, it was shown that IL-10 inhibits the LPS-induced microglial production of TNF-α, IL-1β, lysosomal enzyme activity, NO, superoxide anion, and the chemokine, RANTES *(142,143)*. Ligation of CD40 on microglia induces the production of TNF-α *(144)*. However, upon treatment with IL-10 and TGF-β, CD40-induced production of TNF-α is inhibited *(144)*. Finally, the expression of MHC-II, the class II transactivator transcription factor, and the co-stimulatory molecule, B7-2, induced by IFN-γ or LPS, was shown to be inhibited by IL-10 and/or TGF-β in primary murine microglia

Table 2
Deactivators of Microglia

Deactivators of microglia	Inhibiting effects on microglia	Refs.
IL-10	TNF-α, IL-1β, IL-12, IL-8 production	*(177)*
	Lysosomal enzyme activity/phagocytosis	*(142)*
	Superoxide anion production	*(178)*
	NO production	*(144)*
	RANTES production	*(146)*
	MHC II expression	*(179)*
	IL-8 production	*(75)*
	B7-2 expression	*(145)*
		(143)
TGF-β	TNF-α production	*(142)*
	RANTES production	*(144)*
	MHC-II expression	*(146)*
	Phagocytosis	*(41)*
	IL-8 production	*(178)*
IL-4	MHC-II expression	*(146)*
	Class II transactivate expression	*(41)*
	Phagocytosis	*(178)*
	IL-8 production	
Soluble TNF receptor	IL-12 production	*(180)*
Fractalkine	LPS-induced TNF-α production	*(181)*
Apolipoprotein E	NO production	*(182)*
Lipocortin 1	COX-2 expression	*(183)*
	NO production	
Prostaglandins (i.e., PGE_2)	TNF-α production	*(41)*
PGE receptors (EP_2)	Induce IL-10 production	*(184)*
	NO production	*(185)*
	IL-1β production	*(186)*
	cAMP production	
Vitamin E	Morphological transformation	*(187)*
	LFA-1, VLA-4, and ICAM-1 expression	

(145,146). Another deactivator of microglia is prostaglandin. For example, it was demonstrated *(147)* that prostaglandin E2 selectively inhibits the production of TNF-α (by 95%) and IL-6 in LPS-stimulated microglia. Cyclo-oxygenase-2, pro-IL-1β, or iNOS production was not affected by such a treatment.

7. SUMMARY

Microglia play a central role in the regulation of immune and inflammatory responses taking place within the CNS, by acting as an APC, and by producing a wide variety of immune regulators. Activation of microglia is crucial for microglial functions. Although few mechanisms of microglia activation have been described in vitro, it will be important to eludicate in vivo mechanisms, in order to modulate functions of microglia that are involved in the pathogenesis of CNS diseases.

REFERENCES

1. del Rio-Hortega, P. and Penfield, W. (1927) Cerebral cicatrix. The reaction of neuroglia and microglia to brain wounds. *Bull. Johns Hopkins Hosp.* **41,** 278.
2. Kreutzberg, G. W. (1987) Microglia, in *Encyclopedia of Neuroscience* (Adelman, G., ed.), Birkhauser, Boston, pp. 661.
3. Lawson, L. J., Perry, V. H., and Gordon, S. (1992) Turnover of resident microglia in the normal adult mouse brain. *Neurosci.* **48,** 405.
4. Streit, W. J. (1995) Microglial cells, in *Neuroglia* (Kettenmann, H. and Ransom, B. R., eds.), Oxford University Press, New York, p. 85.
5. Perry, V. H. and Gordon, S. (1988) Macrophages and microglia in the nervous system. *Trends Neurol. Sci.* **11,** 273.
6. Ferrer, I., Bernet, E., Soriano, E., Del Rio, T., and Fonseca, M. (1990) Naturally occurring cell death in the cerebral cortex of the rat and removal of dead cells by transitory phagocytes. *Neuroscience* **39,** 451.
7. Hume, D. A., Perry, V. H., and Gordon, S. (1983) Immunohistochemical localisation of a macrophage-specific antigen in developing mouse retina: phagocytosis of dying neurons and differentiation of microglia cells to form a regular array in the plexiform layers. *J. Cell. Biol.* **97,** 253.
8. Streit, W. J., Walter, S. A., and Pennell, N. A. (1999) Reactive microgliosis. *Prog. Neurobiol.* **57,** 563.
9. Graeber, M. B. and Streit, W. J. (1990) Perivascular microglia defined. *Trends Neurosci.* **13,** 366.
10. Pennell, N. A. and Streit, W. J. (1998) Tracing of fluoro-gold-pre-labeled microglia injected into the adult rat brain. *Glia* **23,** 84.
11. Gehrmann, J., Matsumoto, Y., and Kreutzberg, G. W. (1995) Microglia: intrinsic immuneffector cell of the brain. *Brain Res. Rev.* **20,** 269.
12. Hickey, W. F. and Kimura, H. (1988) Perivascular microglial cells of the CNS are bone marrow-derived and present antigen in vivo. *Science* **239,** 290.
13. Rezaie, P. and Male, D. (1999) Colonisation of the developing human brain and spinal cord by microglia: a review. *Microsc. Res. Technol.* **45,** 359.
14. Mattiace, L. A., Davies, P., and Dickinson, D. (1990) Detection of HLA-DR on microglia in the human brain is a function of both clinical and technical factors. *Am. J. Pathol.* **136,** 1101.
15. McGeer, P. L., Itagaki, S., Tago, S., and McGeer, E. G. (1987) Reactive microglia in patients with senile dementia of Alzheimer's type are positive for the histocompatibility glycoprotein HLA-DR. *Neurosci. Lett.* **79,** 1285.
16. Sedgwick, J. D., Schwender, S., Imrich, H., Dorries, R., Butcher, G. W., and ter Meulen, V. (1991) Isolation and direct characterization of resident microglial cell from the normal and inflamed nervous system. *Proc. Natl. Acad. Sci. USA* **88,** 7438.
17. Hao, C., Richardson, A., and Fedoroff, S. (1991) Macrophage-like cells originate from neuroepithelium in culture: characterisation and properties of the macrophage-like cells. *Int. J. Dev. Neurosci.* **9,** 1.
18. Kurz, H. and Christ, B. (1998) Embryonic CNS macrophages and microglia do not stem from circulating, but from extravascular precursors. *Glia* **22,** 98.
19. Richardson, A., Hao, C., and Fedoroff, S. (1993) Microglial progenitors cells: a subpopulation in cultures of mouse neopallial astroglia. *Glia* **7,** 25.
20. Cuadros, M. A. and Navascues, J. (1998) The origin and differentiation of microglial cells during development. *Progr. Neurobiol.* **56,** 173.
21. Ashwell, K. W. S., Hollander, H., Streit, W., and Stone, J. (1989) The appearance and distribution of microglia in the developing retina of the rat. *Vis. Neurosci.* **2,** 437.
22. Boya, J., Calvo, J., and Carbonell, A. L. (1987) Appearance of microglial cells in the postnatal rat retina. *Arch. Histol. Jpn.* **50,** 223.

23. Alliot, F., Lecain, E., Grima, B., and Pessac, B. (1991) Microglial progenitors with a high proliferative potential in the embryonic and adult mouse brain. *Proc. Natl. Acad. Sci. USA* **88,** 1541.
24. Hickey, W. F., Vass, K., and Lassmann, H. (1992) Bone marrow derived elements in the central nervous system: an immunohistochemical and ultrastructural survey of rat chimeras. *J. Neuropathol. Exp. Neurol.* **51,** 246.
25. Unger, E. R., Sung, J. H., Manivel, J. C., Chenggis, M. L., Blazar, B. R., and Krivit, W. (1993) Male donor-derived cells in the brains of female sex-mismatched bone marrow transplant recipients: a Y-chromosome specific in situ hybridization study. *J. Neuropathol. Exp. Neurol.* **52,** 460.
26. Krall, W. J., Challita, P. M., Perlmutter, L. S., Skelton, D. C., and Kohn, D. B. (1994) Cell expressing human galactocerebrosidase from a retroviral vector repopulate macrophages and central nervous system microglia after murine bone marrow transplantation. *Blood* **83,** 2737.
27. De Groot, C. J. A., Huppes, W., Sminia, T., Kraal, G., and Dijkstra, C. D. (1992) Determination of the origin and nature of brain macrophages and microglia cells in mouse central nervous system, using non-radioactive in situ hybridization and immunoperoxidase techniques. *Glia* **6,** 301.
28. Matsumoto, Y. and Fujiwara, M. (1987) Absence of donor-type major histocompatibility complex class I antigen-bearing microglia in the rat central nervous system of radiation bone marrow chimeras. *J. Neuroimmunol.* **17,** 71.
29. Eglitis, M. and Mezey, E. (1997) Hematopoietic cells differentiate into both microglia and macroglia in the brains of adult mice. *Proc. Natl. Acad. Sci. USA* **94,** 4080.
30. Kitamura, T., Miyake, T., and Fujita, S. (1984) Genesis of resting microglia in the gray matter of mouse hippocampus. *J. Comp. Neurol.* **226,** 421.
31. Cuzner, M. L. (1997) Microglia in health and disease. *Biochem. Soc. Trans.* **25,** 671.
32. Penfield, W. (1928) Neuroglia and microglia. The interstitial tissues of the central nervous system, in *Special Cytology* (Cowdry, P. B., ed.), Paul B. Hoeber, New York, p. 1032.
33. McGeer, P. L. and McGeer, E. G. (1999) Inflammation of the brain in Alzheimer's disease: limitations for therapy. *J. Leuk. Biol.* **65,** 409.
34. Boya, J., Calvo, J., and Prado, A. (1979) The origin of microglia cells. *J. Anat.* **129,** 177.
35. Ling, E. A., Kaur, C., and Wong, W. C. (1982) Light and electron microscopic demonstration of non-specific esterase in ameboid microglial cells in the corpus callosum in postnatal rats: a cytochemical link to monocytes. *J. Anat.* **135,** 385.
36. Murabe, Y. and Sano, Y. (1982) Morphological studies on microglia. V. Microglial cells in the cerebral cortex of the rat with special reference to their possible involvement in synaptic function. *Cell Tissue Res.* **223,** 493.
37. Fujimoto, E., Miki, A., and Mizoguti, H. (1989) Histochemical study of the differentiation of microglial cells in the developing human cerebral hemispheres. *J. Anat.* **166,** 253.
38. Castellano, B., Gonzalez, B., Jensen, M. B., Pedersen, E. B., Finsen, B. R., and Zimmer, J. (1991) A double staining technique for simultaneous staining of astrocytes and microglial cells in vibratome brain sections and astroglial cell cultures. *J. Histochem. Cytochem.* **39,** 561.
39. Ulvestad, E., Williams, K., Bo, L., et al. (1994) HLA- class II molecule (HLA-DR, -DP, -DQ) on cells in the human CNS studies in situ and in vitro. *Immunology* **82,** 533.
40. DeWitt, D. A., Perry, G., Cohen, M., Doller, C., and Silver, J. (1998) Astrocytes regulate microglial phagocytosis of senile plaque cores of Alzheimer's disease. *Exp. Neurol.* **149,** 329.
41. Von Zahn, J., Moller, T., Kettenmann, H., and Nolte, C. (1997) Microglial phagocytosis is modulated by pro- and anti-inflammatory cytokines. *Neuroreport* **8,** 3851.
42. Ford, A. L., Goodsall, A. L., Hickey, F., and Sedgwick, J. D. (1995) Normal adult ramified microglia seperated from other central nervous system macrophages by flow

cytometry sorting. Phenotypic differences defined and direct ex vivo antigen presentation to myelin basic protein-reactive CD^{4+} T cells compared. *J. Immunol.* **154,** 4309.
43. Carson, M. J., Reilly, C. R., Sutcliffe, J. G., and Lo, D. (1998) Mature microglia resemble immature antigen-presenting cells. *Glia* **22,** 72.
44. Havenith, C. E. G., Askew, D., and Walker, W. S. (1998) Mouse resident microglia: isolation and characterization of immunoregulatory properties with naïve CD^{4+} and CD^{8+} T cells. *Glia* **22,** 348.
45. Dickson, D. W., Lee, S. C., Mattiace, L. A., Yen, S. H. C., and Brosnan, C. (1993) Microglia and cytokines in neurological disease with special reference to AIDS and Alzheimer's disease. *Glia* **7,** 75.
46. Aloisi, F., De Simone, R., Columba-Cebezas, S., and Levi, G. (1999) Opposite effects of interferon-γ and prostaglandin E2 on tumor necrosis factor and interleukin-10 production in microglia: a regulatory loop controlling microglia pro- and anti-inflammatory activities. *J. Neurosci. Res.* **56,** 571.
47. Aloisi, F., Ria, F., Penna, G., and Adorini, L. (1998) Microglia are more efficient than astrocytes in antigen processing and in Th1 but not Th2 cell activation. *J. Immunol.* **160,** 4671.
48. Sedgwick, J. D., Schwender, S., Gregersen, R., et al. (1993) Resident macrophages (ramified microglia) of the adult Brown Norway rat central nervous system are constitutively major histocompatibility complex class II positive. *J. Exp. Med.* **177,** 1145.
49. Becher, B. and Antel, J. P. (1996) Comparison of phenotypic and functional properties of immediately ex vivo and cultured human adult microglia. *Glia* **18,** 1–10.
50. Ulvestad, E., Williams, K., Matre, R., Nyland, H., Olivier, A., and Antel, J. P. (1994) Fc receptors for IgG on cultured human microglia mediate cytotoxicity and phagocytosis of antibody-coated targets. *J. Neuropathol. Exp. Neurol.* **53,** 27.
51. Williams, K., Ulvestad, E., and Antel, J. P. (1995) B7/BB1 antigen expression on adult human microglia studies in vitro and *in situ*. *Eur. J. Immunol.* **24,** 3031.
52. Dangond, F., Windhagen, A., Groves, C. J., and Hafler, D. A. (1997) Constitutive expression of costimulatory molecules by human microglia and its relevance to CNS autoimmunity. *J. Neuroimmunol.* **76,** 132.
53. De Simone, R., Giampaolo, A., Giometto, B., Gallo, P., Levi, G., Peschle, C., and Aloisi, F. (1995) The co-stimulatory molecule B7 is expressed on human microglia in culture and in multiple sclerosis acute lesions. *J. Neuropathol. Exp. Neurol.* **54,** 175.
54. Satoh, J., Lee, Y. B., and Kim, S. U. (1995) T-cell costimulatory molecules B7-1 (CD80) and B7-2 (CD86) are expressed in human microglia but not in astrocytes in culture. *Brain. Res.* **704,** 92.
55. Williams, K., Ulvestad, E., Antel, J. P., and McLaurin, J. (1994) Activation of adult human derived microglia by myelin phagocytosis in vitro. *J. Neurosci. Res.* **38,** 433.
56. Matyszak, M. K., Denis-Donini, S., Cltterio, S., Longini, R., Granucci, F., and Ricciardi-Castagnoli, P. (1999) Microglia induce myelin basic protein-specific T cell anergy and T cell activation, according to their state of activation. *Eur. J. Immunol.* **29,** 3063.
57. Perry, V. H., Bell, M. D., Brown, H. C., and Matyszak, M. K. (1995) Inflammation in the nervous system. *Curr. Opin. Neurobiol.* **5,** 636.
58. Munoz-Fernandez, M. A. and Fresno, M. (1998) The role of tumor necrosis factor, interleukin-6, interferon-gamma and inducible nitric oxide synthase in the development and pathology of the nervous system. *Progr. Neurobiol.* **56,** 307.
59. Cuzner, M. L., Gveric, D., Strand, C., Loughlin, A. J., Paemen, L., Opdenakker, G., and Newcombe, J. (1996) The expression of tissue-type plasminogen activator, matrix metalloproteinases and endogeneous inhibitor in the central nervous system in multiple sclerosis: comparison of stages in lesion evolution. *J. Neuropathol. Exp. Neurol.* **55,** 1194.
60. Black, R. A., Rauch, C. T., Kozlosky, C. J., et al. (1997) A metalloproteinase disintegrin that releases tumour-necrosis factor-alpha from cells. *Nature* **385,** 729.

61. Gearing, A. J. H., Beckett, P., Christodoulou, M., et al. (1994) Processing of tumor necrosis factor-α precursor by metalloproteinases. *Nature* **370,** 555.
62. Piani, D., Frei, K., Do, K. Q., Cuenod, M., and Fontana, A. (1991) Murine brain macrophages induce NMDA receptor mediated neurotoxicity by secreting glumatate. *Neurosci. Lett.* **133,** 159.
63. Nicoletti, F., Patti, F., Cocuzza, C., Zaccone, P., Nicoletti, A., Di Marco, R., and Reggio, A. (1996) Elevated serum levels of interleukin-12 in chronic progressive multiple sclerosis. *J. Neuroimmunol.* **70,** 87.
64. El Khoury, J., Hickman, S. E., Thomas, C. A., Loike, J. D., and Silverstein, S. C. (1998) Microglia, scavenger receptors, and the pathogenesis of Alzheimer's disease. *Neurobiol. Aging* **19,** S81.
65. Haga, S., Aizawa, T., Ishii, T., and Ikeda, K. (1996) Complement gene expression in mouse microglia and astrocytes in culture: comparisons with mouse peritoneal macrophages. *Neurosci. Lett.* **216,** 191.
66. Gremo, F., Sogos, V., Ennas, M. G., Meloni, A., Persichini, T., Colasanti, M., and Lauro, G. M. (1997) Features and functions of human microglia cells. *Adv. Exp. Med. Biol.* **429,** 79.
67. Asencio, V. C. and Campbell, I. L. (1997) Chemokine gene expression in the brains of mice with lymphocytic choriomeningitis. *J. Virol.* **71,** 7832.
68. Nishiyori, A., Minami, M., Ohtani, Y., Takami, S., Yamamoto, J., Kawaguchi, N., et al. (1998) Localisation of fractalkine and CXCR1 mRNAs in rat brain: does fractalkine play a role in signaling from neuron to microglia? *FEBS Lett.* **429,** 167.
69. Raine, C. S. (1994) The immunology of the multiple sclerosis lesions. *Ann. Neurol.* **36,** S61.
70. Trapp, B. D., Bo, L., Mork, S., and Chang, A. (1999) Pathogenesis of tissue injury in MS lesions. *J. Neuroimmunol.* **98,** 49.
71. Bo, L., Mork, S., Kong, P. A., Nyland, H., Pardo, C. A., and Trapp, B. D. (1994) Detection of MHC class II-antigens on macrophages and microglia, but not on astrocytes and endothelia in active multiple sclerosis lesions. *J. Neuroimmunol.* **51,** 135.
72. Schonrock, L. M., Kuhlmann, T., Alder, S., Bittsch, A., and Bruck, W. (1998) Identification of glial cell proliferation in early multiple sclerosis. *Neuropathol. Appl. Neurobiol.* **24,** 320.
73. Cannella, B. and Raine, C. S. (1995) The adhesion molecule and cytokine profile of multiple sclerosis lesions. *Ann. Neurol.* **37,** 424.
74. Smith, K. J., Kapoor, R., and Felts, P. A. (1999) Demyelination: the role of reactive oxygen and nitrogen species. *Brain Pathol.* **9,** 69.
75. Smith, M. E., van der Maesen, K., and Somera, F. P. (1998) Macrophage and microglial responses to cytokines in vitro: phagocytic activity, proteolytic enzyme release, and free radical production. *J. Neurosci. Res.* **54,** 68.
76. Smith, M. E. (1993) Phagocytosis of myelin by microglia in vitro. *J. Neurosci. Res.* **35,** 480.
77. Trotter, J., De Jong, L. J., and Smith, M. E. (1986) Opsonization with antimyelin antibody increases the uptake and intracellular metabolism of myelin in inflammatory macrophages. *J. Neurochem.* **47,** 779.
78. Reichert, F., Saada, A., and Rotshenker, S. (1994) Peripheral nerve injury induces Schwann cells to express two macrophage phenotypes: phagocytosis and the galactose-specific lectin MAC-2. *J. Neurosci.* **14,** 3231.
79. Rinner, W. A., Bauer, J., Schmidts, M., Lassmann, H., and Hickey, W. F. (1995) Resident microglia and hematogeneous macrophages as phagocytes in adoptively transferred experimental autoimmune encephalomyelitis: an investigation using rat radiation bone marrow chimeras. *Glia* **14,** 257.
80. Prineas, J. W. and Graham, J. S. (1981) Multiple sclerosis: capping of surface immunoglobulin G on macrophages engaged in myelin breakdown. *Ann. Neurol.* **10,** 149.

81. Prineas, J. W., Kwon, E. E., Cho, E. S., and Sharer, L. R. (1984) Continual breakdown and regeneration of myelin in progressive multiple sclerosis plaques. *Ann. NY Acad. Sci.* **436,** 11.
82. Nyland, H., Mork, S., and Matre, R. (1982) *In-situ* characterization of mononuclear cell infiltrates in lesions of multiple sclerosis. *Neuropathol. Appl. Neurobiol.* **8,** 403.
83. Cammer, W. B., Bloom, R., Norton, W. T., and Gordon, S. (1978) Degradation of basic protein in myelin by neutral proteases secreted by stimulated macrophages: a possible mechanism of inflammatory demyelination. *Proc. Natl. Acad. Sci. USA* **75,** 1554.
84. Gijbels, K., Proost, P., Masure, S., Carton, H., Billiau, A., and Opdenakker, G. (1993) Gelatinase B is present in the cerebrospinal fluid during experimental autoimmune encephalomyelitis and cleaves myelin basic protein. *J. Neurosci. Res.* **36,** 432.
85. Proost, P., Van Damme, J., and Opdenakker, G. (1992) Leukocyte gelatinase B cleavage releases encephalitogens from human myelin basic protein. *Biochem. Biophys. Res. Commun.* **192,** 1175.
86. Sedgwick, J. D., Ford, A. L., Foulcher, E., and Airriess, R. (1998) Central nervous system microglial cell activation and proliferation follows direct interaction with tissue-infiltrating T cell blasts. *J. Immunol.* **160,** 5320.
87. Zhang, M. and Tracey, K. J. (1998) Tumor necrosis factor, in *The Cytokine Handbook* (Thompson, A. W., ed.), Academic, San Diego, p. 517.
88. Hofman, F. M., Hinton, D. R., Johnson, K., and Merrill, J. E. (1989) Tumor necrosis factor identified in multiple sclerosis brain. *J. Exp. Med.* **170,** 607.
89. Selmaj, K., Raine, C. S., Cannella, B., and Brosnan, C. F. (1991) Identification of lymphotoxin and tumor necrosis factor in multiple sclerosis lesions. *J. Clin. Invest.* **87,** 949.
90. Reickmann N. P., Albrecht, M., Kitze, B., Weber, T., Tumani, H., Broocks, A., et al. (1995) Tumor necrosis factor-α messenger RNA expression in patients with relapsing-remitting multiple sclerosis is associated with disease activity. *Ann. Neurol.* **37,** 82.
91. Glabinski, A., Mirecka, M., and Pokoca, L. (1995) Tumor necrosis factor alpha but not lymphotoxin is overproduced by blood mononuclear cells in multiple sclerosis. *Acta Neurol. Scand.* **91,** 276.
92. Probert, L., Akassoglou, K., Pasparakis, M., Kontogeorgos, G., and Kollias, G. (1995) Spontaneous inflammatory demyelinating disease in transgenic mice showing central nervous system-specific expression of tumor necrosis factor-α. *Proc. Natl. Acad. Sci. USA* **92,** 11,294.
93. Taupin, V., Renno, T. T., Bourbonniere, L., Peterson, A. C., Rodriguez, M., and Owens, T. (1997) Increased severity of experimental autoimmune encephalomyelitis, chronic macrophage/microglial reactivity, and demyelination in transgenic mice producing tumor necrosis factor-alpha in the central nervous system. *Eur. J. Immunol.* **27,** 905.
94. Ruddle, N. H., Bergman, C., McGrath, K., Lingenheld, E., Grunnet, M., Padula, S., and Clark, R. (1990) An antibody to lymphotoxin prevents transfer of experimental allergic encephalomyelitis. *J. Exp. Med.* **172,** 1193.
95. Selmaj, K., Papierz, W., Glabinski, A., and Kohno, T. (1995) Prevention of chronic relapsing experimental autoimmune encephalomyelitis by soluble TNF receptor 1. *J. Neuroimmunol.* **56,** 135.
96. Selmaj, K. W. and Raine, C. S. (1995) Experimental autoimmune encephalomyelitis: immunotherapy with anti-tumor necrosis factor antibodies and soluble tumor necrosis factor receptors. *Neurology* **45,** S44.
97. Selmaj, K. W. and Raine, C. S. (1988) Tumor necrosis factor mediates myelin and oligodendrocytes damage in vitro. *Ann. Neurol.* **23,** 339.
98. Louis, J.-C., Magal, E., Takayama, S., and Varon, S. (1993) CNTF protection of oligodendrocytes against natural and tumor necrosis factor-induced death. *Science* **259,** 689.

99. D'Souza, S., Alinauskas, K., McCrea, E., Goordyer, C., and Antel, J. P. (1995) Differential susceptibility of human CNS-derived cell populations to TNF-dependent and independent immune-mediated injury. *J. Neurosci.* **15,** 7293.
100. Butt, A. M. and Jenkins, H. G. (1994) Morphological changes in oligodendrocytes in the intact mouse optic neve following intravitreal injection of tumor necrosis factor. *J. Neuroimmunol.* **51,** 27.
101. Selmaj, K. W., Walczak, A., Mycko, M., Berkowicz, T., Kohno, T., and Raine, C. S. (1998) Suppression of experimental autoimmune encephalomyelitis with a TNF binding protein (TNFbp) correlates with down-regulation of VCAM-1/VLA-4. *Eur. J. Immunol.* **28,** 2035.
102. Selkoe, D. J. (1991) The molecular pathology of Alzheimer's disease. *Neuron* **6,** 487.
103. Meda, L., Cassatella, M. A., Szendrei, G. I., Otvos, L., Baron, P., Villalba, M., Ferrari, D., and Rossi, F. (1995) Activation of microglial cells by β-amyloid protein and interferon-β. *Nature* **374,** 647.
104. Barger, S. W. and Harmon, A. D. (1997) Microglial activation by Alzheimer amyloid precursor protein and modulation by apolipoprotein E. *Nature* **388,** 878.
105. Noda, M., Nakanishi, H., and Akaike, N. (1999) Glutamate release from microglia via glutamate transporter is enhanced by amyloid-beta peptide. *Neuroscience* **92,** 1465.
106. Paresce, D. M., Ghosh, R. N., and Maxfield, F. R. (1996) Microglial cells internalize aggregates of the Alzheimer's disease amyloid β-protein via a scavenger receptor. *Neuron* **17,** 553.
107. Wiley, C. A., Schrier, R. D., Nelson, J. A., Lampert, P. W., and Oldstone, M. B. A. (1986) Cellular localization of human immunodeficiency virus infection within the brains of acquired immune deficiency syndrome patients. *Proc. Natl. Acad. Sci. USA* **83,** 7089.
108. Bagasara, O., Lavi, E., Bobroski, L., Khalili, K., Prestaner, J. P., and Pomerantz, R. J. (1996) Cellular reservoirs of HIV-1 in the central nervous system of infected individuals: identification by the combination of in situ PCR and immunohistochemistry. *AIDS* **10,** 573.
109. Glass, J. D., Fedor, H., Wesselingh, S. L., Selnes, O. A., and McArthur, J. C. (1993) Clinical-neuropathological correlation in HIV-associated dementia. *Neurology* **43,** 2230.
110. Strizki, J. M., Albright, A. V., Sheng, H., O'Connor, M., Perrin, L., and Gonzalez-Scarano, F. (1996) Infection of primary human microglia and monocyte-derived macrophages with acute HIV-1 isolates: evidence of differential tropism. *J. Virol.* **70,** 7564.
111. He, J., Chen, Y., Farzan, M., Choe, H., Ohagen, A., et al. (1997) CCR3 and CCR5 are coreceptors for HIV-1 infection of microglia. *Nature* **385,** 645.
112. Gonzalez-Scarano, F. and Baltuch, G. (1999) Microglia as mediators of inflammatory and degenerative diseases. *Annu. Rev. Neurosci.* **22,** 219.
113. Williams, K., Bar-Or, A., Ulvestad, E., et al. (1992) Biology of adult human microglia in culture: comparisons with peripheral blood monocytes and astrocytes. *J. Neuropathol. Exp. Neurol.* **51,** 538.
114. Wilt, S. G., Milward, E., Zhou, J. M., Nagasato, K., Patton, H., et al. (1995) In vitro evidence for a dual role of tumor necrosis factor-α in human immunodeficiency virus type 1 encephalopathy. *Ann. Neurol.* **37,** 381.
115. Wesselingh, S. L., Takahashi, K., Glass, J. D., McArthur, J. C., Griffin, J. W., and Griffin, D. E. (1997) Cellular localization of tumor necrosis factor mRNA in neurological tissue from HIV-infected patients by combined reverse transcriptase/polymerase chain reaction in situ hybridization and immunohistochemistry. *J. Neuroimmunol.* **74,** 1.
116. Adamson, D. C., Wildemann, B., Sasaki, M., et al. (1996) Immunologic NO synthase: elevation in severe AIDS dementia and induction by gp41. *Science* **274,** 1917.
117. Polazzi, E., Levi, G., and Minghetti, L. (1999) Human immunodeficiency virus type 1 Tat protein stimulates inducible nitric oxide synthase expression and nitric oxide production in microglial cultures. *J. Neuropathol. Exp. Neurol.* **58,** 825.

118. Yagi, R., Tanaka, S., and Koike, T. (1999) Thapsigargin induces microglial transformation from amoeboid to ramified type in vitro. *Glia* **28,** 49.
119. Streit, W. J., Graeber, M. B., and Kreutzberg, G. W. (1988) Functional plasticity of microglia: a review. *Glia* **1,** 301.
120. Amat, J. A., Ishiguro, H., Nakamura, K., and Norton, W. T. (1996) Phenotypic diversity and kinetics of proliferating microglia and astrocytes following cortical stab wounds. *Glia* **16,** 368.
121. Suzumura, A., Sawada, M., Yamamoto, H., and Marunouchi, T. (1990) Effects of colony stimulating factors on isolated microglia in vitro. *J. Neuroimmunol.* **30,** 237.
122. Elkabes, S., DiCicco-Bloom, E. M., and Black, I. B. (1996) Brain microglia macrophages express neurotrophins that selectively regulate microglial proliferation and function. *J. Neurosci.* **16,** 2508.
123. Stroll, G. and Jander, S. (1999) The role of microglia and macrophages in the pathophysiology of the CNS. *Progr. Neurobiol.* **58,** 233.
124. Hartlage-Rubsamen, M., Lemke, R., and Schliebs, R. (1999) Interleukin-1beta, inducible nitric oxide synthase, and nuclear factor-kB are induced in morphologically distinct microglia after rat hippocampal lipopolysaccharide/interferon gamma. *J. Neurosci. Res.* **57,** 388.
125. Chabot, S., Williams, G., and Yong, V. W. (1997) Microglia production of TNF-α is induced by activated T lymphocytes. Involvement of VLA-4 and inhibition by interferonβ-1b. *J. Clin. Invest.* **100,** 604.
126. Chabot, S., Williams, G., Hamilton, M., Sutherland, G., and Yong, V. W. (1999) Mechanisms of IL-10 production in human microglia–T cell interaction. *J. Immunol.* **162,** 6819.
127. Tan, J., Town, T., Paris, D., Placzek, A., Parker, T., Crawford, F., Yu, H., Humphrey, J., and Mullan, M. (1999) Activation of microglial cells cy the CD40 pathway: relevance to multiple sclerosis. *J. Neuroimmunol.* **97,** 77.
128. Chabot, S., Hirashima, H., and Yong, V. W. (1999) VCAM-1 induces NF-κB activation in macrophages, submitted.
129. Kettenmann, H., Banati, R., and Walz, W. (1993) Electrophysiological behavior of microglia. *Glia* **7,** 93.
130. McLarnon, J. G., Xu, R., Lee, Y. B., and Kim, S. U. (1997) Ion channels of human microglia in culture. *Neuroscience* **78,** 1217.
131. Gehrmann, J., Mies, G., Bonnekoh, P., Banati, R. B., Iijima, T., Kreutzberg, G. W., and Hossmann, K. A. (1993) Microglial reaction is the rat cerebral cortex induced by cortical spreading depression. *Brain Pathol.* **3,** 11.
132. Eder, C., Klee, R., and Heinemann, U. (1998) Involvement of stretch-activated Cl^- channels in ramification of murine microglia. *J. Neurosci.* **18,** 7127.
133. Brown, H., Kozlowski, R., and Perry, H. (1998) The importance of ion channels for macrophages and microglia activation in vitro. *Glia* **22,** 94.
134. Walz, W., Ilschner, S., Ohlemyer, C., Banati, R., and Kettenmann, H. (1993) Extracellular ATP activates a cation conductance and a K^+ conductance in cultured microglial cells from mouse brain. *J. Neurosci.* **13,** 4403.
135. Haas, S., Brockhaus, J., Verkhratsky, A., and Kettenmann, H. (1996) ATP-induced membrane currents in amoeboid microglia acutely isolated from mouse brain slices. *Neuroscience* **75,** 257.
136. Illes, P., Norenberg, W., and Gebicke-Haerter, P. J. (1996) Molecular mechanisms of microglial activation. B. Voltage- and purinoreceptor-operated channels in microglia. *Neurochem. Int.* **29,** 13.
137. Ferrari, D., Chiozzi, P., Falzoni, S., Hanau, S., and Di Virgilio, F. (1997) Purinergic modulation of interleukin-1β release from microglial cells stimulated with bacterial endotoxin. *J. Exp. Med.* **185,** 579.

138. Moller, T., Nolte, C., Burger, R., Verkhratsky, A., and Kettenmann, H. (1997) Mechanisms of C5a and C3a complement fragment-induced [Ca^{2+}] signaling in mouse microglia. *J. Neurosci.* **17,** 615.
139. Wang, X., Bae, J. H., Kim, S. U., and McLarnon, J. G. (1999) Platelet-activating factor induced Ca($^{2+}$) signaling in human microglia. *Brain Res.* **842,** 159.
140. Pyo, H., Joe, E.-H., Jung, S., Lee, S. H., and Jou, I. (1999) Gangliosides activate cultured rat brain microglia. *J. Biol. Chem.* **274,** 34,584–34,589.
141. Bhat, N. R. and Sharma, K. K. (1999) Microglial activation by small heat shock protein, alpha-crystallin. *Neuroreport* **10,** 2869.
142. Hu, S., Chao, C. C., Ehrlich, L. C., Sheng, W. S., Sutton, R. L., Rockwold, G. L., and Peterson, P. K. (1999) Inhibition of microglial cell RANTES production by IL-10 and TGF-β. *J. Leukoc. Biol.* **65,** 815.
143. Lodge, P. A. and Sriram, S. (1996) Regulation of microglial activation by TGF-β, IL-10 and CSF-1. *J. Leukoc. Biol.* **60,** 502.
144. Tan J., Town, T., Saxe, M., Paris, D., Wu, Y., and Mullan, M. (1999) Ligation of microglial CD40 results in p44/42 mitogen-activated protein kinase-dependent TNF-a production that is opposed by TGF-β and IL-10. *J. Immunol.* **163,** 6614.
145. Menendez-Iglesias, B., Cerase, J., Ceracchini, C., Levi, G., and Aloisi, F. (1997) Analysis of B7-1 and B7-2 costimulatory ligands in cultured mouse microglia: upregulation by interferon-γ and lipopolysaccharide and downregulation by interleukin-10, prostaglandin E2 and cyclic AMP-elevating agents. *J. Neuroimmunol.* **72,** 83.
146. O'Keefe, G. M., Nguyen, V. T., and Benveniste, E. N. (1999) Class II transactivator and class II MHC gene expression in microglia: modulation by the cytokines TGF-β, IL-4, IL-13 and IL-10. *Eur. J. Immunol.* **29,** 1275.
147. Petrova, T. V., Akama, K. T., and Van Eldik, L. J. (1999) Selective modulation of BV-2 microglial activation by prostaglandins E2. Differential effects on endotoxin-stimulated cytokine induction. *J. Biol. Chem.* **274,** 28,823.
148. Nakamura, Y., Si, Q. S., and Kataoka, K. (1999) Lipopolysaccharide-induced microglial activation in culture: temporal profiles of morphological change and release of cytokines and nitric oxide. *Neurosci. Res.* **35,** 95.
149. Elkabes, S., Peng, L., and Black, I. B. (1998) Lipopolysaccharide differentially regulates microglial trk receptor and neurotrophin expression. *J. Neurosci. Res.* **54,** 117.
150. Spleiss, O., Appel, K., Boddeke, H. W., Berger, M., and Gebicke-Haerter, P. J. (1998) Molecular biology of microglia cytokine and chemokine receptors and microglial activation. *Life Sci.* **62,** 1707.
151. Pyo, H., Chung, S., Jou, I., Gwag, B., and Joe, E. H. (1997) Expression and function of outward K^+ channels induced by lipopolysaccharide in microglia. *Mol. Cells* **7,** 610.
152. Sugaya, K., Chou, S., Xu, S. J., and McKinney, M. (1998) Indicators of glial activation and brain oxidative stress after intraventricular infusion of endotoxin. *Brain Res. Mol. Brain Res.* **58,** 1.
153. Romero, L. I., Tatro, J. B., Field, J. A., and Reichlin, S. (1996) Roles of IL-1 and TNF-α in endotoxin-induced activation of nitric oxide synthase in cultured rat brain cells. *Am. J. Physiol.* **270,** R326.
154. Gottschall, P. E., Yu, X., and Bing, B. (1995) Increased production of gelatinase B (matrix metalloproteinase-9) and interleukin-6 by activated rat microglia in culture. *J. Neurosci. Res.* **42,** 335.
155. Stohwasser, R., Giesebrecht, J., Kraft, R., Muller, E. C., Hausler, K. G., Kettenmann, H., Hanisch, U. K., and Kloetzel, P. M. (2000) Biochemical analysis of proteosomes from mouse microglia: Induction of immunoproteosomes by interferon-γ and lipopolysaccharide. *Glia* **29,** 355.
156. Janabi, N., Mirshahi, A., Wolfrom, C., Mirshahi, M., and Tardieu, M. (1996) Effect of interferon-γ and TNF-α on the differentiation/activation of human glial cells: implication for the TNF alpha receptor 1. *Res. Virol.* **147,** 147.

157. Visentin, S., Agresti, C., Patrizio, M., and Levi, G. (1995) Ion channels in rat microglia and their different sensitivity to lipopolysaccharide and interferon-γ. *J. Neurosci. Res.* **42,** 439.
158. Meda, L., Bernasconi, S., Bonaiuto, C., Sozzani, S., Zhou, D., Jr. Otvos, L., et al. (1996) Beta-amyloid (25-35) peptide and IFN-γ synergistically induce the production of the chemotactic cytokine MCP-1/JE in monocytes and microglial cells. *J. Immunol.* **157,** 1213.
159. Bonaiuto, C., McDonald, P. P., Rossi, F., and Cassatella, M. A. (1997) Activation of nuclear factor-κB by beta-amyloid peptides and interferon-gamma in murine microglia. *J. Neuroimmunol.* **77,** 51.
160. Kust, B. M., Biber, K., van Calker, D., and Gebicke-Haerter, P. J. (1999) Regulation of K^+ channel mRNA expression by stimulation of adenosine A2a-receptors in cultured rat microglia. *Glia* **25,** 120.
161. Moller, T., Kann, O., Verkhratsky, A., and Kettenmann, H. (2000) Activation of mouse microglial cells affects P2 receptor signaling. *Brain Res.* **853,** 49.
162. Prinz, M., Kann, O., Draheim, H. J., Schumann, R. R., Kettenmann, H., Weber, J. R., and Hanisch, U. K. (1999) Microglial activation by components of gram-positive and -negative bacteria: distinct and common routes to the induction of ion channels and cytokines. *J. Neuropathol. Exp. Neurol.* **58,** 1078.
163. Meda, L., Baron, P., Prat, E., Scarpini, E., Scarlato, G., Cassatella, M. A., and Rossi, F. (1999) Proinflammatory profile of cytokine production by human monocytes and murine microglia stimulated with beta-amyloid[25-35]. *J. Neuroimmunol.* **93,** 45.
164. Silei, V., Fabrizi, C., Venturini, G., Salmona, M., Bugiani, O., Tagliavini, F., and Lauro, G. M. (1999) Activation of microglial cells by PrP and beta-amyloid fragments raises intracellular calcium through L-type voltage sensitive calcium channels. *Brain Res.* **818,** 168.
165. Akama, K. T., Albanese, C., Pestell, R. G., and van Eldik, L. J. (1998) Amyloid beta-peptide stimulates nitric oxide production in astrocytes through an NF-κB-dependent mechanism. *Proc. Natl. Acad. Sci. USA* **95,** 5795.
166. Kopec, K. K. and Carroll, R. T. (1998) Alzheimer's beta-amyloid peptide 1-42 induces a phagocytic response in murine microglia. *J. Neurochem.* **71,** 2123.
167. McDonald, D. R., Bamberger, M. E., Combs, C. K., and Landreth, G. E. (1998) Beta-amyloid fibrils activate parallel mitogen-activated protein kinase pathways in microglia and THP-1 monocytes. *J. Neurosci.* **18,** 4451.
168. Pyo, H., Jou, I., Jung, S., Hong, S., and Joe, E. H. (1998) Mitogen-activated protein kinases activated by lipopolysaccharide and beta-amyloid in cultured rat microglia. *Neuroreport* **30,** 871.
169. Pyo, H., Jou, I., Jung, S., and Joe, E. (1999) cAMP potentiates beta-amyloid-induced nitric oxide release from microglia. *Neuroreport* **10,** 37.
170. Van Muiswinkel, F. L., Veerhuis, R., and Eikelenboom, P. (1996) Amyloid beta protein primes cultured rat microglial cells for an enhanced phorbol 12-myristate 13-acetate-induced respiratory burst activity. *J. Neurochem.* **66,** 2468.
171. Combs, C. K., Johnson, D. E., Karlo, J. C., Cannady, S. B., and Landreth, G. E. (2000) Inflammatory mechanisms in Alzheimer's disease: inhibition of beta-amyloid-stimulated proinflammatory responses and neurotoxicity by PPARγ agonists. *J. Neurosci.* **20,** 558.
172. Tan, J., Town, T., Paris, D., Mori, T., Suo, Z., Crawford, F., Mattson, M. P., Flavell, R. A., and Mullan, M. (1999) Microglial activation resulting from CD40-CD40L interaction after beta-amyloid stimulation. *Science* **286,** 2352.
173. Bianca, V. D., Dusi, S., Bianchini, E., Dal Pra, I., and Rossi, F. (1999) Beta-amyloid activates the O-2 forming NADPH oxidase in microglia, monocytes, and neutrophils. A possible inflammatory mechanism of neuronal damage in Alzheimer's disease. *J. Biol. Chem.* **274,** 15,493.

174. Rogove, A. D., Siao, C., Keyt, B., Strickland, S., and Tsirka, S. E. (1999) Activation of microglia reveals a non-proteolytic cytokine function for tissue plasminogen activator in the central nervous system. *J. Cell. Sci.* **112,** 4007.
175. Herms, J. W., Madlung, A., Brown, D. R., and Kretzschmar, H. A. (1997) Increase of intracellular free Ca^{2+} in microglia activated by prion protein fragment. *Glia* **21,** 253.
176. Chang, J. Y. and Liu, L. Z. (1999) Manganese potentiates nitric oxide production by microglia. *Brain Res. Mol. Brain Res.* **68,** 22.
177. Sawada, M., Suzumura, A., Hosoya, H., Marunouchi, T., and Nagatsu, T. (1999) Interleukin-10 inhibits both production of cytokines and expression of cytokine receptors in microglia. *J. Neurochem.* **72,** 1466.
178. Ehrlich, L. C., Hu, S., Sheng, W. S., Sutton, R. L., Rockswold, G. L., Peterson, P. K., and Chao, C. C. (1998) Cytokine regulation of human microglial cell IL-8 production. *J. Immunol.* **160,** 1944.
179. Ehrlich, L. C., Hu, S., Peterson, P. K., and Chao, C. C. (1998) IL-10 down-regulates human microglial IL-8 by inhibition of NF-κB activation. *Neuroreport* **9,** 1723.
180. Becher, B., Dodelet, V., Fedorowicz, V., and Antel, J. P. (1996) Soluble tumor necrosis factor receptor inhibits interleukin-12 production by stimulated human adult microglial cells in vitro. *J. Clin. Invest.* **98,** 1539.
181. Zujovic, V., Benavides, J., Vige, X., Carter, C., and Taupin, V. (2000) Fractalkine modulates TNF-alpha secretion and neurotoxicity induced by microglial activation. *Glia* **29,** 305.
182. Laskowitz, D. T., Matthew, W. D., Bennett, E. R., Schmechel, D., Herbstreith, M. H., Goel, S., and McMillian, M. K. (1998) Endogeneous apolipoprotein E Suppresses LPS-stimulated microglial nitric oxide production. *Neuroreport* **9,** 615.
183. Minghetti, L., Nicolini, A., Polazzi, E., Greco, A., Perretti, M., Parente, L., and Levi, G. (1999) Down-regulation of microglial cyclo-oxygenase-2 and inducible nitric oxide synthase expression by lipocortin-1. *Br. J. Pharmacol.* **126,** 1307.
184. Petrova, T. V., Akama, K. T., and van Eldik, L. J. (1999) Cyclopentenone prostaglandins suppress activation of microglia: down-regulation of inducible nitric oxide synthase by 15-deoxy-delta 12,14-prostaglandin J2. *Proc. Natl. Acad. Sci. USA* **96,** 4668.
185. Caggiano, A. O. and Kraig, R. P. (1999) Prostaglandin E receptor subtypes in cultured rat microglia and their role in reducing lipopolysaccharide-induced interleukin-1β production. *J. Neurochem.* **72,** 565.
186. Levi, G., Minghetti, L., and Aloisi, F. (1998) Regulation of prostanoid synthesis in microglial cells and effects of prostaglandins E2 on microglial functions. *Biochimie* **80,** 899.
187. Heppner, F. L., Roth, K., Nitsch, R., and Hailer, N. P. (1998) Vitamin E induces ramification and downregulation of adhesion molecules in cultured microglial cells. *Glia* **22,** 180.

15

Invasion of Ischemic Brain by Immune Cells

Hiroyuki Kato and Takanori Oikawa

1. INTRODUCTION

The normal central nervous system (CNS) has long been believed to be an immunologically privileged site, because it is isolated from the systemic circulation, and is able to exclude components of the immune system by the blood–brain barrier (BBB) *(1)*. The CNS constitutes a leukocyte-deficient environment with limited regenerative capacity *(2)*. Although recent evidence suggests that the CNS is continuously patrolled by a small number of T-lymphocytes, and that certain cells of macrophage lineage (perivascular cells) are slowly replaced by hematogenous cells *(3)*, this notion is principally true, and other immune cells, such as polymorphonuclear leukocytes (PMNLs) and monocytes, migrate only in response to tissue damage in the CNS. The infiltration of these immune cells into the CNS is a critical step in the evolution of pathological and host defense processes of various neurological diseases, such as cerebral ischemia, hemorrhage, trauma, bacterial or viral infection, and multiple sclerosis *(4)*. In these pathological settings, the infiltration of hematogenous cells is added to the local inflammatory response composed of activation of resident microglia and astrocytes.

Microglia are the immune effector cells in the CNS parenchyma *(5)*. Microglia represent a stable cell pool, and are thought to be in a downregulated form of cells of macrophage lineage. They are considered to take up residence primarily as a single, massive influx during fetal life, and are replenished only rarely by hematogenous cells *(6,7)*. Microglia, which are normally quiescent, may become rapidly activated under various pathological conditions. Activated microglia may develop into brain macrophages or phagocytes. Earlier studies *(8)* have shown that microglia, when stimulated, release a variety of cytotoxic agents that may be important mediators of neuronal injury.

Perivascular cells are the third component of immune cells in the CNS. They have also been called "perivascular microglia," "perivascular macrophages," or "fluorescent granular perithelial" cells, and can be distinguished from resident microglia by their unique association with vessels, lying between brain parenchyma and bloodstream, and by constitutive expression of immunomolecules, such as major histocompatibility complex (MHC) class II antigen (Ag) and ED2, which are not expressed in resting microglia *(9,10)*. Studies using chimeric rats have demonstrated that this population of perivascular cells is regularly replaced by bone marrow cells *(6,11)*.

From: *The Neuronal Environment: Brain Homeostasis in Health and Disease*
Edited by: W. Walz © Humana Press Inc., Totowa, NJ

The inflammatory response in the CNS may recruit these subclasses of immune cells, both intrinsic and extrinsic, depending on the nature of injury. Largely, microglia are the first cells to respond even to subtle pathological stimuli, and can be the sole source of inflammatory cells; when the CNS is severely damaged, and the BBB is breached, migration of exogenous immune cells (leukocytes) may aid or override the lesion. This chapter provides an overview of the inflammatory response and the invasion of immune cells in the brain. Cerebral ischemia is used as a model of brain injury and subsequent inflammation, to focus on how invasion of immune cells is elicited and formed.

2. INFLAMMATORY RESPONSES IN CEREBRAL ISCHEMIA

Inflammatory responses occur within the CNS during disease, and immunological mechanisms are involved in a number of disease processes of the CNS. Diseases that are not primarily inflammatory in nature, such as trauma and ischemia, as well as truly inflammatory diseases, such as viral infection and multiple sclerosis, produce a well-defined response of inflammation *(4)*. Activation and accumulation of leukocytes and cells derived from the mononuclear phagocyte system, including microglia, are the hallmark of this response. There is a close relationship between primary brain injury, production of chemical mediators (cytokines and chemokines), and the cellular inflammatory response. This subheading outlines the inflammatory response in the brain after focal cerebral ischemia in the rat, which is a model of human ischemic stroke. Major components of the response are first, activation of resident microglia, and, second, infiltration of bloodborne leukocytes (Fig. 1). The latter is the major concern of this chapter, and the following subheadings focus on the mechanisms of immune cell invasion into the brain.

2.1. Microglial Activation

Microglia are the first nonneuronal cell to respond to virtually any, even subtle, morphologically nonapparent CNS injury. Their activation is induced rapidly, within minutes after acute trauma and ischemia *(12)*. Brain injury elicits a graded response from microglia, in which the degree of microglial activation reflects the severity of the injury *(13,14)*. The graded activation of microglia consists of morphological changes, from resting microglia with highly ramified processes, to activated microglia with enlarged cell bodies and contracted, stout processes, then to phagocytic microglia with amoeboid morphology, and a stepwise expression of immunomolecules; fully activated microglia express many phenotypic markers and effector molecules that are shared with hematogenous macrophages.

Even sublethal ischemic injury induces microglial activation, although limited. Therefore, microglial activation *per se* is not always pathogenic. When ischemia induces selective neuronal damage, sparing glial cells and without disrupting the BBB (e.g., following brief global ischemia or in peri-infarct zones of focal ischemia), microglial cells transform into macrophages (Fig. 1B), without eliciting the invasion of leukocytes. Microglia may be the sole source of phagocytes in this setting *(15)*. This is similar to the response in the rat facial nucleus to motor neuron degeneration *(16)*. Microglia synthesize and secrete a number of cytokines, and appear to be coordinators of the inflammatory response since their activation may be followed by infiltration of

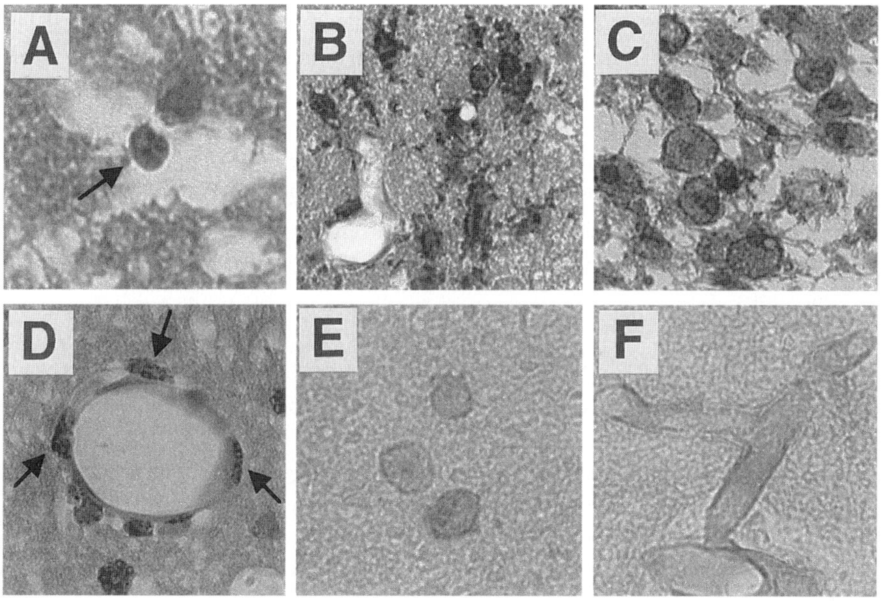

Fig. 1. Inflammatory responses to 1 h of middle cerebral artery occlusion in the rat. **(A)** Infiltration of a polymorphonuclear leukocyte (arrow) in the ischemic core (3 d after ischemia). Hematoxylin and eosin staining. **(B)** Accumulation of amoeboid, phagocytic microglia in the peri-infarct zone (7 d). Immunostaining for a microglia/macrophage marker, MRF-1. **(C)** Accumulation of monocytes/macrophages in the infarct (7 d). MRF-1 immunostaining. **(D)** Activation of perivascular cells (3 d, arrows). Immunostaining for a phagocyte/macrophage marker ED1. **(E)** Infiltration of T lymphocytes in the infarct (7 d). Immunostaining for a T-cell marker CD5. **(F)** Expression of ICAM-1 on endothelial cells in the ischemic core (1 d). Immunostaining for an ICAM-1 marker CD54. Original magnification, ×400.

peripheral immune cells. However, the mechanism by which microglial cells are activated, or how their activation corresponds to the severity of neuronal injury and the appearance of bloodborne cells, is still unknown. Microglia in the CNS, which is beyond the scope of this chapter, is fully described in Chapter 14, and the microglial response in cerebral ischemia has been reviewed elsewhere *(13)*.

2.2. *Invasion of Leukocytes*

Cerebral infarction develops in the center of focal ischemia. In this lesion, all tissue constituents (including glial cells, as well as neurons) may be destroyed, the BBB is breached, and the invasion of leukocytes predominates in the inflammatory response *(17)*. Unlike normal brain microvessels that are clear of leukocytes, brain microvessels from ischemic zones are filled with leukocytes within minutes after ischemia. Many of the leukocytes primarily PMNLs found in vessels within ischemic tissue, are adherent to the endothelium *(18–20)*. This microvascular event may exacerbate the degree of tissue injury; the rheologic effects of sticky leukocytes in the blood vessels (vascular plugging) may interfere with normal perfusion in an already compromised ischemic tissue vascular bed.

Brain injury in the ischemic core evokes not only endogenous brain parenchymal cell damage, but also an exogenous inflammatory response, which includes infiltration of leukocytes *(17,21,22)*. PMNLs begin to infiltrate into the ischemic area 12 h after ischemia, reach a peak at 1–3 d (Fig. 1A), and disappear by 7 d. Monocytes, which are most abundantly seen within an area of infarction, invade massively, as the second wave, after 2–3 d, and cover the entire lesion by 7 d (Fig. 1C). In addition, a significant number of T-lymphocytes invade the area of infarction (Fig. 1E; *17,23,24*). The significance of this T-cell infiltration, which is observed especially in autoimmune diseases, such as experimental autoimmune encephalomyelitis (EAE) *(25,26)*, remains uncertain. The infiltration of leukocytes may aggravate tissue damage by release of cytotoxic products from these activated leukocytes, i.e., by generation and release of oxygen radicals and cytotoxic products that are harmful to the already compromised tissue *(22,27)*. Within an area of infarction, the third component, perivascular cells, are also activated morphologically and phenotypically, and transform into macrophages (Fig. 1D; *28,29*).

The exact nature of the signaling mechanisms in brain inflammation still remains to be elucidated, but undoubtedly involves certain cytokines and chemokines, as well as the expression of adhesion molecules and proteinases that together promote both recruited cell adherence and infiltration and enhanced permeability of brain endothelium. Focal ischemia is a powerful stimulus to elicit genomic responses in the brain, in the form of multiple gene expression. Simultaneously with the expression of immediate early genes and heat shock proteins (for the role of these gene expressions in cerebral ischemia, *see* ref. *30*), is the third wave, which is mostly composed of increased cytokine gene expression, such as tumor necrosis factor (TNF)-α and interleukin (IL)-1β. Chemokines, such as IL-8 and monocyte chemoattractant protein (MCP)-1, are also increased, and are likely to play a role in PMNL and mononuclear cell infiltration. The presence of recruited leukocytes at the site of inflammation is critically dependent on the coordinated expression of adhesion molecules (ligands and receptors) on inflammatory cells and activated endothelial cells (Fig. 1F), respectively.

3. BLOOD–BRAIN BARRIER

The most notable difference between the CNS and other peripheral tissue is the existence of a barrier that separates the CNS tissue from systemic circulation. The strict homeostasis of the CNS environment and the intact barrier are essential for optimal brain functioning. The BBB is impermeable to proteins, ions, many small peptides, amino acids, and drugs, and also appears to rigorously exclude hematogenous cells. The BBB consists of a complex cellular system of endothelial cells, astrocytes, pericytes, perivascular macrophages, and basal lamina. CNS capillaries differ from other capillaries in peripheral organs by the presence of narrow tight junctions, severely restricting the amount of paracellular flux, and the paucity of pinocytotic vesicles, thereby limiting the amount of transcellular flux *(31,32)*. Brain endothelial cells are almost completely surrounded by astrocytic foot processes, and astrocytes induce BBB properties *(33,34)*. Perivascular cells or FGP cells possess a high phagocytic capacity, and are considered to act as a second line of defense *(10)*.

The normal endothelial layer provides a thromboresistant surface that prevents platelet and leukocyte adhesion and activation of any coagulation system. The highly specialized cerebral endothelial cells form a tight barrier, which isolates the brain from immune surveillance, and allows only a few activated T-cells to migrate into the CNS

(3). The low expression of MHC Ags, the low number of Ag-presenting cells in the normal CNS, and the fact that the CNS is not properly drained by a fully developed lymphatic vasculature, make the brain an immunosecluded site *(25,31)*. Even though the immune-privileged status of the CNS limits access of systemic immune cells through the BBB, an immune response can occur in this compartment, with or without a major breach of the BBB. The barrier function of the BBB can change dramatically during various diseases of the CNS. Enhanced BBB permeability is considered to be the result either of opening of tight junctions or of enhanced pinocytotic activity and the formation of transendothelial channels *(31,35)*. The BBB itself may play an active role in mediating the immune response, either by production of chemical mediators or by the expression of adhesion molecules (*see* Subheading 4.6.).

4. CYTOKINES

Cytokines are small proteins with mol wt from 8 to 30 kDa, which possess multiple biological functions *(36)*. Cytokines serve to control a variety of biologic processes, and have a pivotal role in inflammation *(37–39)*. Cytokines appear to be produced primarily in response to external stimuli, and exert their biological effects through specific cell-surface receptors. Cytokines regulate the interaction of immune cells and orchestrate immune responses *(40)*. The proinflammatory cytokines, including IL-1, TNF-α, and IL-6, can influence the function and synthesis of other cytokines, by a complex cytokine network *(41)*. These cytokines are produced by a variety of cells. IL-1β is thought to be produced in the brain by various cellular elements, including endothelial cells, microglia, astrocytes, and neurons *(37)*; TNF-α is produced by neurons, astrocytes, microglia, and leukocytes *(36,42,43)*. Recent evidence suggests that IL-1β, TNF-α, and IL-6 may be key components in the activation and recruitment of leukocytes into the brain *(44)*. TNF-α is known to induce adhesion molecules on endothelial cells and mediates the migration of leukocytes across cerebral endothelial cells *(45)*. TNF-α also regulates the expression of chemokines, such as macrophage inflammatory protein (MIP)-1α and MCP-1 *(46)*.

The cytokine cascade appears to initially involve the release of IL-1 and TNF-α. These cytokines then lead to subsequent production of other proinflammatory cytokines, including IL-6 and IL-8, activation and infiltration of leukocytes, and the production of anti-inflammatory cytokines, including IL-4 and IL-10, which may produce a negative feedback on the cascade *(39)*. The strong evidence in support of this proinflammatory cytokine cascade comes from studies that used agents that block TNF-α and IL-1 *(47,48)*. These antagonist strategies reduce the inflammatory response and improve disease outcome *(47,49)*. Cytokines may influence transport of compounds into the brain by opening the BBB. In vitro studies *(50)* revealed that administration of TNF, IL-1, and IL-6 to monolayers of cerebral endothelial cells leads to an increase in the permeability.

Cerebral ischemia leads to a release of free fatty acids and other proinflammatory lipid metabolites, which in turn promote the expression and release of a proinflammatory cytokine cascade. IL-1β mRNA and protein are induced in focal ischemic tissue of the rat brain, as early as 1 h after ischemia, with a peak within several hours, and remain elevated for a few days *(51,52)*. Concomitant with IL-1β expression, IL-1 receptor and receptor antagonist (IL-1ra) mRNAs are increased *(53)*. Injection of IL-1β into the lateral ventricle of the rat brain, immediately after focal cerebral ischemia, leads to increases in brain edema and infarct volume; injection of anti-IL-1 antibody decreases them *(54)*. Treatment with IL-1ra injected into the rat brain, or peripherally, also decreases the infarct size, edema formation, and PMNL infiltration *(55–57)*. Furthermore, recombinant adenovirus vectors, carrying the human IL-1ra cDNA, reduces brain injury in permanent focal ischemia in the rat *(58)*. Thus, IL-1β appears to be a pathogenic mediator of ischemic brain damage.

A similar early expression of TNF-α following ischemia in the rat has been shown. TNF-α mRNA is detected as early as 1 h, peaks after several hours, and remains elevated for several days *(42,52)*. The TNF receptors, p55 and p75, are also upregulated within 6 and 24 h, respectively *(43)*. Injection of TNF-α into the rat brain, 24 h before focal cerebral ischemia, significantly increases the infarct size, which is reversed by anti-TNF-α antibody *(59)*. These observations suggest that TNF-α mediates ischemic brain injury, but recent studies have reported conflicting findings *(60)*. Neuronal damage, caused by focal cerebral ischemia and epileptic seizures, is exacerbated in mice genetically deficient in TNF receptors *(61)*. There is no difference between the mice that are lacking genes for TNF-α and IL-6 and wild-type animals, concerning the number of infiltrating PMNLs and the degree of BBB dysfunction after traumatic brain injury *(62)*. Therefore, further studies are needed to elucidate the potential effects of cytokines in both processes of damage and repair in CNS inflammation.

5. CHEMOKINES

Chemokines (chemotactic cytokines) are 8–10 kDa proteins with 20–70% homology in amino acid sequences. They have been classified into four subfamilies (C, CC, CXC, and CX_3C) defined by spacing of cysteines in a highly conserved motif, and act through specific high- and low-affinity receptors that belong to the superfamily of serpentine G-protein-coupled receptors *(63,64)*. Chemokines play an essential role for the infiltration of leukocytes into the brain. They are thought to provide the directional cues for the movement of leukocytes in development, homeostasis, and inflammation *(65)*. They may provide the signals that convert the low-affinity, selectin-mediated leukocyte-endothelial cell interaction into high-affinity, integrin-mediated interaction that leads to extravasation of leukocytes (*see* next subheading).

The dramatic increase in the secretion of chemokines during inflammation results in the selective recruitment of leukocytes into the tissue of inflammation. The capacity to precisely control the movement of inflammatory cells suggests that various chemokines and their receptors act in concert with other cytokines to cause tissue infiltration, by increasing the circulating pool of a given leukocyte, upregulating particular adhesion molecules, as well as increasing leukocyte responsiveness to a chemokine. The main stimuli for chemokine production are early proinflammatory cytokines, such as IL-1 and TNF-α, bacterial products, such as lipopolysaccharide, and viral infection *(63)*. Intrahippocampal injections of human recombinant chemokines in mice confirmed their role in leukocyte recruitment in brain parenchyma; MCP-1 is a potent chemoattractant specific for monocytes, and IL-8 and MIP-2 provoke a dramatic PMNL recruitment associated with a destruction of the BBB, which can be attenuated by prior depletion of leukocytes *(66)*.

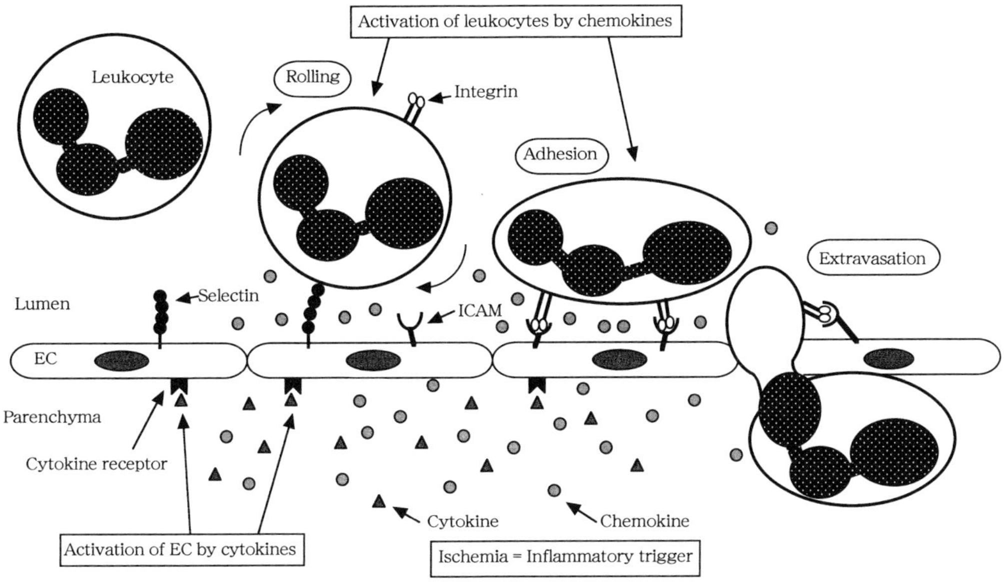

Fig. 2. A simplified model of leukocyte infiltration into ischemic brain. Endothelial cells adjacent to a site of inflammation (ischemia) are stimulated by cytokines (TNF and IL-1) and express the adhesion molecules, E-selectin, P-selectin, and ICAM-1. Leukocytes rolling on the endothelium in a selectin-mediated process are exposed to chemokines for a sufficient time to be programmed for activation, including that of integrins (Mac-1 and LFA-1) on leukocytes. Firm adhesion is caused by specific receptor–ligand interactions between integrins on leukocytes and ICAMs on ECs, and finally transendothelial migration (extravasation) takes place.

A systemic increase of IL-8 mRNA expressing mononuclear cell and IL-8 levels in plasma from patients with ischemic stroke, suggest that IL-8 could be involved in recruiting blood PMNLs to the site of cerebral ischemia *(67)*. PMNL infiltration into the area of cerebral infarction in the rat is preceded by an increased expression of cytokine-induced neutrophil chemoattractant (CINC), a member of the IL-8 family, in the brain (with a peak at 12 h) and the serum *(68,69)*. IL-8 and CINC are produced by endothelial cells and/or leukocytes in response to cytokines, TNF-α and IL-1β *(70)*. Expression of MCP-1 (with a peak between 12 h and 2 d) precedes the massive infiltration of hematogenous monocytes into ischemic tissue of the rat *(71)*. MCP-1 mRNA is initially expressed in astrocytes surrounding the ischemic tissue, and subsequently in infiltrating macrophages and reactive microglia in the infarcted tissue *(72)*.

6. LEUKOCYTE ADHESION AND MIGRATION

An important prerequisite for an inflammatory response to occur is the trafficking of hematogenous leukocytes from the blood to the site of inflammation, through the endothelial wall. In this process, adhesion and extravasation of leukocytes occur in response to molecular changes on the surface of blood vessels that signal brain injury. There are three major steps for the infiltration of leukocytes: rolling, firm adhesion, and transendothelial migration *(73–75)*. All the processes are facilitated by an interaction of specific ligands that are constitutively expressed on leukocytes and receptors on the surface of endothelial cells which are inducible by inflammatory stimuli (Fig. 2; Table 1).

Table 1
Adhesion Molecules Involved in Leukocyte-Endothelium Interactions

Molecules	Expressed on	Ligand and target cell
Selectins		
P-selectin	EC, platelets	Carbohydrate on leukocytes, EC
E-selectin	EC	Carbohydrate on leukocytes
L-selectin	PMNL, monocytes, lymphocytes	Carbohydrate on EC
Integrins		
CD11a/CD18 (LFA-1)	PMNL, monocytes, lymphocytes	ICAM-1, ICAM-2 on EC
CD11b/CD18 (Mac-1)	PMNL, monocytes	ICAM-1, iC3b on EC
CD11c/CD18 (p150,95)	PMNL, monocytes	iC3b on EC
VLA-4	Monocytes, lymphocytes	VCAM-1 on EC
Ig superfamily		
ICAM-1	EC and various cells	LFA-1, Mac-1 on leukocytes
ICAM-2	EC	LFA-1 on leukocytes
VCAM-1	EC	VLA-4 on leukocytes

EC, endothelial cells; PMNL, polymorphonuclear leukocytes.

The first stage (rolling) is mediated by a family of adhesion molecules called "selectins," which are glycoprotein molecules found on leukocyte and endothelial cell surfaces, and recognize carbohydrates on target cells. Selectins consist of P-selectin and E-selectin on endothelial cells and L-selectin on leukocytes, and are upregulated during inflammation *(76)*. Leukocytes flowing in the blood make the first contact with the vasculature, and reduce the velocity in the bloodstream near or at the site of injury, then roll to a site where the leukocytes ultimately leave the bloodstream for the lesion. The light binding of leukocytes to the vessel wall is strengthened under the influence of chemokines that are released from the site of injury. On activation, firm adherence of leukocytes to the endothelial lining (sticking) is mediated by the binding of a leukocyte membrane glycoprotein receptor complex, termed β_2-integrin, to its endothelial ligand intercellular adhesion molecule (ICAM)-1, ICAM-2, and vascular cell adhesion molecule (VCAM)-1 *(77,78)*. The integrin family is, for the most part, expressed constitutively and solely on leukocytes. The CD18 integrin complex consists of three heterodimers, which all share a common β-subunit (CD18), and are distinguished from each other by distinct α-subunits. The three α-subunits are termed "leukocyte function-associated antigen-1" (LFA-1, or CD11a, present on all leukocytes), Mac-1 (CD11b, present mostly on PMNLs and monocytes), and p150,95 (CD11c, present on neutrophils and monocytes). ICAMs are members of the immunoglobulin supergene family. ICAM-1 is broadly expressed on a variety of cell types at low levels, and binds to LFA-1 and Mac-1, but ICAM-2 is expressed only on endothelial cells and leukocytes and is recognized only by LFA-1 *(79,80)*. ICAM-2 is constitutively expressed, but ICAM-1 is inducible by proinflammatory cytokines, such as IL-1 and TNF-α; VCAM-1 is not constitutively expressed, but is upregulated by these cytokines *(81)*. In the final stage, the leukocyte adhering to the endothelium begins to crawl between endothelial cells and leaves the bloodstream to enter the brain tissue (transendothelial migration).

The appearance of leukocytes in ischemic tissue has previously been considered to represent a pathologic response to existing injury. Recent evidence suggests that leukocytes may also be directly involved in the pathogenesis and extension of ischemia-reperfusion injury. Two proposed mechanisms of leukocyte involvement in ischemia-reperfusion injury are, first, direct microvascular plugging by leukocytes after endothelial adhesion *(19,82)*, and, second, transendothelial migration of leukocytes with subsequent CNS tissue infiltration and neurotoxic injury. Thus, leukocytes may physically obstruct microvessels *(18,19,83)*, release vasoconstrictor mediators *(84,85)*, cause dysfunction of the vasoreactivity of cerebral arteries *(86)*, injure endothelial cells directly, or migrate into the brain tissue, where they may damage parenchymal cells, through the release of cytotoxic enzymes, oxygen free radicals, and products of the phospholipid cascade *(87,88)*. PMNLs are the major source of oxygen radicals in reperfusion, after focal cerebral ischemia in the rat *(27)*. Adhesion of leukocytes to microvascular endothelium is essential for either of these mechanisms.

Thus, the leukocyte migration, from the bloodstream to the tissue through the endothelium, is a critical step of tissue inflammation. Adhesion molecules are expressed on endothelial cells soon after permanent and transient focal cerebral ischemia in the rat, and precede the adhesion and migration of leukocytes into the ischemic tissue. The expression parallels that of chemokines, and follows that of cytokines, such as TNFα and IL-1β, which are powerful upregulators of leukocyte-endothelium adhesion molecules. ICAM-1 mRNA expression is upregulated after focal cerebral ischemia in the rat, within a few hours, and ICAM-1 on endothelial cells is increased during the first week *(23,24,89–92)*. Similar upregulation of VCAM-1 and P- and E-selectins after ischemia have also been reported *(89,90,93–96)*.

7. ANTILEUKOCYTE INTERVENTION STUDIES

The importance of cellular adhesion molecules for leukocyte trafficking is now apparent. Some of the strongest evidence supporting the involvement of leukocytes in ischemia-reperfusion injury comes from therapeutic studies in experimental models, showing that CNS injury is reduced by prevention of leukocyte adhesion and infiltration (for comprehensive review, *see* ref. 97). Efforts have been centered on PMNLs since these are the first to arrive at the brain parenchyma within hours after the insult *(20,98,99)*.

Administration of anti-neutrophil serum leads to a 80–95% reduction of circulating PMNLs. Depletion of circulating PMNLs, in transient focal cerebral ischemia in the rat, reduces postischemic brain edema formation and the size of cerebral infarction *(21,100)*. Treatment with antibodies, directed against the CD11b–CD18 complex on PMNLs and monocytes of rats subjected to transient focal ischemia, leads to a significant reduction in infarct volume and PMNL infiltration *(101–104)*. Blocking of the corresponding ligand on endothelial cells, ICAM-1, has also shown a striking protective effect *(105,106)*. ICAM-1 knockout leads to a striking reduction of infarct volume after transient focal ischemia, compared to wild-type mice *(107,108)*. However, treatment with antileukocyte antibodies is ineffective in permanent focal ischemia models *(106,109)*, and neutropenia is not beneficial in transient global cerebral ischemia that leads to selective neuronal damage, instead of to infarction *(110–112)*.

Thus, leukocytes clearly contribute to ischemic brain damage in focal cerebral ischemia followed by reperfusion, if the ischemic insult is not too severe. The following may explain the success of antileukocyte strategies in transient focal cerebral ischemia. The expression of leukocyte adhesion molecules on endothelial cells appears to be triggered by reperfusion; reperfusion allows a large number of leukocytes to reach affected areas, and to induce tissue damage; microvascular plugging by leukocytes interferes with the reperfusion of the ischemic tissue in transient focal ischemia. In permanent focal ischemia or brief global cerebral ischemia, the contribution of leukocyte-mediated injury may be overridden by other mechanisms of brain damage. The role of monocytes in the setting of cerebral ischemia is still unresolved, and more work is needed to separate potential beneficial effects on tissue repair from destructive processes.

Although initial brain inflammation can contribute to the degree of brain damage after injury, anti-inflammatory interventions, to limit the degree of damage, have been shown to interfere with nervous regeneration and recovery *(113)*. Better regeneration occurs in the CNS associated with more marked inflammation *(114)*, and activated macrophage and microglial facilitation of neuronal plasticity/recovery after injury is apparently associated with secretion of neurotrophic factors from macrophages *(115)*. The inflammation that occurs in response to injury in the CNS appears to serve multiple purposes. Although certain strategies may be required to intervene early in brain inflammation to reduce injury and neurodegeneration, other intervention may be necessary to facilitate repair and recovery of regeneration processes after CNS injury *(116)*.

8. CONCLUSION

Cerebral ischemia induces inflammatory responses including the invasion of immune cells into the brain, which may have a major impact on the extent of tissue damage and subsequent recovery. Inflammation occurs with some delay after the onset of ischemia, and, therefore, this component may be a promising target for therapeutic intervention that extends the time window for treatment considerably, since early recruitment of patients is a major obstacle in stroke therapy. In addition, an increased understanding of inflammatory and immunological mechanisms offers great potential for the development of new anti-inflammatory strategies in stroke (for the inflammatory component of a disease that is not primarily inflammatory in nature), and other truly inflammatory disorders of the CNS. One of the important questions that remains to be answered is whether the inflammatory responses only expand the acute ischemic damage, as suggested by many reports that showed beneficial effects of antileukocyte therapies, or whether these reactions are necessary for proper wound healing at the chronic stage. If inflammation is indeed an important contributor to ischemic brain injury, effort should be taken to maximize therapeutic efficacy and minimize side effects, by suppressing cytotoxicity and enhancing neurotrophic properties.

REFERENCES

1. Fabry, Z., Raine, C. S., and Hart, M. N. (1994) Nervous tissue as an immune compartment: the dialect of the immune response in the CNS. *Immunol. Today* **15,** 218–224.
2. Streilein, J. W. (1995) Unveiling immune privilege. *Science* **270,** 1158–1159.
3. Hickey, W. F. (1999) Leukocyte traffic in the central nervous system: the participants and their roles. *Semin. Immunol.* **11,** 125–137.

4. Stoll, G. and Jander, S. (1999) The role of microglia and macrophages in the pathophysiology of the CNS. *Progr. Neurobiol.* **58,** 233–247.
5. Streit, W. J., Graeber, M. B., and Kreutzberg, G. W. (1988) Functional plasticity of microglia: a review. *Glia* **1,** 301–307.
6. Hickey, W. F., Vass, K., and Lassmann, H. (1992) Bone marrow derived elements in the central nervous system: an immunohistochemical and ultrastructural survey of rat chimeras. *J. Neuropathol. Exp. Neurol.* **51,** 246–256.
7. Lawson, L. J., Perry, V. H., and Gordon, S. (1992) Turnover of resident microglia in the normal adult mouse brain. *Neuroscience* **48,** 405–415.
8. Banati, R. B., Gehrmann, J., Schubert, P., and Kreutzberg, G. W. (1993) Cytotoxicity of microglia. *Glia* **7,** 111–118.
9. Graeber, M. B., Streit, W. J., and Kreutzberg, G. W. (1989) Identity of ED2-positive perivascular cells in rat brain. *J. Neurosci. Res.* **22,** 103–106.
10. Mato, M., Ookawara, S., Sakamoto, A., Aikawa, E., Ogawa, T., Mitsuhashi, U., et al. (1996) Involvement of specific macrophage-like lineage cells surrounding arterioles in barrier and scavenger function in brain cortex. *Proc. Natl. Acad. Sci. USA* **93,** 3269–3274.
11. Hickey, W. F. and Kimura, H. (1988) Perivascular microglia are bone marrow derived and present antigen in vivo. *Science* **239,** 290–292.
12. Morioka, T., Kalehua, A. N., and Streit, W. J. (1991) The microglial reaction in the rat dorsal hippocampus following transient forebrain ischemia. *J. Cereb. Blood Flow Metab.* **11,** 966–973.
13. Kato, H. (1997) Microglia: Inflammatory markers in stroke, in *Neuroinflammation: Mechanism and Management* (Wood, P. L., ed.), Humana Press, Totowa, pp. 91–107.
14. Kato, H. and Walz, W. (2000) The initiation of the microglial response. *Brain Pathol.* **10,** 137–143.
15. Kato, H., Kogure, K., Araki, T., and Itoyama, Y. (1995) Graded expression of immunomolecules on activated microglia in the hippocampus following ischemia in a rat model of ischemic tolerance. *Brain Res.* **694,** 85–93.
16. Streit, W. J. and Kreutzberg, G. W. (1988) Response of endogenous glial cells to motor neuron degeneration induced by toxic ricin. *J. Comp. Neurol.* **268,** 248–263.
17. Kato, H., Kogure, K., Liu, X. H., Araki, T., and Itoyama, Y. (1996) Progressive expression of immunomolecules on activated microglia and invading leukocytes following focal cerebral ischemia in the rat. *Brain Res.* **734,** 203–212.
18. Hallenbeck, J. M., Dutka, A. J., Tanishima, T., Kochanek, P. M., Kumaroo, K. K., Thompson, C. B., Obrenovitch, T. P., and Contreras, T. J. (1986) Polymorphonuclear leukocyte accumulation in brain regions with low blood flow during the early postischemic period. *Stroke* **17,** 246–253.
19. Schmid-Schönbein, G. W. (1987) Capillary plugging by granulocytes and the no-reflow phenomenon in the microcirculation. *Fed. Proc.* **46,** 2397–2401.
20. Garcia, J. H., Liu, K. F., Yoshida, Y., Lian, J., Chen, S., and del Zoppo, G. L. (1994) Influx of leukocytes and platelets in an evolving brain infarct (Wistar rat). *Am. J. Pathol.* **144,** 188–199.
21. Matsuo, Y., Onodera, H., Shiga, Y., Nakamura, M., Ninomiya, M., Kihara, T., and Kogure, K. (1994) Correlation between myeloperoxidase-quantified neutrophil accumulation and ischemic brain injury in the rat: effects of neutrophil depletion. *Stroke* **25,** 1469–1475.
22. Kochanek, P. M. and Hallenbeck, J. M. (1992) Polymorphonuclear leukocytes and monocytes/macrophages in the pathogenesis of cerebral ischemia and stroke. *Stroke* **23,** 1367–1379.
23. Schroeter, M., Jander, S., Witte, O. W., and Stoll, G. (1994) Local immune responses in the rat cerebral cortex after middle cerebral artery occlusion. *J. Neuroimmunol.* **55,** 195–203.

24. Jander, S., Kraemer, M., Schroeter, M., Witte, O. W., and Stoll, G. (1995) Lymphocytic infiltration and expression of intercellular adhesion molecule-1 in photochemically induced ischemia of the rat cortex. *J. Cereb. Blood Flow Metab.* **15,** 42–51.
25. Wekerle, H., Linnington, G., Lassmann, H., and Meyermann, R. (1986) Cellular immune reactivity within the CNS. *Trend Neurosci.* **9,** 271–277.
26. Stoll, G., Müller, S., Schmidt, B., van der Meide, P., Jung, S., Toyka, K. V., and Hartung, H. P. (1993) Localisation of interferon-γ and Ia-antigen in T cell-line mediated experimental autoimmune encephalomyelitis. *Am. J. Pathol.* **142,** 1866–1975.
27. Matsuo, Y., Kihara, T., Ikeda, M., Ninomiya, M., Onodera, H., and Kogure, K. (1995) Role of neutrophils in radical production during ischemia and reperfusion of the rat brain: effect of neutrophil depletion on extracellular ascorbyl radical formation. *J. Cereb. Blood Flow Metab.* **15,** 941–947.
28. Brierley, J. B. and Brown, A. W. (1982) The origin of lipid phagocytes in the central nervous system: I. The intrinsic microglia. *J. Comp. Neurol.* **211,** 397–406.
29. Kato, H. (1999) The role of glial and inflammatory reactions in cerebral ischemia, in *Maturation Phenomenon in Cerebral Ischemia III* (Ito, U., Fieschi, C., Orzi, F., Kuroiwa, T., and Klatzo, I., eds.), Springer-Verlag, Berlin, pp. 135–141.
30. Kogure, K. and Kato, H. (1993) Altered gene expression in cerebral ischemia. *Stroke* **24,** 2121–2127.
31. de Vries, H. E., Kuiper, J., de Boer, A. G., van Berkel, T. J. C., and Breimer, D. D. (1997) The blood-brain barrier in neuroinflammatory diseases. *Pharmacol. Rev.* **49,** 143–155.
32. Rubin, L. L. and Staddon, J. M. (1999) The cell biology of the blood-brain barrier. *Annu. Rev. Neurosci.* **22,** 11–28.
33. Brightman, M. (1991) Implication of astroglia in the blood-brain barrier. *Ann. NY Acad. Sci.* **633,** 343–347.
34. Abbott, N. J., Revest, P. A., and Romero, I. A. (1992) Astrocyte-endothelial interaction: physiology and pathology. *Neuropathol. Appl. Neurobiol.* **18,** 424–433.
35. Abbott, N. J. (2000) Inflammatory mediators and modulation of blood-brain barrier permeability. *Cell. Mol. Neurobiol.* **20,** 131–147.
36. Rothwell, N. J. and Hopkins, S. J. (1995) Cytokines and the nervous system. II: actions and mechanisms of action. *Trends Neurosci.* **18,** 130–136.
37. Rothwell, N. (1991) Functions and mechanism of IL-1 in the brain. *Trends Pharmacol. Sci.* **12,** 430–435.
38. Hartung, H. P., Jung, S., Stoll, G., Zielasek, J., Schmidt, B., Archelos, J. J., and Toyka, K. V. (1992) Inflammatory mediators in demyelinating disorders of the CNS and PNS. *J. Neuroimmunol.* **40,** 197–210.
39. Clark, W. M. (1997) Cytokines and reperfusion injury. *Neurology* **49(Suppl. 4),** S10–S14.
40. Arai, K., Lee, F., Miyajima, A., Miyatake, S., Arai, N., and Yokota, T. (1990) Cytokines: coordinators of immune and inflammatory response. *Annu. Rev. Biochem.* **59,** 783–836.
41. Benveniste, E. N. (1992) Inflammatory cytokines within the central nervous system: source, function, and mechanism of action. *Am. J. Physiol.* **263,** C1–16.
42. Liu, T., Clark, R. K., McDonnell, P. C., Young, P. R., White, R. F., Barone, F. C., and Feuerstein, G. Z. (1994) Tumor necrosis factor-α expression in ischemic neurons. *Stroke* **25,** 1481–1488.
43. Botchkina, G. I., Meistrell III, M. E., Botchkina, I. L., and Tracey, K. J. (1997) Expression of TNF and TNF receptors (p55 and p75) in the rat brain after focal cerebral ischemia. *Mol. Med.* **3,** 765–781.
44. Sei, Y., Vitkovic, L., and Yokoyama, M. M. (1995) Cytokines in the central nervous system: regulatory roles in neuronal function, cell death and repair. *Neuroimmunomodulation* **2,** 121–133.
45. Sedgwick, J. D., Riminton, D. S., Cyster, J. G., and Korner, H. (2000) Tumor necrosis factor: a master-regulator of leukocyte movement. *Immunol. Today* **21,** 110–113.

46. Tessier, P. A., Naccache, P. H., Clark-Lewis, I., Gladue, R. P., Neote, K. S., and McColl, S. R. (1997) Chemokine network in vivo: involvement of C-X-C and C-C chemokines in neutrophil extravasation in vivo in response to TNF-α. *J. Immunol.* **159,** 3595–3602.
47. Ohlsson, K., Bjork, P., Bergenfeldt, M., Hageman, R., and Thompson, R. C. (1990) Interleukin-1 receptor antagonist reduces mortality from endotoxin shock. *Nature* **348,** 550–552.
48. McCarthy, P. L., Abhyankar, S., Neben, S., Newman, G., Sieff, C., Thompson, R. C., Burakoff, S. J., and Ferrara, J. L. (1991) Inhibition of interleukin-1 by interleukin-1 receptor antagonist prevents graft-versus-host disease. *Blood* **78,** 1915–1918.
49. Aiura, K., Gelfand, J. A., Burke, J. F., Thompon, R. C., and Dinarello, C. A. (1993) Interleukin-1 (IL-1) receptor antagonist prevents Staphylococcus epidermis-induced hypotension and reduces circulating levels of tumor necrosis factor and IL-1β in rabbits. *Infect. Immun.* **61,** 3342–3350.
50. de Vries, H. E., Blom-Roosemalen, M. C., van Oosten, M., de Boer, A. G., van Berkel, T. J., Breimer, D. D., and Kuiper, J. (1996) The influence of cytokine on the integrity of the blood-brain barrier in vitro. *J. Neuroimmunol.* **64,** 37–43.
51. Liu, T., McDonnell, P. C., Young, P. R., White, R. F., Siren, A. L., Hallenbeck, J. M., Barone, F. C., and Feuerstein, G. Z. (1993) Interleukin-1β mRNA expression in ischemic rat cortex. *Stroke* **24,** 1746–1751.
52. Wang, X., Yue, T. L., Barone, F. C., White, R. F., Gagnon, R. C., and Feuerstein, G. Z. (1994) Concomitant cortical expression of TNF-α and IL-1β mRNAs follows early response gene expression in transient focal ischemia. *Mol. Chem. Neuropathol.* **23,** 103–114.
53. Wang, X., Barone, F. C., Aiyar, N. V., and Feuerstein, G. Z. (1997) Interleukin-1 receptor and receptor antagonist gene expression after focal stroke in rats. *Stroke* **28,** 155–162.
54. Yamasaki, Y., Matsuura, N., Shozuhara, H., Onodera, H., Itoyama, Y., and Kogure, K. (1995) Interleukin-1 as a pathogenic mediator of ischemic brain damage in rats. *Stroke* **26,** 676–681.
55. Relton, J. K. and Rothwell, N. J. (1992) Interleukin-1 receptor antagonist inhibits ischaemic and excitotoxic neuronal damage in the rat. *Brain Res. Bull.* **29,** 243–246.
56. Relton, J. K., Martin, D., Thompson, R. C., and Russell, D. A. (1996) Peripheral administration of interleukin-1 receptor antagonist inhibits brain damage after focal cerebral ischemia in the rat. *Exp. Neurol.* **138,** 206–213.
57. Garcia, J. H., Liu, K. F., and Relton, J. K. (1995) Interleukin-1 receptor antagonist decreases the number of necrotic neurons in rats with middle cerebral artery occlusion. *Am. J. Pathol.* **147,** 1477–1486.
58. Betz, A. L., Yang, G. Y., and Davidson, B. L. (1995) Attenuation of stroke size in rats using an adenoviral vector to induce overexpression of interleukin-1 antagonist in brain. *J. Cereb. Blood Flow Metab.* **15,** 547–551.
59. Barone, F. C., Arvin, B., White, R. F., Miller, A., Webb, C. L., Willette, R. N., Lysko, P. G., and Feuerstein, G. Z. (1997) Tumor necrosis factor-α. A mediator of focal ischemic brain injury. *Stroke* 1233–1244.
60. Shohami, E., Ginis, I., and Hallenbeck, J. M. (1999) Dual role of tumor necrosis factor-α in brain injury. *Cytokine Growth Factor Rev.* **10,** 119–130.
61. Bruce, A. J., Boling, W., Kindy, M. S., Peschon, J., Kraemer, P. J., Carpenter, M. K., Holtsbergm, F. W., and Mattson, M. P. (1997) Altered neuronal and microglial responses to excitotoxic and ischemic brain injury in mice lacking TNF receptors. *Nature Med.* **2,** 788–794.
62. Stahel, P. F., Shohami, E., Younis, F. M., Kariya, K., Otto, V. I., Lenzlinger, P. M., et al. (2000) Experimental closed head injury: analysis of neurological outcome, blood-brain barrier dysfunction, intracranial neutrophil infiltration, and neuronal cell death in mice deficient in genes for pro-inflammatory cytokines. *J. Cereb. Blood Flow Metab.* **20,** 369–380.

63. Baggiolini, M., Dewald, B., and Moser, B. (1994) Inerleukin-8 and related chemotactic cytokines: CXC and CC chemokines. *Adv. Immunol.* **55,** 97–179.
64. Baggiolini, M., Dewald, B., and Moser, B. (1997) Human chemokines: an update. *Annu. Rev. Immunol.* **15,** 675–705.
65. Luster, A. D. (1998) Chemokines: chemotactic cytokines that mediate inflammation. *N. Engl. J. Med.* **338,** 436–445.
66. Bell, M. D., Taub, D. D., and Perry, V. H. (1996) Overriding the brain's intrinsic resistance to leukocyte recruitment with intraparenchymal injections of recombinant chemokines. *Neuroscience* **74,** 283–292.
67. Kostulas, N., Kivisakk, P., Huang, Y., Matusevicius, D., Kostulas, V., and Link, H. (1998) Ischemic stroke is associated with a systemic increase of blood mononuclear cells expressing interleukin-8 mRNA. *Stroke* **29,** 462–466.
68. Liu, T., Young, P. R., McDonnell. P. C., White, R. F., Barone, F. C., and Feuerstein, G. Z. (1993) Cytokine-induced neutrophil chemoattractant mRNA expressed in cerebral ischemia. *Neurosci. Lett.* **164,** 125–128.
69. Yamasaki, Y., Matsuo, Y., Matsuura, N., Onodera, H., Itoyama, Y., and Kogure, K. (1995) Transient increase of cytokine-induced neutrophil chemoattractant, a member of the interleukin-8 family, in ischemic brain areas after focal ischemic in rats. *Stroke* **26,** 318–323.
70. Strieter, R. M., Kunkel, S. L., Showell, H. J., Phan, S. H., Ward, P. A., and Marks, R. M. (1989) Endothelial cell gene expression of a neutrophil chemotactic factor by TNF-α, LPS, and IL-1β. *Science* **243,** 1467–1469.
71. Wang, X., Yue, T. L., Barone, F. C., and Feuerstein, G. Z. (1995) Monocyte chemoattractant protein-1 messenger RNA expression in rat ischemic cortex. *Stroke* **26,** 661–665.
72. Gourmala, N. G., Buttini, M., Limonta, S., Sauter, A., and Boddeke, H. W. (1997) Differential and time-dependent expression of monocyte chemoattractant protein-1 mRNA by astrocytes and macrophages in rat brain: effects of ischemia and peripheral lipopolysaccharide administration. *J. Neuroimmunol.* **74,** 35–44.
73. Kishimoto, T. K. and Rothlein, R. (1994) Integrins, ICAMs and selectins: role and regulation of adhesion molecules in neutrophil recruitment to inflammatory sites. *Adv. Pharmacol.* **25,** 117–169.
74. Springer, T. A. (1994) Traffic signals for lymphocyte recirculation and leukocyte emigration: the multistep paradigm. *Cell* **76,** 301–314.
75. Clark, W. M. and Zivin, J. A. (1997) Antileukocyte adhesion therapy: preclinical trials and combination therapy. *Neurology* **49(Suppl. 4),** S32–S38.
76. Fassbender, K., Mössner, R., Motsch, L., Kischka, U., Grau, A., and Hennerici, M. (1995) Circulating selectin- and immunoglobulin-type adhesion molecules in acute ischemic stroke. *Stroke* **26,** 1361–1364.
77. Smith, C. W., Rothlein, R., Hughes, B., Mariscalco, M. M., Rudloff, H. E., Schmalstieg, F. C., and Anderson, D. C. (1988) Recognition of an endothelial determinant for CD 18-dependent human neutrophil adherence and transendothelial migration. *J. Clin. Invest.* **82,** 1746–1756.
78. Argenbright, L., Letts, L., and Rothlein, R. (1991) Monoclonal antibodies to the leukocyte membrane CD18 glycoprotein complex and to intercellular adhesion molecule-1 inhibit leukocyte-endothelial adhesion in rabbits. *J. Leukoc. Biol.* **49,** 253–257.
79. Smith, C. W., Marlin, S. D., Rothlein, R., Toman, C., and Anderson, D. C. (1989) Cooperative interactions of LFA-1 and Mac-1 with intercellular adhesion molecule-1 in facilitating adherence and transendothelial migration of human neutrophils in vitro. *J. Clin. Invest.* **83,** 2008–2017.
80. Petruzzelli, L., Takami, M., and Humes, H. D. (1999) Structure and function of cell adhesion molecules. *Am. J. Med.* **106,** 467–476.

81. Kallmann, B. A., Hummel., V., Lindenlaub, T., Ruprecht, K., Toyka, K. V., and Rieckmann, P. (2000) Cytokine-induced modulation of cellular adhesion to human cerebral endothelial cells is mediated by soluble vascular cell adhesion molecule-1. *Brain* **123,** 687–697.
82. Mori, E., del Zoppo, G. L., Chambers, D., Copeland, B. R., and Arfors, K. E. (1992) Inhibition of polymorphonuclear leukocyte adherence suppresses no-reflow after focal cerebral ischemia in baboons. *Stroke* **23,** 712–718.
83. Hatchell, D. L., Wilson, C. A., and Saloupis, P. (1994) Neutrophils plug capillaries in acute experimental retinal ischemia. *Microvasc. Res.* **47,** 344–354.
84. Faraci, F. M., Lopez, A. G., Breese, K., Armstrong, M. L., and Heistad, D. (1991) Effect of atherosclerosis on cerebral vascular responses to activation of leukocytes and platelets in monkeys. *Stroke* **22,** 790–796.
85. Mugge, A., Heistad, D. D., Padgett, R. C., Waack, B. J., Densen, P., and Lopez, J. A. (1991) Mechanisms of contraction induced by human leukocytes in normal and atherosclerotic arteries. *Circ. Res.* **69,** 871–880.
86. Akopov, S., Sercombe, R., and Seylaz, J. (1994) Leukocyte-induced acute endothelial dysfunction in middle cerebral artery in rabbits. Response to aggregating platelets. *Stroke* **25,** 2246–2252.
87. Weiss, S. J. (1989) Tissue destruction by neutrophils. *N. Engl. J. Med.* **320,** 365–376.
88. Welbourn, C. R., Goldman, G., Paterson, I. S., Valeri, C. R., Shepro, D., and Hechtman, H. B. (1991) Pathophysiology of ischaemia reperfusion injury: central role of the neutrophil. *Br. J. Surg.* **78,** 651–655.
89. Wang, X. and Feuerstein, G. Z. (1995) Induced expression of adhesion molecules following focal brain ischemia. *J. Neurotrauma* **12,** 825–832.
90. Okada, Y., Copeland, B. R., Mori, E., Tung, M.-M., Thomas, W. S., and del Zoppo, G. J. (1994) P-selectin and intercellular adhesion molecule-1 expression after focal brain ischemia and reperfusion. *Stroke* **25,** 202–211.
91. Matsuo, Y., Onodera, H., Shiga, Y., Shozuhara, H., Ninomiya, M., Kihara, T., Tamatani, T., Miyasaka, M., and Kogure, K. (1994) Role of cell adhesion molecules in brain injury after transient middle cerebral artery occlusion in the rat. *Brain Res.* **656,** 344–352.
92. Zhang, R. L., Chopp, M., Zalonga, C., Zhang, Z. G., Jiang, N., Gautam, S. C., et al. (1995) The temporal profiles of ICAM-1 protein and mRNA expression after transient MCA occlusion in the rat. *Brain Res.* **682,** 182–188.
93. Wang, X., Yue, T. L., Barone, F. C., and Feuerstein, G. Z. (1995) Demonstration of increased endothelial-leukocyte adhesion molecule-1 mRNA expression in rat ischemic cortex. *Stroke* **26,** 1665–1669.
94. Haring, H. P., Berg, E. L., Tsurushita, N., Yagaya, M., and del Zoppo, G. (1996) E-selectin appears in nonischemic tissue during experimental focal cerebral ischemia. *Stroke* **27,** 1386–1392.
95. Jander, S., Pohl, J., Gillen, C., Schroeter, M., and Stoll, G. (1996) Vascular cell adhesion molecule-1 mRNA is expressed in immune-mediated and ischemic injury of the rat nervous system. *J. Neuroimmunol.* **70,** 75–80.
96. Zhang, R., Chopp, M., Zhang, Z., Jiang, N., and Powers, C. (1998) The expression of P- and E-selectins in three models of middle cerebral artery occlusion. *Brain Res.* **785,** 207–214.
97. Härtl, R., Schürer, L., Schmid-Schönbein, G. W., and del Zoppo, G. J. (1996) Experimental antileukocyte interventions in cerebral ischemia. *J. Cereb. Blood Flow Metab.* **16,** 1108–1119.
98. Clark, R. K., Lee, E. V., Fish, C. J., White, R. F., Price, W. J., Jonal, Z. L., Feuerstein, G. Z., and Barone, F. C. (1993) Development of tissue damage, inflammation and resolu-

tion following stroke: an immunohistochemical and quantitative planimetric study. *Brain Res. Bull.* **31,** 565–572.

99. Zhang, R. L., Chopp. M., Li, Y., Zalonga, C., Jiang, N., Jones, M. L., Miyasaka, M., and Ward, P. A. (1994) Anti-ICAM-1 antibody reduces ischemic cell damage after transient middle cerebral artery occlusion in the rat. *Neurology* **44,** 1747–1751.
100. Shiga, Y., Onodera, H., Kogure, K., Yamasaki, Y., Yashima, Y., Syozuhara, H., and Sendo, F. (1991) Neutrophil as a mediator of ischemic edema formation in the brain. *Neurosci. Lett.* **125,** 110–112.
101. Clark, W. M., Madden, K. P., Rothlein, R., and Zivin, J. A. (1991) Reduction of central nervous ischemic injury in rabbits using leukocyte adhesion antibody treatment. *Stroke* **22,** 877–883.
102. Chen, H., Chopp, M., Zhang, R. L., Bodzin, G., Cehn, Q., Rusche, J. R., and Todd III, R. F. (1994) Anti-CD11b monoclonal antibody reduces ischemic cell damage after transient focal cerebral ischemia in rat. *Ann. Neurol.* **35,** 458–463.
103. Chopp, M., Zhang, R. L., Chen, H., Li, Y., Jiang, N., and Rusche, J. R. (1994) Postischemic administration of an anti-Mac-1 antibody reduces cell damage after transient middle cerebral artery occlusion in rats. *Stroke* **25,** 869–876.
104. Zhang, Z. G., Chopp, M., Tang, W. X., Jiang, N., and Zhang, R. L. (1995) Postischemic treatment (2-4h) with anti-CD11b and anti-CD18 monoclonal antibodies are neuroprotective after transient (2h) focal cerebral ischemia in the rat. *Brain Res.* **698,** 79–85.
105. Clark, W. M., Lauten, J. D., Lessov, N., Woodward, W., and Coull, B. M. (1995) The influence of antiadhesion therapies on leukocyte subset accumulation in central nervous system ischemia in rats. *J. Mol. Neurosci.* **6,** 43–50.
106. Zhang, R. L., Chopp, M., Li, Y., Zalonga, C., Jiang, N., Jones, M. L., Miyasaka, M., and Ward, P. A. (1995) Anti-ICAM-1 antibody reduces ischemic cell damage after transient but not permanent MCA occlusion in the Wistar rat. *Stroke* **26,** 1438–1443.
107. Connolly, E. S., Winfree, C. J., Springer, T. A., Naka, Y., Yan, S. D., Stern, D. M., et al. (1996) Cerebral protection in homozygous null ICAM-1 mice after middle cerebral artery occlusion: role of neutrophil adhesion in the pathogenesis of stroke. *J. Clin. Invest.* **97,** 209–216.
108. Soriano, S. G., Lipton, S. A., Wang, Y. F., Xiao, M., Springer, T. A., Guitierrez-Ramos, J. C., and Hickey, P. R. (1996) Intercellular adhesion molecule-1-deficient mice are less susceptible to cerebral ischemia-reperfusion injury. *Ann. Neurol.* **39,** 618–624.
109. Garcia, J. H., Liu, K. F., and Bree, M. P. (1996) Effects of CD11b/18 monoclonal antibody on rats with permanent cerebral artery occlusion. *Am. J. Pathol.* **148,** 241–248.
110. Grögaard, B., Schürer, L., and Gerdin, B. (1989) Delayed hypoperfusion after incomplete forebrain ischemia in the rat: the role of polymorphonuclear leukocytes. *J. Cereb. Blood Flow Metab.* **9,** 500–505.
111. Schürer, L., Grögaard, B., Gerdin, B., and Arfors, K. E. (1990) Effects of neutrophil depletion and superoxide dismutase on postischemic hypoperfusion of rat brain. *Adv. Neurol.* **52,** 57–62.
112. Schürer, L., Grögaard, B., Gerdin, B., Kempski, O., and Arfors, K. E. (1991) Leukocyte depletion does not affect post-ischemic nerve cell damage in the rat. *Acta Neurochir.* **111,** 54–60.
113. Hirschberg, D. L., Yoles, E., Belkin, M., and Schwartz, M. (1994) Inflammation after axonal injury has conflicting consequences for recovery of function: rescue of spared axons is impaired but regeneration is supported. *J. Neuroimmunol.* **5,** 9–16.
114. Guest, J. D., Rao, A., Olson, L., Bunge, M. B., and Bunge, R. P. (1997) The ability of human Schwann cell grafts to promote regeneration in the transected nude rat spinal cord. *Exp. Neurol.* **148,** 502–522.

115. Batchelor, P. E., Liberatore, G. T., Wong, J. Y., Porritt, M. J., et al. (1999) Activated macrophages and microglia induce dopaminergic sprouting in the injured striatum and express brain-derived neurotrophic factor and glial cell line-derived neurotrophic factor. *J. Neurosci.* **19,** 1708–1716.
116. Lazarov-Spiegler, O., Rapalino, O., Agranov, G., and Schwartz, M. (1998) Restricted inflammatory reaction in the CNS: a key impediment to axonal regeneration? *Mol. Med. Today* **4,** 337–342.

Index

H

I

O

P